THEORY OF SELF-ADAPTIVE
CONTROL SYSTEMS

THEORY OF SELF-ADAPTIVE CONTROL SYSTEMS

Proceedings of the Second IFAC Symposium on
the Theory of Self-Adaptive Control Systems
September 14-17, 1965
National Physical Laboratory
Teddington, England

Edited by
P. H. Hammond
Autonomics Division
National Physical Laboratory

Springer Science+Business Media, LLC 1966

Library of Congress Catalog Card No. 66-12631

© 1966 Springer Science+Business Media New York
Originally published by International Federation of Automatic Control
Düsseldorf, Germany in 1966
Softcover reprint of the hardcover 1st edition 1966
ISBN 978-1-4899-6157-0   ISBN 978-1-4899-6289-8 (eBook)
DOI 10.1007/978-1-4899-6289-8

PREFACE

The thirty-seven papers in this volume cover a wide field of research activities concerned with the analysis and synthesis of self adaptive, optimal and multi-level control systems. They represent the papers submitted to the 2nd IFAC Symposium on "The Theory of Self Adaptive Control Systems" which was organized by the Society of Instrument Technology at the request of the United Kingdom Automation Council acting on behalf of the Theory Committee of IFAC under the chairmanship of Prof. John G. Truxal. The meeting was held at the National Physical Laboratory, Teddington, England from September 14-17, 1965 by kind permission of the Director of the Laboratory, Dr. J. V. Dunworth.

The Symposium was attended by 150 people from 17 countries. The sessions were recorded and the discussions were subsequently summarized by scientific reporters. These summaries are believed to correctly convey the essence of the points discussed. Written discussion remarks, submitted after the Symposium, are reproduced in full, together with the author's reply where appropriate. The grouping of the papers into symposium sessions has been rearranged slightly in the Proceedings to accommodate papers which were submitted for post publication only.

The help of all the organizations and individuals who prepared and organized the Symposium, gave papers and contributed to discussions, recorded and edited these discussions and helped with the editing of the Proceedings is gratefully acknowledged.

P. H. HAMMOND
Editor

Teddington, Middlesex
May, 1966

ACKNOWLEDGMENTS

SPONSORING ORGANIZATIONS

INTERNATIONAL FEDERATION OF AUTOMATIC CONTROL (IFAC)

President: Prof. J. F. Coales
United Kingdom

Hon. Secretary: Dr. G. Ruppel
IFAC, Postfach 10 250
Dusseldorf,
West Germany

UNITED KINGDOM AUTOMATION COUNCIL (UKAC)

President: Prof. J. F. Coales
United Kingdom

ORGANIZING SOCIETY

SOCIETY OF INSTRUMENT TECHNOLOGY

President: S. S. Carlisle
Past President: Prof. G. D. S. MacLellan

HOST ESTABLISHMENT

The National Physical Laboratory
Teddington, Middlesex, England

Director: Dr. J. V. Dunworth

ORGANIZING COMMITTEE

Chairman: Prof. J. H. Westcott (UK)

Prof. J. G. Balchen (Norway)
P. A. N. Briggs (UK)
Capt. T. W. E. Dommett (UK)
P. H. Hammond (UK)
Prof. G. D. S. MacLellan (UK)

Secretary: R. M. Wilde (UK)

Cmdr. A. A. W. Pollard (UK)
J. F. M. Scholes (UK)
Prof. J. G. Truxal (USA)
Prof. A. A. Voronov (USSR)

SESSION CHAIRMEN

Prof. J. G. Balchen (Norway)
S. L. H. Clarke (UK)
P. H. Hammond (UK)
G. D. S. MacLellan (UK)
Dr. V. J. Rutkovsky (USSR)
Prof. J. H. Westcott (UK)

REPORTING AND EDITORIAL SUPPORT COMMITTEE

P. A. N. Briggs (UK)
J. M. C. Clark (UK)
I. G. Cumming (UK)
Dr. W. Fincham (UK)

Dr. A. T. Fuller (UK)
Dr. A. D. G. Hazlerigg (UK)
D. Q. Mayne (UK)
I. H. Rowe (UK)

Staff of the Machine Translation and Systems Research Group, Autonomics
Division, National Physical Laboratory.

CONTENTS

PARAMETER ESTIMATION AND MODELLING TECHNIQUES

MODEL REFERENCE AND ADAPTIVE CONTROLLERS

ADAPTIVE CONTROLLERS WITH PARAMETER IDENTIFICATION OR PREDICTION

OPTIMAL AND MULTILEVEL CONTROL SYSTEMS

CONTROL USING PERTURBATION TECHNIQUES

ADAPTIVE AND OPTIMAL CONTROLLERS USING STOCHASTIC THEORY

A CLASS OF LEARNING CONTROL SYSTEMS
USING STATISTICAL DECISION PROCESSES

K. S. Fu
School of Electrical Engineering
Purdue University
Lafayette, Indiana

ABSTRACT

The basic concept of learning control systems is introduced. Synthesis of learning control systems using statistical decision theory is discussed. Learning is needed if the a priori information is unknown or incompletely known in an adaptive system. The controller will establish the necessary information for control during the system's operation. Two types of learning, with external supervision and without external supervision, have been described. Several learning schemes, including both parametric and nonparametric approaches, have been proposed. Problems of possible computational difficulties and the convergence of learning processes are also discussed.

INTRODUCTION

Recently the concept of introducing learning to automatic control systems has been proposed.[1-3] A learning control system may be considered as a system which modifies its control law (or policy) to maintain a good performance as a result of its experience, in the presence of unpredictable changes of environment. From this viewpoint, a learning (control) system is certainly also adaptive in the sense of conventional adaptive control systems.[1-4] Moreover, a learning system will be benefited from its past experience to improve its performance. A conventional adaptive control system is designed to modify itself in the face of a new environment so as to optimize its performance. However, if the changes of environment are unpredictable, the amount of a priori information required for the system's adaptive action is usually not completely available. In some instance, the environment may vary so fast that the system cannot maintain optimum performance through adaptive action. It is in these types of situation that a learning system is preferable. A learning control system is designed to recognize familiar patterns in a situation, to gain more a priori information as time proceeds, and then from its past experience, to react in an optimum manner toward the best performance.

Several approaches have been proposed for the study of learning control systems.[5,6] This paper is concerned with the application of statistical decision theory to learning control systems. The statistical decision theory has been recently applied to optimal and adaptive control systems.[7-11] However, most works assume the probability distributions of the system's random parameters are known a priori. The decision of selecting a particular control policy is made on the basis of the a priori information using either Bayes or minimax criterion. In the case that the required probability distributions are unknown or only partially known, the application of learning process to the control system is proposed. The controller will establish its knowledge about the required probability distributions, and the decision of selecting a proper control policy is made on the basis of the established (or the learned) probability distributions of random parameters. The statistical knowledge about the random parameters is updated after each stage of decision so the system performance is improved as the learning process proceeds.

Superior numbers refer to similarly-numbered references at the end of this paper.

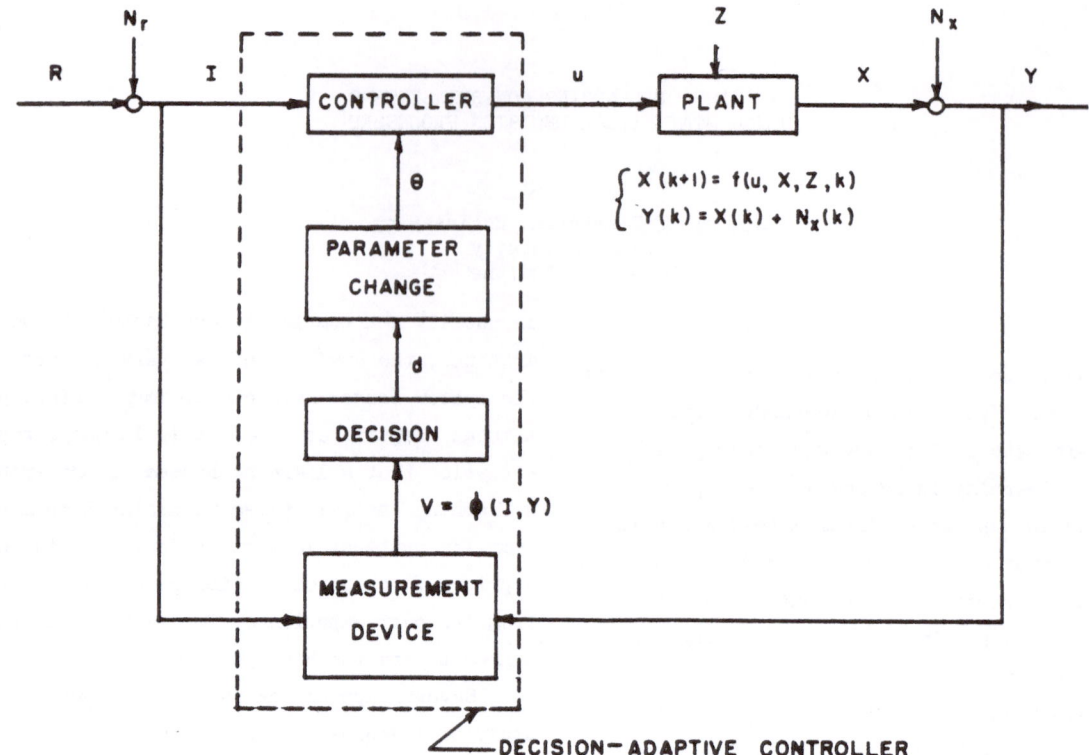

Fig. 1. A DECISION—ADAPTIVE CONTROL SYSTEM

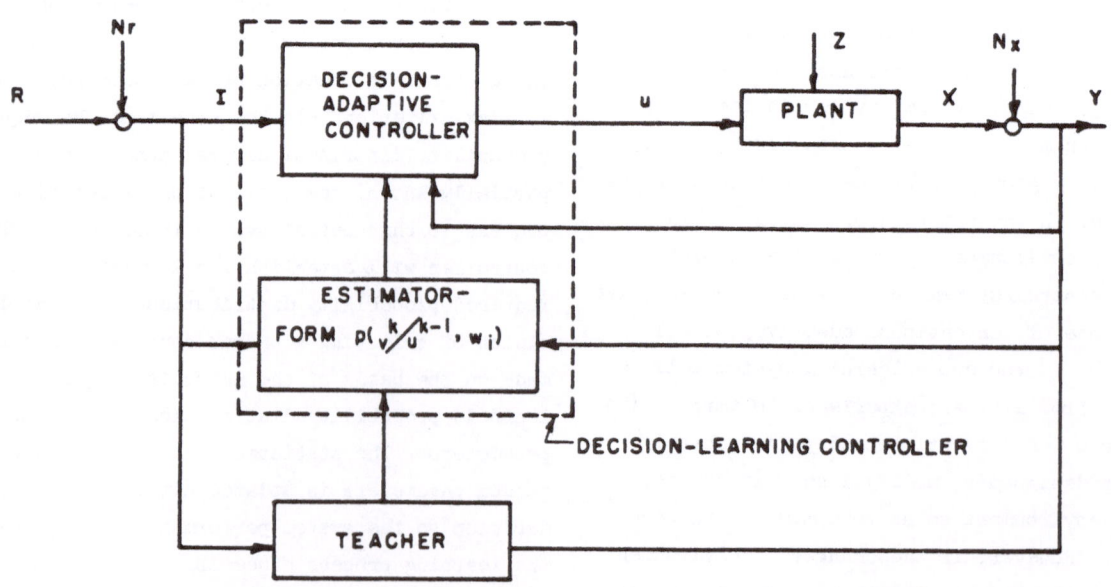

Fig. 2. A DECISION —LEARNING CONTROL SYSTEM WITH
EXTERNAL SUPERVISION

DECISION-ADAPTIVE CONTROL SYSTEMS

A block diagram of a decision-adaptive control system is shown in Fig. 1. A concise interpretation of the decision theoretic approach to the control problem may be obtained by considering the statistical decision process as a zero-sum two-person statistical game. The two players are referred to as the decision-adaptive controller and the plant-environment. A non-negative weighting function measures the loss incurred by the decision-adaptive controller as a consequence of each possible decision in selecting a proper control policy. The aim of the controller is to minimize the average (or the maximum) loss in selecting control policies.[12,13] The operation of the decision-adaptive controller may be summarized as follows: (1) Take measurements of the observable plant states Y and the system's input I at discrete time intervals. Each measurement vector $V = (Y,I)$ or $\phi(Y,I)$, a function of Y and I, can be considered as a point in the n-dimensional measurement space Ω_V. The Ω_V-space will be partitioned into m regions, where m is the number of all possible admissible control policies (either m possible control signals u_i or m quantized regions of controller parameters Θ_i). (2) Based on the set of measurements $V(k)$ taken at k^{th} time instant compute the multivariate likelihood function $p\{V(1), \ldots, V(k)/\omega_j\}$, $i = 1, \ldots, m$, where ω_i is the ith partitioned region in the Ω_{V^k}-space called the ith control situation. The partition of the Ω_V-space is made such that, with respect to ω_i, the ith control policy u_i (or Θ_i) is the optimum policy. (3) With a priori probabilities $P(\omega_i)$, $i = 1, \ldots, m$, make a decision on the proper selection of control policy in the sense of minimizing the average loss (Bayes decision rule). Let $L(\omega_i, u_j)$ represent the loss incurred by the controller if the control policy u_j is decided when ω_i is the actual control situation,

$$v^k = \{V(1), \ldots, V(k)\}$$

and $$u^k = \{u(1), \ldots, u(k)\}$$

The Bayes decision rule calls for minimizing the average loss

$$R_k = \sum_{i=1}^{m} P(\omega_i) \, r_k(\omega_i) \qquad (1)$$

where $$r_k(\omega_i) = \int_{\Omega_{V^k}} L(\omega_i, u_j) \, p(v^k/u^{k-1}, \omega_i) dv^k$$

For zero-one loss function, the Bayes decision rule is to select u_j if

$$P(\omega_j) \, p(v^k/u^{k-1}, \omega_j) \geq P(\omega_i) \, p(v^k/u^{k-1}, \omega_i)$$

for all $i \neq j$ (2)

or $P(\omega_j) \, p\{V(k)/v^{k-1}, u^{k-1}, \omega_j\} \, p(v^{k-1}/u^{k-1}, \omega_j) \geq$

$$P(\omega_i) \, p\{V(k)/v^{k-1}, u^{k-1}, \omega_i\} \, p(v^{k-1}/u^{k-1}, \omega_i)$$

for all $i \neq j$ (3)

In the case of using Bayes sequential decision procedure the controller may use the dynamic programming approach to decide the optimum control policy - a backward computation procedure.[15] A forward procedure suggests that the controller computes the generalized sequential probability ratios[16]

$$U_k(v^k/u^{k-1}, \omega_i) = \frac{p(v^k/u^{k-1}, \omega_i)}{\left[\prod_{j=1}^{m} p(v^k/u^{k-1}, \omega_j)\right]^{1/m}} ,$$

$$i = 1, \ldots, m \qquad (4)$$

for the ith control situation and at the kth measurement. Compare $U_k(v^k/\omega_i)$ with the stopping bound $A(\omega_i)$ where

$$A(\omega_i) = \frac{1 - e_{ii}}{\left[\prod_{j=1}^{m} (1-e_{ij})\right]^{1/m}} \qquad (5)$$

and e_{ij} = the probability of deciding u_i if the jth control situation is true. If $U_k(v^k/u^{k-1}, \omega_i) < A(\omega_i)$, the decision d_i is rejected; otherwise an additional measurement is requested. The last decision left is accepted as the final decision. For $m = 2$, the procedure reduces to the standard Wald's sequential probability ratio test. The feature of sequential decision procedure is that the error level can be controlled by slightly sacrificing the average number of measurements.

The error probabilities e_{ij} can be specified a priori. The controller examines measurements $V(j)$ as they accumulate instead of taking fixed number of measurements as in the classical fixed-sample decision procedure.

It is noted that in the formulation of decision-adaptive control systems, the function of the decision-adaptive controller can be interpreted as a pattern recognition process. Each control situation can be considered as a pattern class of the plant-environment. The controller tries to recognize the present plant-environment pattern or control situation and to apply a proper control policy (u_i or Θ_i) optimum with respect to that situation. The decision on a proper control policy is to minimize the average loss due to misrecognition (or the probability of misrecognition in zero-one loss case) of plant-environment patterns.

LEARNING IN DECISION-ADAPTIVE CONTROL SYSTEMS

In using statistical decision procedures to control the plant-environment when environmental changes are unpredictable, the controller often does not have complete a priori information for its operation. Thus the controller will encounter some difficulties in computing accurate conditional density functions $p(V^k/u^{k-1}, \omega_i)$ which will directly affect the result of decision. In order to improve the performance of the controller under the condition of insufficient a priori information, the learning process is introduced into the operation of the controller. The controller will estimate all unknown a priori information from its experience. The decision of selecting a proper control policy (or the recognition of control situation) is made on the basis of the estimated information. The learning scheme used by the controller is designed to formulate the statistical decision problem in terms of a present measurement and a sequence of learning measurements from the controller's previous experience. The controller's previous experience includes the measurements previously taken and the classifications of these measurements into

control situations. Let, at the kth time instant, $V(k) = F_i(k) + N(k)$ for the ith control situation where

$$F_i = \begin{bmatrix} X_i \\ R_i \end{bmatrix}, \quad \text{and} \quad N = \begin{bmatrix} N_x \\ N_r \end{bmatrix}$$

Suppose that the controller is presented with the measurement $V(k+1)$ after it has taken k learning measurements $V(1), \ldots, V(k)$. The decision as to which control policy the controller should choose now is a function of $V(1), \ldots, V(k)$ as well as $V(k+1)$. The origin of the $V(j)$, $j = 1, \ldots, k$, is from previous measurements to be classified. After each successive V is measured information is provided to the controller about the classification of the measurement V. Depending upon whether this information about the classification of V is provided by an external source, the learning scheme can be divided into two cases, namely, learning with external supervision and learning without external supervision.

DECISION-LEARNING CONTROL SYSTEMS WITH EXTERNAL SUPERVISION

In the case of learning with external supervision, the information concerning the correct classification of successive measurement is provided by an external aid - a "teacher" (Fig. 2). The set of learning measurements V^k may be partitioned according to the ω_i's to which they correspond. It is convenient to assume that learning measurements from one control situation do not provide information about the statistical structure of other control situations. The learning measurements V^k may be considered as k learning measurements from ω_i. Suppose that, as is usually the case, V^k and the event $V \in \omega_i$ are statistically independent, then

$$p\{V^{k+1}/u^k, \omega_i\} = p\{V(k+1)/V^k, u^k, \omega_i\}\, p\{V^k/u^k, \omega_i\}$$

$$= p\{V(k+1)/V^k, u^k, \omega_i\}\, p\{V^k/u^k\} \qquad (6)$$

The learning process for establishing $p\{V(k+1)/V^k, u^k, \omega_i\}$ is given by[17]

$$p\{V(1)/\omega_i\} \to p\{V(2)/V(1),u(1),\ \omega_i\} \to \cdots$$

$$\to p\{V(k+1)/V^k,u^k,\omega_i\}$$

where $u(1) = u_i, \ldots, u(k) = u_i$.*

The nonincreasing property of the average loss as the number of learning measurements increases for learning with external supervision can be easily proved.[14] This property implies that the controller can improve its performance, or learn, as more experience, or more learning measurements are provided.

If the form of the density function $p\{V(k+1)/V^k,\omega_i\}$, $i = 1, \ldots, m$, is known a priori but a parameter vector (or a set of parameters) α_i, $i = 1, \ldots, m$, unknown, the value of α_i can be learned by using the above suggested procedure. Initially, α_i may be assumed as a random variable with a priori density function $p(\alpha_i)$.

$$p\{V(k+1)/V^k,\ \omega_i\} =$$

$$\int_{\Omega_{\alpha_i}} p\{V(k+1)/V^k,\alpha_i,\omega_i\} \cdot p(\alpha_i/V^k,\omega_i)\ d\alpha_i \qquad (7)$$

For given α_i, $p\{V(k+1)/V^k,\alpha_i,\omega_i\}$ is known. Thus the problem reduces to the computation of $p(\alpha_i/V^k)$. By Bayes' theorem,

$$p(\alpha_i/V^k) = \frac{p(V^k/\alpha_i)\ p(\alpha_i)}{\int p(V^k/\alpha_i)\ p(\alpha_i)\ d\alpha_i} \qquad (8)$$

In order to simplify the computation, $p(\alpha_i)$ should be chosen such that the form of $p(\alpha_i/V^k)$ is of the same family as $p(\alpha_i)$ (reproductive property). In this case, with a reproducible a priori density function $p(\alpha_i)$, the structure of the "Estimator" (Fig. 2) remains the same; only the parameter α will change with k.[16-18]

If α_i is not fixed but unknown, but is time-varying and unknown, the density function $p(\alpha_i)$ can no longer be assumed stationary. In the case

of slowly time-varying mean vector α_i, a learning scheme has been introduced to "track" the statistical property of α_i.[17] For the case of a normally distributed V with time-varying mean vector α_i (not necessarily be slowly varying), a learning scheme to estimate the mean vector has been developed. The stationary component and the time-varying component of the mean vector can be separately learned and the learned parameters converge to the true parameter values.[20]

If the form of the density function $p\{V(k+1)/V^k,\omega_i\}$, $i = 1, \ldots, m$, is also unknown, a nonparametric estimation technique must be used. Several approaches have been suggested for this purpose.[13,21,22] The product extension method has been proposed as a means of approximating higher order density functions.[13] The structure which lower order density functions impose on the higher order ones was considered in terms of an information content measure of agreement between densities. The information content measure of agreement between densities also results in a convenient means of choosing between alternative product forms as estimates of a density function. Aizerman et.al., have proposed the application of potential function method to the estimation of unknown density functions.[21] The unknown density function is approximated by a weighted sum of a set of orthonormal functions. An algorithm for learning with external supervision has been suggested and its convergence proved. Sebestyen has proposed to estimate an unknwon density function for ω_i based on the assumption that the partitioned region ω_i can be approximated by the union of circular sets representing equiprobable contours of normal processes.[22] Each circular set is established based on actually occurred learning measurements (called adaptive sample set). Mathematically, consider ω_{i1}, ω_{i2}, $\ldots$, ω_{ij}, $\ldots$ are the sample sets in ω_i, then

$$P(\omega_i)\ p(V^{k+1}/\omega_i) = P(\omega_i,\ V^{k+1})$$

$$= \sum_{j=1}^{N_i} P(\omega_{ij})\ p(V^{k+1}/\omega_{ij}) \qquad (9)$$

where N_i is the number of sample sets in ω_i after $k+1$ measurements. For simplicity, the normal processes for each sample set are assumed to have independent components v_q and equal variances. Thus

$$p(V^{k+1}/\omega_{ij}) = \frac{1}{(\sigma\sqrt{2\pi})^{n(k+1)}} \exp\left\{-\frac{1}{2\sigma^2}\sum_{q=1}^{n(k+1)}[v_q - m_q(\omega_{ij})]^2\right\} \quad (10)$$

The radius of circular sets are, in general, pre-assigned. However, it can be adjusted, if necessary, during the learning process.[19]

DECISION-LEARNING CONTROL SYSTEMS WITHOUT EXTERNAL SUPERVISION

In the previous discussion of learning control systems using statistical decision processes, the learning measurements V^k are assumed to be all correctly classified by an external aid - "teacher". In practical situations, it may not be always possible that a "teacher" can be provided for the system during its complete period of operation. A "teacher" may be present only during the "learning period". However, the "learning period" must be finite, it is certainly hoped that the system can also learn during the actual "operating period". The learning process performed during the "operating period" in the absence of a "teacher" is called learning without external supervision. Two schemes are proposed for decision-learning control systems without external supervision. The first scheme is that the controller correlates each measurement V with all control situations. The second scheme is that the controller uses its own decision to direct the learning process.

In order to correlate the measurements $V(1), \ldots, V(k)$ of unknown classifications with each control situation ω_i, $i = 1, \ldots, m$, there are totally m^k possible situations which must be considered. Let the m^k possible partitions of $V(1), \ldots, V(k)$ be $\Delta_1, \ldots, \Delta_{m^k}$, then

$$p\{V(k+1)/V^k, \omega_i\} = \sum_{j=1}^{m^k} p\{V(k+1)/V^k, \omega_i, \Delta_j\} P(\Delta_j/V^k, \omega_i)$$

or

$$p\{V(k+1)/V^k, \omega_i\} = \frac{1}{p(V^k, u^k/\omega_i)}\sum_{j=1}^{m^k} p\{V(k+1)/V^k, \omega_i, \Delta_j\}\, p(V^k/\omega_i, \Delta_j) P(\Delta_j/\omega_i) \quad (11)$$

It is noted from Eq. (11) that the computation of $p\{V(k+1)/V^k, \omega_i\}$ will grow exponentially as k increases. That is, m^k computations are required for the optimum utilization of k learning measurements. For this reason it does not seem practical to use the scheme for large values of m and k. Several recursive schemes have been developed to avoid this difficulty.[16,23] Fu and Chen have proposed to use a function modified from the a posteriori function of ω_i defined as

$$\hat{p}(\alpha_i/V^k) = \frac{P(\omega_i)p\{V(k)/V^{k-1}, \alpha_i\}p(\alpha_i/V^{k-1})p(V^{k-1}/\omega_i)}{\sum_{j=1}^{m} P(\omega_j)\int p\{V(k)/V^{k-1}, \alpha_j\}p(\alpha_j/V^{k-1})p(V^{k-1}, \omega_j)\, d\alpha_j} \quad (12)$$

It is noted that $\int \hat{p}(\alpha_i/V^k)\, d\alpha_i$ = a posteriori probability of the ith control situation. It can be shown that, as $k \to \infty$, $\hat{p}(\alpha_i/V^k) \to \delta(\alpha_i - \alpha_{io})$ where α_{io} is the true parameter value of α_i.

For the case that the controller used its own decisions to direct the learning process, several decision schemes can be applied. The controller may use its own decision in terms of Bayes decision procedure discussed above. In this case, the adaptive action of the controller is actually used as a guide for the learning process. The learning measurements taken by the controller are classified on the basis of the decision made from the estimated probability density functions up to the present. It is hoped that, as the estimated density function for each ω_i approaches the corresponding true density function, the decisions will approach optimum Bayes decisions. Since the classifications by the controller itself can hardly be error-free, in general, the convergence of the learning process cannot be guaranteed. However, if sufficient a priori information has been established either as a pre-specified knowledge about the plant-environment or through a process of learning with external supervision (training), the error probability of the controller's own decisions will be very small. The learning process may converge even in the absence of external supervision. This situation is even obvious if the controller uses a sequential decision procedure to make its own decision since the error probabilities can be pre-specified in the sequential case. For $m = 2$, the requirements on the initial knowledge of measurement distributions for each control situation ω_i, $i = 1, 2$, or on the pre-specified error probabilities can be determined such that the learning process will converge.[20]

The second approach for the controller using its own decisions is to apply a search technique to determine the best control policy. The basic block diagram of the scheme is shown in Fig. 3. Since the controller does not have a complete a priori knowledge about the plant-environment, the index of performance surface to be searched is in general unknown or incompletely known. The search procedure selected for this purpose must have the feature that very little a priori information is required. Under this condition, the decision can be formulated on the basis of randomized control policies, i.e.,

$$r_k(\omega_i) = \sum_{j=1}^{m} \int_{\Omega_{V^k}} L(\omega_i, u_j)$$

$$p(V^k/u^{k-1}, \omega_i) \, P\{u_j(k)/V^k, u^{k-1}\} \, dV^k \quad (13)$$

In order to establish the information about $P\{u_j(k)/V^k, u^{k-1}\}$, the reinforcement technique becomes a good choice.[19,24] It is essentially a random search procedure with Markov properties. The features of the reinforcement technique are: (1) less a priori information required, (2) false (local) extrema avoided, (3) simplicity

in computation, and (4) convergence sometimes even with nonstationary searched surfaces. In using the randomized control policies with reinforcement techniques, the controller first evaluates the system's performance. From the system's performance measure, the type and the amount of reinforcement on each of the probabilities $P\{u_j(k)/V^k, u^{k-1}\}$, $j = 1, \ldots, m$ are determined. Each control policy is initially considered as equally probable for every control situation or they are associated with different initial probabilities on the basis of a priori information. These probabilities associated with each control policy, $P\{u_j(k)/V^k, u^{k-1}\}$, will be modified through the reinforcement technique after $V(k)$ is observed. Corresponding to a particular control situation ω_i, in a limit, the probability of the optimum control policy will approach one with probability one. However, in this case, the decision made by the controller is, in general, biased due to the reason that the controller may not have the chance to try all possible admissible control policies. In order to eliminate this disadvantage, an alternative approach has been suggested. The decision on $u(k)$, $k = 1, 2, \ldots$, is generated completely on the basis of a set of subjective probabilities $P\{u_j(k)/V^k, u^{k-1}\}$. The set of subjective probabilities is modified through an algorithm of reinforcement after each measurement.[*]

CONCLUSIONS AND FURTHER REMARKS

An informal introduction of learning control systems has been presented. Particularly, the application of statistical decision theory to the synthesis of learning control systems has been discussed. In general, learning processes may be classified into two classes, namely, learning with external supervision and learning without external supervision. Relationships between learning

---

* A detailed discussion of this approach will be reported in a separate paper "An Algorithm for Learning Without External Supervision and Its Application to Learning Control Systems" by Z. J. Nikolic and K. S. Fu.

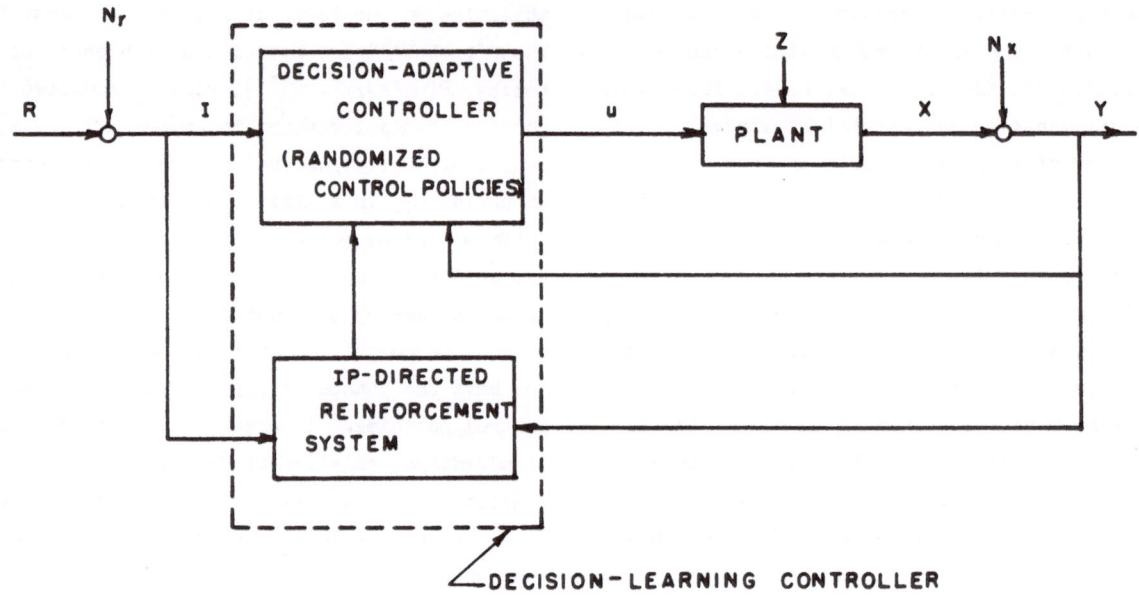

Fig. 3. A DECISION-LEARNING CONTROL SYSTEM
WITHOUT EXTERNAL SUPERVISION

control and pattern recognition have been demonstrated. Several learning schemes have been described and compared. The learning controller will establish the necessary information about the plant-environment during its operation such that the system's performance will be improved as the learning process proceeds.

The increase of dimensionality of v^k and u^k as k increases is a problem which should be seriously considered in practical computations. However, any additional information about the plant-environment may help to reduce the difficulty of this problem. For example, if the function f only relates $X(k+1)$ to $X(k)$ and $u(k)$ or, furthermore, the plant is memoryless, the trouble due to the increase of dimensionality in computation will be greatly simplified. The performance of the system learning with external supervision is bounded by the system's performance of which all the necessary a priori information is known. The performance of the system learning without external supervision is bounded by the performance of the system learning with external supervision. In general, the rate of learning is much slower in the case of learning without external supervision than that with external supervision. If the a priori knowledge is sufficiently good, the scheme of learning using controller's own Bayes decision procedure seems a practical one. The computations and the memory capacity involved in the learning scheme are comparatively much reduced although the convergence of the process usually can not be guaranteed. The algorithm of reinforcement for the controller learning with randomized control policies is, in general, not unique. The problem of obtaining an optimum algorithm for maximum learning rate is certainly an interesting and important one.

ACKNOWLEDGMENT

This work was supported in part by the National Science Foundation Grant GP-2183.

The author would also like to thank Mr. Z. J. Nikolic for his helpful discussions.

REFERENCES

(1) E. Mishkin and L. Braun, Jr., ed., _Adaptive Control Systems_, Chapter 1, McGraw-Hill, New York, 1961.

(2) J. E. Gibson, _Nonlinear Automatic Control_, Chapter 11, McGraw-Hill, New York, 1963.

(3) J. T. Tou, _Modern Control Theory_, Chapter 7, McGraw-Hill, New York, 1964.

(4) L. A. Zadeh, "On the Definition of Adaptivity," Proc. IEEE, Vol. 51, No. 3, March, 1963.

(5) J. E. Gibson, K. S. Fu, et. al., "Philosophy and State of the Art of Learning Control Systems," Rept. TR-EE 63-7, School of Electrical Engineering, Purdue University, Lafayette, Indiana, November, 1963.

(6) K. S. Fu, "Learning Control Systems," COINS Symposium, June, 1963, Evanston, Illinois, Published in _Computer and Information Sciences_, Spartan Books, Washington, D. C., 1964.

(7) J. G. Truxal and J. J. Padalino, "Decision Theory," Chapter 15 of _Adaptive Control Systems_ edited by Mishkin and Braun, McGraw-Hill, 1961.

(8) J. C. Hsu and W. E. Meserve, "Decision-Making in Adaptive Control Systems," Trans. IRE on Automatic Control, pp. 24-32, January 1962.

(9) Y. Sawaragi, Y. Sunahara, and T. Nakamizo, "Decision Making in Adaptive Control Systems with Random Inputs," Parts I, II, and III, Tech. Repts. of the Engr. Res. Inst., Kyoto University, 1963.

(10) A. A. Feldbaum, "One Class of Self-Organizing Systems with Dual Control," Automatika i Telemekhanika, Vol. 25, No. 4, April, 1964.

(11) M. Aoki, "On Control System Equivalents of Some Decision Theoretic Theorems," Journal of Mathematical Analysis and Applications, January, 1965.

(12) D. Blackwell and M. A. Girshick, _Theory of Game and Statistical Decisions_, Wiley, New York, 1954.

(13) K. S. Fu and J. A. Lebo, "On the Selection of Decision Criteria and the Estimation of Probabilities in Pattern Recognition," Rept. TR-EE 64-16, School of Electrical Engineering, Purdue University, Lafayette, Indiana, Sept. 1964.

(14) K. S. Fu, "A Statistical Approach to the Design of Pattern Recognition and Learning Systems," Cybernetics, No. 2, 1962.

(15) R. Bellman, R. Kalaba, and D. Middleton, "Dynamic Programming, Sequential Estimation and Sequential Detection Processes," Proc. Nat. Acad. Sci., Vol. 47, 1961, pp. 338-341.

(16) K. S. Fu and C. H. Chen, "A Sequential Decision Approach to Problems in Pattern Recognition and Learning," Proc. Third Symposium on Adaptive Processes, IEEE Special Publication, Sept. 1964.

(17) N. Abramson and D. Braverman, "Learning to Recognize Patterns in a Random Environment," Trans. IRE on Information Theory, Sept., 1962.

(18) D. G. Keehn, "Learning the Mean Vector and Covariance Matrix of Gaussian Signals in Pattern Recognition," Tech. Rept. No. 2003-6, Stanford, Electronics Laboratories, Feb. 1963.

(19) M. D. Waltz and K. S. Fu, "A Computer-Simulated Learning Control System," IEEE International Convention Record, Part 1, 1964.

(20) K. S. Fu and C. H. Chen, "Sequential Decisions, Pattern Recognition and Machine Learning," Rept. TR-EE 65-6, School of Electrical Engineering, Purdue University, Lafayette, Indiana, April, 1965.

(21) M.A. Aizerman, E. M. Braverman, and L. I. Rozoner, "The Probability Problem of Pattern Recognition Learning and the Method of Potential Functions," Automatika i Telemekhanika, Vol. 25, No. 6, Sept. 1964.

(22) G. S. Sebestyen, "Pattern Recognition by an Adaptive Process of Sample Set Construction," Trans. IRE on Information Theory, Vol. IT-8, Sept. 1962.

(23) S. C. Fralick, "The Synthesis of Machines which Learn without a Teacher," Tech. Rept. No. 61308-9, Stanford Electronics Laboratory, April, 1964.

(24) R. W. McLaren and K.S. Fu, "Synthesis of Learning Systems Operating in an Unknown Random Environment," International Conference on Microwave, Circuit Theory and Information Theory, Sept. 7-12, 1964, Tokyo, Japan.

Discussion

Prof. J.H. Westcott (UK), Prof. R.J.A. Paul (UK) and Mr. P.C. Young (UK) asked the author to comment on the rate of convergence, the amount of storage required and the sensitivity of the learning control system to changes in system statistics.

Prof. Fu replied that convergence is fast when the learning is supervised. For learning without supervision, the first scheme proposed in the paper converges very slowly and requires a large memory; the second scheme is a good practical solution as only simple computation is required, but it is difficult to prove convergence. In a few simple cases it has been found that convergence depends on the a priori information. The third scheme converges, but the convergence is slow because a global random search is required to find the true minimum. If the system statistics are non-stationary it is nevertheless possible, in many cases, to express the non-stationary statistics - for example a non-stationary mean - as a polynomial of time; the parameters of this polynomial could then be estimated using the methods outlined in the paper.

Mr. T. Horrocks (UK) questioned the applicability of the minimum principle in situations where the environment is not intelligent. This prompted the following contribution by Prof. L.E. Zachrisson (Sweden):-

Note on the non-necessity of saddlepoints in games against nature

Main Argument

$\{u\}$ denotes my set of strategies

$\{v\}$ denotes nature's set of strategies

The loss function is $L(u, v)$

I wish to minimise $L(u,v)$ where <u>I do not know</u> what nature has selected. If I choose a strategy u the worst that can happen (but it can happen) is that nature chooses $v = v'(u)$ so that $L(u,v)$ is maximised and becomes:

$$L(u,v'(u)) = \max_{v} L(u,v) .$$

If I wish to minimise the worst that can happen I have to choose $u = u''$ such that:

$$L'' = L(u'',v'(u'')) = \min_{u} \max_{v} L(u,v) .$$

With this choice of u my loss will never exceed L'' (but it might be equal to L''):

$$L(u'',v) \leqslant \max_{v} L(u'',v) = L(u'',v'(u'')) = L'' .$$

As far as I can see the question as to whether the equality in the (true) relation:

$$\min_{u} \max_{v} L(u,v) \geqslant \max_{v} \min_{u} L(u,v)$$

is ever satisfied need never arise. (Of course, equality $\Longleftrightarrow$ existence of saddle-paint).

ELABORATION

The min-max strategy implies that nature knows my choice, and that I act to minimise the worst that can happen. The max-min strategy implies that I know nature's choice and nature actively opposes my desire to minimise L by choosing itself such a strategy that the min value will be as large as possible.

Min-max = max-min implies that I cannot do any better by knowing nature's choice (if nature plays rationally). But as I do not know nature's choice there is no need to consider the situation.

USEFULNESS OF MIXED STRATEGIES

Nevertheless, it is true that if:

$$\text{min-max} > \text{max-min}$$

I can perform better by using mixed strategies. In this case the loss-function is:

$$\bar{L}(F,v) = \int L(u,v) \, dF(u) .$$

If the distribution that places all probabilities max at the point u'' is denoted by F'' , then:

$$\bar{L}(F'',v) = L(u'',v)$$

and

$$\min_{F} \max_{v} \bar{L}(F,v) \leqslant \max_{v} \bar{L}(F'',v) = \max_{v} L(u'',v) = L'' .$$

If nature has already performed its mixing then I can, by making an appropriate mixing myself, limit my loss to max-min instead of min-max.

But, as I see it, this circumstance does not contradict the statement that the min-max problem is meaningful in its own right.

13

ADMISSIBLE ADAPTIVE CONTROL

by H. Kwakernaak
Department of Applied Physics
Technological University
Delft, Netherlands

ABSTRACT

Ignorance concerning system operation or parameter values often makes it impossible to optimize
a control system. From statistical decision theory the notion of admissible control is borrowed.
It is shown how admissible controllers may be obtained. The selection of suitable admissible control policies is discussed.

1. MOTIVATION

The relation between optimal and adaptive control is not always well-understood. Roughly, optimal control is control which is "better" than
any other possible manner of control. In order to
judge how well various control systems perform
relative to one another, some performance criterion
must be introduced. Precisely what is meant by
adaptive control is not always clear; usually
this notion implies that the control is modified
according to changing circumstances. If the control is modified, this is obviously done to improve the performance of the system. When the modification is introduced so as to obtain the most
possible improvement, we clearly have a case of
optimization.

If a controller is to be designed which must
cope with system or environment changes which
cannot be foreseen at all, not even in a statistical fashion, the concept of optimality breaks
down, however (Zadeh[1]), for reasons which will be
explained. It will be discussed how in such a
case the notion of optimal control must be replaced by that of admissible control, that is,
control which cannot be improved simultaneously
for all possible realizations of the changing
conditions.

2. A CONTROL PROBLEM PROTOTYPE

For the sake of definiteness, we shall employ
the following conception of a control problem. A
discrete-time system with state at time n denoted
by x_n and input denoted by u_n is described by a
given state transition law

$$x_{n+1} = f(x_n,u_n,w_n,\theta,n) \tag{1}$$

where $[w_n]$ is a sequence of independent random
variables and θ a parameter which happens to be of
interest. It is possible to observe at each instant n a quantity y_n which is specified by the
known relation

$$y_n = g(x_n,w_n,\theta,n). \tag{2}$$

All quantities mentioned may or may not be vector-
valued. The discrete-time description is chosen
for convenience, but the following discussion
applies to continuous-time systems just as well.

The system is run during the period n = 0,
1,...,N. It is assumed that a controller is installed which as time proceeds assigns values to
the input u_n according to some rule. At the end
of a run the performance of the interconnection
of system and controller is measured by the number

$$\sum_{n=0}^{N} L(x_n,u_n,w_n,n). \tag{3}$$

We shall think of this quantity as having the
value of cost, so that the smaller it is, the
better the system has performed.

3. PERFORMANCE FUNCTIONS

In the foregoing we have introduced a parameter θ upon which the operation of the system
depends. We have mentioned this parameter explicitly because we shall now assume that the
value this parameter assumes is not known beforehand. In fact we shall use the notion of parameter in a very wide sense: θ will also be
thought to represent our ignorance concerning the
precise operation of parts of the system or the
statistical properties of the random processes
involved.

At the end of a single run the performance
of the system is judged by the outcome of the number (3). This outcome is determined by the particular realization of the random processes $[w_n]$
during that run, the value of the parameter θ,
and the control which is employed. We shall specify the controller in the following manner.
Clearly at the instant n all information on which
to base the choice of the input u_n is contained
in the sequence of observations $[y_0,y_1,...,y_n]$
and the knowledge of the previously applied inputs
$[u_0,u_1,...u_{n-1}]$. Therefore, a function

Superior numbers refer to similarly-numbered references at the end of this paper.

$$u_n = k_n(y_o, y_1, \ldots, y_n; u_o, u_1, \ldots, u_{n-1}) \qquad (4)$$

will be called the control law k_n; following Bellman[2] a sequence of control laws $\pi = \{k_o, k_1, \ldots, k_N\}$ will be called the policy π. The implementation of a policy will be called a controller.

To compare the relative merits of controllers before the system is run, it is usual to take the mathematical expectation of the cost over the stochastic variables. We therefore define the function

$$V(\pi, \theta) = E\left(\sum_{n=o}^{N} L(x_n, u_n, w_n, n)\right) \qquad (5)$$

It is clear that the policy π and the parameter θ appear as independent variables of this function. For fixed π we shall call $(V_\pi, .)$ the performance function of the policy π.

4. ADMISSIBLE CONTROL

If the value of the parameter θ were known, the performance of any given controller would be specified by the single number $V(\pi, \theta)$; by selecting the policy which minimizes this number an optimal solution of the control problem is obtained. Since θ is assumed not to be known, however, another criterion must be devised to select a controller. To this end, we borrow from statistical decision theory (Weiss[3]) the notion of admissibility. We define: An admissible policy is a policy which cannot be improved for all values of θ simultaneously. More precisely, if is an admissible policy, there is no other policy π' such that

$$V(\pi', \theta) \leqq V(\pi, \theta) \quad \text{for all } \theta \qquad (6)$$

$$V(\pi', \theta) < V(\pi, \theta) \quad \begin{array}{l}\text{for at least one} \\ \text{value of } \theta\end{array}$$

Clearly, if nothing is known concerning the value of θ, no one can object to using an admissible policy since no other policy can be guaranteed to yield better performance.

5. BAYES CONTROL

In the preceding discussion we assumed that so little is known regarding the parameter θ that not even a probability distribution can be assigned to it. Suppose, however, that $\beta(\theta)$ is a function defined on the range of parameter θ which has all properties of a probability distribution function. Then for fixed π we can take the expectation of $V(\pi, \theta)$ with respect to this distribution; we shall write this expectation in the generalized form

$$\int V(\pi, \theta) \, d\beta(\theta) \qquad (7)$$

Such a function $\beta(\theta)$ will be called an a priori distribution. For any given a priori distribution we can look for the policy π which minimizes the quantity (7); such a policy will be called a Bayes policy with respect to the a priori distribution $\beta(\theta)$.

It has to be kept in mind, however, that the function $\beta(\theta)$ should not be thought of as an actual probability distribution of θ since θ may not even be a stochastic variable.

6. ADAPTIVE CONTROL

When an a priori distribution for θ is assumed and the corresponding Bayes controller is sought, the problem has been made into a problem of stochastic or Bayesian optimal control theory. It is well-known from the work of Wald[4], Bellman[2] and Feldbaum[5] that finding the optimal policy for a stochastic control problem falls apart into two phases. At a particular instant n, first the a posteriori distribution for the state of the system and the parameter θ must be found; we shall denote this distribution somewhat loosely as

$$p(x_n, \theta \mid y_o, y_1, \ldots, y_n; u_o, u_1, \ldots, u_{n-1}) \qquad (8)$$

The a posteriori distribution at time n can be found from the a posteriori distribution at time (n-1) through the use of Bayes' rule; hence the name Bayes policy. The solution of the problem is considerably simplified if there exists a quantity of low dimension z_n which completely specifies the a posteriori distribution at time n; naturally this quantity is a function of $y_o, \ldots, y_n$, $u_o, \ldots, u_{n-2}$ and u_{n-1}. We shall call such a quantity, if it exists, the information state of the system.

The second phase in the solution of a stochastic control problem consists of finding the optimal control action which must be based upon the a posteriori distribution. If there exists an information state z_n the optimal control law takes the form

$$u_n = k_n(z_n) \qquad (9)$$

The actual solution of the optimization problem can be carried out with the help of dynamic programming[2]. For examples the reader is referred to the literature.

We shall follow Bellman[2] and consider this procedure of first gathering and evaluating new information and then deciding the best action on the basis of this information as characteristic for adaptive control. The first phase can also be viewed as identification and the second as control proper.

7. BAYES POLICIES AND ADMISSIBLE POLICIES

There is a very important result from statistical decision theory[3,4] which will loosely be stated as follows: Let π be a policy which is

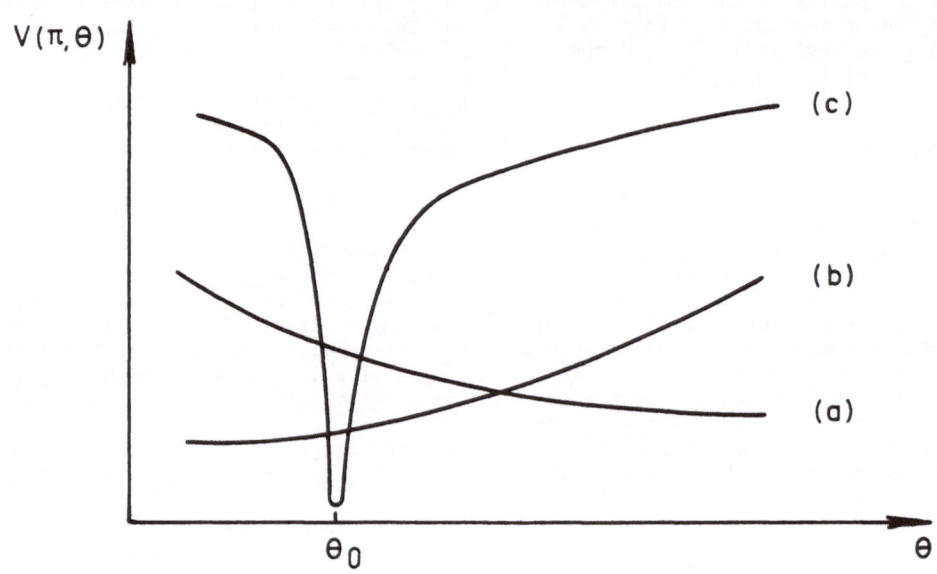

FIGURE 1: PERFORMANCE FUNCTIONS OF
VARIOUS ADMISSIBLE CONTROLLERS

Bayes with respect to an a priori distribution $\beta(\Theta)$. Then if π is unique, it is admissible. If π is not unique, it is admissible if $\beta(\Theta)$ assigns nonzero probability or probability density to all values of Θ.

The proof of this statement is not very difficult, and can be found in textbooks on statistical decision theory (Weiss[3]).

8. ADMISSIBLE ADAPTIVE CONTROL

Since we are interested in finding admissible controllers, for us the most important conclusion of the preceding section is that admissible controllers can be obtained by assuming a suitable a priori distribution on Θ and determining the control policy which is optimal for the resulting Bayes problem. Such a policy is adaptive as defined in Section 6 since it exhibits the features of identification and control proper. Combining these ideas, we have now developed the notion of admissible adaptive control, that is, control which (i) takes into account uncertainties in the system model, and (ii) has the property that it cannot be improved upon simultaneously for all circumstances.

In this notion of admissible adaptive control two prevailing trends in modern control theory, namely that of designing the best possible control system and that of making the system adapt to changes in its environment are brought together. A further discussion is given in the following sections.

9. SELECTION OF ADMISSIBLE CONTROL POLICIES

From the theorem outlined in Section 7 it follows that admissible policies can be found by assigning a suitable a priori distribution to the parameter Θ and constructing the corresponding Bayes controller. In the choice of the a priori distribution one has great freedom although several aspects need be contemplated. One consideration is that it is reasonable to incorporate in the choice of $\beta(\Theta)$ what a priori information one has on Θ. If for one reason or another certain values of Θ are considered more "likely" than other values, the distribution $\beta(\Theta)$ should assign more weight to this region. Among statisticians the legitimacy of this procedure is still subject to furious discussion (Savage[6]), but we shall not go into this here.

Another consideration is that of mathematical convenience. The Gaussian distribution is very popular for this reason, especially in connection with linear systems and quadratic performance functions. A mathematically often even more convenient choice of the a priori distribution is to assign probability one to a particular value Θ_0 of the parameter Θ.

No matter what a priori distribution is chosen, however, it should be kept in mind that the eventual design must be judged from the entire behavior of the performance function for all values of Θ. Curves (a) and (b) of Figure 1 illustrate possible behaviors of performance functions of different admissible controllers. It is up to the designer to decide which controller he prefers. Curve (c) illustrates what could be the performance function of an admissible controller which is obtained by optimizing with respect to an a priori distribution which assigns probability one to the parameter value $\Theta = \Theta_0$. It is clear that a curve like the one sketched is not a very good performance function due to the sensitivity to small differences in the parameter value.

There has been one major attempt in statistical decision theory to remove the arbitrariness in selecting an admissible policy by introducing the minimax principle. That is, for a given policy π, the performance function $V(\pi,\Theta)$ is reduced to a single number $M(\pi) = \sup_\Theta V(\pi,\Theta)$, where Θ varies over its entire permissible range. A minimax policy is then defined as a policy for which $M(\pi)$ is minimized. Such a policy has the property that the upper bound which holds for the performance function is the lowest that can be achieved for any policy. From examples and applications it appears, though, that as often as not the minimax principle yields policies with unfavorable properties, due to the fact that in obtaining them one has to reckon with the most adverse and unreasonable circumstances. Moreover minimax policies are difficult to find.

An encouraging result, however, is that systems which are operated over long periods of time may exhibit the phenomenon that the control ultimately becomes quite independent of the assumed a priori statistics due to the accumulation of fresh information which completely overrides the initial assumption (Aoki[7]). In such a case there are few problems because the choice of the a priori distribution is of little importance.

10. TIME-VARYING PARAMETERS

The problem of compensating for a single time-varying parameter denoted by a_n can be fitted into the present set-up by taking the parameter Θ to be $\Theta = (a_0, a_1, \ldots, a_N)$. A convenient stochastic model for obtaining admissible controllers is the drift model whereby the variations in a_n are described by the relation

$$a_{n+1} = a_n + w_n \qquad (10)$$

Here $[w_n]$ is a sequence of Gaussian independent random variables with identical distributions and zero mean. The variance is chosen in accordance with the expected variations of the parameter.

Another method which is often proposed for adaptive control is the following: (i) measure the value of the parameter at time n, (ii) assume that the parameter remains at this value during the rest of the control period, and (iii) compute a control action which is optimal on the basis of this assumption. This procedure is repeated at all instants of time.

It is clear that this procedure will result in the use of an optimal policy if the value of the parameter remains constant during the entire control period. In this case one cannot improve on the control of the system; therefore viewed with respect to all possible realizations of the time-varying parameter the procedure constitutes an admissible policy.

11. THE ADMISSIBILITY OF AN EXISTING DESIGN METHOD

A design philosophy which is fairly well-established by now is the optimal filtering and control theory prompted by the work of Wiener. This method can be examined in the light of the present discussion as to what extent it yields admissible control policies. For given second-order properties of the random processes involved, Wiener's linear controller is optimal if the processes are Gaussian. Consequently the linear controller is admissible even if the exact distributions are not known.

12. RESUME

An adaptive control system is defined as a control system which exhibits the features of (i) identification in the form of the determination of a posteriori statistics and (ii) control proper. Ignorance concerning the system operation or parameter values often precludes the determination of an optimal controller; to allow a criterion of choice between different controllers from statistical decision theory the notion of admissible control is borrowed. Admissible control is defined as control which cannot be improved for all realizations of the unknown parameters simultaneously.

Admissible adaptive controllers can be obtained as the solution of a stochastic optimization problem by assigning to the unknown parameters a fictitious probability distribution. The eventual evaluation of any chosen controller, however, should be based on the consideration of the entire performance function of the controller, with the unknown parameter as the independent variable. The final selection of a controller is left to the designer: Admissibility is no guarantee for acceptability.

It is the purpose of this paper to add to the current insight into adaptive and optimal control theory rather than to suggest specific design methods. In the opinion of the author one of the most important conclusions is that relaxing the requirement of optimality to that of admissibility will often result in a more realistic approach to control problems.

REFERENCES

(1) Zadeh, L.A., "What is Optimal?", I.R.E. Trans. on Inf. Th., 4, p. 3 (1958).
(2) Bellman, R. E., Adaptive Control Processes: A Guided Tour. Princeton Univ. Press, Princeton, N. J., 1961.
(3) Weiss, L., Statistical Decision Theory, McGraw-Hill, New York, N. Y., 1961.
(4) Wald, A., Statistical Decision Functions, Wiley, New York, N. Y., 1950.
(5) Feldbaum, A. A., "Dual Control Theory," Pts. I through IV, Automation and Remote Control 21, Nos. 9, 11, 22, Nos. 1, 2 (1960, 1961).
(6) Savage, L. J. "The Foundations of Statistics Reconsidered," Proc. Fourth Berkeley Symposium on Mathematical Statistics and Probability, Vol. I, p. 575-586, Univ. of Calif. Press, Berkeley and Los Angeles, Calif., 1961.
(7) Aoki, M., "On Some Convergence Questions in Bayesian Optimization Problems." I.E.E.E. Trans. on Aut. Contr., 10, p. 180-182 (April, 1965).

DISCUSSION

Dr. O. L. R. Jacobs (UK) pointed out that in section 3 of the paper Θ is stated to be an unknown but constant parameter, but in section 10 it is stated that Θ might be non-stationary or even a Markov process. He asked how these two statements are to be reconciled.

Dr. Kwakernaak replied that if there is a time varying parameter a, then Θ, the set of unknown parameters, is:

$$\Theta = (a_1 \ldots\ldots a_n)$$

If the parameter is stochastic it can be incorporated as one of the variable in the state equations.

Prof. C. J. D. M. Verhagen (Netherlands) noted that $\beta(\Theta)$ is stated to have all the properties of a probability distribution function yet is not in fact such a function. He asked how, if a concept is defined by the sum of its properties, such a distinction can be maintained?

Dr. Kwakernaak admitted that this question is difficult to answer since it is a question of philosophy. He preferred not to call $\beta(\Theta)$ a probability distribution because Θ is not a random variable.

A GRADIENT METHOD FOR DETERMINING OPTIMAL CONTROL OF NON-LINEAR STOCHASTIC SYSTEMS

by D.Q. Mayne
Control Section
Imperial College of Science
and Technology, London.

ABSTRACT

This paper describes a feasible method for estimating the gradient of the cost function with respect to the parameters of the controller in order to improve or, possibly to optimise the control of non-linear discrete-time stochastic systems. When the disturbances have few discrete values a gradient method for determining the optimal control sequences for each possible outcome of disturbances is presented. For more complex problems a Monte Carlo method is proposed.

INTRODUCTION

This paper discusses the possibility of improving and, in certain cases, of optimising non-linear stochastic control systems. The formulation of the problem is wide enough to cover the adaptive control problem since the unknown parameters can be included in an augmented state-vector. Florentin[1] discusses an example where the probability density of the state-vector, conditional on the available observations, can be expressed in terms of a small set of variables, called the sufficient statistics. This conditional density, is, in effect, the 'state' of the stochastic control problem, and has been called the hyperstate. A stochastic difference or differential equation, describing the future evolution of the sufficient statistics, can often be formed. It is theoretically possible to obtain the optimal control as a function of these statistics, and this has been done by Florentin[1] using dynamic programming, a technique which cannot be used for more complex problems. Kushner[2] shows how the problem may be formulated when the hyperstate must be expressed as a probability density, derives equations describing the future evolution of this conditional density, and shows formally how the optimal control may be obtained, using dynamic programming techniques, in terms of the current conditional density. The problem is considerably more complex since the hyperstate is now a density function.

Since the dynamic programming technique, though conceptually simple and invaluable in formulating the problem, involves excessive storage, it seems necessary to resort to gradient techniques. Because the future evolution of the hyperstate, given the current hyperstate, is described by stochastic equations, it is necessary, even with gradient techniques, to reduce the dimensionality of the problem. One way of doing this is to replace the continuous random disturbances by random disturbances which have a finite number of discrete states and which occur only at discrete instants of time. This will be called the sampling method. Alternatively Monte Carlo techniques may be used to generate several possible realisations of the hyperstate trajectory. The gradient of the cost with respect to the initial control for each trajectory may be used to estimate the required improvement to the initial control. Successive approximation methods, using both approaches, are described in the paper.

Since this paper was first written a paper[3] of direct relevance by Kwakernaak, who seems to have been the first to apply Monte Carlo techniques to stochastic control problems, has appeared. Some material in the original paper has been deleted in an attempt to present results, which, apart from some introductory remarks, complement those of Kwakernaak. In particular, a different technique of evaluating gradients is used, and a useful technique of variance reduction is presented.

FORMULATION OF THE PROBLEM

Consider a non-linear system whose state at time k is x_k. The evolution of the system is described by the difference equation:

$$x_{k+1} = f_k(x_k, u_k, \xi_k) \qquad (1)$$

where u_k is the control vector at time k and

Superior numbers refer to similarly-numbered references at the end of this paper.

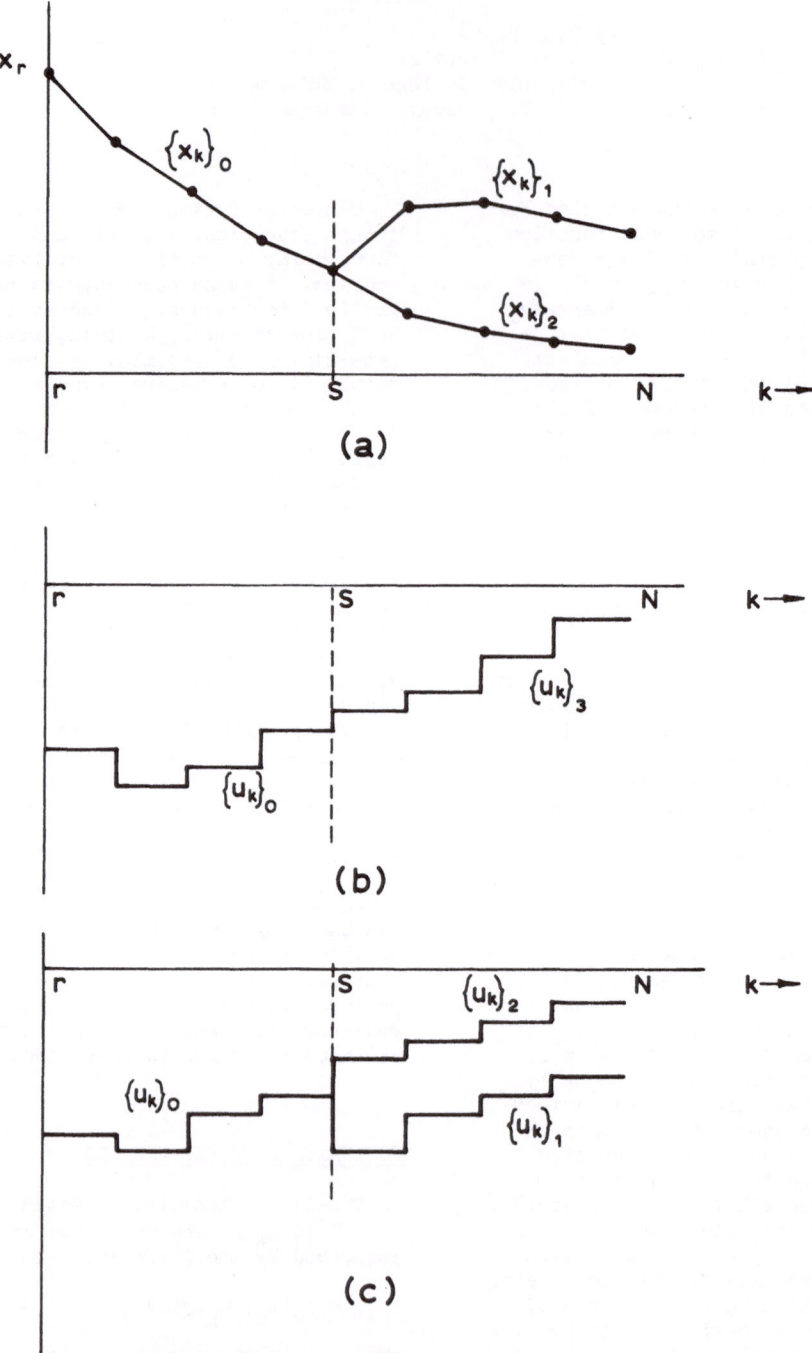

Fig. 1

$\{\xi_k\}$ is a sequence of independent random vector variables. At time k a noisy observation y_k is made:

$$y_k = h_k(x_k, \eta_k) \qquad (2)$$

y_k may be a vector and $\{\eta_k\}$ a sequence of independent random vector variables. The process begins at time 1. At time r, the objective is to minimise V_r the expected value of the future cost:

$$V_r = E\left[\sum_{k=r}^{N-1} L_k(x_k, u_k) + L_N(x_N)\right] \qquad (3)$$

Let s_k be a vector composed of the previous observations and inputs:

$$s_k^T = (y_k^T, u_{k-1}^T, \ldots\ldots y_2^T, u_1^T, y_1^T) \qquad (4)$$

This has been called the information state. At time r the hyperstate is the conditional density $p(x_r/s_r)$. In order to perform the expectation in equation (3) it is necessary to estimate the future densities of x conditional on the current information state s_r and the sequence of proposed future controls. If we can specify these future densities by a set of parameters z_k, then at time r, given s_r, it is possible to estimate the future densities from a stochastic difference equation:

$$z_{k+1} = f_k^*(z_k, u_k, \xi_k, \eta_k) \qquad (5)$$

where the boundary value z_r is known from $p(x_r/s_r)$. A simple example is given by Florentin[1]. If the expectation in equation (3) is performed with respect to the density of x_k specified by z_k and then with respect to the density of z_k, it is possible to write:

$$V = E \sum_{k=r}^{N-1} L_k^*(z_k, u_k) + L_N^*(z_N) \qquad (6)$$

where the expectation in equation (6) is taken over all possible realisations of z_k given z_r. The optimisation operation, using dynamic programming, gives the optimum control laws $u_k^0 = g_k^0(z_k)$. Equations (5) and (6) specify the problem considered in this paper, which is mainly concerned with feasible methods of calculating expectations in order to evaluate gradients of the cost function to improve or, possibly, to optimise the control. Note that in obtaining equation (6) an expectation has already been performed. If this is difficult to do analytically, the methods proposed in this paper may be extended to this operation as well, as proposed by Kwakernaak (3)

for the initial density $p(x_r/s_r)$. In this case equations (1), (3) and (5) would be employed. Equations (1) and (3) being similar to equations (5) and (6) will be used in the sequal, x being regarded as the hyperstate, to avoid complicating the notation.

SAMPLING METHOD

One way of simplifying the expectation operation is to replace continuous distributions by discrete distributions, so that, for example, at time s

$$P(\xi_s = a_1) = p_1$$
$$P(\xi_s = a_2) = 1 - p_1 = p_2 \qquad (7)$$

The control problem, given distributions of this sort, is of interest in its own right; the question as to whether a reasonable approximation to the original problem can be made in this way will not be examined here. It is plausible that the method improves on the commonly adopted procedure of obtaining the control at time k by optimising ignoring future disturbances.

Consider the system described by equations (1) and (3) where x is now regarded as the hyperstate of an original problem. Consider initially a simplified problem where:

$$\xi_k = 0, \; k \neq s$$

and let the distribution ξ_s be specified by equation (7). Thus x will vary as shown in Figure 1(a) during the interval $[r, s-1]$. During the intervals $[s, N]$ the value of x will depend on the value of the disturbance ξ_s. The two realisations of the trajectory corresponding to $\xi_s = a_1$ and $\xi_s = a_2$ will be labelled 1 and 2 respectively and the label 0 will be used for the initial $(k \leqslant s-1)$ part of the trajectory. Thus $[x_k]_0$ denotes x_k where $k \leqslant s-1$, $[x_k]_1$ denotes x_k where $k \geqslant s$ and $\xi_s = a_1$. If future values of the hyperstate will not be available the optimal control sequence could be specified as in Figure 1(b). If future values of the hyperstate will be available, the only new information occurs at time s when the value of ξ_s is ascertained. A control sequence appropriate to each value of ξ_s must be chosen leading to the specification shown in Figure 1(c). Note x_s has a single value, but u_s has two values $[u_s]_1$ and $[u_s]_2$. It is for this reason that the division between the trajectory 0, on the one hand, and the

21

trajectories 1 and 2, on the other hand, is put between $k = s$ and $k = s - 1$. The problem of optimal control, given that the hyperstate will be updated, is to choose the sequences $\{u_k\}_0$, $\{u_k\}_1$ and $\{u_k\}_2$ to minimise V where:

$$V = V_o + p_1 V_1 + p_2 V_2$$

$$V_o = \sum_{k=r}^{s-1} L_k(x_k, u_k)$$

$$V_1 = \left[\sum_{k=s}^{N-1} L_k(x_k, u_k) + L_N(x_N) \right]_1$$

$$V_2 = \left[\sum_{k=s}^{N-1} L_k(x_k, u_k) + L_N(x_N) \right]_2$$

The problem is now amendable to gradient techniques. Equations for the first variation method for improving a control sequence $\{u_k\}$ for deterministic discrete-time systems are given below. Assume that the nominal control sequences $\{u_k\}_j$, $j = 0, 1, 2$, have been chosen (Figure 1(c)), and the trajectories corresponding to these control sequences have been generated. Then along any of the trajectories 0, 1 and 2 the variable V_{x_k} can be generated by the backwards recursive relation:

$$V_{x_k} = \frac{\partial H_k}{\partial x_k} \tag{9}$$

$$H_k(x_k, u_k, V_{x_k}) = L_k(x_k, u_k) + \left\langle V_{x_{k+1}}, f_k(x_k, u_k, \xi_k) \right\rangle \tag{10}$$

where $\langle , \rangle$ denotes the inner product of two vectors and where all variables are evaluated on the appropriate trajectory. V_{x_t} is a variable such that, if x_t is changed by an amount δx_t keeping the controls and disturbances unaltered but allowing x_k, $k > t$, to change in accordance with the dynamic equations of the system, then the first order change δV in cost will be:

$$\delta V = \left\langle V_{x_t}, \delta x_t \right\rangle$$

Equation (10) is the adjoint equation for discrete-time systems. In our problem, the variables $\left[V_{x_k} \right]_0$, $\left[V_{x_k} \right]_1$, $\left[V_{x_k} \right]_2$ are such that:

Due to $[\delta x_k]_0$, $\delta V = \left[\left\langle V_{x_k}, \delta x_k \right\rangle \right]_0 = $
$$\triangleq \left\langle \left[V_{x_k} \right]_0, \left[\delta x_k \right]_0 \right\rangle \tag{11a}$$

Due to $[\delta x_k]_1$, $\delta V_1 = \left[\left\langle V_{x_k}, \delta x_k \right\rangle \right]_1 = $
$$\triangleq \left\langle \left[V_{x_k} \right]_1, \left[\delta x_k \right]_1 \right\rangle \tag{11b}$$

Due to $[\delta x_k]_2$, $\delta V_2 = \left[\left\langle V_{x_k}, \delta x_k \right\rangle \right]_2 = $
$$\triangleq \left\langle \left[V_{x_k} \right]_2, \left[\delta x_k \right]_2 \right\rangle \tag{11c}$$

The boundary conditions for $\left[V_{x_k} \right]_1$ and $\left[V_{x_k} \right]_2$ are given by:

$$\left[V_{x_N} \right]_1 = \left[\frac{\partial L_N(x_N)}{\partial x_N} \right]_1$$
$$\left[V_{x_N} \right]_2 = \left[\frac{\partial L_N(x_N)}{\partial x_N} \right]_2 \tag{12}$$

The boundary value for $\left[V_{x_k} \right]_0$ can be found as follows: consider the first order change δV in the expected cost due to a change δx_s.

Due to δx_s, $\delta V = p_1 \left[\left\langle \frac{\partial L_s(x_s, u_s)}{\partial x_s}, \delta x_s \right\rangle \right]_1$
$$+ p_2 \left[\left\langle \frac{\partial L_s(x_s, u_s)}{\partial x_s}, \delta x_s \right\rangle \right]_2$$
$$+ p_1 \left[\left\langle V_{x_{s+1}}, \frac{\partial f_s(x_s, u_s, \xi_s)}{\partial x_s} \delta x_s \right\rangle \right]_1$$
$$+ p_2 \left[\left\langle V_{x_{s+1}}, \frac{\partial f_s(x_s, u_s, \xi_s)}{\partial x_s} \delta x_s \right\rangle \right]_2$$

During the interval $[s, s + 1]$ $[u_s]_1 \neq [u_s]_2$

Whence:
$$\left[V_{x_s} \right]_0 = p_1 \left[V_{x_s} \right]_1 + p_2 \left[V_{x_s} \right]_2 \tag{13}$$

where:
$$\left[V_{x_s} \right]_1 = \left[\frac{\partial H_s}{\partial x_s} \right]_1, \quad \left[V_{x_s} \right]_2 = \left[\frac{\partial H_s}{\partial x_s} \right]_2 \tag{14}$$

Once V_{x_k} is evaluated for each trajectory the variation δV due to δu_k can be evaluated. For a deterministic system:

Due to δu_k, $\delta V = \left\langle \dfrac{\partial H_k}{\partial u_k}, \delta u_k \right\rangle$

For the stochastic system being considered the total change δV in expected cost due to changes $\left[\delta u_k \right]_j$, $j = 0, 1, 2$, is:

$$\delta V = \left[\sum_{k=1}^{s-1} \left\langle \frac{\partial H_k}{\partial u_k}, \delta u_k \right\rangle \right]_0$$
$$+ p_1 \left[\sum_{k=s}^{N-1} \left\langle \frac{\partial H_k}{\partial u_k}, \delta u_k \right\rangle \right]_1 \qquad (15)$$
$$+ p_2 \left[\sum_{k=s}^{N-1} \left\langle \frac{\partial H_k}{\partial u_k}, \delta u_k \right\rangle \right]_2$$

The routine for improving the control sequence is, therefore:

$$\left[\delta u_k \right]_0 = -\varepsilon \left[\frac{\partial H_k}{\partial u_k} \right]_0$$
$$\left[\delta u_k \right]_1 = -\varepsilon p_1 \left[\frac{\partial H_k}{\partial u_k} \right]_1 \qquad (16)$$
$$\left[\delta u_k \right]_2 = -\varepsilon p_2 \left[\frac{\partial H_k}{\partial u_k} \right]_2$$

where ε is a suitably chosen small positive number. When the hyperstate is not updated $\{u_k\}_1 = \{u_k\}_2$. Label this part of the control sequence as $\{u_k\}_3$.

Then:

$$\left[\delta u_k \right]_0 = -\varepsilon \left[\frac{\partial H_k}{\partial u_k} \right]_0$$
$$\left[\delta u_k \right]_3 = -\varepsilon \left[p_1 \left[\frac{\partial H_k}{\partial u_k} \right]_1 + p_2 \left[\frac{\partial H_k}{\partial u_k} \right]_2 \right] =$$
$$= -\varepsilon \left[E \frac{\partial H_k}{\partial u_k} \right] \qquad (17)$$

Example

Consider the simple example:

$$x_{k+1} = x_k + u_k + \xi_k, \quad x_1 = 5$$
$$V = E \left[\sum_{k=1}^{4} u_k^2/2 + x_5^2/2 \right]$$
$$\xi_k = 0, \; k \neq 3$$
$$P(\xi_3 = \pm 1) = \tfrac{1}{2}$$

Let the nominal control sequence be:

$$u_1, \; u_2, \; \left[u_3\right]_1, \; \left[u_4\right]_1, \; \left[u_3\right]_2, \; \left[u_4\right]_2$$

Then:

$$\left[x_5\right]_1 = 6 + u_1 + u_2 + \left[u_3\right]_1 + \left[u_4\right]_1$$
$$\left[x_5\right]_2 = 4 + u_1 + u_2 + \left[u_3\right]_2 + \left[u_4\right]_2$$
$$H_k = u_k^2/2 + V_{x_{k+1}} (x_k + u_k + \xi_k)$$

The recursive relation for V_x is:

$$V_{x_k} = \frac{\partial}{\partial x_k} \left[V_{x_{k+1}} (x_k + u_k + \xi_k) \right] = V_{x_{k+1}}$$

where $V_{x_N} = x_N$

Whence:

$$\left[V_{x_5}\right]_1 = \left[V_{x_4}\right]_1 = \left[V_{x_3}\right]_1 =$$
$$= 6 + u_1 + u_2 + \left[u_3\right]_1 + \left[u_4\right]_1$$

$$\left[V_{x_5}\right]_2 = \left[V_{x_4}\right]_2 = \left[V_{x_3}\right]_2 =$$
$$= 4 + u_1 + u_2 + \left[u_3\right]_2 + \left[u_4\right]_2$$

$$\left[V_{x_3}\right]_0 = \tfrac{1}{2}\left[V_{x_3}\right]_1 + \tfrac{1}{2}\left[V_{x_3}\right]_2 =$$
$$= 5 + u_1 + u_2 + \tfrac{1}{2}\Big\{\left[u_3\right]_1 + \left[u_4\right]_1$$
$$\left[u_3\right]_2 + \left[u_4\right]_2 \Big\}$$

$$\left[V_{x_1}\right]_0 = \left[V_{x_2}\right]_0 = \left[V_{x_3}\right]_0$$

$$\therefore \quad \delta u_1 = -\varepsilon \frac{\partial H_1}{\partial u_1} = -\varepsilon \frac{\partial}{\partial u_1} \left[\tfrac{1}{2} u_1^2 + \left[V_{x_2} \right]_0 \times \right.$$

$$\left. \left[x_1 + u_1 + \xi_1 \right] \right] = -\varepsilon \left[u_1 + \left[V_{x_2} \right]_0 \right]$$

i.e.

$$\delta u_1 = -\varepsilon \left[5 + 2u_1 + u_2 + \tfrac{1}{2} \left\{ \left[u_3 \right]_1 + \left[u_4 \right]_1 + \left[u_3 \right]_2 + \left[u_4 \right]_2 \right\} \right]$$

$$\delta u_2 = -\varepsilon \left[5 + u_1 + 2u_2 + \tfrac{1}{2} \left\{ \left[u_3 \right]_1 + \left[u_4 \right]_1 + \left[u_3 \right]_2 + \left[u_4 \right]_2 \right\} \right]$$

$$\left[\delta u_3 \right]_1 = -\frac{\varepsilon}{2} \left[6 + u_1 + u_2 + 2 \left[u_3 \right]_1 + \left[u_4 \right]_1 \right]$$

$$\left[\delta u_4 \right]_1 = -\frac{\varepsilon}{2} \left[6 + u_1 + u_2 + \left[u_3 \right]_1 + \left[2u_4 \right]_1 \right]$$

$$\left[\delta u_3 \right]_2 = -\frac{\varepsilon}{2} \left[4 + u_1 + u_2 + 2 \left[u_3 \right]_2 + \left[u_4 \right]_2 \right]$$

$$\left[\delta u_4 \right]_2 = -\frac{\varepsilon}{2} \left[4 + u_1 + u_2 + \left[u_3 \right]_2 + 2 \left[u_4 \right]_2 \right]$$

Optimality is achieved when:

$$u_1 = u_2 = -1$$
$$\left[u_3 \right]_1 = \left[u_4 \right]_1 = -\frac{4}{3}$$
$$\left[u_3 \right]_2 = \left[u_4 \right]_2 = -\frac{2}{3}$$

when the values of δu become zero. When no observations are made, the nominal control sequence is u_1, u_2, u_3, u_4 and:

$$\delta u_1 = -\varepsilon \left[5 + 2u_1 + u_2 + u_3 + u_4 \right]$$

$$\delta u_2 = -\varepsilon \left[5 + u_1 + 2u_2 + u_3 + u_4 \right]$$

$$\delta u_3 = -\varepsilon \left[5 + u_1 + u_2 + 2u_3 + u_4 \right]$$

$$\delta u_4 = -\varepsilon \left[5 + u_1 + u_2 + u_3 + 2u_4 \right]$$

Optimality is achieved when:

$$u_1 = u_2 = u_3 = u_4 = -1$$

Extensions

Using the rule for evaluation of V_{x_k} where trajectories unite, the method can be easily extended to deal with random disturbances having more than two discrete levels, and occurring at more than one instant of time. The number of trajectories increases alarmingly as k increases. However, for this simplified problem, the optimal value of the current control u_r can be found by repeating the algorithm routine until

$$\left[\frac{\partial H_k}{\partial u_k} \right]_j \quad j = 0 \ldots \ldots J, \text{ is zero.}$$

If desired the second variation method can be used to hasten convergence in the final stages.

MONTE CARLO TECHNIQUES

Consider again the original problem of finding an improvement δu_r if the control action u_r for a given state x_r where $\{\xi_k\}$ is a sequence of independent vector random variables of known distribution - i.e., a change δu_r is sought which will reduce the cost V_r defined by equation (3) and where the expectation is taken over all possible realisations of the sequence $\{\xi_k\}$. A particular realisation $\{\xi_k\}_j$ of this sequence is generated using a random number generator which has the necessary statistical properties. The state variable and control action sequences $\{x_k\}_j$ and $\{u_k\}_j$ corresponding to this realisation are generated using the system difference equation (1) and sub-optimal control laws

$$u_k = g_k (x_k), \quad k = r \ldots \ldots N - 1$$

The sequence $\{V_{x_k}\}_j$ is then calculated in reverse time using equations (9) and (10) together with the boundary condition:

$$\left[V_{x_N} \right]_j = \left[\frac{\partial L_N (x_N)}{\partial x_N} \right]_j$$

Hence

$$\left[\frac{\partial H_r}{\partial u_r} \right]_j$$

can be calculated for this realisation of $\{\xi_k\}$. The first order change $\left[\delta V_r \right]_j$ in $\left[V_r \right]_j$

24

due to a first order change δu_r in u_r where $[V_r]_j$ is the cost of this j^{th} realisation is:

$$[\delta V_r]_j = \left\langle \left[\frac{\partial H_r}{\partial u_r}\right]_j, \quad \delta u_r \right\rangle$$

This is one estimate of the expected value of δV_r

$$\delta V_r = E\left[\delta V_r\right]_j = \left\langle E\left[\frac{\partial H_r}{\partial u_r}\right]_j, \quad \delta u_r \right\rangle$$

The optimal value of u_r given the subsequent control laws must satisfy:

$$E\left[\frac{\partial H_r}{\partial u_r}\right]_j = 0$$

If several realisations $j = 1 \ldots\ldots J$ are generated then an improved estimate of $E\left[\frac{\partial H_r}{\partial u_r}\right]_j$ is given by $\overline{\left[\frac{\partial H_r}{\partial u_r}\right]_j}$ where the bar denotes the average of the J estimates $\left[\frac{\partial H_r}{\partial u_r}\right]_j$

If this estimate is sufficiently large compared with its standard deviation, u_r can be improved by the algorithm

$$\delta u_r = -\varepsilon \overline{\left[\frac{\partial H_r}{\partial u_r}\right]_j}$$

Repeated application of this algorithm yields the optimal value of u_r for the state x_r given the subsequent non-optimal control laws $g_k(x_k)$, $k = r+1 \ldots\ldots N-1$.

This procedure is satisfactory for improving the control action for a given state. Determination of optimal control for a stochastic system however requires the determination of optimal control laws which specify the control action for each point (x, k) in phase space. This is not practical. However, in many cases it should be possible to specify the control laws by a finite set of parameters which can then be estimated:

$$u_k = h_k (x_k, \theta_k), \quad k = 1 \ldots\ldots N-1$$

where $h_k(. , .)$ is a known function and θ_k is a set of parameters. The simplest example is the 'open-loop' controller:

$$u_k = \theta_k, \quad k = 1 \ldots\ldots N-1$$

A controller which is linear over the 'spread' of the sequence $\{x_k\}_j$ would probably be suitable for many applications; for the case when u_k is a scalar variable this can be expressed as:

$$u_k = \theta_k + \left\langle \beta_k, x_k \right\rangle$$

where θ_k is a scalar and β_k a vector.

Monte Carlo techniques can be used to estimate the parameters of these simple controllers as follows. Given an initial state x_1 a set of J sequences $\{x_k\}_j$ and $\{u_k\}_j$ are generated. The control actions are obtained using a nominal set of parameters for each control law h_k. The J sequences $\{V_{x_k}\}_j$ are then evaluated and hence the sequences $\left\{\frac{\partial H_k}{\partial u_k}\right\}_j$.

For the open-loop controller:

$$[\delta u_k]_j = \delta\theta_k$$

so that $$[\delta V_1]_j = \sum_{k=1}^{N-1} \left\langle \left[\frac{\partial H_k}{\partial u_k}\right]_j, \delta\theta_k \right\rangle$$

where $[\delta V_1]_j$ is the first order variation in the cost $[V_1]_j$ of the jth trajectory due to variations $\delta\theta_k$ in θ_k, $k = 1 \ldots\ldots N-1$. Thus the algorithm for improving θ_k is:

$$\delta\theta_k = -\varepsilon \overline{\left[\frac{\partial H_k}{\partial u_k}\right]_j}$$

For the linear feedback controller where u_k is a scalar:

$$[\delta u_k]_j = \left[\delta\theta_k + \left\langle \delta\beta_k, [x_k]_j \right\rangle\right]$$

and, arguing as above, it can be shown that the algorithm for improving θ_k and β_k is:

$$\delta\theta_k = -\varepsilon \overline{\left[\frac{\partial H_k}{\partial u_k}\right]_j}$$

$$\delta\beta_k = -\varepsilon \overline{\left[\frac{\partial H_k}{\partial u_k} \cdot x_k\right]_j}$$

Similar algorithms can be derived for more complex control laws.

Reduction of Variance

One feature of the naive method used above is that in its application to linear dynamic systems which have additive disturbances which have symmetric distributions and zero mean and where the cost function is quadratic the estimate of

$$E \frac{\partial H_k}{\partial u_k}$$

still has finite variance for a finite number of samples. It would be desirable to eliminate this sampling variance for linear systems so that if the method is applied to non-linear problems a non-zero sampling variance would result only if the system is non-linear over the 'spread' of the sequences $\{x_k\}_j$.

Let the linear system equations be:

$$x_{k+1} = Ax_k + Bu_k + C\xi_k$$

and the initial state is x_1 at $k = 1$. Assume that a nominal linear controller has been chosen

$$u_k = G.x_k$$

A particular realisation $\{\xi_k\}_{j-}$ of the disturbance sequence is chosen and the sequences $\{x_k\}_{j-}$, $\{u_k\}_{j-}$ determined. It is easily shown that:

$$\left[x_k\right]_{j-} = \bar{x}_k + \left[\eta_k\right]_{j-}$$

where $\bar{x}_k$ is the value of x_k that would be obtained if the disturbances were removed and $\left[\eta_k\right]_j$ is a linear combination of $\xi_1 \ldots \ldots \xi_k$. If $\left[V_k\right]_{j-}$ is defined as:

$$\left[V_k\right]_{j-} = \sum_{k=1}^{N-1} \left[\tfrac{1}{2}\langle x_k, Qx_k\rangle + \tfrac{1}{2}\langle u_k, Ru_k\rangle\right] + \tfrac{1}{2}\langle x_N, Qx_N\rangle$$

and V_k as:

$$V_k = E\left[V_k\right]_{j-}$$

Then:

$$\left[V_{x_k}\right]_{j-} = Q\left[x_k\right]_{j-} + A^T\left[V_{x_{k+1}}\right]_{j-}$$

where $\left[V_{x_N}\right]_{j-} = Q\left[x_N\right]_{j-}$

It is easily seen that:

$$\left[\frac{\partial H_r}{\partial u_r}\right]_{j-} = \left[\frac{\partial H_r}{\partial u_r}\right]_0 + \mathcal{L}\left[\left[\xi_1\right] \ldots \ldots \left[\xi_{N-1}\right]_{j-}\right]_v$$

where $\left[\frac{\partial H_r}{\partial u_r}\right]_0$ is the value of $\frac{\partial H_r}{\partial u_r}$ when the disturbances are removed and $\mathcal{L}$ denotes a linear function. If the antithetic variable method (4) is used, another sequence $\{\xi_k\}_{j+}$ of the disturbance is generated where:

$$\left[\xi_k\right]_{j+} = -\left[\xi_k\right]_{j-}$$

If $\left[\frac{\partial H_r}{\partial u_r}\right]_{j+}$ is now evaluated using the same procedure, then:

$$\left[\frac{\partial H_r}{\partial u_r}\right]_{j+} = \left[\frac{\partial H_r}{\partial u_r}\right]_0 - \mathcal{L}\left[\left[\xi_1\right]_{j-} \ldots \ldots \left[\xi_{N-1}\right]_{j-}\right]$$

The estimator $\left[\frac{\partial H_r}{\partial u_r}\right]_j$ where:

$$\left[\frac{\partial H_r}{\partial u_r}\right]_j = \tfrac{1}{2}\left[\frac{\partial H_r}{\partial u_r}\right]_{j-} + \tfrac{1}{2}\left[\frac{\partial H_r}{\partial u_r}\right]_{j+} = \left[\frac{\partial H_r}{\partial u_r}\right]_0$$

thus has zero variance as required.

Example

Using the antithetic variate method in a non-linear scalar problem specified by:

$$x_{k+1} = x_k + f(u_k) + \xi_k$$

$$V_r = \sum_{k=r}^{N-1} \left[\tfrac{1}{2}x_k^2 + \tfrac{1}{2}u_k^2\right] + \tfrac{1}{2}x_N^2$$

where $f(\ .\ )$ is a saturating type function:

$$f(u_k) = u_k, \quad |u_k| \leq 1$$

$$f(u_k) = 2 - \exp(1 - u_k), \quad u_k > 1$$

$$f(u_k) = -2 + \exp(u_k + 1), \quad u_k < -1$$

it was found that the estimated variance of

$$\frac{\overline{\delta H_r}}{\delta u_r}$$

was reduced by a factor of 50 - 5000 compared with naive Monte Carlo while the system was operating in the non-linear region. In the linear region the factor, theoretically infinite, was found to be in the range 10^{10} - 10^{14}.

REFERENCES

1. FLORENTIN, J. J., Optimal, Probing, Adaptive Control of a Simple, Bayesian System. Journal of Electronics and Control, Vol. 13, No. 2, p. 165, August 1962.

2. KUSHNER, H. J., On the Dynamical Equations of Conditional Probability Density Functions with Applications to Optimal Stochastic Control Theory. Journal of Mathematical Analysis and Applications, Vol. 8, No. 2, p. 332, April 1964.

3. KWAKERNAAK, H., On-line iterative optimisation of stochastic control systems. Automatica, Vol. 2, No. 3, p. 195, January 1965.

4. HAMMERSLEY, J. M. and HANDSCOMB, D. C., Monte Carlo Methods. Methuen and Co. Ltd. 1964.

Discussion

Prof. J.L. Douce (UK) quoted the example following equation 17 where u_k is assumed to be constant over two intervals. He asked whether this is generally true.

Mr. Mayne replied that the nominal sequence was chosen to be constant. It was not assumed that the solution would be of this form.

SUMMARY OF DISCUSSION

PREDICTION OF THE OPTIMAL CONTROL SEQUENCE

by V. N. Novoseltsev
USSR National Committee on Automatic Control
Moscow, USSR

Consider the discrete-time system described by the equation:

$$x_k = f_k(x_o;\ u_o, u_1 \ldots u_{k-1};\ h_o, h_1, \ldots h_{k-1}) \qquad \ldots (1)$$

where x_i is the state of the system at time i, u_i is the control signal and h_i is a random disturbance. x_o is the initial state. The performance index of the system is:

$$E[Q] = E\left[\sum_{i=0}^{N} F_i\,(x_i, u_i)\right]. \qquad \ldots (2)$$

Let I_k denote the information state at time k. I_k comprises the prior information and the information obtained by measurement of the system at times $0, 1, \ldots k-1$. Let $\bar{u}^{(k)}$ denote the control <u>sequence</u> $(u_k^{(k)}, u_{k+1}^{(k)}, \ldots u_n^{(k)})$ obtained by minimising $E(Q/I_k)$ at time k. Thus $\bar{u}_k^{(k)}$ is that sequence of controls $(u_k, u_{k+1}, \ldots u_N)$ obtained from performing the following minimisation:

$$\min_{u_k \cdots u_N} \left[E\left\{ \sum_{i=k}^{N} F_i\,(x_i, u_i)/I_k \right\} \right]. \ldots (3)$$

This minimisation is first performed at time 0 and $\bar{u}_o^{(o)}$ is obtained. The optimal control at time 0 is the <u>first</u> member of this sequence, $u_o^{(o)}$. At time 1, I_1 is obtained, the minimisation repeated yielding $\bar{u}_1^{(1)}$ and the first member of this sequence, $u_1^{(1)}$, is the optimal control at time 1. Proceeding in this way $\bar{u}_k^{(k)}$ is obtained at time k and the first member of this sequence, $u_k^{(k)}$, is the optimal control at time k. It can be shown that dynamic programming would yield the same optimal control sequence $(u_o^{(o)}, u_1^{(1)}, \ldots u_N^{(N)})$.

EXAMPLE

To simplify the numerical solution a continuous time system is used to illustrate the procedure of determining the optimal control sequence. The system is described by the following equations:

$$x(t) = \mu.u(t) \qquad \ldots (4)$$
$$y(t) = x(t) + h(t)$$

μ is an unknown gain but the prior mean and variance of μ are known to be m and σ_μ^2 respectively. The control $u(t)$ can be measured perfectly, $y(t)$ is a noisy measurement of the state $x(t)$. $h(t)$ is white noise of variance $s.\delta(t)$. $x^*(t)$ is the desired output. For simplicity put $x^*(t) = \text{const} = a$. The performance index is:

$$E[Q] = E\left[\int_0^T [x(t) - x^*(t)]^2\, dt\right]. \quad \ldots (6)$$

To determine $u^o(0)$, the optimal control at $t = 0$, we have to optimise $E(Q|I_o)$ with respect to the <u>function</u> $u(t)$, $0 \leqslant t \leqslant T$; $u^o(0)$ is the value of the optimal function at $t = 0$. Let $\hat{x}(t|I(0))$ denote $E(x(t)|I(0))$ and $\sigma_x^2(t|I(0))$ denote $\mathrm{Var}\,(x(t)|I(0))$. It is easily shown[1] that:

$$\hat{x}(t|I(0)) = m\,u(t) \qquad \ldots (7)$$

$$\sigma_x^2\,(t|I(0)) = \cfrac{u^2(t)\,\sigma_\mu^2}{1 + \dfrac{\sigma_\mu^2}{S}\,z(t)} \qquad \ldots (8)$$

where:
$$z(t) = \int_0^t u^2(\tau)\, d\tau. \qquad \ldots (9)$$

Thus:

$$E(Q|I(0)) = \int_0^T \left[(\hat{x}(t|I(0))-a)^2 + \sigma_x^2(t|I(0))\right] dt$$

$$= \int_0^T \left[(m\sqrt{z} - a)^2 + \cfrac{z\,\sigma_\mu^2}{1 + \dfrac{\sigma_\mu^2}{S}\,z}\right] dt \quad \ldots (10)$$

since:
$$\dot{z} = u^2 . \qquad \ldots(11)$$

Equation (10) must be optimised with respect to the *function* $u(t)$, i.e. with respect to the *function* $z(t)$, $0 \leq t \leq T$, where $z(0) = 0$ and $z(T)$ is free. If $u(t)$ is differentiable, $z(t)$ is twice differentiable and classical calculus of variations can be used. Let $F(z, \dot{z})$ denote the integrand of equation (10). The Euler equation is:

$$\dot{F} - \dot{z} \frac{\partial F}{\partial \dot{z}} = \text{const}$$

i.e.
$$z^\circ(t) = c^2 t \qquad \ldots(12)$$

and
$$u^\circ(t) = c \qquad \ldots(13)$$

where $z^\circ(t)$, $u^\circ(t)$ are optimal functions. The value of c is given by

$$\frac{\partial}{\partial c} \left[(mc - a)^2 + \frac{\sigma_\mu^2 c^2}{1 + rc^2} \right] = 0 \qquad \ldots(14)$$

where
$$r = \frac{\sigma_\mu^2 T}{S} . \qquad \ldots(15)$$

The optimal value of c is therefore given by:

$$m^2 r^2 c^5 - mar^2 c^4 + (2m^2 r - \sigma_\mu^2 r)c^3 - 2marc^2$$
$$+ (m^2 + \sigma_\mu^2)c - ma = 0 \qquad \ldots(16)$$

where
$$r \ll 1 , \quad u^\circ = \frac{ma}{m^2 + \sigma_\mu^2}$$

when
$$r \gg 1 , \quad u^\circ = \frac{a}{m}$$

$r \gg 1$ means that $\text{Var}(x(T) \mid I(0)) \ll 1$ and is referred to as the complete information level.

The solution defined by equation (16) can also be used at time t, if m and σ_μ^2 are replaced by $M(t)$ and $D_\mu(t)$ where:

$$M(t) = E(x(t) \mid I(t)) \qquad \ldots(17)$$

$$D_\mu(t) = \text{Var}(x(t) \mid I(t)) . \qquad \ldots(18)$$

It has been shown[1] that:

$$M(t) = \frac{M(0) + \int_0^t y(\tau) u(\tau) \, d\tau}{1 + \frac{D_\mu(0)}{S} \int_0^t u^2(\tau) d\tau} \qquad \ldots(19)$$

$$D_\mu(t) = \frac{D_\mu(0)}{1 + \frac{D_\mu(0)}{S} \int_0^t u^2(\tau) d\tau} . \qquad \ldots(20)$$

Figure (1) shows for $t = 0$ the optimal control $u^c = c$ as a function of the time left to go, T, when $m = a = \sigma_\mu^2 = S = \text{const} = 1$. When T is very small, $r \ll 1$, and $u^\circ \to 0.5$. With larger T u° increases above the 'complete information level' in order to 'probe' the system. When T is very large sufficient time is available for information gathering, no extra probing is required, and the control decreases to the complete information level.

REFERENCE

1. A. A. Feldbaum. "The Principles of Optimal Control System Theory" Moscow, 1963.

Dr. D. Q. Mayne (UK) commented as follows on this contribution:-

The procedure proposed in these remarks, while it may be useful in some applications, is not optimal. The procedure proposes the prediction, at time k and using the information 'state' I_k at time k, of a future optimal control sequence which is 'open-loop', i.e. the acquisition of future data is ignored in the prediction otherwise the future control values

$$u^{(k)}_{k+r} , \qquad r \geqslant 1 ,$$

would have to be specified in feedback form as functions of the (future) information 'states' I_{k+r}. For a lucid discussion of this point see S. E. Dreyfus "Some types of Optimal Control of Stochastic Systems", J. Siam Ser. A. Control vol. 2, Na 1, Jan. 1963, pp.131-140.

Dr. Novoseltsev replied as follows:-

It is obvious that the conciseness of the contribution has produced misunderstanding of some of its aspects. As a matter of fact the **aquisition** of future data is not ignored when optimal prediction is to be performed.

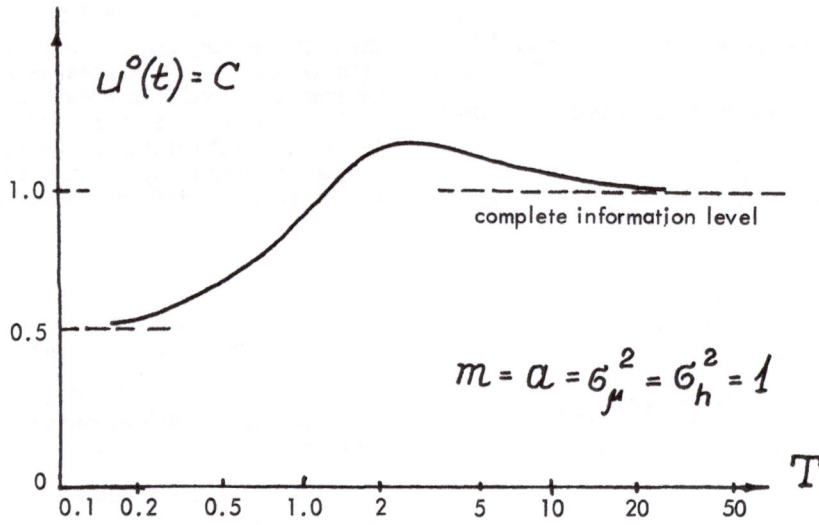

complete inf. level is $u^o_{compl.} = \dfrac{a}{m} = 1$

Figure 1

For detailed consideration take the example given in the contribution for Gaussian distributions. The **a**quisition of future data is now the measurement of y_j, $(j = 0,1,...,i)$.

Consider vectors $\vec{u}_r = u_0, u_1, ..., u_r$, $\vec{y}_r = y_0, ..., y_r$, etc. Then for a closed-loop system when $\vec{y}_i$ is known, the a posteriori distribution $P(\mu)$ is

$$P(\mu \mid \vec{u}_i, \vec{y}_i) = C_i P_o(\mu) \prod_{j=0}^{i} P(y_j \mid \mu, u_j) =$$

$$= C_i \exp\left\{ - \frac{(\mu - m_i)^2}{2\sigma_i^2} \right\}. \qquad ... (1)$$

where

$$m_i = \frac{\left(\frac{\sigma_\mu}{\sigma_h}\right)^2 \sum_{j=0}^{i} u_j y_j + m}{1 + \left(\frac{\sigma_\mu}{\sigma_h}\right)^2 \sum_{j=0}^{i} u_j^2} \qquad ... (2)$$

and

$$\sigma_i^2 = \frac{\sigma_\mu^2}{1 + \left(\frac{\sigma_\mu}{\sigma_h}\right)^2 \sum_{j=0}^{i} u_j^2}. \qquad ... (3)$$

$E(\mu)$ for the i-th step can be found as follows:

$$E(\mu \mid I_o) = \frac{I}{(2\pi)^{(i+1)/2} \sigma_h^{i+1} \sqrt{2\pi}\sigma_\mu} \int_{-\infty}^{\infty} \cdots \int_{-\infty}^{\infty}$$

$$\frac{(\sigma_\mu/\sigma_h)^2 \sum_{j=0}^{i} u_j(\mu u_j + h_j) + m}{I + (\sigma_\mu/\sigma_h)^2 \sum_{j=0}^{i} u_j^2} \times$$

$$\times \prod_{j=0}^{i} \exp\left\{ - \frac{h_j^2}{2\sigma_h^2} \right\} \exp\left\{ - \frac{(\mu-m)^2}{2\sigma_\mu^2} \right\} dh_o dh_1 ... dh_i \, d\mu = m. \qquad ... (4)$$

For the open-loop system, i.e. with no measurements of y_i the conditional probability would be given by the next equation instead of (1)

$$P(\mu \mid \vec{u}_i) = \int_{\Omega(\vec{y}_i)} P(\mu \mid \vec{u}_i, \vec{y}_i) \, P(\vec{y}_i) \, d\vec{y}_i$$

$$= \int_{\Omega(\vec{y}_i)} \frac{P(\mu)P(\vec{y}_i \mid \mu)}{P(\vec{y}_i)} \, P(\vec{y}_i) \, d\vec{y}_i = P(\mu) \qquad ... (5)$$

and there would be no information accumulation at all. In the open-loop system it would be $E(\mu) = m$, $D(\mu) = \sigma_\mu^2$ for all future i. Then $E(x_i \mid I_o)$ (open loop) $= u_i m$, $D(x_i \mid I_o)$ (open-loop) $= u_i^2 \sigma_\mu^2$.

All these formulae are valid for any step, not only for $k = 0$.

As for the future optimal control sequence at the k-th step, it is clear that it will not be used in reality for at the next $(k+I)$-th step the new prediction of control sequence will be found, based upon I_{k+1} instead of I_k. For this reason there is no need to specify future $u_{k+r}^{(k)}$ in feedback form, $u_{k+r}^{(k)}$ are simply the numerical parameters used to find the actual control $u_k^{opt} = u_k^{(k)}$ at every k-th step:

$$u_k^{opt} = u_k^{(k)} \left(\vec{u}_{k-1}^{(k-1)}, \quad \vec{y}_{k-1} \right).$$

The last equation is the feedback form of optimal control.

ADAPTIVE CONTROL IN BIOLOGICAL SYSTEMS

SELF-ADAPTIVE RETINAL PROCESSES AND THEIR IMPLICATIONS TO ADAPTIVE CONTROL

by Dr. Jan J. Kulikowski
Institute of Automatic Control
Polish Academy of Sciences
Warsaw, Poland

ABSTRACT

An analysis of two-dimensional optical images (or spatial distributions of other physical quantities) with various signal-to-noise ratios requires an optimum balance between integration (or averaging) to detect weak signals, and differentiation to increase contrasts. This is connected with sensitivity-selectivity balance.

Based on some physiological observations this paper provides certain models of retinal receptive fields which adapt to various conditions. These models called adaptive centric operators have a relatively small number of inputs. Nevertheless, a mapping system containing them achieves optimization in picture processing for various conditions.

INTRODUCTION

Visual perception is concerned with the reception of spatial images in time. This process is performed by several levels of hierarchical organization and each level is likely to be optimised within certain limits. The lowest level of the visual signal processing occurs in the retina and appears to involve a goal seeking task. Models of it may find applications as pre-processors which adapt to various conditions reducing the amount of data flow in an optimal way.

The analysis by the nervous system of two-dimensional images projected on the retina over a period of time requires both sufficient sensitivity to perceive the signals and selectivity to distinguish details. Within the limits of the system sensitivity and selectivity are in opposition therefore an optimum balance should be found. It should also be noted that space and time are interchangeable in communication systems having a given channel capacity, therefore the spatial selectivity can be increased at the expense of the temporal one.

This paper deals mainly with a balance between sensitivity and spatial selectivity.

PHYSIOLOGICAL DATA

An elementary unit of the first level is the retinal receptive field shown in Fig. 1a.

There are connections from receptors converging on one ganglion cell (through some intermediate cells, not shown in Fig. 1). However, the receptive fields consist of antagonistic parts: excitatory "on-centres" and inhibitory "off-surroundings" and vice versa. This two-centre system is evidently more advanced (in the evolutionary sense) than such simple contrast-detecting networks with lateral inhibition as are found in the primitive eye of Limulus (Fig. 1b). The larger the size of the centre, the greater is the sensitivity and the smaller is the selectivity. The main units - ganglion cells are threshold elements and their output is a frequency of a constant amplitude pulse train. Normally, in response to a step input signal the pulse frequency increases, but after some time decreases exhibiting a passive adaptation to a steady state. However some experimental data indicate that there must be other principles underlying this behaviour, e.g.

1. The sizes of the centres are modified by luminance levels to optimal values[1] and changes in selectivity have been observed.[2]

2. Some mammalian retinae in which a spontaneous pulse activity (i.e. without signals) or an initial frequency exists show additional phenomena, e.g.

    a. The average pulse frequency can decrease below the spontaneous activity level with an increase in luminance, as shown in Fig. 2a (Ref. 3).

    b. The interval histogram distributions of pulses become multimodal and these further modes caused by luminance are spaced apart at intervals of the main mode (Fig. 2b); this suggests that some pulses are deleted.[4]

All the facts cannot be explained by simple linear models and a highly non linear feedback system must be postulated.[5]

ASSUMPTIONS

1. Only the central and foveal areas of the retina are considered.

"Superior numbers refer to similarly-numbered references at the end of this paper"

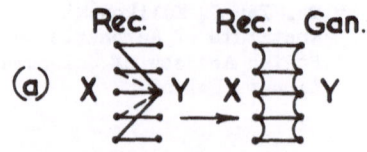

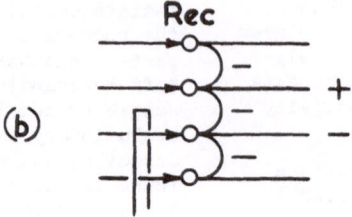

Fig. 1.(a) Scheme of a receptive field
(b) and a contrast-detecting network

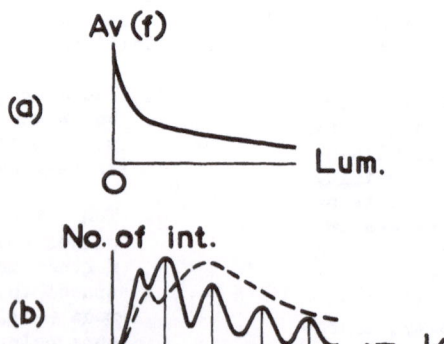

Fig. 2. The average (a) and
instantaneous (b) frequency of pulses

2. There is a feedback system, but feedforward is not discussed.

3. The number of input signals (x), not necessarily independent of each other, are chosen to be equal to the number of outputs (y); this means that any input is assumed to be a set of signals from several receptors; in man the receptor-to-ganglion ratio is 1:1 but in other mammals usually higher.

4. The system considered here of alternately on-centre and off-centre cells with corresponding antagonistic surroundings has greater possibilities than networks with lateral inhibition, however, for brevity, only on-centre cells are to be described, because the mathematical method is the same in both cases.

DEFINITIONS

A set of input signals (x) may be described by giving a reference level equal to the average signal intensity and deviations from this level. In this set only adjacent signals of different intensity supply new information and may be called independent optical signals. The reference level is sometimes not relevant.

Decorrelation is the transformation of input signals such that the number of output signals approaches the number of independent optical signals. Thus, some redundancy may be eliminated. On the other hand for signals of minimum width (fitting the receptor width) and minimum duration (equalling the reaction time) only those of sufficient intensity, (Fig. 3) to bring the product of those three variables above the threshold of absolute detectability will produce a response. Signals of smaller intensity than this threshold may be conditionally detected provided that either spatial or temporal summation takes place. In other words it may be possible with signals of more than elementary width and duration to accept a lower intensity threshold if the centre of the receptive field is matched in size with the optical signal.

Adaptive decorrelation is a signal transformation in which the decorrelation differs according to signal intensity; signals of lower intensity are to a greater extent spatially and temporally summated whilst those of greater intensity are differentiated leaving only the independent signals. The former process increases sensitivity; the latter selectivity. Thus, there is an interplay of two parallel processes: summation (integration of weak signals) and differentiation (contour enhancement).

These functions may be performed approximately by a receptive field, because signals within its centre are integrated and all the field performs a double (off-on-off) spatial differentiation. Varying the size of the on-centre would simulate adaptive decorrelation.

Since separate control of spatial and temporal summation is difficult, the mutal relationship between them remains to be chosen, so that only spatial summation needs to be controlled.

1. Temporal summation does not depend directly on the spatial one (i.e. the on-centre size); the summation time may be, for example, dependent on a cell output (y) decreasing with an increase in y.

2. Summation time is inversely proportional to the centre size; this may facilitate detection of a class of weak signals, the geometrical size of which is matched with the centre size.

3. Summation time is directly proportional to the centre size; this provides the lowest detectability threshold when weak signals are expected to be small in size and also short in duration, but both temporal and spatial selectivity are poor.

Further considerations are mainly concerned with the latter (simplest) case.

MATHEMATICAL MODELS

Fig. 4a-1 shows the on-centre receptive field. Because of its symmetry it can be described by a one-dimensional function which indicates a contribution of various inputs of various distances (d) from the centre to the output. As shown in Fig. 4a-2, the excitatory area of positive "A" which spreads over the whole of the receptive field in darkness, decreases with an increase in luminance.

Provided that the receptive field of one ganglion cell consists of a large number of inputs, the function (Fig. 4a-2) may be described in terms of two Gaussian distributions.

$$\text{In darkness} \quad A_d = A \, e^{-(d/d_1)^2}$$

$$\text{and output} \quad y_n = \sum_{-d}^{+d} A_d x_{nd}$$

$$\text{where} \quad d_1 = \text{constant}$$

$$x_{nd} = \text{input, distant "d" from "n"}$$

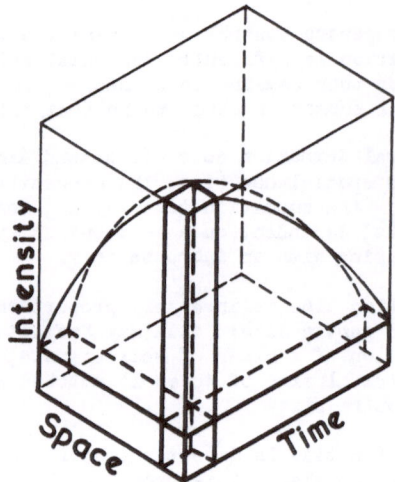

Fig. 3. Detectability threshold volume.

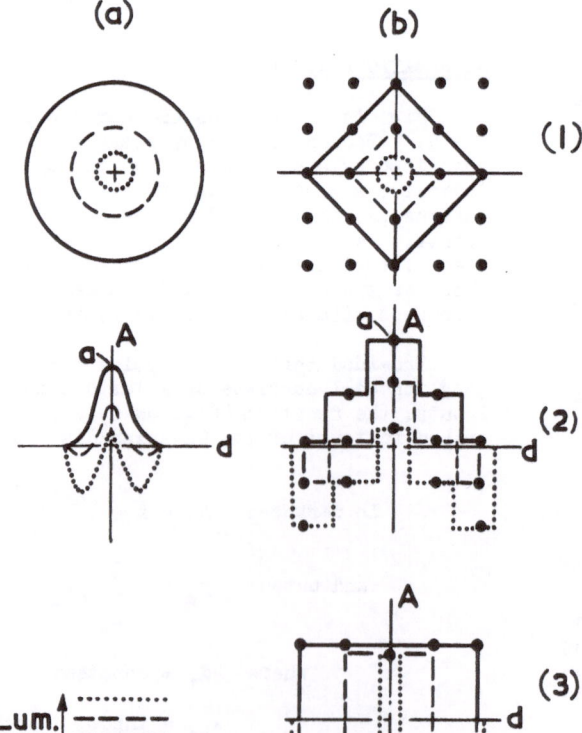

Fig. 4. Receptive field transformation functions

38

For higher luminance levels a feedback (B) occurs

$$A_{df} = A e^{-(d/d_1)^2} - w(y) B \; e^{-(d/d_2)^2}$$

where $d_2 > d_1$

$w(y)$ = spatial weighting function.

In practice, however, only a small number of elements can be used. Therefore a simplified model, called an adaptive centric operator (Fig. 4b-1) which may be described by matrix equations is used. First realization of the operator is given by the equation

$$
\begin{bmatrix} \cdots \\ y_{n-1} \\ y_n \\ y_{n+1} \\ \cdots \end{bmatrix}
=
\begin{bmatrix}
\cdots & \cdots & \cdots & \cdots & \cdots & \cdots & \cdots \\
a-2d & a-d & a & a-d & a-2d & 0 & 0 \\
0 & a-2d & a-d & a & a-d & a-2d & 0 \\
0 & 0 & a-2d & a-d & a & a-d & a-2d \\
\cdot & \cdot & \cdot & \cdot & \cdot & \cdot & \cdot
\end{bmatrix}
-
$$

$$
\begin{bmatrix}
\cdots & & \\
w(y_{n-1}) & 0 & 0 \\
0 & w(y_n) & 0 \\
0 & 0 & w(y_{n+1}) \\
\cdot & \cdot & \cdot
\end{bmatrix}
\begin{bmatrix}
\cdots & & \\
b\,b\,b\,b\,b\,0\,0 \\
0\,b\,b\,b\,b\,b\,0 \\
0\,0\,b\,b\,b\,b\,b \\
\cdot & \cdot & \cdot
\end{bmatrix}
\begin{bmatrix} \cdots \\ x_{n-1} \\ x_n \\ x_{n+1} \\ \cdots \end{bmatrix}
$$

or $\underline{Y} = (\underline{A} - \underline{W}\,\underline{B})\,\underline{X} = \underline{A}_f\,\underline{X}$

The transformation function matrices $\underline{A}$ and $\underline{A}_f$ are non-singular because of the assumption 3, and their elements are concentrated about the main diagonal. The elements of $\underline{A}$ may be chosen to decrease with the distance from the diagonal (Fig. 4b-2), while elements of $\underline{B}$ are constant, each row being multiplied by a spatial weighting function of the nearest outputs, e.g. $w(y_n) = c y_{n-1} + y_n + c y_{n+1}$ (where $c < 1$). The latter operation causes inhibition of the neighbouring cells. If this weighting function is constrained the equation has stable solution when a method of successive iterations is used. Some of the results obtained for coefficients a=1; b=0.03; c=05; d=0.5, when a mapping system is a linear array of thirty receptors, are quoted in Table 1.

Three kinds of input signals are examined; uniform over all the mapping system, square wave modulation and a separate point. In response to the uniform signals the output value increases rapidly, but after that it decreases exponentially. For square waves and points the output signals tend to values denoting excitation (+) and inhibition (-). This enhances contrasts, while the uniform background component only speeds up this process. Apparently, the greater the signal intensity the stronger the first reaction and the quicker the settling down. This means also a longer temporal summation for weak signals.

Table 1

Input signals	(x)	1 it.	3 it.	5 it.
uniform	3	6	0.06	0.0006
square wave	0	0.5	0.25	0.09
	1	1.5	1.22	1.00
	1	1.5	1.22	1.00
	0	0.5	0.25	0.09
	1	2.5	0.5	-0.002
	2	3.5	1.44	0.58
	2	3.5	1.44	0.58
	1	2.5	0.5	-0.002
	2	4.5	-0.005	-0.11
	3	5.5	0.65	0.16
	3	5.5	0.65	0.16
	2	4.5	-0.005	-0.11
	0	2.5	-1.13	-0.59
	5	7.5	2.19	0.89
	5.	7.5	2.19	0.89
	0	2.5	-1.13	-0.59
point	0	6	-0.12	-0.56
	12	12	2.9	1.39
	0	6	-0.12	-0.56

MULTI-DECISIONAL MODEL

Further simplification may be obtained with a multi-decisional model having a small number of states (Fig. 5b). Within any of the states the model in linear. When outputs are greater than certain threshold values, rows of the matrix $\underline{A}$ are switched to new values, as in Fig. 5b-2 (broken and dotted lines). There are two possibilities for the strategy of switching:

1. all rows are switched in the same way within a limited region of interaction, e.g. five rows as in Fig. 5b.

2. firstly, the row which corresponds to the greatest output is switched, and then neighbouring rows are switched to values of lower orders.

The next problem is to choose a decision criterion which should maximize the performed index. This should take into account

1. the operator sensitivity, defined as a sum of positive factors ($S = \Sigma a_{d+}$),

2. the selectivity coefficient ($e = \Delta a_d / a$) limited by the summation centre,

   that is to say $\int_{t}^{t+T} \left[e^2(t) + \lambda s^2(t) \right] dt = \text{max.}$

Provided that the coefficient λ is constant, the generalised function of performance can be determined (in more general cases λ can be changed by higher-level orders). Depending on the intensity of the output signals the matrix $\underline{A}$ can be switched appropriately by a decision element. The decision element can also classify signals according to their signal-to-noise ratio

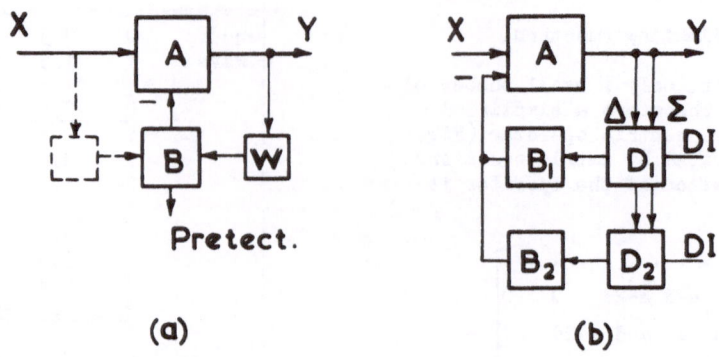

Fig.5. Simplified block diagrams.

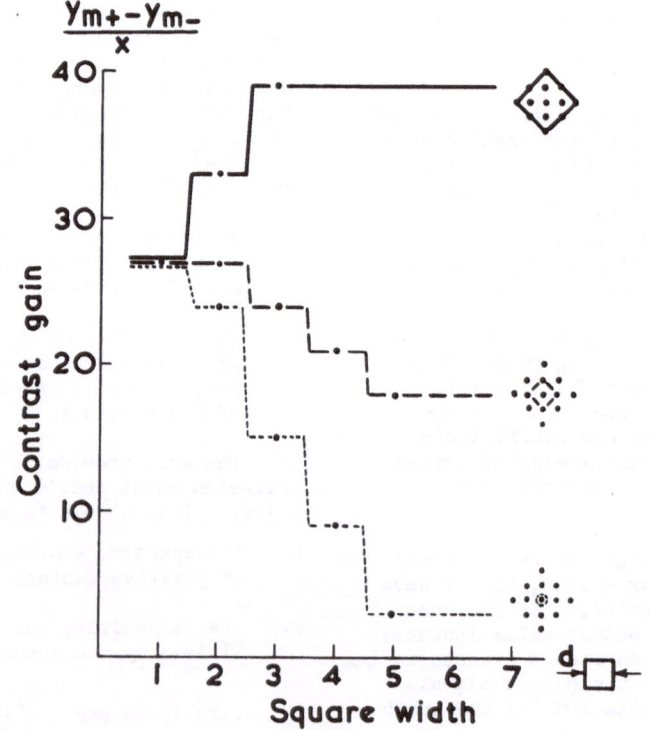

Fig. 6. Contrast gain

40

if it has two opposite inputs, one for the sum of positive signals (Σy^2_n) and the other for the difference in neighbouring signals ($\Sigma(y_n - y_{n+1})^2$). The latter operation may be performed by using the contrast network (Fig. 1b). In any case, the decision process should possess some hysteresis to ensure stability, that is

	Increase	Decrease	Dec. Region
$\lambda\Sigma y^2_n - \Sigma(y_n - y_{n\pm1})^2$	>P	<P'<P	1
	>Q>P	<Q'<Q	2
	>R>Q	<R'<R	3

The first component of the above relationship is a function of signal energy while the other is a measure of high-frequency spectrum energy. The latter quantity contains a considerable amount of noise, the spectrum of which is usually wider than that of the signals. Thus if this component exceeds some threshold value the operators are switched toward extended summation.

The way of switching the matrix $\underline{A}$ must be chosen, and this can be done, e.g. by changing the weighting coefficients (a_d) so that their relative position is unchanged (Fig. 4b-2). A two-dimensional mapping system containing such operators gives different contrasts (i.e. differences between maximum positive and negative outputs $y_{m+}-y_{m-}$) for different sizes of optical signals (Fig. 6). When the summation centres are small (dotted line) small spots evoke stronger responses than large ones. The large summation areas (continuous line) process larger spots with considerable gain while the smaller spot signals are processed with relatively less gain, and, in addition to this, are a little blurred, according to the shape of the transformation function (Fig. 4b-2).

Much greater gain can be achieved with another way of switching (Fig. 4b-3) in which the central areas are extended at the same value of the weighting coefficients as in the centre. In such a case, however, it is impossible to recognise on the map sizes of spots smaller than the central area size.

Summing up, the multi-decisional model seems to be more promising, because the decision process can be directly expressed in terms of the performance index, and the centric operators have been simulated with relative ease. In addition to this, it is easy to make the matrices $\underline{A}$ asymmetrical and have eccentric operators which provide non-symmetrical processing of the image within the mapping system. This enables the determination of contour position and eventually direction of contour movement.

DISCUSSION

The models described can explain some non-linear phenomena in the retinae. Owing to non-linearity in the system the average frequency of

the group of cells may decrease with an increase in luminance. The multi-modal distribution of pulse intervals may be explained as a result of changes in summation areas. In darkness all receptors within the receptive field evoke the cell pulses in a random way. When illuminated, however, the distant receptors develop inhibition in the cell, which is equivalent to deletion of pulses. This results in a multi-modal distribution. It should be noted that the processes of matching the sizes of summation centres are common for the vertebrates and their influence may be observed even in higher levels of signal processing. However, the retinae of lower vertebrates also possess highly specialized receptive fields, e.g. the frog's retina has four kinds of cells, detecting different features: edges, moving curves, changing contrasts, and dimming, which have been modelled previously.[6] This means that irreversible coding is more influential than reversible. As is well known, one can detect more features if a pattern recognition machine has several layers, e.g. the first detecting only the simplest features, etc. The retinae of mammals have more universal cells detecting only general features such as transients, contrast gradients, and sometimes direction, and this is the process of reversible coding, in which very little essential information is lost. Adaptation of the sizes of the summation centres depends, in fact, on the signal-to-noise level. In the half-light, there is no redundancy, signals are weak, and therefore spatial and temporal correlation is needed. This is effected by an increase in size of the summation centres and also by a temporal summation. For higher luminance levels, signals should be decorrelated; sizes therefore decrease and the summation time is shorter because of rapid negative feedback. This enables the steady state component of signals to be sorted out and directed to another brain centre (pretectum) controlling the size of the pupil. Further levels of the visual channel deal with signals having redundancy optimally reduced, and these levels can detect more complicated features.

One can conclude that systems for the measurement and identification of complex processes could contain adjustable elements described as adaptive centric operators in addition to well known feature detectors, which are non-adjustable.

In any case adaptive operators can be associated with centric operators of the feature-detecting system. It seems reasonable to make the succeeding operators very selective in order to "clean up" patterns which may be blurred because of summation processes in the retina. In biological systems this is done by the laterial geniculate body.

CONCLUSION

The processes described ensure optimal

adaptive decorrelation of signals, i.e. the
steady-state components of signals are sorted
out, and contrasts and transients are detected.
The models called adaptive centric operators
may be useful as pre-processors in optical
signal processing or after some modifications
in a transformation of state vectors.

ACKNOWLEDGMENTS

The final version of this paper was prepared during the author's stay at the National
Physical Laboratory.   The author wishes to
express his gratitude to Dr. A.M. Uttley,
Superintendent of the Autonomics Division, for
permission to continue this research.   The
author thanks also to Mr. P.H. Hammond,
Dr. A. Hazlerigg, Mr. P.A.N. Briggs and
Miss M. Longden for helpful discussions.   A
personal grant from the British Council is
gratefully acknowledged.

REFERENCES

1.   Glezer, V., Kosteyanets, N., "Changes in
     the Effective Size of the Receptor Field",
     Biofizika, Vol. 6, No. 6, 1961.

2.   Robson, J., Campbell, F., "A Threshold
     Contrast Function for the Visual System".
     The Physiological Basis for Form Discrim-
     ination Symp. at Brown University,
     January 23-24, 1964.

3.   Arduini, A., in Brain Mechanisms,
     Elsevier, Amsterdam, 1963, p. 184.

4.   Kosak, W. et al., "Analysis of Maintained
     Firing Patterns", Aust. J. Sci. Vol. 25,
     p.102.

5.   Kulikowski, J., in Control Problems of
     Large Systems" (in Polish) P.A.N.,
     Warsaw, 1964, p.164.

6.   Herscher, M., Kelly, T., "Functional
     Electronic Model of the Frog Retina",
     IEEE Trans., Vol. MIL-7, No. 2/3, 1963

Discussion

Mr. P.H. Hammond (UK) asked the author to
give some details of the experimental techniques
which led to the conclusion that the size of the
centres are modified to optimum values with
variation of luminance level.

In reply Dr. Kulikowski referred to the
experiments of V. Glezer (ref. 1 of the paper).
On the human subject, the size of the centre was
estimated by giving stimuli of different 'square
size'. For each test, the luminance was adjusted

to maintain constant the total flux passing
through the square.

The results show that at low levels of
luminance, the summation area extends over the
whole of the receptive field. For example, in
a darkened room, only objects having strong
contours can be distinguished. As the luminance
increases, the summation area decreases in size
and more detail of the object can be distinguished.
The system therefore optimizes to the signal
available contrary to the type of system which
only accepts or rejects an image.

ADAPTIVE FUNCTIONS OF MAN IN VEHICLE CONTROL SYSTEMS

Y. T. Li,* L. R. Young** and J. L. Meiry***
Department of Aeronautics and Astronautics
Massachusetts Institute of Technology
Cambridge, Massachusetts USA

ABSTRACT

Inability of human pilots to introduce adequate adaptation of their control provided much of the motivation for the development of automatic adaptive control systems. The rapid change in vehicle system characteristics in high performance aircraft, which may climb from sea level to extremely high altitudes in minutes, required automatic adaptive control to relieve the burden on the operator.

This paper examines the principles and compositions of existing automatic adaptive control systems and on these bases the human adaptive as well as primary control functions are analyzed.

In general, human outshines the automatic system with his huge capacity of open loop or programmed control; but he lacks the capacity and speed for making on-line computations needed in the operation of active continuous adaptive system. Human can also perform some passive type or very simple active type adaptation but would require the assistance of a computer to perform complicated active adaptation.

This limitation is responsible for the domination of mechanized systems for adaptive control. There are advantages attributable to computer-assisted human adaptive control system considering the huge capacity of his open loop adaptation. Visual or some other form of multi-input display becomes a necessary medium when computer-man coupling is to be made effectively. Considering man's remarkable ability of pattern recognition, this task may be in many cases easier than the coupling of a computer with an automatic actuator when a complicated function is to be recognized and manipulated.

INTRODUCTION

To be called adaptive, a control system must exhibit a degree of flexibility to perform satisfactorily under a wide range of environmental conditions. In many respects all adaptive control systems are in a sense attempts to imitate the remarkable adaptability of human beings. In the introduction to a book entitled Adaptive Control Systems, Truxal stated that "adaptive control has been viewed as the instrumentation realization of a prime characteristic of the human being in a control task."[8] The various terms currently used to describe adaptive control concepts such as learning, hill-climbing, optimal, self-organizing, self-repairing, etc., all reflect the desire of the adaptive system designer to imbue his system with characteristics of human intelligence.

A dramatic demonstration of human adaptability was the recent incident in which PAA pilot Charles Kimes successfully controlled, rerouted and landed his Boeing 707 in California after losing one wing tip and an outboard engine. Obviously, further understanding of the nature of manual adaptive control would be of significant value to the overall field of automatic adaptive control.

Ironically it is not the versatility of human adaptive control but rather its limitations which have led to the great surge in development of automatic adaptive control systems during the past few years. In particular, adaptive flight control systems were required in high performance aircraft to provide the human pilots with adequate handling qualities over a wide range of environmental conditions. In this and many other situations the objective of the adaptive control system is to reduce the range of adaptation required of the human operator, dealing mostly with continuous changes of operating conditions rather than single shot events. Although maneuvers required in adapting to continuous change of operating conditions are often less complex than those pertaining to a single shot change of events, the associated dynamic

---

*Professor of Instrumentation
**Assistant Professor of Aeronautics & Astronautics
***Assistant Professor of Aeronautics & Astronautics

Superior numbers refer to similarly-numbered references at the end of this paper.

problems of the former are nevertheless more challenging. For one thing, the system has a primary operating function which involves certain dynamic characteristics so that the additional change of operating parameters makes the system behave as a continuous time-varying system. In addition, the system behavior must be identified and modified quickly enough to overcome the effect on performance of the continuously changing dynamic behavior. A simple example of this type of situation is illustrated by the change of the handling quality of a high-performance aircraft as it climbs from sea level to the rarefied upper atmosphere, requiring considerable adaptation in the control system to maintain satisfactory response at all altitudes.

Unfortunately current studies of adaptive control systems still suffer from the lack of universal quantitative measures of adaptivity. For a simple sinusoidal function one can use the sensitivity function to assess the variation of component frequency response. However, attempts to define meaningful measures of adaptivity in terms of transient response have not been very conclusive. This lack of quantitative measurement of adaptivity also reflects the lack of a clear-cut method of classification of the types of adaptive control system. For the current study, the comparison of man and machines in adaptive control will be made with emphasis placed upon the type of logic used for system identification and fast adaptive response.

CLASSIFICATION OF CONTINUOUS ADAPTIVE CONTROL SYSTEMS

Generally speaking the objective of an adaptive control system is to maintain the dynamic behavior of the handling quality of a control system within desirable tolerance while the operating system parameters undergo some changes. An active adaptive system is one which accomplishes this objective by making active adjustments of the parameters of the controller.* A properly designed controller should allow adequate range of adjustment to counteract the variation of the operating system parameter. One obvious adaptive control logic is to program the adjustment to the controller parameter as a function of the operating system parameters or the environmental conditions which influence those parameters. This is generally known

---

*In a more general sense, an active system is one which introduces a drive function to compensate for a deviation of operating condition. The drive function may be introduced by a modification of the controller parameter or by introducing a separate function. The former is more commonly done.

as an open loop adaptive control system, with the advantage of being the fastest scheme among all types of adaptive control systems and the major disadvantage of the difficulties associated with measurement of the dominant operating parameters or the environmental conditions.

While active systems may seem to be the sure way to improve the adaptability of a control system, it must be recognized that most practical feedback control systems are inherently adaptive to some degree. Consequently, improvement of system adaptability is possible by choice of the system configuration without the use of active adjustment of the controller parameters. Systems with this approach may be called passive adaptive systems.

For people familiar with control principles the ultimate scheme to assure the realization of any system performance is "feedback", which involves the generation of a set of suitable signals to completely define the actual performance. The deviation of the measured actual performance from some desired performance is then used to control an actuating device to drive the adjustable parameters in a direction to reduce this deviation until it becomes zero or a minimum.[1,4] Whereas this feedback logic applies really for systems whose performances are in the form of measurable physical quantities, it becomes increasingly difficult when the desired performances are complicated analytical definitions, as in the case of most adaptive systems. Furthermore, the performance function of a dynamic system, by its very definition, must be evaluated over a sufficiently long time interval, hence the response of this type of adaptive system tends to be slow. Nevertheless, the concept of using a precisely defined performance index in feedback arrangement represents the ultimate adaptive control system. It is slow, but in principle it can yield any desired degree of adaptability. Systems which follow this logic will be called active adaptive control systems with precise performance indices.

One of the objectives of continuous adaptive control is to overcome quickly the adverse effect of rapid changes in the environmental conditions. Consequently, despite the general advantage of active feedback sysgems, one must often sacrifice the exact adaptation achieved with precisely defined performance indices in exchange for speed and, possibly, simplicity. The design process of this type of adaptive system usually begins with the concept of a well-defined performance index which is progressively simplified and made less precise in an effort to gain speed of adaptation. This type of system is therefore called active adaptive control with inferred performance indices.

The four types of adaptive control systems classified above and repeated below cover most manual and automatic systems:
  a) Open loop adaptive control system
  b) Passive adaptive control system
  c) Active adaptive control with precisely defined performance indices
  d) Active adaptive control with inferred performance indices

Obviously each of the four types may be subdivided into various species. Furthermore many practical systems may incorporate the concept of more than one type. However, for the study of the inherent capability of man as compared with that of machine, an examination may be made along each of these four types.

Open Loop Adaptive Control

Open loop adaptive control, as illustrated in Fig. 1, relies upon knowledge of the effects of various environmental disturbances on the vehicle characteristics and measurement of these disturbances in order to compensate for the expected effects. The primary control may be either open loop or closed loop as indicated in Fig. 1, with one or more distinct variables fed back, but no performance index is transmitted and fed back for the adjustment of vehicle parameters. Open loop adaptive control is the most rapid adaptive control available. By using prior study to anticipate the change of vehicle characteristics one obviates the necessity for time-consuming calculation of performance indices. Men are capable of exceedingly complex open loop adaptive control, as evidenced by the numerous athletic performances carried out in open loop manner. In balancing a motorcycle at various speeds, a different amount of "steering effort" is needed. An experienced rider can apparently make the open loop continuous adjustment of the "steering effort" as a function of speed. Similar open loop control is associated with the ability of automobile drivers to adapt to different road conditions. By reducing speed on a winding road, he is in effect introducing an open loop parameter adjustment to meet a change in environmental conditions, in this case the input spectrum.

In a mechanized system, the distinction between an open loop and a closed loop control scheme is clear. In the case of human control, however, the distinction is not quite so clear because at any one time he may use open and closed loop control together. In the beginning of a learning period, feedback dominates and the response is slow. As he gains more and more experience, open loop control will take

hold and the response will be faster. Often he may have to use some feedback to regain his open loop operation during the "warm-up" period. 3 For instance, in changing from one type of airplane to another, a pilot uses his previously acquired set of control parameters associated with each type of aircraft. He may supplement this strictly open loop adaptation by the use of a few evaluation commands to check on the aircraft performance when he feels a little rusty.

Passive Adaptive Control

Passive adaptive control refers to the inherent ability of all control systems to maintain their transfer characteristics relatively invarient without active adjustment of the controller parameters. For a control system with a primary feedback loop, the static performance of the primary output is fully adaptive. Active drive signals can be generated to reduce the static deviation to zero, through an infinite gain preceding the point in the loop where the control parameter is varied. Dynamic behavior is also somewhat adaptive in a passive sense if only the low frequency portion of the dynamic response is of interest since the high static gain introduced by the primary feedback loop remains quite effective over the low frequency response. Extension of the high loop gain requirement into a higher range of frequencies poses a problem which is generally incompatible with requirements for stability in systems incorporating a typcially low pass vehicle. One interesting scheme for overcoming this difficulty in passive adaptive systems is the use of a pre-filter,5 as shown in Fig. 2. The purpose of this filter is to block the high frequency component of the input signal from entering the feedback system, thereby allowing the use of more lead and more gain in the compensation to improve the adaptation to medium range frequencies. The use of the pre-filter eliminates the excessive overshoot which might otherwise occur with discontinuities in the input. The net effect of this scheme is an exchange of better adaptability at a sacrifice of the overall system response.

Humans appear to be capable of establishing a passive adaptive control system and in particular using the pre-filter technique. For example, in driving a car around a corner or in flying an airplane, one can maintain effective and smooth control over the vehicle at a variety of speeds providing one can relax by ignoring the higher frequency components and high transient rates of the input. This "relaxing" attitude is effectively a mental "filter" necessary to avoid operator-induced-oscillations.

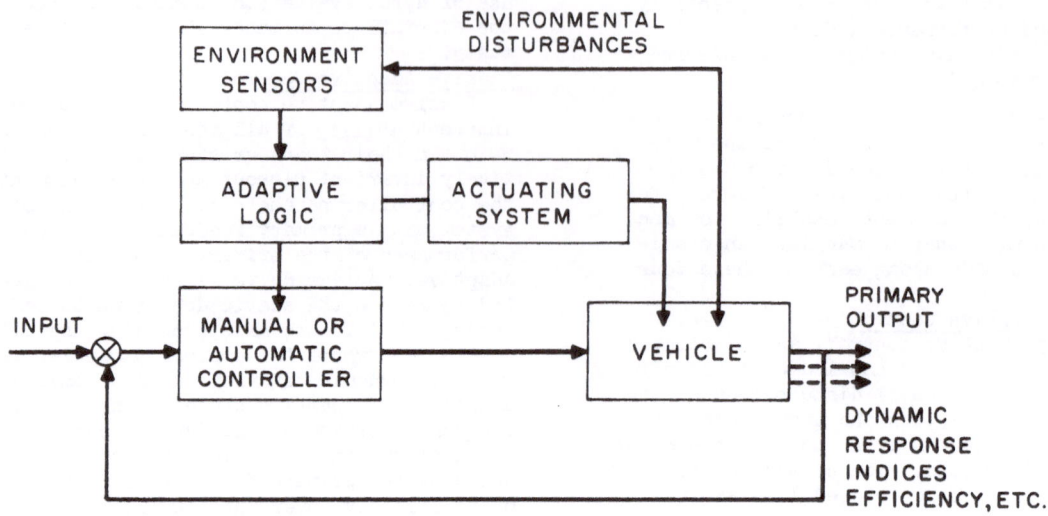

FIG. 1   "OPEN LOOP" ADAPTIVE CONTROL

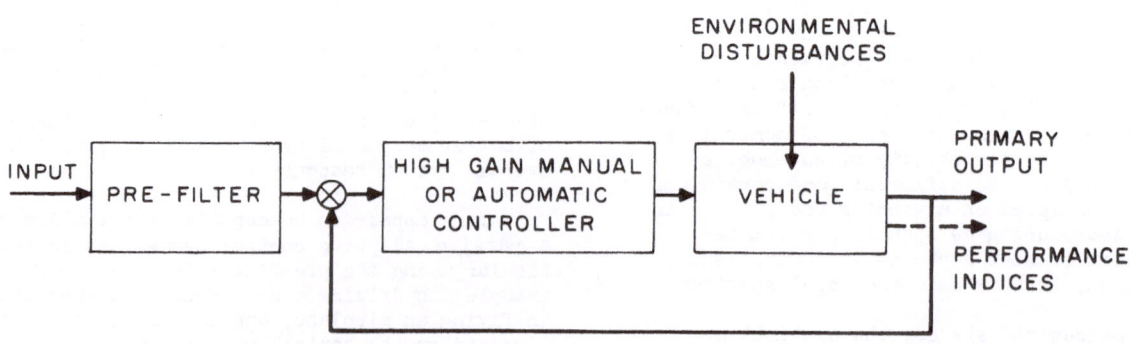

FIG. 2   PASSIVE ADAPTIVE SYSTEM

Active Adaptive Control with Precise Performance Indices

Conceptually one of the most straightforward techniques for adaptive control is through on-line minimization or nulling of some precisely defined performance index. By continuously calculating the performance index, and by knowing the proper adjustments of control parameters in order to reduce this index, the overall system may theoretically be maintained at a prescribed level or at its optimum with respect to that performance index. Though conceptually simple, this form of adaptive control often is quite complicated in its mechanization. The principle is illustrated in Fig. 3, showing the necessity for two feedback loops, a primary loop covering the control of the output; and a secondary feedback loop which adjusts control parameters to minimize some precisely defined performance index. The techniques used to generate a performance index or set of performance indices adequate to describe the dynamic property of a system is usually known as process identification. In particular for on-line adaptive control the process identification must be made available as a set of measurable signals. Most known schemes for this purpose involve the utilization of the fact that the cross correlation function between the primary input and the primary output is the weighting function of the system when the input is white Gaussian noise. However, systems based upon this principle are in general complicated and slow. Humans are incapable of on-line computation as required by this type of adaptive control systems. On the other hand, if the performance indices are slowly computed by a machine, it becomes a trivial problem to close the adaptive loop either by man or by machine assuming that an effective display system is available for man and a set of control signals are available for the machine.

Active Adaptive Control with Inferred Performance Indices

Since the generation of a set of signals to represent the exact desired performance may be quite involved and is often very slow, the first concern in the design of an active adaptive control system is to find an easily calculated performance index which is somehow related to the desired performance. Thus, for example, a function of the mean square error between the output of a system and the output of a model,9 with acceptable performance may be used as an inferred performance index for adaptive control because it is easily calculated and generally related to the ideal performance. There is some possibility that man may retain a mental image of a proper

system response to his control action and make some modification of his action when the response deviates, thus performing a function similar to a model reference adaptive control system.13 As a rule, however, men are unable to compute mentally even the rather simply defined performance indices required for this type of active adaptive control. His adaptive action, without computer aid, is more likely to be "open loop" based upon earlier experience than active feedback control.

While the model reference type of inferred performance index is simple to generate, the mean square operation is still a time consuming process. This mean square operation is, in effect, a smoothing process dictated by the definition of the performance index.2 If the smoothing process is modified and eliminated in part,2, then the response time will be faster but the original requirement of the inferred performance index is no longer met. The minimum amount of smoothing needed in the generation of a performance index for acceptable performance depends upon the pattern of the primary input signal, the configuration used for the generation of the performance index, the desired response speed and amount of adaptation. The design of the system configuration is guided by engineering intuition, but the final overall performance can best be evaluated by the simulation with a computer. Thus the final configuration of this type of adaptive control as illustrated in Fig. 4 may be quite simple but the design can only be realized though trial and error process. For this reason, similar operation is still beyond the ability of humans unless the computation part is done for him.

Another group of automatic fast response adaptive control systems incorporates a relay type non-linear device, 5, which induces limit cycle behavior which in a fact serves as test input. The limit cycle frequency (in some instances it may be of the order of 1000 c.p.s.) is generally higher than the frequency of the control signal, and the limit cycle amplitude is related to the parameters of the system. For this reason the limit cycle amplitude and frequency can be observed and used as an inferred performance index to control an adjustable parameter. (In principle, the non-linear property of the relay has an inherent tendency to lower the effective gain in response to an increase in limit cycle amplitude and thus achieve a measure of adaptivity without additional parameter control. This effect tends to maintain the limit cycle amplitude and system loop gain stationary despite changes in other gains around the control loop.).

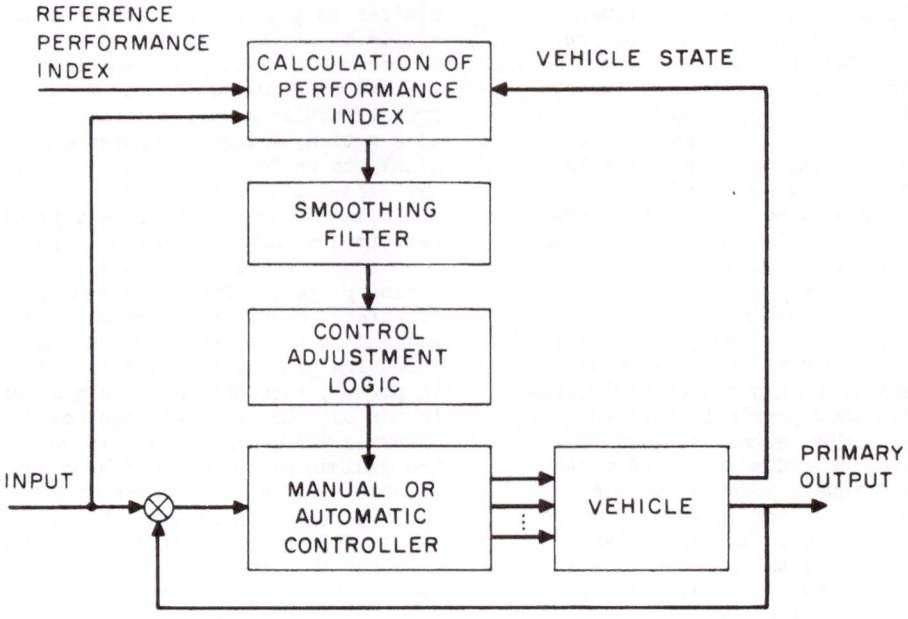

FIG. 3   ACTIVE ADAPTIVE CONTROL WITH PRECISE PERFORMANCE INDICES

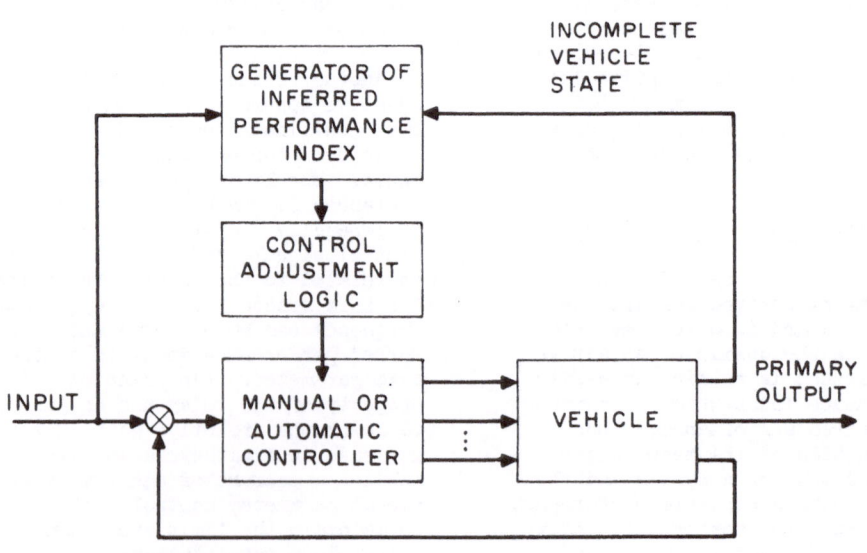

FIG. 4   ACTIVE ADAPTIVE CONTROL WITH
INFERRED PERFORMANCE INDICES

When used in an automatic active adaptive control system, the high frequency limit cycle component of the output signal may be filtered out and its amplitude or frequency used to adjust lead or gain of a compensation device. Men can usually recognize the limit cycle frequencies in the output and can use this information to adjust control system parameters (mostly his own action) in an adaptive fashion. Furthermore for many difficult control tasks, humans tend to operate in a bang-bang manner, thereby generating system limit cycle behavior without the necessity for any external non-linearity. By observing the limit cycle amplitude he can adjust the size and timing of the command pulses he delivers to the system in order to stabilize the system output with an acceptable level of limit cycle amplitude. In addition to the amplitude and frequency of the limit cycle, men are capable of performing simple pattern recognition on these high frequency components of the error and using this information for adaptive control. For example, diverging high frequency oscillation is quickly observed and the corrective control may be taken after two or three cycles, whereas sluggish response of the system as evidenced by sudden decay of the limit cycle is also immediately apparent to the human operator. 11

In summary, a human operator is inferior to automatic adaptive controllers for on-line calculation of even simple performance indices. On the other hand, humans have a very large storage capacity for open loop control programs and are quite capable of performing bang-bang type adaptive control. It would appear the exploitation of man's superior ability in the last two aspects may be further developed with the aid of on-line computation and proper tie-in between machine and man through effective display. Consideration along this direction is discussed in the following part of this paper.

HUMAN SENSORS AND THEIR FUNCTION IN ADAPTIVE VEHICLE CONTROL

The block diagram of Fig. 5 illustrates the functional positions of the eye and the inner ear balance mechanisms (vestibular system) in vehicle control. The eye can observe indications displayed on instruments or the view of the outside world with respect to his vehicle. The vestibular system responds only to vehicle motions, not to input commands, and is primarily sensitive to vehicle rates over most of the frequency range commonly of interest in control. The block marked "Logic" refers to the control and compensation fucntions of the central nervous system and is shown as also receiving inputs from other sources than the visual and vestibular system. These include visual information on other than the primary input variables, as well as tactile and audio information on the system state.

Visual Systems

Information is conventionally transmitted to pilots primarily through visual displays, and virtually all the research on engineering characteristics of human operators has been conducted with visual displays. The eyes have the ability to furnish very fine discrimination of displacement on displays and can detect approximately one minute of arc angular displacment between two dots, or a few seconds of arc displacement between two lines in a vernier type measurement. They are chiefly used, however, for their ability to sense information for complex pattern recognition; such as reading numbers on a dial, interpreting the state of a system from an integrated attitude display, finding the runway lights of an airport amidst a great variety of other city lights, etc. The visual channel is somewhat poorer at estimating velocities of moving objects on a display and is almost valueless in estimating accelerations. In addition, the limited central viewing area permits the eye to take in accurate information on only one small area at a time, and displays of information on several dials must be assimilated sequentially by eye with scanning movement.

Since knowledge of output rate may be essential for stabilization of an unstable vehicle as well as for fast active adaptive control as discussed above, it is often found necessary to display rate of change of the vehicle output as well as the output itself. An obvious example is the display of aircraft rate of climb as well as altitude, but it is doubtful that separate dials are the appropriate display configuration for control of difficult high order systems.

For stabilization and control of systems with short characteristic time, such as the "short period mode" of an aircraft, fast response instruments and integrated display become essential. The necessity for integrated visual presentation of velocity information may be illustrated by reference to a series of experiments recently performed in our laboratory. In the first part of this experiment only system error was displayed and the simulated vehicle being controlled was made more difficult by increasing the first order lag cascaded to a double integration. In such cases, the operators tended to use their proportional control stick in a bang-bang

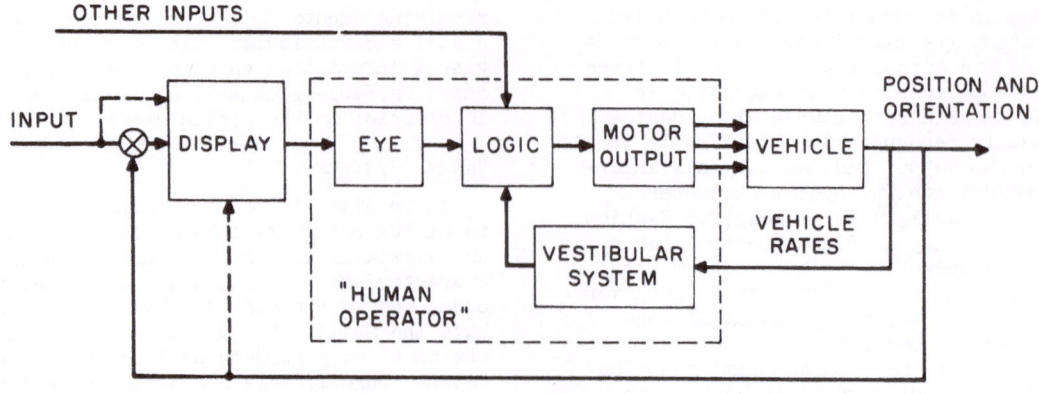

FIG. 5   THE HUMAN SENSORS IN VEHICLE CONTROL

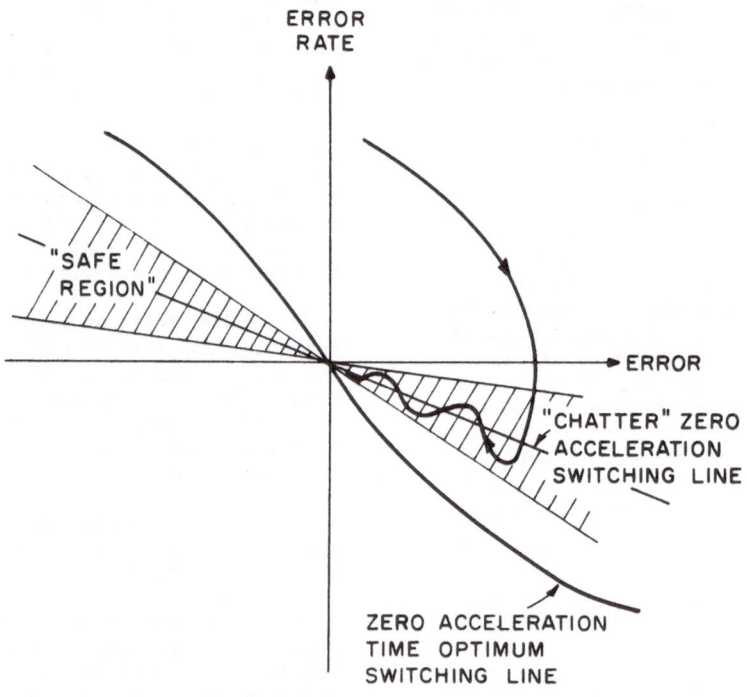

FIG. 6   PHASE PLANE DISPLAY WITH "SAFE REGION"

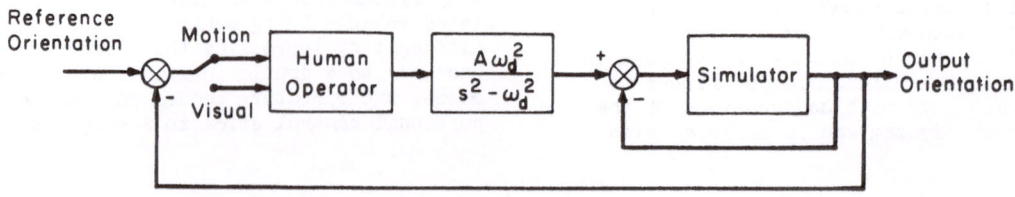

FIG.7 CONTROL OF INVERTED PENDULUM WITH VISUAL OR MOTION FEEDBACK

50

manner. Analysis of the operator's response in the phase plane yields a collection of switching points lying mostly in the second and fourth quadrants, showing how the operator bases his control on error rate derived from the visual display of error. With visual display of error, an experienced operator can maintain marginal control of the simulated system with a lag of no more than one second.

As a next step, the operator was shown a phase plane display with a moving spot indicating error and error rate on the horizontal and vertical axis respectively, as seen in Fig. 6. 10 By observing error rate the operator can now make some estimation of the acceleration of the error. With this information, he can readily guide the displayed spot from a large initial deviation into a "safe region" in the second and fourth quadrant of the phase plane. Once in the safe region, similar tactics will allow him to keep the spot within this region while bringing it toward the origin with a few repeated switchings.

In general, for systems higher than second order, a switching surface in state space is necessary to complelely define a control law. For a third order system the three dimensional switching surface may be projected onto the phase plane, and for a fixed value of acceleration leads to a single line in the phase plane. For example, the "zero acceleration time optimum switching line" shown in Fig. 6 is the intersection of the three dimensional time optimum switching surface with the two dimensional phase plane. Similarly, other switching logic may be characterized by the corresponding zero acceleration switching lines on the phase plane.

Those zero acceleration switching lines lying closer to the vertical than that of the optimum switching logic will result in full swing oscillatory mode or unstable behavior. On the other hand, switching surfaces with the zero acceleration switching lines lying closer to the horizontal will result in a slower response which "chatters" in toward the origin.

In manual control of a third order system using a phase plane display, considerable uncertainty exists in the estimation of the instantaneous acceleration, the knowledge of the switching surface, and the timing of the operator's control reversals. To allow for these uncertainties and still avoid overshoot or instability, we have deliberately chosen the control logic corresponding to the chatter mode. The corresponding zero acceleration switching line must carry sufficient lead angle over the optimum zero acceleration

switching line to allow for these uncertainties. The uncertainty region centered around the chosen zero acceleration switching line is therefore the "safe region." In operation, this region is used as a target area to which the operator brings the displayed spot. With this arrangement an average operator can easily control a simulated system with three cascaded integrations. This example is not intended to stress the merits of the phase plane display, for it has serious drawbacks in requiring two axes for control of one output. However, it is an illustration of a technique for using integrated display ideas to replace simple dials and assist the human's visual system is gathering rate information.

The Vestibular Sensors

As indicated in the block diagram of Fig. 5 the human vestibular motion sensing apparatus will be stimulated whenever the vehicle moves. Generally the outputs of the vestibular system in terms of perceived motion have been considered only in their negative aspects, as they interfere with successful control performance by pilots. Bizarre or prolonged motion as sensed by the vestibular system can result in motion sickness, aviator's vertigo, or momentary confusion in interpretation of visual displays. Most of these phenomena can be explained in terms of the dynamic response and threshold levels of the vestibular sensors.

Realizing that the vestibular sensors provide some indication of vehicle motion, we have concentrated on the positive aspects of the use of motion cues in manual control of vehicles. The semicircular canals behave like overdamped angular accelerometers and provide angular velocity information over the frequency range 0.1 to 10 radians per second about all body axes. The otoliths are overdamped linear acceleration sensors furnishing information on linear velocity over the frequency range 0.1 to 1.5 radians per second. 7

A set of experiments designed to illustrate the ability of men to use vestibular as well as visual inputs in vehicle control is illustrated in the block diagram of Fig. 7. The task was to control the orientation of a moving base flight simulator about the roll axis to maintain its axis of symmetry vertical. 12 The vehicle dynamics were those of an inverted pendulum, and the divergence frequency ω was permitted to vary. Larger values of ω, indicating the movement of the unstable pole further into the right half plane (or control of a shorter pendulum) proved more difficult to control. Comparison was made of man's ability to control the system using (1) only visual cues while sitting outside the simulator, (2) only motion cues while sitting in the simulator with no visual display, and (3)

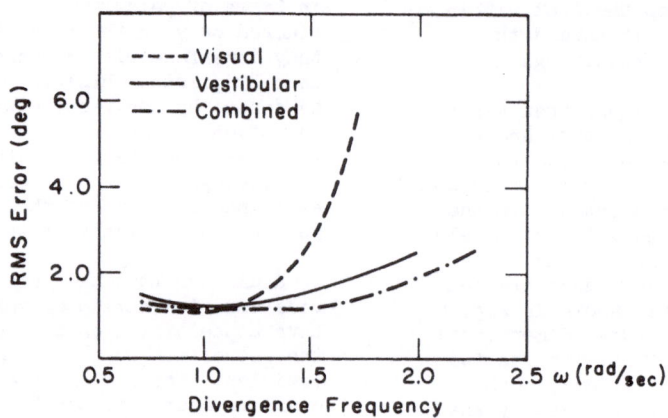

FIG. 8   RMS ERROR FOR CONTROL
OF INVERTED PENDULUM

both visual and motion cues, placed in the simulator with the hood off. As seen in Fig. 8, the RMS error increased with difficulty of the task (increasing divergence frequency) for all conditions; however the addition of motion cues permitted subjects to maintain control of the simulator at higher divergence frequencies which had proven uncontrollable with the visual cues alone. Analysis of the operator's control behavior in terms of switching line slopes for a bang-bang model explains this phenomenon by the operator's ability to generate the required increased lead compensation at higher frequencies when motion cues were available. Notice further that at very low divergence frequencies, when the system is quite easy to control and rate compensation is less important, the pure visual display mode had the lowest error. This result is also compatible with the known ability of the visual system to detect very small deviations in position.

This experiment, as well as others, confirms our belief that the manual controller relies heavily upon inputs received through the motion sensors in situations where considerable lead or rate compensation is necessary for adequate performance. This finding tends to explain the number of cases of vehicle simulation wherein fixed base tests were found to be uncontrollable; however motion base simulation and the actual flight tests proved the vehicle dynamics to be quite flyable.

In the adaptive control situation, with changing vehicle characteristics, it seems clear that the vestibular system plays an important role in early identification of these changes. By sensing rate and acceleration, the motion sensors provide an indication of changes in vehicle state long before the visually observed position errors have had a chance to build up to non-negligible values. Thus the first indication of an unexpected automobile skid on ice is the abrupt sidewise acceleration rather than the observed change in heading of the car. Furthermore, since the vestibular system furnishes a second and independent measurement of vehicle motion, it can in some circumstances be used in parallel with the visual input to detect malfunctions, particularly in the visual display system.

CONCLUSIONS

Ever since the launching of man-in-space programs, the question of whether the role of men during flight should be primarily a pilot or passenger has been hotly debated. This question cannot be answered in a general way,

for each phase of manual control involves consideration of the human's limitation and requirement for special training, continuous attention and control effort. All these burdens should therefore be considered together with other human factors associated with a particular mission in order to arrive at an optimum manual operation loading condition.

On the other hand, manual controls are economical in some aspects, and parallel manual and automatic control does provide reliability through redundancy.

It is not the purpose of this paper to recommend manual control as a substitute for automatic adaptive control for any specific mission. On the other hand, this paper does indicate some of the capabilities and limitations of man for making continuous adaptation - a function higherto is dominated by inanimate systems.

In summary, man is limited in his capacity for continuous computation and estimation of the rate of a signal. However, he is endowed with huge capacity for receiving visual and non-visual inputs. He possesses or can learn an enormous amount of programmed control logic and exhibits a high degree of flexibility in his modes of manual maneuvering. All these limitations and capabilities apply to his control or the primary loop or the adaptive loop. For this reason, it would seem to be arbirary to let man handle the primary loop manually while insisting that automatic systems must be used to handle the adaptive loop. A more logical approach would be an analysis of the action of man in performing all phases of control and to provide assistance to him in the specific area of his limitation.

The simplest and by far the most common early method for treating the problem of changing vehicle characteristics was by placing the bulk of the active adaptive control burden on the man. An alternate arrangement was to make the control system automatically adaptive in an attempt to present the pilot with the same handling qualities under all environmental conditions. This leaves the man a relatively simpler primary tracking task to perform but places a severe burden on the automatic adaptive system. When the environmental conditions vary over wide ranges and complicated patterns in view of the human capabilities and limitations discussed above, a third man-machine adaptive configuration is suggested, as shown schematically in Fig. 9. Man can act most efficiently in his adaptive role when he is only lightly loaded with the primary tracking task, and for this reason a considerable amount of passive vehicle stabilization is

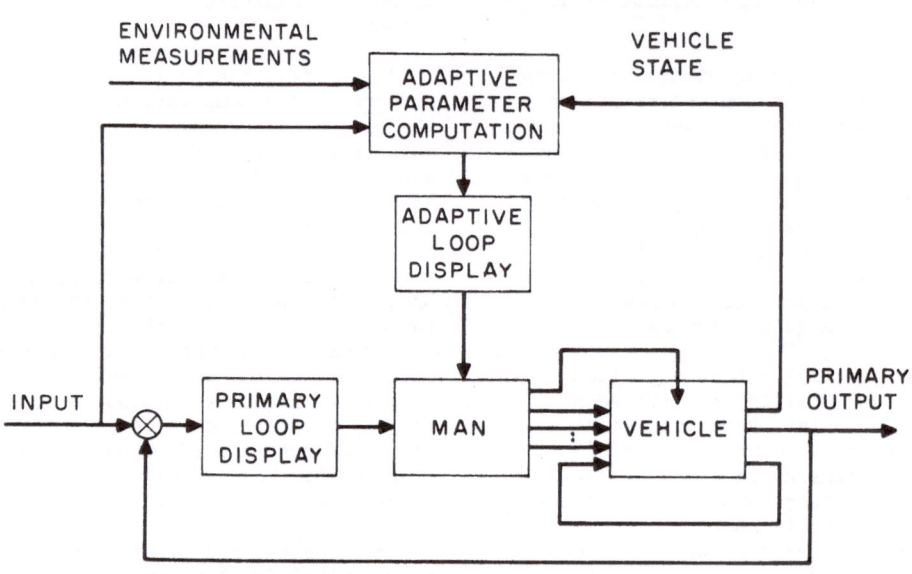

FIG. 9   A MAN-MACHINE ADAPTIVE CONFIGURATION

required. Man's weakness in computation of complex performance indices and his inability to make accurate measurements of the environment lead to the use of automatic assistance in these roles. The block labeled "Adaptive Parameter Computation" is indicated to assist man in the measurement and computation required for the identification portion of adaptive control, and a separate block labeled "Adaptive Loop Display" is shown to communicate the information on the characteristics of the system to the human operator. Note that the information coming through the adaptive loop display is not primary control variables but rather is related to a variety of functions relevant to vehicle characteristic identification. The nature of the adaptive loop display can take advantage of man's ability to recognize complex patterns. Thus, for instance, the characteristics of the vehicle dynamics might be displayed by means of the system step response, even though no step inputs are actually given to the real system, thereby enabling the operator to identify rise time, overshoot, damping and static gain relative to a reference step response in a rather straightforward manner. This seemingly simple task for man is in fact quite difficult to instrument to yield appropriate actuating signals for automatic control. It is felt that with this type of configuration the man and the machine each contribute their best to the overall system performance; the machine in performing stable rapid closed loop control and in measuring and computing a variety of indices and displaying them to the man, and the man in using his training and experience together with a number of computer indices and vehicle state variables fed back to him to weigh all the information and decide upon the appropriate adaptive adjustments.

To fully utilize the joint man-machine capabilities, more effective display systems must be developed. The few examples of man-vehicle adaptive control illustrated in this paper serve only as an introduction to this area of research.

ACKNOWLEDGMENT

This paper and the related work were sponsored by NASA under Grant No. NsG 577.

REFERENCES

(1) Draper, C. S. and Li, Y. T., Principles of Optimizing Control Systems and an Application to the Internal Combustion Engine, American Society of Mechanical Engineers.

(2) Hagen, K. E., The Dynamic Behavior of Parameter-Adaptive Control Systems, Sc. D. thesis, Massachusetts Institute of Technology, June 1964.

(3) Krendel, E. S. and D. T. McRuer, Servo Approach to Skill Development. J. Franklin Institute, 269:24, 1960.

(4) Li, Y. T., Optimalizing Systems for Process Control, Instruments, Volume 25, No. 1, 72-77, No. 2, 190-193, No. 3, 324-359. January, February, March 1962.

(5) Li, Y. T. and Vander Velde, W. E., Philosophy of Non-linear Adaptive Systems, Proceedings of the First International Congress of the International Federation of Automatic Control.

(6) Li, Y. T. and H. P. Whitaker, Performance Characterization for Adaptive Control Systems, Proceedings of the First International Symposium on Optimizing and Adaptive Control.

(7) Meiry, J. L., The Vestibular System and Human Dynamic Space Orientation, Doctoral thesis, Massachusetts Institute of Technology, June 1965.

(8) Mishkin, E. and L. Brown (editors), Adaptive Control Systems, McGraw-Hill, 1961.

(9) Osborn, P. V., Whitaker, H. P., and Kezer, A., New Developments in the Design of Model Reference Adaptive Control Systems, Institute of Aerospace Sciences Paper 61-39, January, 1961.

(10) Platzer, H. L., A Non-linear Approach to Human Tracking. The Franklin Institute, Interim Technical Report No. I-2490-1.

(11) Young, L. R., D. M. Green, J. I. Elkind and J. A. Kelly. The Adaptive Dynamic Response Characteristic of the Human Operator in Simple Manual Control. IEEE Transaction on Human Factors in Electronics, Sept. 1964. Vol. HFE, pp. 6-13.

(12) Young, L. R. and J. L. Meiry, Bang-Bang Aspects of Manual Control in High Order Systems. IEEE Transactions on Automatic Control. Vol.AC-10,No.3, pp336-341. July 1965

(13) Young, L. R. and L. Stark, Biological Control Systems - A Critical Review and Evaluation. Developments in Manual Control, NASA CR-190, March 1965.

Discussion

Dr. M.V. Meerov (USSR) asked whether the adaptability of manual control systems with high loop gain had been investigated. Prof. Li replied that in the cases where the high gain system took the form of relay control, that better adaptability had been achieved.

The criterion used to describe the 'adaptibility' was the sensitivity function of the system. This was defined as the percentage change in performance (in the frequency domain) expressed as a percentage of the change in disturbance. Results show that the higher the gain, the better is the sensitivity function. Asked about the effectiveness of a display in the form of a phase plane portrait of error and error rate given as a single dot on a CRT screen, Prof. Li stated that an operator having no knowledge of phase plane took ¾ hour to learn to control a system consisting of a first order lag with two integrators. If, however, the concept of phase plane had been explained beforehand, the training period was reduced to four minutes.

Mr. P. H. Hammond (UK) recalled the work of Birmingham (ref. 1) who used a special form of single dimensional display. The signal displayed was a composite function of the error and weighted proportions of its derivatives. He asked the authors to compare their method with that of Birmingham. Prof. Meiry in reply said that this type of display (termed 'quickening') had the deficiency that information regarding position was lost. The original use of quickening had been concerned with a submarine diving, where the vessel dynamics were represented essentially by 2 integrators plus 2 second order systems. To position the submarine, a depth dial was still required. Results showed that an operator could increase his performance by an additional integrator when given a phase plane display instead of a quickening type display, i.e. he could control a system consisting of 3 integrators plus a first order lag of 1 second.

Prof. R. J. A. Paul (UK) queried the applicability of theoretical studies to practice. It was stated that whereas a controller might be satisfactory on a simulation of an aircraft, a frequent experience was that (in practice) as the stiffness of an aircraft increased through the flight envelope, what had been a tolerable limit cycle became a drastic oscillation on the pilot. This was due to the structural hysteresis and dead zone effects. During the development of the X15 aircraft (in which use has been made of the limit cycle of the high gain system) a great deal of instrumentation had to be installed to ensure that the aircraft was stable over the complete flight envelope. On the basis of the total experience and data obtained, it was concluded that if this had been available some years ago, an adaptive controller would not have been needed.

Written contribution to the discussion
Mr. B. R. Gaines (UK)

At the Cambridge University Psychological Laboratory we have been studying the control strategy of the human operator on high-order systems. A typical system to be controlled or stabilized is three cascaded variable lags with joystick input and compensatory display. In following a low-frequency sinusoidal input the operator uses apparently proportional control at short lags (less than half second) and switching-mode control at long lags (greater than four seconds). At intermediate lags there are transitions between these modes and an error/error-rate diagram shows 'bursts' of ellipsoidal motion.

Controlling the autonomous system with no input at high lags the operator uses the switching-mode control described in the paper, but is not able to dead-beat the system and maintains an ellipsoidal limit-cycle in the phase-plane. With practice on a fixed lag the operator may eliminate this and so our overall system has a loop which varies the lag to maintain the limit cycle. Part of the effect of practice is to increase the lead or anticipation; the usual stategy is to switch when the velocity is zero but initially this is done with reaction-time lag and eventually with anticipation. The 'feel' of the system with input compared with the autonomous one is that the former is more 'sluggish', presumably because it is less under operator-control.

We have found it very important that the operator has kinesthetic feedback from the position of the control. His actions have little immediate effect on the display and spring-loading of the control not only gives him memory of these but also a zero-reference. We also have put in visual aids to error-rate estimation, not through X-Y display, but instead by showing a short line (in place of a spot) whose inclination to the vertical is proportional to error rate. This cue resembles looking down at a car on a road and is very effective.

The authors replied as follows. The experiments at Cambridge support our findings that human operators tend to track continuously in "easy" systems and exhibit bang-bang control when faced with a high order controlled element requiring considerable lead for stabilization. As pointed out in Ref. 12 of the paper, the generation of lead compensation by the operator requires mental computation of the integral of control stick output; this computation being considerably simpler with pulse or bang-bang operation than with continuous control. Of course the switching lines or switching planes for the operator in bang-bang control must be discontinuous because of his reaction time and consequently the existence of limit cycles is to be expected with high lag systems. These limit cycles apparently provide the controller with significant information concerning the characteristics of the overall system, and may be an important cue in manual adaptive control.

REFERENCE

1. Birmingham, H. P. and Taylor, F. V. - A human engineering approach to the design of man-operated continuous control systems. Naval Research Laboratory, Washington D.C., NRL Report 4333, April 1954.

AUTONOMIC CONTROL OF RESPIRATION

by Dr. W.F. Fincham and Dr. I.P. Priban,
National Physical Laboratory,
Teddington,
Middlesex, England

ABSTRACT

Two aspects of the biological processes controlling respiration in man are discussed in this paper. The first describes how the properties of the blood are used as the basis for a hill-climbing type control system. Deviations from the optimum are corrected by changes in the rate of ventilation of the lungs so that ventilation keeps pace with metabolism. The second aspect considered is the minimization of the energy expenditure for breathing. This involves the neuromuscular part of the respiratory system which selects, for each breath, both the duration and the volume change in the lungs, as well as specifying which muscles are activated.

INTRODUCTION

In most of the recent work on the modelling of biological control systems, an important feature has been the inclusion of reference or 'desired' values at which the system attempts to maintain its outputs. Although the performance of such models has been similar to that of the real system modelled under certain idealised conditions of operation, we feel that the inclusion of several reference levels is unrealistic. In fact, there is no experimental evidence to suggest that biological reference levels exist. If a system is to attain close control of its output variables, even when some of the system parameters are subject to large changes, it is likely that its operation can be explained in terms of the theory of self-adaptive systems. We believe that the respiratory control system is an example of such a system, and in this paper a model is proposed which provides a basis for a theoretical and experimental study.

BASIS FOR CONTROL

Biological processes proceed at optimum rates only if the physical and chemical state of their environments are closely controlled. The state is determined mainly by the values of variables within the blood, such as temperature, pH (e.g. acidity), the partial pressures of carbon dioxide and oxygen (pCO_2 and pO_2) and the level of saturation of haemoglobin with oxygen. The chemical state is, therefore, dependent upon the relative exchange rates of CO_2 and O_2 with respect to the blood. In the body there are two major sites where these gas exchanges occur.

The first site is that of the tissue cells of the body where the process of interest is that of metabolism. This can be represented by the energy balance equation:

$$\text{Food (chemical energy)} + O_2 =$$

$$\begin{array}{c}\text{Heat}\\\text{Energy}\end{array} + \begin{array}{c}\text{stored}\\\text{chemical}\\\text{energy}\end{array} + \begin{array}{c}\text{work}\\\text{energy}\end{array} + CO_2 + H_2O$$

The important aspect is that O_2 is consumed and CO_2 produced. The CO_2 is transported via the circulation to the lungs, the second site, where it is exchanged for O_2. The oxygenated blood then returns to the tissues.

The exchange of gases between the circulation of the lungs and the gas in the lungs is mainly a passive process. The ventilation of the lungs requires the expenditure of energy in the respiratory muscles of the chest wall and diaphragm. If the chemical environment of the tissues is to be maintained at its optimum state for changing levels of metabolism, then the level of ventilation must be controlled.

The two aspects of respiratory control described in the paper are

(1) The control of the level of ventilation

(2) The mechanism by which the energy expenditure in the respiratory muscles it kept at the minimum consistent with the level of ventilation.

CHEMICAL CONTROL OF VENTILATION

For an accurate description of the chemical state of the blood, at least the five variables mentioned above must be considered. However, for the purposes of this investigation a more convenient representation is used, which highlights the functional characteristics of blood relevant to this study, in particular of haemoglobin. For a given sample of haemoglobin, the ordinate of the curve shown in figure 1 represents the gas exchange ratio that can be achieved without producing changes in the pH or pCO_2. In order to show a two-dimensional curve, the abscissa has been labelled as a composite function of pH and pCO_2. The curve, based on data obtained by other workers[1,2], has a peak value of about 0.67, whereas a typical value for the metabolic gas exchange ratio (Em) is 0.76. This means that as blood passes through

Superior numbers refer to similarly-numbered references at the end of this paper

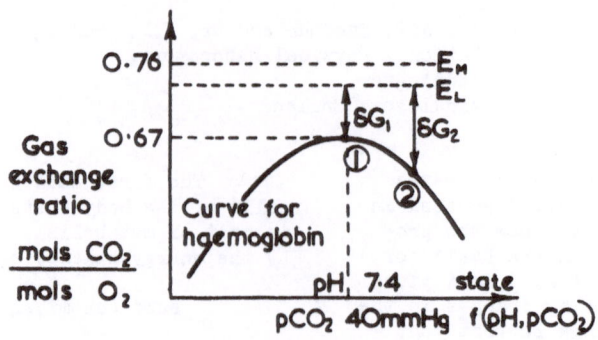

Fig. 1. Characteristics of Respiration

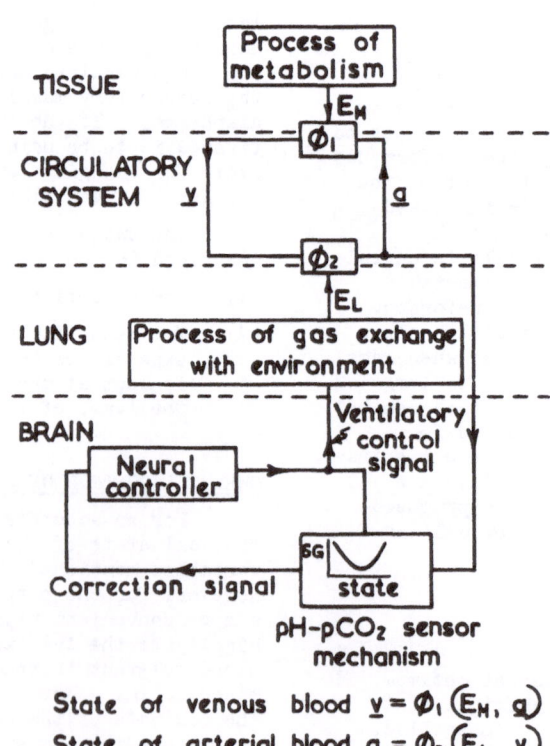

State of venous blood $\underline{v} = \phi_1 (E_M, \underline{a})$

State of arterial blood $\underline{a} = \phi_2 (E_L, \underline{v})$

Fig. 2. Respiratory Gas Control System

the tissues, haemoglobin can only partly buffer against the tendency for a change in the state of the blood, and therefore both pH and pCO_2 will change.

Because breathing is a rhythmical event producing alternating gas flow into and from the lungs, there are cyclic variations in the pCO_2 and pO_2 values of the gases in the lungs. This, in turn, causes variations in the gas exchange rate ratio between the blood and the gas phase and so induces a fluctuating component in the values of pCO_2, pO_2 and pH of the arterial blood leaving the lungs. The mean values of these variables will depend on the composition of blood entering the lungs and on the total volumes of CO_2 and O_2 exchanged.

The shape of the characteristic shown in fig. 1 is utilised in the following way. If the mean state of the blood is at point 1, then gas exchange in the lungs (E_L) can be accomplished with the smallest variation about the mean state. If, however, the blood is at state 2, then for the same E_L, the variation in state will be larger. For the two states considered, the relative magnitudes of these variations would be a function of the ratio $\delta G_1/\delta G_2$.

The data given in ref. 3 indicates that a change in pO_2 alters the flanks of the curve, but has little effect on the lateral position of the peak.

Anatomical and physiological findings suggest that a sensor mechanism exists which is capable of responding to these fluctuations (or perturbations) in pCO_2, pH or both. If this is accepted, then a complete optimizing system can be postulated (fig. 2).

The principle of control is that ventilation is adjusted in the direction that will minimize the magnitude of the perturbation signal. When the perturbation signal is at a minimum, the mean values of pH and pCO_2 are those shown in fig. 1. These are, in fact, the values found in normal man. Further, at these values the level of ventilation will be just sufficient to keep pace with metabolism.

The 'control signal' connection to the sensor is necessary so that the sensor output is independent of the amplitude and frequency of breathing and dependent only on the state of the blood passing through it.

CONTROL OF THE RESPIRATORY MUSCLES

The level of ventilation may be expressed in terms of two components:

Level of ventilation $(litre/min)$ $\dot{V} =$

Frequency of breathing $(breaths/min)$ f $\times$

Swept volume (tidal volume) $(litre/breath)$ V_T

Theoretical studies have shown that for each value of $\dot{V}$, there is only one value for f at which the energy expended in the respiratory muscles will be at a minimum[4].

At a fast frequency of breathing and a small tidal volume, high input power is required to produce the relatively larger changes in velocity of muscle shortening. Because the 'dead space' volume of the airways has to be cleared before any useful gas exchange can occur, increased energy is required to compensate for the relative reduction in the effective gas exchange volume (tidal volume - dead space volume). Alternatively, at large tidal volumes, large movements are required for longer durations. The power required to distend the chest wall and enlarge the lungs rises steeply as the tidal volume increases.

The natural frequency of breathing is found to agree with the predicted frequency at which the power necessary for a particular level of ventilation is at a minimum. This finding is fundamental to the hypothesis that breathing is controlled so that energy expenditure is minimized.

An additional aspect is that many of the muscles used for breathing may also be used to cause movement of the body or to maintain posture. Therefore, it seems reasonable to hypothesise that, if certain muscles are involved in producing, say, movement, then other less loaded muscles are used to cause breathing.

The muscles are stimulated by impulses from a neural network in the brain, and the function of this composite neuromuscular system is:

1. to regulate the period and the tidal volume of each breath so that the average energy expended for breathing is at a minimum;

2. to select those muscles of respiration which can shorten most easily and thus effect ventilation while developing only minimum tension.

STRUCTURE AND FUNCTION OF RESPIRATORY NEUROMUSCULAR ASSEMBLY

In the chest wall, two sets of main muscle may be defined according to their function. For example, shortening of the 'inspiratory set' causes the ribs to be raised and increases the volume within the rib cage. Anatomically, the principle difference between these sets is the angle of inclination of the muscle relative to the spinal pivot point of the ribs. Fig. 3 shows a typical component of this assembly. Nervous activity from the brain excites the synaptic units. Unit G is a junction in the neural pathway to the muscle spindle. The spindle is a hybrid of muscle fibres and specialised nerve endings. Unit G stimulates the fibres, whereas the nerve endings form a region which is sensitive to the tension

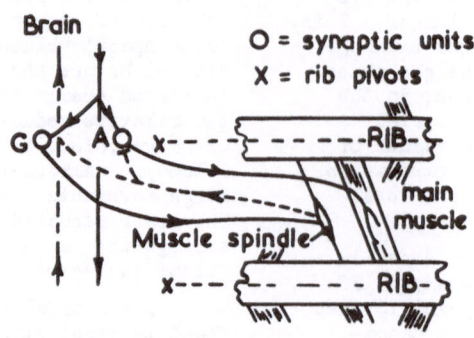

Fig. 3. Component of the Respiratory Neuromuscular System

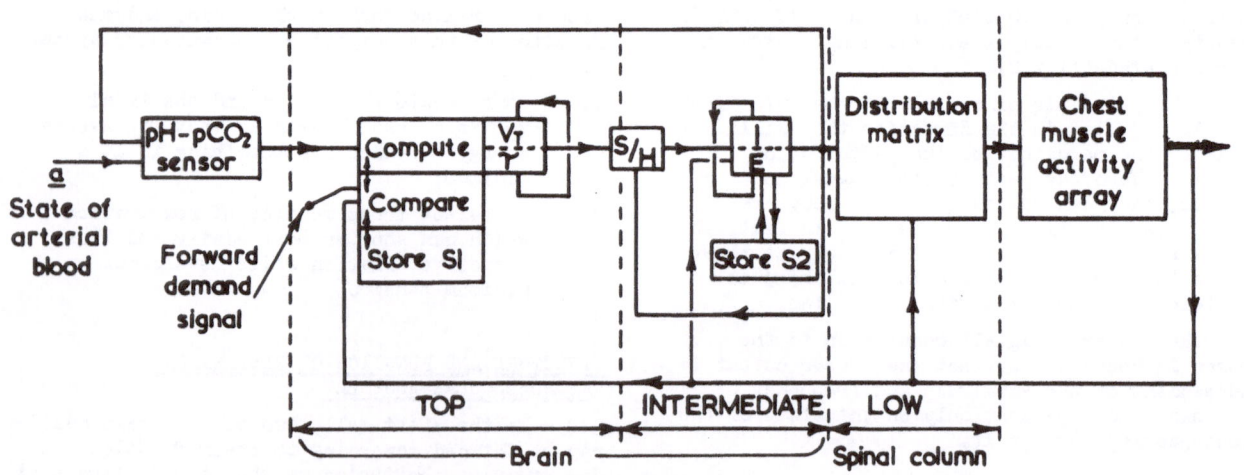

Fig. 4. Neuromuscular Control System

60

developed within the spindle. The output of the receptor is therefore determined by the length of the spindle relative to its excitation. The excitation passing from unit A to the main muscle is determined by the level of main excitation from the brain and by the signal from the spindle. The latter is a feedback signal in the sense that it is a measure of the ability of the main muscle to shorten and so displace the ribs by an amount equivalent to the excitation from the brain. The spindle may influence conditions at other local type A synaptic units and also sends information to the brain.

DESCRIPTION OF CONTROL SYSTEM

The neuromuscular control system has, for convenience, been divided into three levels of control. To the top level is attributed the most important aspect of control, which is the 'optimal design' of each breath in terms of its duration and tidal volume. In the remaining levels, the full temporal and spatial pattern of the activity to the respiratory muscles is determined (fig. 4).

BRAIN RESPIRATORY ACTIVITY (TOP AND INTERMEDIATE CONTROL LEVELS)

The inputs to the neurones in the brain that are associated with respiration, are the signals representing the ventilatory drive (the magnitude of the perturbation signal from the $pH-pCO_2$ sensor) and information from the muscle spindles. At the top level of control this information is used to determine the direction in which the level of ventilation should be adjusted and to determine the appropriate values of tidal volume and breath duration. To do this, the top level requires a knowledge of what previous adjustments have achieved in terms of increased or decreased energy expenditure. The implication is that storage and logical decision processes are involved in order to select appropriate sets of cause and effect and so allow the control to function satisfactorily. For example, the top level controller must decide whether a change in the chemical state of the blood is the result of a change in the level of metabolism or in the level of ventilation. The output from the top level control is therefore the predicted best values of tidal volume V_T and breath duration τ. In fig. 4 the block labelled τ/V_T and its associated feedback represents the mechanism which ensures that the product $(1/\tau)$ x V_T matches the ventilatory demand.

If the flow of air through the mouth or free movement of the chest wall is restricted experimentally, no change is apparent in the duration of that particular breath. This result is assumed to arise because a sample/hold device (S/H) exists between the top and intermediate levels of control. It permits changes in the optimal design of a breath to be recognised only at the instant of completion of the present breath, that is, at the end of an expiration.

At the intermediate level of control the detail of the breath is increased in terms of the temporal and some spatial distribution of activity to be sent from the brain. The design is based on feedback from the muscle spindles. The feedback signal can be interpreted in terms of how close previous optimal designs were with respect to the present true optimum. The true optimum is continually changing because of the changing environment and of interaction with other biological sub-systems not directly involved in the control of respiration. By this means, a characteristic pattern of a breath is built up for each individual. In essence, each individual has a typical waveform of gas flow, as measured at the mouth, which is determined by his particular physiological or pathological state.

A principal component of the respiratory control centre in the brain is in reality a network of cells that exhibit phasic activity. This activity is in phase with breathing. A portion of the cells are active during inspiration and the remainder during expiration (blocks I and E, fig. 4). The feedback loop around I and E indicates oscillatory behaviour in which only one cell group is instantaneously active. The store S2 allows the build-up of the optimal pattern. The process in the intermediate level allows some relatively rapid adjustment to the breath design, in terms of flow rate, providing that any discrepancies are received early in the execution of a breath. For example, if the flow rate during inspiration were briefly impeded soon after starting the breath, then some compensation might result during the same phase of breathing.

Additional information regarding the model structure of the intermediate level has come from breath-holding experiments. In these tests subjects were required to hold their breath for periods of time ranging between 0.5 and 1.5 seconds, and on average about 0.8 seconds. This period was chosen on the basis that the test would cause only negligible changes in ventilatory demand signal. The result of the pilot tests show that, after an initial period, the subjects responded in identical ways. The duration of that phase of a breath during which the 'hold signal' appeared was extended by a period equal to that of the 'hold' time. Also, the volume of gas inspired, or expired, was very close to that expected if no hold signal had occurred. It would seem that the voluntary control of the breath only interrupts the execution of the optimal pattern.

The output from the brain, that is from the intermediate level controllers, consists of bursts of frequency modulated neural pulses flowing in several selected channels down the spinal column. The duration of the burst determines the duration of the breath. The instantaneous activity within a channel represents the optimal design of energy expenditure in terms of the firing of a particular set of respiratory muscles.

SPINAL CORD ACTIVITY

The processes in the spinal cord are described in terms of a low level controller. Nerves leave the spinal cord at different levels and pass to the muscles. The levels are interconnected and so one level may influence the instantaneous rate of impulses passing to chest muscles at other levels. The conditions at the synaptic units (type A) determine which will be stimulated. The low level controller thus acts as a distribution switch or matrix which routes the activity from the brain to those muscles which can shorten most easily. The final spatial and temporal distribution of activity to the respiratory muscles is achieved at this level of control.

CHEST MUSCLE ACTIVITY

In fig.4, the block labelled 'chest muscle activity array' represents the spatial arrangement of muscle units, the value of each element in the array being a measure of the electrical activity in that muscle unit. As each unit is activated, mechanical forces are generated. The net effect of muscle activation is to produce a volume change in the thorax. The muscle spindle (fig. 3), associated with each muscle unit, forms part of a loop whose purpose is to ensure that the change in length of the muscle unit is equivalent to the activity reaching it. The cumulative output of the spindles can also be interpreted in terms of a measure of the energy expenditure above what is considered, by the top level control, to be optimal.

SUMMARY OF CONTROL SEQUENCE

An analysis of a series of breaths shows a recurring variation in breath duration every 3.6 breaths[5]. If this modulation were the result of a purely random process, the expected cycle length would be 3.0 breaths. The modulation is interpretated as part of a search process in which both frequency and tidal volume vary cyclically to enable selection of the mean values at which the average energy expenditure in the muscles is at a minimum. This search process is assumed to be inherent in the τ/V_T assembly of the top level controller. The feedback from the low level enables the high level to discriminate between an increased expenditure resulting from a 'controlled' variation in τ and V_T and an increase resulting from other unspecified disturbances. The criterion used in allocating the pattern of activity to the muscles is that of continuously seeking the best performance index, defined as the highest number of units of ventilation per unit energy expenditure. The feedback paths from the low level to the higher level controllers allow the overall process to select the best values of τ and V_T even if mechanical loading on the respiratory system is changed. For example, a change in posture is usually accompanied by an invol-untary change in respiratory frequency. The change is fast which suggests that the output of the top level controller can be changed rapidly by signals from other areas in the nervous system. This signal is labelled 'forward demand' and acts as a transient feed forward signal to allow for future possible increases in energy expenditure, such as might be involved in exercise.

It is the local loop between the spinal column and respiratory muscles which determines the final detailed route of activation in the distribution unit of the spinal column. This process appears to involve the continuous appraisal of the muscle tension relative to muscle electrical activity and involves a low amplitude 10 c/s component.

Only two aspects of the control of respiration are reported in this paper. Other aspects are the regulation of airway diameter and the control of the distribution of the circulation. These have been discussed elsewhere[6].

ACKNOWLEDGMENT

Support for this work by a Medical Research Council grant, and the provision of facilities at the National Physical Laboratory, are gratefully acknowledged.

REFERENCES

(1) Roughton, F.J.W., *Handbook of Physiology*, Section 3, Ed. Fenn, W.O. and Rahn, H. Amer. Physiol. Soc. 1964.

(2) Margaria, R., A Mathematical Treatment of the Blood Dissociation Curve for Oxygen. *Clin. Chem.* Vol. 9, No. 6, 1963.

(3) Naeraa, N., Strange Petersen, E. and Boye, E., The Influence of Simultaneous, Independent Changes in pH and Carbon Dioxide Tension on the 'in vitro' Oxygen Tension-Saturation Relationship of Human Blood., *Scand. J. Clin. Lab. Invest.* Vol. 15, No. 2, 1963.

(4) Mead, J., Control of Respiratory Frequency. *J. Appl. Physiol.*, Vol. 15, No. 3, 1960.

(5) Priban, I.P., An Analysis of some Short-term Patterns of Breathing in Man at Rest. *J. Physiol.*, Vol. 166, pp 425-434, 1963.

(6) Priban, I.P., and Fincham, W.F., Self-Adaptive Control and the Respiratory System. *Nature*, Vol. 208, No. 5008, 1965.

Discussion

Dr. J.F. Meredith (UK) asked whether an elaborate mechanism was necessary to minimize the energy expended in breathing as it was unlikely that this amounted to more than 10% of the energy expenditure in the body.

Dr. Priban answered that unless all the
biological processes functioning simultaneously
were working efficiently, the overall energy
expenditure in the body would be much higher.
It should be remembered that the work done
in ventilating the lungs could increase by
tenfold during heavy exercise. Under
such conditions, unless the energy expenditure
were optimized, exercise could be continued
only for very brief periods of time.

METABOLIC CONTROL IN THE MAMMALIAN MICROCIRCULATION

by A. S. Iberall and S. Z. Cardon
General Technical Services, Inc.
Yeadon, Pennsylvania

ABSTRACT

Adaptive control by the autonomic system is complex in that conflicting demands of all autonomic functions of all the tissues of the body must be satisfied. The microcirculation is discussed for illustration, particularly those activities related to metabolic control. It is proposed that the level of local metabolism is determined by the oxygen supply which is controlled by microcirculation flow mechanisms. Some experimental studies of one aspect of these flow mechanisms are described, the rate of flow of red blood cells in the capillaries. It is further noted that the design architecture of the microcirculation is a result of adaptive growth control. A model for the structure of the microcirculation is proposed.

While attention to adaptive control in man has been generally directed toward control activities in the conscious state it is likely that the more sophisticated adaptive control activities are those of the autonomic system. In fact, adaptive control in the conscious state appears to be limited to a very small number of adaptions at any one time. On the other hand, the autonomic system appears to be constantly regulating and controlling a myriad of activities throughout the entire system. Supplies of oxygen, sugar, and other requirements of the individual cells are supplied in accordance with need, most often in the face of conflicting demands from the different organs, tissues, and individual cells. It is likely that much more can be learned about complex adaptive control by study of the autonomic system. A case in point is the microcirculation in which all the mass transfer and heat transfer functions of the circulating blood takes place.

For example, one function of the microcirculation in mammals is to provide a flux of oxygen under adequate partial pressure to its perfused tissues. In some tissue one could suppose that this might be supplied by passive diffusion through fixed resistances. However, in actively consuming tissue, an active control mode is required to supply oxygen to meet demands, frequently unanticipated demands. Since essentially the same control modes of action are used for a very wide range of ambient excitations independent of the excitation type or mode coupling, it may be considered an adaptive control system. (Similar comment would undoubtedly apply to many of the other supply functions of the microcirculation, possibly to the resistance control and temperature regulation functions. To add to the complexity, many of the functions and regulatory mechanisms are undoubtedly coupled. This discussion however, will be confined to the metabolic control functions.) The skeletal muscles is an apt example of consuming tissue. The following model is proposed as fitting physiological facts for this system.

The muscles must form a prime mover engine, in that for any time schedule of body displacement (illustratively, a mental schedule of constant speed) the loads on the muscles are indeterminate. The muscle engines therefore must always be in an idling state, ready to respond with output horsepower to meet load demands. Such engine cycles are generally run by control of the thermodynamic process that supplies the energy transformation. In the present instance it is a chemical oxidation process and one would expect that the fuel flow, the oxygen flow, or the level of an inhibiting combustion product could be the factor controlling the rate of energy transformation. It is clear that the tissue is amply supplied with fuel (sugar) at regulated levels, and the major combustion product, carbon dioxide, is absorbed in an ample sink. The oxygen appears to be supplied in a more limited fashion with little oxygen storage in the system. Furthermore, the level of oxygen in the blood drops sharply in the blood's passage through the skeletal muscle under load. Thus, it appears that the most plausible control element is the oxygen supply. It is proposed that the local flux of oxygen to the muscle tissue is controlled by the microcirculation functioning in an active control mode. The reasoning leading to this proposal is the following.

From our own earlier experimental work[1], it appears that the highest frequency of local power control is confined within a band approximating one cycle per 1-2 minutes. We have loosely referred to this as a 100 second cycle. This cycle is found in the local skin temperature, internal temperature, ventilation and metabolic rate, and in heart rate. From the work of Lewis[2], and Krogh[3], and various comments culled from the microcirculation conferences of the past ten years, it appears likely that similar cycles are present in the flows in the microcirculation. The active

Superior numbers refer to similarly-numbered references at the end of this paper.

oxygen exchange elements -- from effective oxygen exchange area considerations -- are the capillaries. From blood flow considerations, they are essentially impeded (for example, it is commonly stated that the capacity of the capillaries is so great that if all were flowing at the same time, all the blood in the body would be contained in them). Thus, the 100 second power cycle likely represents an effective intermittent opening and closing of the capillaries in this time domain. In actuality, the intermittent oxygen flow through capillaries may consist only of an intermittent red blood cell flow (there is some experimental indication for this which will be discussed below) and not serum flow since it is the red cells which are the main oxygen carriers in the blood. Thus the flow variation may not necessarily be due to mechanical opening and closing. It is likely that the overall cycling process is recognized in physiology in at least selected tissue as vasomotion.

The process of equivalent opening and closing is not random, as evidenced by locally coherent thermal power cycles appearing at every point in and on the body (as demonstrated in fluctuations in skin and internal temperature). Since the thermal cycles from point to point are not highly correlated, there is apparently little central control of the local resistance beds. However, the resistances of larger aggregates are not indeterminate as is indicated by cyclic coherence in oxygen consumption at the mouth, i.e., a 100 second cycle is present in ventilation and metabolic rate. Thus the local opening and closing cycle appears quasi-coordinated into a non-linear limit cycle that propagates like a ring oscillator through the system.

The muscles and microcirculation supply elements, the capillaries, are not acted upon in a more rigidly scheduled firing order familiar to prime mover engineering. Instead there appears to be a loose but well coordinated statistically determined firing order, linked from region to region, which is continuously adjusting to the power demands of the system. This means that for a given activity level characterized by a velocity pattern for the whole body, the power demand is continuously determined by the local force loads on the system.

The system resistance as a whole has a determinate slowly varying mean state, with individual regions indeterminate and available to adjustment to a best fit of conflicting inputs. Thus the system operates generally in a resistance control mode in the microcirculation. The nature of the control algorithm can only be guessed at. It may involve subtle cues of fatigue, coordination by electrical signals and chemical byproduct concentrations.

For example, we have proposed[1] a hypothetical mechanistic scheme in which the local capillary is normally closed but opens for a limited time when a specific signal is presented. The muscle fibers, the transforming motor element, on the other hand it is proposed, will use up all the oxygen at whatever rate the surrounding tissue permits. These two actions can create the conditions for a local instability.

Specifically, an elementary unit of the system may be defined as consisting of a local group of muscle fibers, which are sufficient in extent to represent not only the electrical firing from a single nerve fiber, but of all those adjacent motor units that cooperate to produce a coordinated nervous wave (see Adrian[4] for discussion of such cooperative motor units activity). This thus serves to define, even if vaguely, a motor element that can be fully coordinated in an electrical sense, from the level of intermittent individual muscle fiber discharge up to saturated and coordinated waves of muscle activity. In prototype, the motor element may be viewed as disposed in a cylindrical form. The muscle fibers in such a bundle alternate with capillaries (see, for example, Krogh[3]). These capillaries are connected in a network shunting from oxygen rich, high pressure arterial blood to oxygen poor, low pressure venous blood. On the approach side, the arterial side, the capillary networks are supplied by small arteries, the arterioles, which possess a muscle motor unit, a sphincter, which by opening and closing, can control flow to the capillaries. The predominant pressure drop takes place at the arteriole level, nominally through the arteriole sphincters, though there may be other additional closure mechanisms. It is likely that total blood flow is controlled by sphincters, but the red blood cell fraction may be otherwise controlled, likely electrically. The stage is thus set for control action of the oxygen flow. (In earlier descriptive form, it has been discussed by Krogh[3].)

There is a flux of blood axially along the capillaries, among the muscle fibers. There is in addition an oxygen flux from the capillaries out to the tissue and thence to the active muscle fibers. This forms a flux stream orthogonal to the blood flux stream. Now, according to the present hypothesis the muscle fibers are an active sink that will utilize oxygen at whatever rate its boundary permits. Thus, the muscle fibers do not determine the rate governing reaction, which must be determined instead by a 'diffusion' rate across the intervening tissue geometry. However, the capillaries are active. They are not all full of red cells. In fact, most of them are free of, or low in, red blood cell flow. (Following Krogh, it is common to regard that most of the capillaries are considered to be closed at any time.) Thus, as an equilibrium force balance, one might expect that with a given number of effective open capillaries, a certain cross-channel equilibrium concentration of oxygen would exist in the tissue, ranging from a nominal 19% oxygen concentration in arterial blood down to near zero for interstitial fluids. Depending on the geometry, this would represent a specific equilibrium oxygen flux into the muscles, and thus its existing equilibrium oxidation rate. There would similarly be an equilibrium counterflux of returning carbon dioxide. There could be more than one model of the equilibrium point, depending on the nature of the boundary equilibrium at the muscle fibers and as determined by the operating points of the metabolic reactions, i.e., the distribution would tend toward an equilibrium following each change in distribution of effectively open capillaries.

The one major open element in this description is the effective number of open capillaries. The question is what makes them open or close, or since they are normally closed, what makes them open for limited times, and in particular, what forms the 100 second nominal timing phase? It is suggested that a component that enters into the cross-channel flux is responsible, although this component may be carried in the axial channel flow. It is likely neither oxygen nor carbon dioxide, since these substances acting only passively would come to spatial and temporal equilibrium. It is also not likely any other passive intermediary, since this would tend to equilibria. It cannot be the oxidation kinetics, since these are too fast and also can only tend toward equilibrium. Therefore, as Krogh also sought, there must be some switching, or escapement element for opening or closing capillaries. It is not ruled out that either a second escapement or a spring return is used for the opposing action. Krogh understood the problem quite well and it motivated his search for a vasodilating agent which he designated as agent X.

Our search for the escapement has suggested adrenaline for the most plausible agent. It is a powerful vasodilator in muscle and its time of action seems correct. It is reputedly a calorigenic agent as well. The problem is to work out a suitable hypothetical model for the local 100 second cycle and then more difficult to demonstrate experimentally the accuracy of the model. Some possibilities are the following:

1. There is a centrally produced cyclic stream of adrenaline into the blood from the adrenals. This may be delivered into the blood at a near 100 second rate to form a predominant timing phase. The pulsed adrenaline wave in the blood can regularize a 'twinkling' in the capillary number by being used up, or tied up by the number of capillaries that open on signal from it. The rate of adrenaline flow could determine the twinkling rate of capillaries and therefore the oxygen flow.

2. A second possibility is that the adrenaline is maintained in the blood at very slowly changing levels. Its flow rate is still used to open a definite number of capillaries. However, the two minute cycle is locally determined. Specifically, in a closed capillary, the red cell supply has dropped appreciably and the muscle

fibers are leaching the oxygen from the surrounding tissue. The timing phase is the time constant associated with this diffusive sink (similar to drainage from a single water well in a field). As the oxidation reaction goes on, its combustion by-products pile up as carbon dioxide and various intermediaries. One of these, perhaps carbon dioxide or lactic acid building up to a threshold level, then triggers the capillary to open, recharging the capillary with blood and the region with oxygen.

3. Somewhat more specifically, Lundholm[5] has suggested that adrenaline acts to increase the lactic acid production in the cells with a resulting increased diffusion into the surrounding tissue, and that it is the lactic acid in the tissue fluid which produces the vasodilation. It is also possible that adrenaline acts simply to increase the cell oxidation rate, which would have the same effect. Interestingly enough, other metabolic by-products have been similarly postulated to mediate blood flow in active organs. (Berne, for example, has proposed adenosine for the heart.) Thus, it is possible that there are different agents for the different organs.

However, our key thought, that regardless of the local mechanism for control of blood flow, it is the oxygen choke that controls the metabolism level, has generally been well regarded.

On the problem of the nature of blood flow in the microcirculation, we recently found some experimental indication that the red cell flow rate in capillaries is oscillatory in nature. Using techniques and equipment developed (and generously provided) by Dr. E. Bloch in his laboratory at Western Reserve University, the mesentery of the frog was examined. The mesentery was exposed on a microscope stage and the magnified image projected onto a screen containing an aperture. A photocell was mounted behind the aperture and its output connected through shaping and amplifying circuitry to a count rate meter, thus making it possible to determine relative rates at which red cells were passing through the particular small vessel under observation.

In choosing capillaries for investigation, the exposed mesentery was scanned by starting from an artery and following it to the smallest connecting vessels to the veins. Such vessels selected for observation were generally ones which had a rapid lively flow, were of small diameter, and could clearly be seen to connect an arteriole to a venule. For example in one experiment, the diameter of the capillary was about 11 microns, that of the feeding arteriole, 23 microns, and the draining venule, 22 microns. In another experiment, the capillary diameter was 19 microns and its length, 290 microns. Other capillaries observed were up to 22 microns in diameter.

In the initial experiment, the rates of flow

in counts per second were noted at 20 second intervals for a total period of 8000 seconds. These rates were then plotted against time. An examination of the resulting curve showed that the red blood cell flow was undulating with two main cycles, at about 80 seconds and 320 seconds, and an indication of a slower cycle at about 1200 seconds. The flows varied from a minimum of 70 cells per second to over 520 cells per second.

The experiments were repeated with the rate meter connected to a recorder thus producing a continuous record for periods up to three hours. This continuous record showed a faster cycle of approximately 25 seconds in addition to cycles of 80 seconds, 350-450 seconds, and of about 1400 seconds. The flows varied from momentary periods of zero flow to well over 500 cells per second. (These data will be presented in detail in a subsequent report.)

One experiment was also run on the rat's lung. The flows in the areas being observed closed off after only brief periods thus indicating the lung was not in good shape. Despite this shortcoming, there was some indication of oscillatory cell flow in this mammalian tissue.

These preliminary experiments indicate that there are regulatory mechanisms for red cell flow operating at the same time domain as the heat production mechanisms in the mammal. Thus, it is suggestive that heat production mechanisms are similar in frogs and mammals even though the thermoregulation is decidedly different. This fits the hypothesis that the mammalian thermoregulation is essentially accomplished by regulation of the heat loss, and that the heat production is an independent factor determined by the activity of the various tissues and the body as a whole.

An oscillatory flow in red cells has been demonstrated apparently independent of mechanical opening and closing of a particular capillary. Further, it is likely that the oxygen is supplied intermittently to the tissue through an intermittent red cell flow, and that this is the determinant of the 100 second cycle in metabolism. Many more observations are considered necessary, in various mammalian tissue and in more normal operation before this conclusion can be considered final. Work is continuing in our laboratory designed to probe at the control mechanisms.

It must be noted that a 100 second cycle takes place outside the muscles and other heat producing tissues. We have previously suggested that such a cycle primarily represents a mechanism for heat production and regulation in the muscles. This data suggests that this cycle is related to microcirculation flow generally whether in muscle for heat production or in other tissue for other purposes in which heat production will be incidental. In the heat production tissue, the cycle would be reflected in heat and

temperature cycles and it is for this cycle that the adrenaline trigger is proposed. In other tissues, it may be reflected in those fluxes, material or energy, which may be involved. In the latter case, the trigger need not be adrenaline. (For example, histamine may be involved in skin reactions.)

For a single flow distribution system for the entire body that permits local cyclic demand according to local power need likely requires that it be an adaptive control system in more than one sense. Up to this point, we have discussed the control in the 100 second time domain. There is evidence for regulation and control functions at other time domains, faster and slower.

At a faster time domain, there is an axial jitter in the red cell flux, likely related to what Krogh[3] called plasma skimming. At a rate of several cycles per second, there is a fluctuation in the number of blood cells per second passing a point in a given capillary. (This has been viewed in Bloch's movies of the microcirculation - see[6] for still illustrations, or see[7] for a popular view of the microcirculation, and[8] for references on the microcirculation.) At a branch point, the fluctuations appear in the division of cells between the two branches, although in longer time the stream will tend to divert from one branch to the other. It is likely that these slower processes represent Krogh's plasma skimming, and one aspect of the 100 second cycle. It may be that this adaptive control cycle is built up from the higher frequency blood cell jitter, and thus not to be found in actual sphincter controlled opening and closing of arterioles. At present it has not been possible to distinguish among the various processes.

It seems clear that the high frequency jitter is mediated by electrical forces. The 100 second cycle appears more likely mediated by an electrochemical or more properly, a hormonal-electrical mechanism that monitors the red blood cell flux in a quantized fashion. It thus appears that the individual cell embarks on a trip through a capillary net very much like a ball in a pin-ball machine. The 'pins' are electrical-chemical interactions at the wall. The walls tend to act like electrical valves. In addition, pursuing the pin ball analogue, there may be sectional closure switchings, also electrically mediated, that more formally control passages or constrictions. It is a combination of such elements that provide an as yet unknown algorithm for an adaptive control that meters the oxygen, tied to the red blood cell carriers, to the local muscles. One surmises that an electrical feedback system furnishes information on regional flow settings.

At a slower time scale, there is evidence that there exists an adaptive zonal control over the blood flow. This may be viewed as follows: The muscle motor elements fire only with a determinate statistical number per unit time, i.e., their number 'twinkles' at a 100 second time domain. While such action appears stochastic, its propagation is not. There are conflicting demands of motor units in the vicinity that must be satisfied, and then there are the conflicting demands among major systems, i.e., the heart, liver, brain, kidney, GI tract, skeletal muscles, and the skin. To satisfy the systemic circulations, it appears likely that there is another ring oscillator mode of control in the nervous system, at the level of the hypothalamus, that controls a near seven minute blood flow division among these various circulations. It appears plausible that the division is affected by resistance, probably at the level of the arteriole sphincters. It is likely that the control of blood flow is affected through this seven minute cycle. The flow shifts are most prominent in the circulation to the skin, and are probably known elsewhere as well (such as in the kidney). Seven minute temperature cycles at the hypothalamus correlated with metabolism have suggested the likelihood of deep body temperature control as taking place at the hypothalamus. Thus it may be that the ring division of blood flow takes place on the basis of providing a control of heat exchange (see experimental data in[9]). Any zone thus has a mean blood flow, and mean oxygen flow provided to it, which, at slow operative levels, is sufficient to supply zonal oxygen need.

In addition to these adaptions of cyclic supply according to conflicting functions and modalities, there is yet another longer time adaptive growth control of the architecture of the distribution system to consider. One might question whether this can be considered adaptive control especially within the context of this symposium. Yet, it surely is when one considers that there is a viability in living systems which is characterized by a reversible adaption of distribution system numbers and sizes in time. From this point of view, growth adaption is certainly adaptive control to various modalities of living.

In the case of the microcirculation, or more generally also the arterial supply, growth control is reflected in the developmental architecture of the system. To describe the flow distribution system requires a geometric-topological architectural model. The need for such a model was originally stressed by Young (1809). As reported by Aperia[10], Young considered "that the system of vessel ramifications... follows on the whole the same laws throughout the entire organism... After a certain independent course every vessel ramifies into two branches of the next stage. Thereby, a simple course for the length as well as for the cross-section of the different stages is stated. Corresponding pieces of the same stage of ramification show the same physical conditions."

In[1], a model was proposed that fits the anatomy, namely that a main arterial branch develops embryologically down the 'center' of each func-

tional unit (the limbs are viewed as an excellent prototype), and that a topologically similar prototype is supplied at each subsequent level of subdivision. The prototype topological element appears to be an elongated cylinder which is then divided into m sectors and n parallel slices. This subdivision into m n segments quickly transforms the large elongated cylinder into m n small pseudo-cylinders which can then be similarly subdivided in turn. This satisfies the topological needs for this sytem in providing it with flow distribution of a sufficient number of elements. The subdivision proceeds from the aorta with an internal diameter in the range of centimeters, to a normal large artery in the range of a centimeter, on down to arterioles in a range near 30 microns. A major essential property appears to be that the mean velocity is approximately constant in all arteries (actually it changes slowly with size) which tends to fix the diameter at each subdivision. It is conceivable that the tubes grow in size to fulfill this condition by some sort of acoustic signal. (Although biologists may commonly think in terms of hydrodynamic erosion, it is more likely that vibrational 'noise' acting on the local cellular wall structure may be a growth stimulus.) However, the oxygen flux geometrically demands a proportioning between the diameter of the blood vessel required to supply the oxygen flux and the diameter per unit length of supplied tissue so that there is an apparent conflict between a first power law of artery diameter and length of the subdivision required for geometric similarity, and two-thirds power law required topologically by gas distribution requirement. This is resolved in a fashion that would be understandable to the more practical minded plumber.

Tube runs are used in (i.e., they grow to) fairly customary lengths for each diameter, suitable for each level of subdivision. Any needed adjustments are made by branching around the main topological tree. Thus the tube runs which have not changed their length much between branches can be represented by a gross scatter (in a logarithmic presentation) around a first power law of variation of diameter with length of runs[11]. A set of design laws for the distribution system can follow that will permit growth or degradation of supply tubes which can adapt to local use. Need for change might be illustrated in a demand for mild modification of the distribution system at various levels in response to a new pattern of muscular exercise.

ACKNOWLEDGMENT

Support for this work under National Aeronautics and Space Administration Contract No. NASW-1066 is gratefully acknowledged.

REFERENCES

(1) Iberall, A., ASME J. Basic Eng. 82 (1) 92,
    103; 82 (3), 513, 1960; Cardon, S., NASA
    CR-141, CR-214, CR-129, 1964, 1965.

(2) Lewis, T., Heart 13, 27, 1926; 15, 177,
    1929-31; Brit. Med. J. 785, 837, 869, 1941.

(3) Krogh, A., THE ANATOMY AND PHYSIOLOGY OF
    CAPILLARIES, Hafner, 1959.

(4) Adrian, E., THE MECHANISM OF NERVOUS ACTION,
    U. of Penna., 1932.

(5) Lundholm, L., Acta Scand. Phys. 39, Suppl.
    133, 3, 1956.

(6) Bloch, E., Am. J. Anat. 110, 125, 1962.

(7) Zweifach, B., Sci. Amer., Jan 1959, 2.

(8) Wiedeman, M., Chapt 27 in Handbook of
    Physiology, Circulation, Sect. 2, Vol. 2,
    1963.

(9) Benzinger,T.,Kitzinger,C., in TEMPERATURE -
    ITS MEASUREMENT AND CONTROL, Vol. 3, Pt. 3,
    Reinhold, 1963.

(10) Aperia, A., Skand. Arch. Physiol. 83, Suppl.
     16, 3, 1940.

(11) Suwa, N., et al, Tokoku J. Exper. Med. 29,
     169, March 1963.

Discussion

Dr. I.P. Priban (UK) asked whether
micro-circulation was in any way under the
control of the nervous system or was the
control purely hormonal.

Dr. Cardon answered that component frequencies of several cycles per second had been
observed and these were likely to be directly
associated with the nervous system, whereas the
lower frequencies could be related to chemical
and hormone action. He considered that the
85 second cycle length was a normal process
cycle associated with metabolism or ventilation.

THE METHOD OF ADAPTIVELY CONTROLLED PSYCHOLOGICAL LEARNING EXPERIMENTS

Gordon Pask and G.L. Mallen
System Research Ltd.,
Richmond, Surrey,
England.

The research reported in this paper has been supported by the Air Force Office of Scientific Research under Contract AF 61(052)640 through the European Office of Aerospace Research U.S.A.F.

1. BACKGROUND DATA

1.1 Introductory Comments

In this paper we discuss the learning process that takes place when a man acquires a "Structured Skill" in adaptively controlled conditions. The phrase "Structured Skill" refers to a skill that satisfies some rather involved axioms 1,2,3,4,5, but for the present purpose we shall interpret a Structured Skill as a skill that is sufficiently well defined to act as the framework for a control procedure and that can be reduced to homogeneous components or subskills. The argument is illustrated by laboratory experiments involving a rather limited class of structured skills but other work indicates that our conclusions are generally applicable.

One practical reason for controlling the learning process is to mediate instruction. Thus, there is a class of control procedures that aimsto maximise the rate of learning, and if these also satisfy certain additional criteria (such as leading to a terminal condition in which the subject is dealing with a real life state of affairs) they are plausibly dubbed teaching procedures (in which case the control mechanism is a teaching device 6,7,8,9,10,11,12). The present discussion is almost completely confined to procedures of this sort. There are, however, several other reasons why we aim to control learning. We may wish to test a student's capability. Vygotsky[13] was probably the first to point out that in order to discover how a man learns it is essential to ascertain how much co-operative assistance he needs in order to learn and it is evident that the determination of this quantity entails some sort of control procedure that gives or witholds co-operative assistance as a function of suitable performance measures 14. Further, there is a whole gamut of experimental situations (in psychophysics, perceptual discrimination, and other fields) wherein the subject certainly learns although his learning is irrelevant to the experimental enquiry[15, 16].

In order to achieve constancy (to approximate stationary state measurement) it is possible to compensate for changes due to learning by the automatic adjustment of salient features in the subject's environment. This type of adjustment is generally feasible if the experimental situation is as well defined as a structured skill.

1.2 Conversational Paradigm

The paradigm case for any control of the learning process is a conversation between a pair of participants. One of these acts in the capacity of a subject, and the other acts as a control mechanism[17,18,19,20]. If the conversation is specialised into a tutorial the subject becomes a "student" and the control procedure becomes a "teaching" procedure. Similarly, if it assumes the specialised form of an interview, then the objective behind the control process is the determination of one or more quantities characterising the subject. In the first place we shall point out several important properties of any ordinary and verbal conversation.

(1) The whole affair is symbolic.

(2) The conversation is a symbolic game that usually ends in a compromise.

(3) This game is partially competitive (in the sense that one participant aims to impose his point of view upon the other and, at any rate in a tutorial, there is an inbuilt dominance pattern). It is also partially co-operative in the sense that even the dominat participant must render his discourse intelligible.

(4) Since the subject learns the other participant must learn and adapt his mode of discourse in order to maintain rapport with him. As a minimum requirement he must render his discourse novel enough or varied enough to maintain the subject's interest and attention and at the same moment he must provide discourse that is intelligible.

(5) Conversations are conducted in an open

Superior Nos. refer to similarly numbered references at the end of paper ended type of language that is also capable of representing many levels of discourse. The lowest level of discourse is usually confined to the statement of questions and replies (the questions may be conceived as problems and the replies as solutions for these problems). Higher levels of discourse are concerned with instructions, statements of interpretation, and assertions of preference. In particular, if we are anxious to view the interaction between the participants as "communication" then it is necessary to introduce considerations of the "meaning" of the communicated data (in fact the many level open ended language is an adequate vehicle for the discourse merely because it includes level assigning phrases like "I mean that ....." where the "...." is some interpretation for another phrase that is referred to).

1.3 Laboratory Arrangements.

Although the laboratory conversation is characterised by properties (1),(2),(3), (4),(5), it differs from a real life conversation in several ways.

In the first place the symbols used in a laboratory control situation are non-verbal events (like pressing a response button selected from an alphabet of alternative buttons or illuminating a stimulus signal lamp selected from an alphabet of signal lamps) in contrast to the verbal signals used in a real life conversation. Next, the learning and adapting participant who aims to control the conversation is conveniently replaced by an adaptive control mechanism (abbreviated to an A.C.M.) which is, in fact, programmed on the special purpose computer shown in FIG.1.

Since the denotation of the non-verbal symbols used in the laboratory is unambiguous the various levels of discourse (which, in a real life situation, are represented in terms of a single language) must be separated. Each level of discourse in the laboratory system is associated with a given level of language (thus the stimulus signs denoting problems are kept distinct from the signs denoting properties of a performance and the response signs denoting solutions are kept distinct from the signs denoting preferences). These levels of language, called L^0, L^1, ... are real systems for communication. However, it should be emphasised that only a pair of levels of stratified open ended language are needed, from the logical viewpoint, in order to comprehend the entire communication system in a canonical form (namely a stratified and open ended object language and a similar descriptive metalanguage) 21, 22, 25, 26.

The convention of distinguishing between levels of discourse rather than reducing the framework to its logical and canonical form yields a picture in which the A.C.M. is an hierarchically organised adap-tive system in the sense of Tarjan [23] and Ashby [24], and it is possible to apply the criteria of stability that are discussed by Tarjan to the interaction between the subject and the A.C.M. In this hierarchical organisation, each level of the hierarchy corresponds to a single level of discourse such as L^0, L^1, ... Depending upon our interpretation of the model, interaction between the subject and the A.C.M. may, FIG.2, or may not FIG. 3, be allowed to occur at all of the levels concerned. We examine systems wherein the interaction is legally limited to L^0 in 2.1, 2.2, 2.3, 2.4, 2.5, and 2.8 and systems characterised by more than one level of discourse in 2.6.

1.4 Compensated Learning.

At this point it is instructive to consider the coupled man/machine system from a point of view that is more familiar in Control Engineering where, given a process that "learns" the first step in controlling it is to associate it with another system (in fact the A.C.M.) in such a fashion that the coupled system (of process interacting with its A.C.M.) is stable and does not "learn". The issue is discussed by Pun [33] and others. In all cases an A.C.M. able to achieve this condition must contain a model representing the process. The form of this model varies according to the type of control algorithm. In the case of a system designed according to the Bellman optimality principle, for example, this is a sequential estimation model.

This argument from control engineering can be applied with equal cogency in the case of a conversation. For all of the systems we shall examine, the conversational participant represented by the A.C.M. maintains a coupled system, the conversation, which does not learn and it does this by the expedient of allowing the subject to learn (or even of encouraging the subject to learn) and compensating for his learning. The form of estimation we carry out is crude but the structural features of the model are fairly elaborate. In particular, as above, it may be pertinent to admit interaction only in L^0 or to arrange for interaction at many levels of discourse.

In the former case our assumptions about the other-than structural characteristics of the subject can be restricted to those in 1.6 and in 1.7 and can be alternatively stated as the purely statistical assumption that man is a selforganising system in the sense of von Foerster 53, 54.

Thus using the term game (as we used it in 1.2(2) and 1.2(3)) the control strategy of the A.C.M. in L^0 assumes that man is a statistical creature and plays a statistical game with him.

But suppose that the coupled system cannot be stabilised in this fashion (in the weak sense of Tarjan that parametric changes cannot be compensated) (and that the assumpt-

Fig. 1. Special purpose computor.

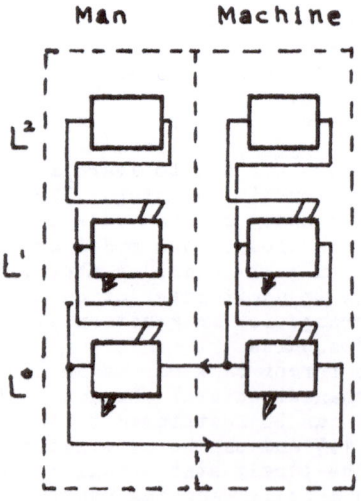

Fig. 2. Man–machine interaction at L^0 level only.

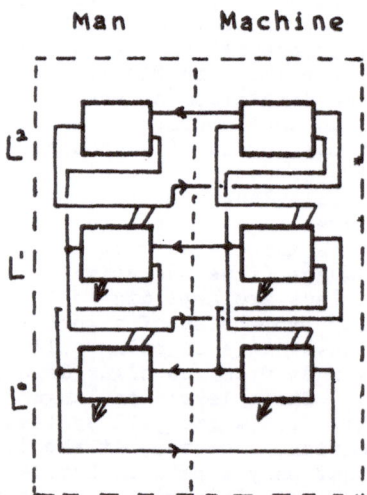

Fig. 3. Man–machine interaction at all levels.

ions in 1.6 and 1.7 are consequently inadequate. We may, of course, amend the model
and add further statistical constraints.
However, since our access to the subject is
limited, we probably could not estimate the
parameters of a more elaborate model in the
interval allowed for interaction. Fortunat-
ely there is another possibility, namely
(by extending Tarjan's construction) to
adjust the higher order parameters of the
process directly. As the process is a
subject, this entails interaction between
the subject and the A.C.M. at many levels
of discourse. Further, the stabilisation
of this interaction involves a different
game in the sense of 1.2(2) and 1.2(3)
wherein a different sort of assumption is
made about the subject. Specifically, in the
metasystems of 2.6.   in which higher
order parameters of the subject are adjusted
the A.C.M. is designed to indulge in a
rational game in L^1 on the assumption that
the subject aims to maximise his degree of
control over the joint or coupled system.
27, 28, 29.

1.5 Cybernetic Model

As suggested by FIG.2 and FIG.3 the
A.C.M. is based on a Cybernetic Model for
the subject that is structured as another,
Tarjan-like or Ashby-like, hierarchical
organisation. It has been argued in
previous publications that a structure of
this type is also the least elaborate
Cybernetic model for the learning process
in man (which is compatible with the viewpoint adopted by Piaget and [30] Vygotsky).
In particular, the components or subskills
of a skill are partially separable sub-
systems of a control system responsible for
the entire skill, these components are
hierarchically organised to provide for the
entire performance, and the learning process
is a business of building up this
hierarchical organisation. Thus a description of this organisation (or, in psychological terms, of this conceptual structure)
is isomorphic with one of the TOTE hierarchies proposed by Miller, Gallanter and
Pribram [31] or Miller and Chomsky [32] and thus
(since a TOTE hierarchy is an "Artificial
Intelligence" programme) with a suitable
"Artificial Intelligence", the organisation
of the subskills is closely related to the
learning systems examined by Koestler [35] and
others 37,38,39,40,41,42,43. Finally,
using arguments of the sort advanced by
Koestler [35] Newell [34], and Pask [26], it is
possible to demonstrate that each of the
subsystems is analogous to any other and
also to the entire system. If, like
Newell [34] for example, we regard one of the
subsystems as a "problem solver" any other
subsystem is a "problem solver" and so is
the entire system. Without this condition,
the concatenation of subsystems would not
be legitimate. However, it is certainly

not the case that one subsystem solves the
same kind of problems as another nor is it
the case that the entire system is solving
the same kind of problems as its subsystems.
44, 45, 48, 50, 51, 52.

What the construction does accomplish is
a reduction of learning to problem solving
whenever the coupled man/machine system is
stable. Thus the initial subsystem solves
problems, denoted by stimuli, and achieves
solutions, denoted by responses, as a result
of applying operation sequences. The next
level of subsystems accepts "problems" that
are differences between properties of the
initial operation sequences and solves these
"problems" by producing novel operation
sequences that are added to the available
operation sequences repertoire. 44,45.

1.6 Structural Assumptions.

It will be sufficient to provide
detailed structural assumptions for the case
of a single subskill of a structured skill
and to discuss these in an informal fashion.
A more formal account is available in 1,2, 3,
46, 47.

The act of performing a subskill is
conceived as the correct solution of a
sequence of similar problems selected from a
certain class that are denoted by unsimplif-
ied stimuli $x^* \in X^*$ spaced an interval Δt
apart. The subject (who, as in 1.5, is viewed as an artificial intelligence system) is
provided with the description of a class of
correct algorithms (in particular, amongst
other things, a correct solution function, Ω^*
is specified so that $y^+ = \Omega^*(x^*)$ is a
correct response denoting a correct solution
and is selected from a set Y of response
states y) In this manner it is possible to
argue that any sequence $\hat{x}^*$ of length τ
stimuli is representative of the problem
class denoted by the stimuli x^*. Learning,
in the spirit of 1.5 is conceived as the
construction of correct operation processes
from operations available to the subject
which satisfy Ω^* and are described by
algorithms that are members of the original
class of correct algorithms.

When learning is complete, any operation
process might be partially applied to
x^* to yield a partial solution of x^* and
conversely when learning is partially complete there will be partially constructed
correct operation processes that are only
capable of providing the solution for
partially solved problems that can be
defined as lying at a certain distance from
a solution. We need to assert that, at a
given trial n the subject is no more uncertain
about the solution of a problem at a distance
R than he is about a problem at a distance
R + 1 from its solution and that learning
reduces the uncertainty associated with a
problem at a given distance from a solution.

Next it is necessary to define simplified stimuli that are derived from $x^* \in X^*$ by a simplifying procedure Φ_μ of degree μ which is so defined that if X_μ is the set of stimuli x that are derived from all x^* in X^* by applying Φ_μ then each $x \in X_\mu$ denotes a class of partially solved problems (at the same <u>distance</u> from their solution) and with the properties indicated below.

(1) If $\hat{x}^*$ is representative $\hat{x}$ is representative for $\hat{x} \in X_\mu$.

(2) $\hat{x} = \Phi_\mu(x^*) \subset X_\mu$

(3) $x = \Phi_0(x^*) = x^*,$

$x = \Phi_{\mu \max}(x^*) = y^+ = \Omega^*(x^*) \in Y$

(4) $X_\mu = \bigcup_{x^* \in X^*} \Phi_\mu(x^*), \quad X^* = X_0,$

$X_{\mu \max} = \Omega^*(X^*) = \Omega(X_\mu)$

Hence changing μ changes the expected uncertainty of the subject. It is often convenient to use the converse of this rank ordering of stimulus subsets, X_η where we define $\eta = 1 - \mu/\mu_{\max}$ as the converse rank ordering "difficulty" (in the sense of less "simplified").

To complete the structural assumptions we define the correct response characteristics function as $\sigma(x,y) = 1$ if $x,y \in X, \Omega(X)$ and 0 if not and the correct response measure, ρ, assuming the value $\rho(n)$ at the nth trial, which is the average value of $\sigma(x,y)$ over a sequence $\hat{x}$ of τ stimuli $\hat{x} = \Phi_\eta(x^*)$.

1.7 Subject Assumptions.

Unlike a well programmed computor a conscious subject <u>must</u> maintain a certain rate of mentation. Further, he is liable to the effects of <u>overload</u> and (due to the fact that he has a minimum rate of mentation) it is possible that a given body of data (the stimuli deemed to be relevant) will fail to occupy his attention, in which case he will attend to irrelevant data. In order to bring the image of 1.5 into conformity with 1.2 it is necessary to stipulate that the subject has conditionally accepted the conditions (I) and (II) below.
(I) He attends to a field of attention determined by the name, i, of a subskill and the class of algorithms cited in 1.6 with a domain consisting of problems, or partially solved problems, denoted by the stimuli $x \in X$ in 1.6, - and -

(II) He prefers to achieve goal satisfying solutions $y^+ = \Omega(x)$ given stimulus x, and rejects operation processes that lead to other solutions .

It is evident that the whole of the argument depends (like any other argument about man) upon conditions (I) and (II). Conversely these conditions are acceptable

if and only if $\hat{x}$ is so chosen that the stimuli occupy the subject's field of attention but do not induce overload. Hence a prerequisite for stable interaction between the subject and an A.C.M. is that the A.C.M. chooses $\hat{x}$ so that (I) and (II) <u>can</u> be accepted by the subject. However, <u>if</u> these conditions are satisfied, it becomes possible to make a number of assumptions about any subject. In particular (1) Since, from 1.6 any $\hat{x}$ derived from $\hat{x}^*$ is representative, the A.C.M. need only select a sequence $\hat{\eta}$ or $\hat{\mu}$ to determine the sequence $\hat{x}$ (given an arbitrarily programmed $\hat{x}^*$) for $\hat{x} = \Phi_\mu(\hat{x}^*)$. (2) Also from 1.6, the value of $\rho(n)$ calculated over $\hat{x}$ is an indication of the subject's uncertainty regarding his choice of y (since he has agreed to accept the goal defined by Ω). In particular, $\rho(n) = 0$ if a sequence $\hat{x}$ with $x(n)$ as its terminal member denotes unintelligible problems, and, in this case, the subject is overloaded.
(3) The difference $\Delta\rho(n) = \rho(n) - \rho(n-\tau)$ is an index of learning and if $\Delta\rho(n) > 0$ the subject is <u>not</u> overloaded although he may be maximally loaded. If he is maximally loaded by $\hat{x}$ he is not under-loaded by $\hat{x}$. (4) The <u>loading</u> is increased by an increase in η and decreased by a decrease in η and this relation is preserved for all n such that $T \geq n \geq 0$ where T is the trial at which a criterion level of performance is achieved.
(5) There is a value of η such that $\hat{x} \subset X_\eta$ is intelligible and $\Delta\rho > 0$.
1.8 Learning Model.
According to the learning model we present in this paper, the acquisition of a subskill is correlated with the development (in the subject) of an organisation that realises and embodies the code for this subskill. The A.C.M. interacts in a partially cooperative fashion with the subject to induce this organisation (the term "induction" being used in its biological sense). The subskill code is an instructional statement at the level of discourse L^4 which prescribes relevance and determines a class of problem solving algorithms which satisfy the L^0 goal.
We assume that this code is embodied at the outset, so that acceptance of 1.7(1) and 1.7(II) is not vacuous. However, the organisation that realises this code and reconstructs it, is not available to the novice.
If 1.7(1) and 1.7(11) are satisfied, the alphabet of L^0 is well defined and it is thus possible to compute information measures over this alphabet for sequences x,y. Further any consistent adaptation is goal directed. In the present model we represent the adaptive changes of learning which occur when the code realising "organisation" is built up in terms of changes in a pair of information measures namely H,

74

the conditional response entropy given x, and I, the maximum value of H, which is assumed to be a monotonic function of η, with I=0 if $\eta = 0$ and I = I_{max} if $\eta = \eta_{max}$.

As we argue later, and in consonance with 1.7(1) 1.7(2) and 1.7(3), the value of $\Delta\rho(n)$ is an index of $-dH/dt$ at the nth trial or, since trials are spaced apart by intervals Δt, at the instant $t=n\Delta t$ and if the value of $\rho(n)$ is changed, maximised, or minimised we shall also argue that an index $\Delta\rho(n)$, or in some conditions, $\rho(n)$ is a sufficient input to an A.C.M. that stabilises the learning process in the subject. The output of the A.C.M. is interpreted, from our assumption of monotonicity, as variation in the maximum value I, of the entropy which is also an intuitively reasonable way of talking about a variation in "difficulty".

The constraints embodied in the subskill learning model to some extent reflect 1.7(5) and 1.7(4). They are (1) there is a maximum value of H, say H_{max} beyond which the subject finds problems unintelligible and is overloaded. (2) There is a minimum value of H, say H_{min} below which the problems posed by stimuli fail to occupy his attention. (3) For all n less than T (or for all t less than $T\Delta t$) if $H_{max} > H > H_{min}$ then $0 > dH/dt$, but $dH/dt > 0$ if $H > H_{max}$.

(4) For all n less than T (or all t less than $T\Delta t$) $dH/dI > 0$ if $H_{max} > H > H_{min}$.

(5) The subject acts as a self-organising system, in the sense of von Foerster 53,54, so that if R, the redundancy $R = 1-H/I$, used as a measure of organisation then $0 < dR/dt$ until $t = T\Delta t$, providing that the A.C.M. makes suitable adjustments of I. The model presented is the simplest that satisfies these constraints. It represents the subject as a self organising system with a limited mentation or data processing capability. Further, for the particular class of skill that is considered, we introduce the empirically supported assumption that dR/dt is maximised if the condition $H = H_{max}$ is maintained (in other words, if the terminal criterion requires $I_t = I_{max}$ and $H_{min} = H_t$ then T is minimised, if this condition is satisfied).

For structured skills that consist of several jointly rehearsed subskills, we must at least, comprehend the effects of transfer of training between the acquisition of the several subskills and a form of interference between them which is due to a limit upon mentation of any sort, i.e. with respect to any of these subskills.

2. **Mathematical Model of the Coupled System.**

2:1 **Single subskill model.**
   From the constraints set out in 1:8(1)...

....(5), we know that for learning to occur, the subject's uncertainty H must satisfy the condition $H_{min} < H < H_{max}$. Now suppose, at some stage in the learning process that the subject is presented with a problem whose difficulty is indexed by η and hence measured by I. Let the subject's initial uncertainty regarding the correct response be H_0 where $H_{min} < H_0 < H_{max}$, then we assume that the subject's learning rate $-dH/dt$ is proportional to 1) a subject characteristic parameter k, 2) the amount by which his initial uncertainty exceeds $H_{min} = (H_0 - H_{min})$. 3) His remaining uncertainty H so that $\frac{dH}{dt} = -k(H_0 - H_{min})H$ whence $H = H_0 \exp(-k(H_0 - H_{min})t)$.......(1).

From 1:8 we have that I is a monotonic function of η and is interpretable as a difficulty measure with minimum $\eta = 0$ and a maximum $\eta = \eta_{max}$. From these assumption, in our model, I_{max} is the maximum difficulty associated with the subskill, so that in general $H \leqslant I \leqslant I_{max}$. The condition that an increase in I, ΔI results in an increase of H, ΔH (note however, that this does not necessarily imply a change in H_0 in equation (1)) gives the constraint cited in 1.8(4). We assume a linear relationship so that $\Delta H / \Delta I = a > 0$ a-constant and for simplicity let a=1 then
$$\Delta I = \Delta H \quad \ldots\ldots\ldots\ldots(2)$$
Initially let I=H=0 at t=0 and let I be increased by the A.C.M. till $I=H=H_{min}$, at $t=t_0$. At this point learning can begin.

In 1:6 we defined a sampling interval Δt. A sequence of stimuli is presented at constant η (or I) during this period, variables are measured and control action taken at the end of each sampling period i.e. at times
$$\Delta t, \ 2\Delta t, \ 3\Delta t \ldots\ldots\ldots\ldots T\Delta t.$$
We introduce the control rule in the following form
$$I_{n\Delta t} = I(n-1)\Delta t + \alpha\Delta I \text{ if } H_{n\Delta t} \leqslant H_{(n-1)\Delta t}.$$
....(3a).
$$I_{n\Delta t} = I(n-1)\Delta t - \beta\Delta I \text{ if } H_{n\Delta t} > H_{(n-1)\Delta t}....$$
....(3b).
Where α and β are control parameters and $H_{n\Delta t}$ is the value of H at $n\Delta t$ before control action is taken. Considering now the behaviour of this basic model we see that at $t = t_0, H = H_{min}$ and therefore from equation (1) learning cannot occur.
   Thus $H_{t_0 + \Delta t} = H_{t_0}$

∴ From (3) $I_{t_0 + \Delta t} = I_{t_0} + \alpha\Delta I$

But From (2) $\Delta H = \Delta I$ therefore, $H_{t_0 + \Delta t}$ becomes $H_{t_0} + \Delta H = H_{min} + \alpha\Delta I$. Thus, during the period $t_0 + \Delta t$ to $t_0 + 2\Delta t$, learning can and does occur and the subject's

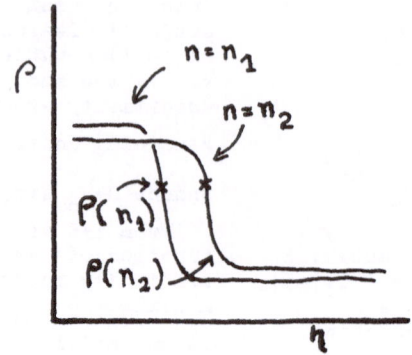

Fig. 4. Illustration of overload concept.

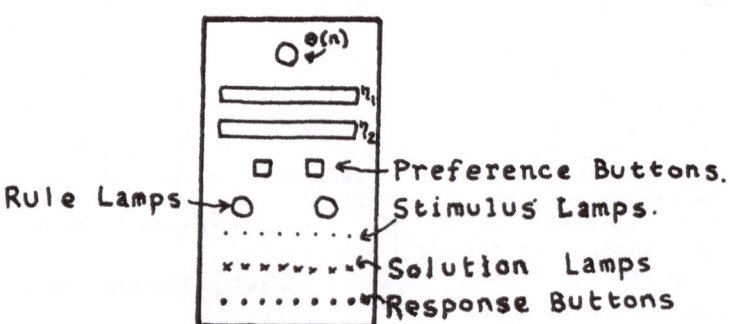

Fig. 5. Subjects display for two subskill system.

uncertainty is reduced to

$$H_{t_o+2\Delta t} = H_{t_o+\Delta t} \exp(-k(H_{t_o+\Delta t}-H_{min})\Delta t)$$
$$= H_{o(t_o+\Delta t)} \exp(-k\alpha\Delta I \Delta t)$$

where $H_{o(t)}$ = value of H at beginning of sampling period t. Hence the decrease in H due to learning during an interval t is given by $-\Delta H_{(t)} = H_{o(t)} (1-\exp(-k(H_{o(t)}-H_{min})\Delta t)$
If this process is allowed to continue then a point is reached at $t=t_1$ at which the reduction in H due to learning is exactly compensated by the increase in $I, \propto \Delta I$ due to control action i.e.,

$$\alpha\Delta I = H_{o(t_1)} (1-\exp(-k(H_{o(t_1)}-H_{min})\Delta t)....(4)$$

This is an equilibrium condition at which the value of $H_{o(t)}$ is constant or $dH/dt = 0$ while I increases to I_{max} at $t=t_2$, after which I is constant.

If the terminal criterion for a subskill is that $\eta_{T\Delta t} = \eta_{max}$ hence $I_{T\Delta t} = I_{max}$ and $H_{o(T\Delta t)} = H_{min}$ then the time taken to reach this terminal condition is $T\Delta t = t_2 + t_d$. Where t_d = time required for H to decay from $H_{o(t_2)}$ to H_{min} at constant $I=I_{max}$

$$= \frac{\log H_{o(t_2)} - \log H_{min}}{k(H_{o(t_2)} - H_{min})} \quad \text{from (1)}$$

It is thus evident, since this latter part of the process is uncontrolled, i.e. $I = I_{max}$ all $t > t_2$, that $T\Delta t$ is minimised if t_2-t_1 is minimised.

Now the RHS of equation(4) is maximised if $H_{o(t_1)} = H_{max}$. That is, the optimum value of the discrete difficulty increase $(\alpha\Delta I) op = H_{max} (1-\exp(-k(H_{max}-H_m)\Delta t)....(5)$
and under this condition the interval t_2-t_1 is minimised. Hence to maximise learning rate $\alpha\Delta I$ is set equal to $(\alpha\Delta I)$ op. Given the above condition $H_{o(t)}$ reaches the equilibrium value H_{max} at t_1 and remains there till $t=t_2$ after which it decays to H_{min} i.e.

$$\frac{d}{dt} H_{o(t)} = 0 \quad t_1 \leq t \leq t_2$$

Now in fact, the observable variables are the difficulty levels and response entropies. These are related to I and H, as stated in 1.7 and 1.8 and we shall now develop these relationships.
Given 1.7(1) the L^o alphabet for the computation of H and I is the set X,Y. Although the absolute value of I is unknown the definitions in 1.7 and 1.8 ensure that $I = I(\eta)$ and that for all n such that $o < n \leqslant T$ they ensure that...

$$I(\eta_2) > I(\eta_1) \quad \text{if}$$
$$\eta_2 > \eta_1, \quad \text{so that}$$
$$\frac{\Delta I(\eta)}{\Delta \eta} > 0.............(6)$$

Now H can be partitioned as $H=H(X)=H(Y|X)$ wherein the term $H(X)$ is the stimulus entropy that depends only on the choice of $\hat{x}$ and is thus constant and the term $H(Y|X)$ is the conditional entropy

$$-\sum_j p(y_j|x) \log p(y_j|x) \quad \text{for } j=1,2,...m$$

indicating uncertainty about the response selection when a stimulus $x \subset X$ is specified. Since the stimulus entropy is constant

$$\frac{dH}{dt} = \frac{d}{dt} H(Y|X)............(7)$$
$$= -\frac{d}{dt} p(y_j|x) \log p(y_j|x).....(8)$$

Now given 1.7(II) and given that Ω in 1.7 is a one to one correspondence, the learning process admitted by 1.7 and 1.8 leads to the increase in only one of the terms $p(y_j | x)$, namely the term $p(y+| x) = p(\Omega(x)|x)$ and on average, the remaining terms decrease to compensate for this change. Hence we can infer from (8) that

$$\frac{d}{dt} H(Y|X) = \frac{dH}{dt} > 0 \equiv \frac{d}{dt} p(y^+/x) < 0...(9a)$$

$$\frac{dH}{dt} = 0 \equiv \frac{d}{dt} p(y^+| x) = 0 ....(9b)$$

$$\frac{dH}{dt} < 0 \equiv \frac{d}{dt} p(y^+| x) > 0 ....(9c)$$

But from 1.7 the response proficiency average ρ, is defined as a monotonic function of the frequency with which $y = y^+ = \Omega(x) = \Omega(x^*)$ and consequently replacing probabilities by the corresponding estimates of frequency of occurrence, we obtain from (9) above the relation.

$$\frac{dH}{dt} > 0 \equiv \Delta\rho < 0 ....(10a)$$

$$\frac{dH}{dt} = 0 \equiv \Delta\rho = 0 ....(10b)$$

$$\frac{dH}{dt} < 0 \equiv \Delta\rho > 0.............(10c)$$

Thus we can substitute $\Delta\rho$ for $-\frac{dH}{dt}$
also, from(6), we can substitute $\Delta\eta$ for ΔI. The discontinuity in the η v ρ curves shown in Fig.4 illustrates the concept of an overload point. These curves show that there is a difficulty level, at any stage in the learning process, above which the problems are unintelligible, so that ρ is very low. In this case, H is high, according to the postulate that H trends are indicated by values of ρ.

Returning to consider, more fully, the implication of equation(5) we note that at equilibrium

$$\frac{dI}{dt} = \frac{\alpha\Delta I}{\Delta t}$$

77

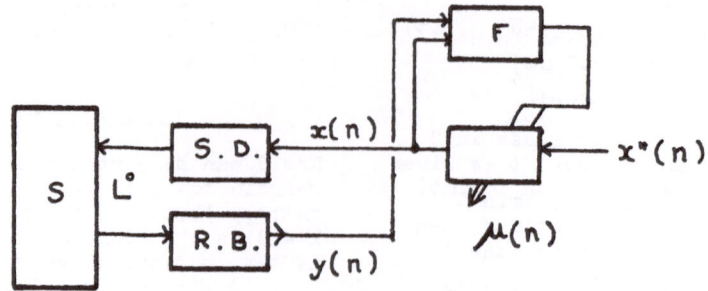

Fig. 6. Single subskill control mechanism.

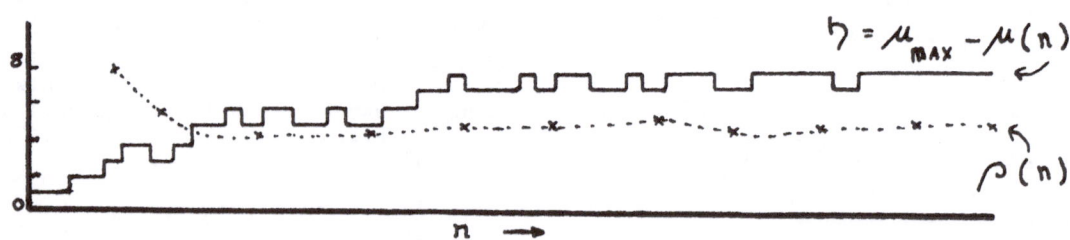

Fig. 7. Single subject response – first-order system.

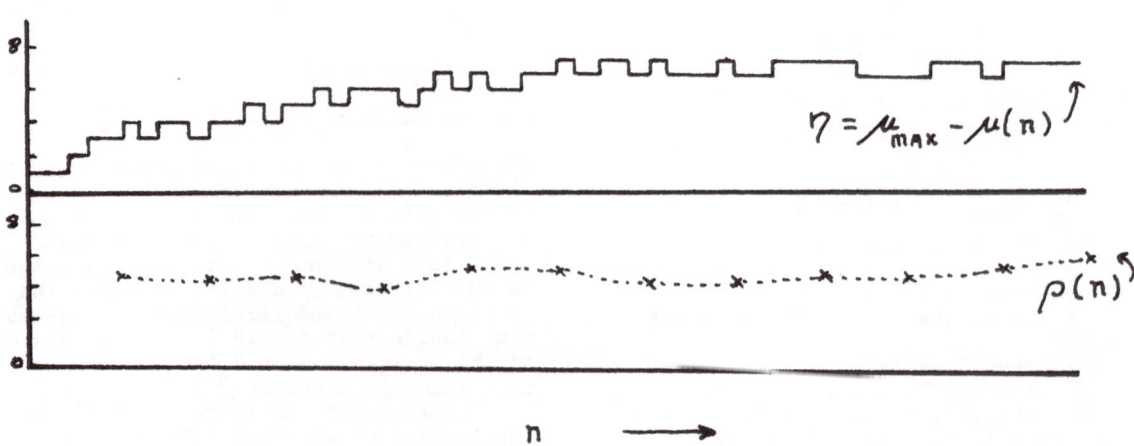

Fig. 8. Single subject response – first-order system – with ρ_0 adjustment.

and by taking the linear approximation to equ (1) i.e. $H = -k(H_o - H_{min})\,t$

whence $\dfrac{dH}{dt} = -k(H_o - H_{min})$

we note that $\dfrac{dI}{dt} = \dfrac{-dH}{dt}$ at equilibrium

and also $\dfrac{d}{dt} H_{o(t)} = 0$ at equilibrium

Hence $\dfrac{d}{dt} H_{o(t)} = \dfrac{dI}{dt} + \dfrac{dH}{dt}$ ....(11)

so that the equation describing the behaviour of the coupled subject - ACM system for a single subskill is

$$\dfrac{d}{dt} H_{o(t)} = \dfrac{dI}{dt} - k\,(H_{o(t)} - H_{min})\ldots.(12)$$

Where $\dfrac{dI}{dt}$ is determined by the control rule (3). We can interpret equ (12) as follows: If due to some external interference, the subject begins to learn more quickly at $H_{o(t)} = H_{max}$ then $\dfrac{dH}{dt}$ goes more negative

and hence $\dfrac{dH_o}{dt}$ goes negative. The

A.C.M. now selects a new $\dfrac{dI}{dt}$ to compensate.

This is the adaptive control problem of selecting a new trajectory in the I,t coordinate system, and for a single subskill this is done by varying the parameter α in equ (3).

2.2 Single Subskill Procedures.

The strategy recommended by the above class of learning model can be embodied in the decision rule F, of a control mechanism of the form of Fig 6, where S.D. represents the stimulus display, R,B.the response board and S the subject. F, can be interpreted as a translation of equ (3) in 2.1 from I, H terms into η, ρ terms so that

$F_1(\eta.\rho) = \Delta\eta(n) = +1$ if $\rho(n) > \rho_o$

(unless $\eta = \eta_{max}$ when $\Delta\eta = 0$)

$\Delta\eta(n) = 0$ if $\rho(n) = \rho_o$

$= -1$ if $\rho(n) < \rho_o$

(unless $\eta = \eta_o$ when $\Delta\eta = 0$)

Where ρ_o is a criterial level of performance chosen (from averages over the empirical data from many subjects) to maximise the expected rate of learning.
In the experiments concerned S.D. and R.B. are realised in terms of the apparatus in Fig 5. Only one rule Ω, is used for these experiments, and so the η and θ indicators are unused. The rule Ω relates the stimulus lamps to the rule lamps.
In consonance with our previous account a stimulus, x, is a configuration of lamps, its degree of simplification determining the number of lamps in the configuration and the amount of cueing data, regarding Ω (x), that is delivered.

A typical response curve for such a system is shown in Fig 7, and according to our assumptions the smoothed value of η should correspond to I. Similarly, from the arguments of 1.8 and 2.1 $\rho(n)$ t is related to $H_{o(t)}$.

In Table 1 we present data to support our assumption that the learning rate is maximised (or $T\Delta t$ is minimised) by maximising the stable value of H_o (minimising the stable value of $\rho(n)$). In order to confirm that this strategy is not too far amiss in the case of an individual subject the value of ρ_o has been varied, by an additional control loop, to maximise I and this leads to the response shown in Fig 8. In Table 1 conditions A,B, and C are associated with increasing values of ρ_o in the control rule F. As ρ_o is increased these mistake percentages are rank ordered $TA < TB < TC$ as in Table 1 and the equivalent ρ values are similarly rank ordered.

TABLE 1
Subskill Stabilisation Data
Number of trials, T, to reach criterion performance, using three different values of ρ_o

$\rho_o = .5$	$\rho_o = .6$	$\rho_o = .7$
T_A	T_B	T_C
181	287	375
203	291	388
213	305	431
237	362	490
240	391	497
251	418	532
268	419	537
286	440	569
294	469	592
301	483	623

Using Jonckheere's Trend Test $S/\sigma = 4.7$ The null hypothesis is rejected in favour of the predicted trend $T_C > T_B > T_A$ at the .01% level of significance ($P < .0001$)
The important point is that $T_C > T_B > T_A$ so that condition A with smallest ρ_o is more effective than B or than C.

2.3 Multiple Subskills.

Most structured skills consist of many related subskills. Of the many possible types of dependence that occur we shall only consider the types known in psychology as positive transfer of training and conditional interference. Positive transfer of training occurs when some of the problems posed by a pair of subskills can be solved by the same methods and conditional interference occurs when the performance of one subskill inter- feres with the joint performance of another until a higher order concept has been acquired (in the case we shall consider the

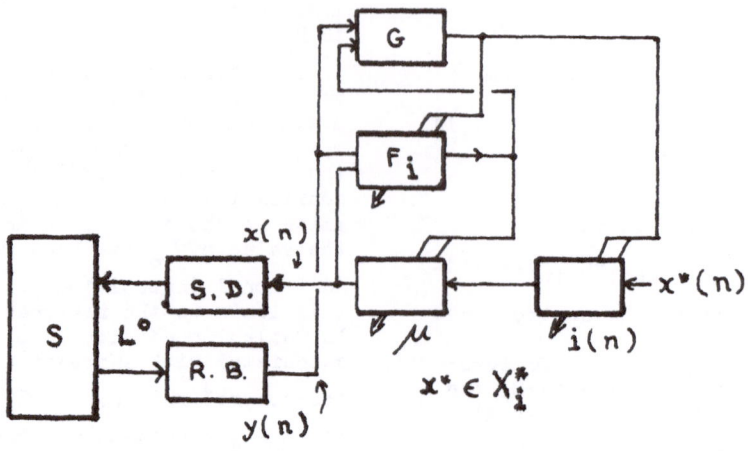

Fig. 9. Multiple subskill control mechanism.

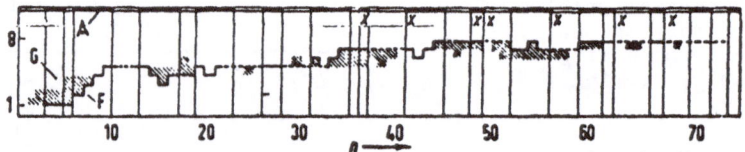

Fig. 10. Characteristics of second-order system with i(n) chosen by machine. $0.7 > \rho\,(n) > 0.55$ for $350 > n$. A=i (n). F=μ_{max} $-\mu_{\beta \bullet}$ (n). G=$\mu_{max} - \mu_{\alpha}$ (n);

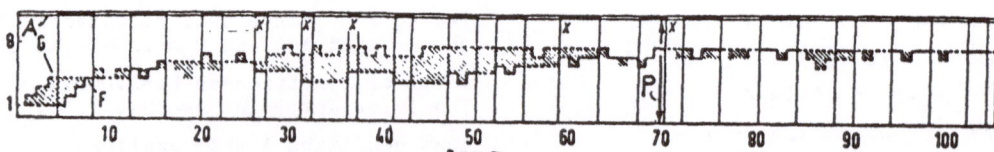

Fig. 11. Characteristics of second-order system with i(n) chosen by subject. $0.7 > \rho\,(n)$ > 0.55 for $500 > n$. A =i (n). F=$\mu_{max} - \mu_{\beta}$ (n). G=$\mu_{max} - \mu_{\alpha}$ (n).

acquisition of this procedure is contingent upon <u>attempting</u> joint rehearsal). The component subskills of the structured skill are called i = 1, 2, ...m.

The single subskill model developed in 2.1. is characterised by the time constant Δt, the sampling period, insofar as control action at time t is determined by the behaviour of the system during the preceding interval $t - \Delta t \rightarrow t$. However, when such sub-systems are coupled, as they are when a structured skill is being taught, then the control time constant, or memory, of the coupled system requires to be much longer, because control action at time t may now be dependent on the state of a subskill last rehearsed some long time previously.

Because of the different order of time constants involved we make the approximation that the subsystems are in their steady states, i.e. the subsystem parameters change very slowly in comparison with the subsystem time constant Δt. Thus, if the complete system is stable the subsystems are in dynamic equilibrium.

a. Positive Transfer of Training

Suppose there are 'm' subskills and that we know, 'a priori', the interactions among them. That is we are given an inter-action matrix

$$z = (z_{ij}) \text{ where } z_{ij} \begin{array}{l} > 0 \quad i \neq j \\ = 0 \quad i = j \end{array}$$

The elements of this matrix indicate the amount by which the achieved state, I_i, of the ith subskill aids learning rate in the jth subskill.
Consider the case when m = 2 and

$$z = \begin{bmatrix} 0 & a \\ b & 0 \end{bmatrix} \quad a, b > 0$$

let us assume that the subject has been brought to equilibrium by the A.C.M. on each component subskill before the condition of joint rehearsal is imposed. Let $t = t_1$ be the earliest time at which both subskills are at equilibrium with $H_0 = H_{max}$ in both cases. Let $t = t_2$ be the earliest time at which $I_i = (I_{max})_i$ for all i, i = 1, 2. Thus from equation 12 we get two coupled equations specifying the new dynamic conditions for the equilibrium of the two subsystems. These are -

$$\frac{dI_1}{dt} = k_1((H_{max})_1 - H_{min}) + bI_2 \dots\dots(13a)$$

$$\frac{dI_2}{dt} = k_2((H_{max})_2 - H_{min}) + aI_1 \dots\dots(13b)$$

Since t_2 is determined by the lowest average $\frac{dI_j}{dt}$ it is evident, under the joint

rehearsal condition that $t_2 - t_1$ is minimised, or learning rate maximised, if

$$\frac{dI_1}{dt} \quad \text{and} \quad \frac{dI_2}{dt}$$

are maintained as nearly equal as possible. Hence if we assume, for simplicity, that
$$k_1 = k_2 \text{ and } (H_{max})_1 = (H_{max})_2$$
we require to minimise

$$|dI_1/dt - dI_2/dt| = \min |bI_2 - aI_1|$$

and this is 0 when $I_2 = aI_1/b$. So the optimal control policy should aim to maintain

$$\frac{dI_2}{dt} = \frac{a}{b} \frac{dI_1}{dt}$$

This policy can be effected either a) by varying the microscopic difficulty steps $(\propto \Delta I)_1$ and/or $(\propto \Delta I)_2$
or b) by varying the relative number of times each subskill is rehearsed (in other words, by varying the probability $P_1(= 1-P_2)$ that subskill 1 is next rehearsed).

b). Conditional Interference

The simplest form of inference is the following:
Learning one subskill inhibits the learning of all the others by equal amounts. So that if the total inhibition from the ith subskill is C, then all other subskills are inhibited by a factor $\frac{c}{m-1}$ thus equation 13 for dynamic equilibrium now becomes

$$\frac{dI_1}{dt} = k_1((H_{max})_1 - H_{min}) + bI_2 - \frac{c}{2-1} \frac{dI_2}{dt}\dots(14a)$$

$$\frac{dI_2}{dt} = k_2((H_{max})_2 - H_{min}) + aI_1 - \frac{c}{2-1} \frac{dI_1}{dt};\dots(14b)$$

For simplicity put

$$k = k_1((H_{max})_1 - H_{min}) = k_2((H_{max})_2 - H_{min})$$

and substituting 14a into 14b and vice versa we get

$$\frac{dI_1}{dt} = \frac{k}{1+c} + \frac{bI_2 - acI_1}{1 - c^2} \dots\dots\dots(15a)$$

$$\frac{dI_2}{dt} = \frac{k}{1+c} + \frac{aI_1 - bcI_2}{1 - c^2} \dots\dots\dots(15b)$$

Equation 15a, b, are now conditions for dynamic equilibrium of the coupled system. Note that $\frac{dI_1}{dt}$ is reduced by a factor

depending on C. Now again $t_2 - t_1$ is minimised for
$$\min \left| \frac{dI_1}{dt} - \frac{dI_2}{dt} \right| \quad \text{but}$$

$$\min \left| \frac{dI_1}{dt} - \frac{dI_1}{dt} \right| = \min \left| \frac{aI_1}{1-c} - \frac{bI_2}{1-c} \right|$$
$$= 0 \text{ when } I_2 = \frac{a}{b} I_1$$

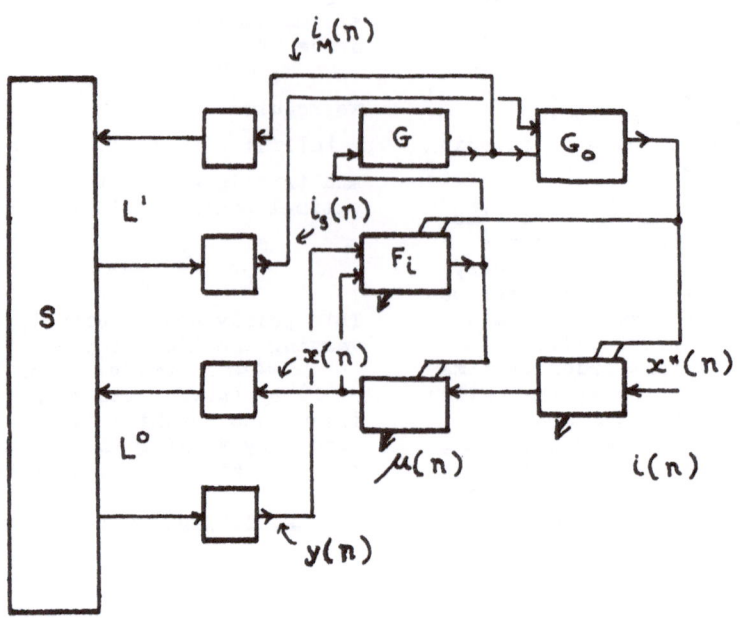

Fig. 12. Higher-order control mechanism.

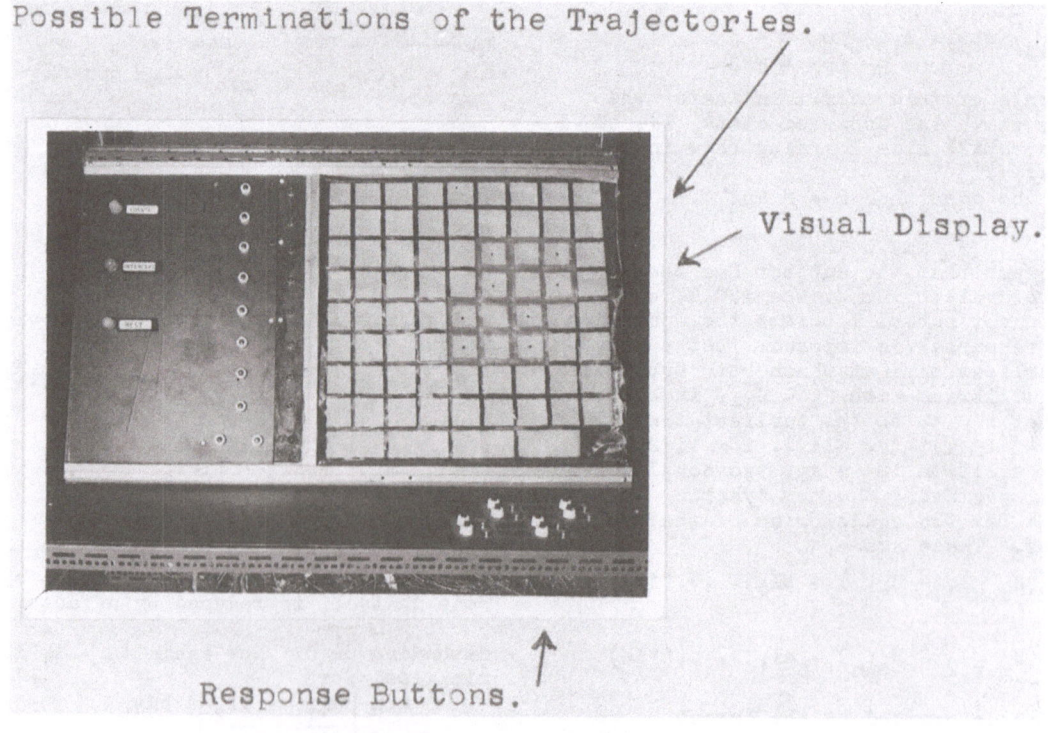

Fig. 13. Trajectory display console.

Hence the optimal control strategy is again to maintain $\dfrac{dI_2}{dt} = \dfrac{a}{b}\ \dfrac{dI_1}{dt}$. So for

<u>symmetrical</u> interference the optimal machine control strategy is determined solely by the positive transfer matrix.

2.4. <u>Types of Concept</u>

Though the above model cannot comprehend concept acquisition explicitly, we can represent the symptoms of the acquisition process as the variation of some of the model parameters. To do this a distinction between concept types becomes useful, a Type 0 concept being associated with the L^0 vocabulary and a Type 1 concept with the higher level L^1 vocabulary.

Thus when the subject becomes aware, in the multiple skills case, that the subskills are related and that <u>joint</u> rehearsal of the subskills is necessary, that is he becomes aware, in an abstract sense, of the existence and structure of the positive transfer matrix, we can say he has acquired a Type 0 concept. Analogously, as indicated in 2.3., the acquisition of a Type 1 concept alters the conditional interference matrix and so we would expect a sudden change in C (equ. 14) to be symptomatic of the Type 1 concept acquisition process.

2.5. <u>Multiple Subskill Procedures</u>

The strategy recommended for instructing a structured skill composed of several subskills which can be learned if and only if a higher order concept is acquired, requires a pair of decisionrules F and G, that can be embodied in the control mechanism in Fig.9. The output of G is here interpreted as the relative frequency of rehearsal of subskill Ω_i, i = 1 or 2 (though as in 2.1. it may also be interpreted as changing the values of $(\alpha \Delta I)_{op}$ which leads to similar experimental results). In the case of control of the relative frequency of rehearsal the pair of rule lamps shown in Fig.5. are used to indicate the chosen rule that is applicable for a given block of trials. A typical response curve for this coupled system appears in Fig. 10. The area between η_1 and η_2 indicates the extent to which the subject acquires a higher order concept and it can be shown that G, the policy of 2.3. minimises this area.

2.6. <u>Higher Order Systems</u>

So far we have subscribed to the organisation of Fig. 2. It is now pertinent to ask what advantages are to be obtained using a higher level of discourse, such as L^1, to realise the organisation in Fig. 3. In practice, the L^1 modality provides the subject with indications of the η_i values

(as in Fig. 4.) and allows him to express preferences for the rehearsal of either the Ω_1 or the Ω_2 subskill after any block of trials (the preference buttons are indicated in Fig. 5.). The naive interpretation of an L^1 communication modality allows the subject unfettered control of the value of i (tempered only by the restriction that Ω_1 and Ω_2, must always be rehearsed and that they must be rehearsed with <u>equal</u> frequency at t = T Δ t). The result of such an interpretation is the response curve shown in Fig. 11. The learning rate is far less (the value of TΔt is far greater) than it is in the case of Fig. 10., where i is chosen by the rule G.

Further, and not surprisingly, the area between η_1 and η_2 is much greater. The fact is that L^1 is not the modality of a <u>control</u> loop when it is interpreted in this fashion. To render it <u>competent</u> we must make the degree of control that the subject is allowed to exert into a function of his success in satisfying the L^0 goals. This is achieved by the "metasystem" in Fig.12., wherein Θ is a function $0 \leqslant \Theta \leqslant 1$ that is defined as $(\eta_1/\eta_{max})(\eta_2/\eta_{max})$. At t=0 the value of Θ = 0 and at t = t_2 its value is equal to 1. In Fig. 12., G produces a "machine rule", preference i_m for a value of i at t = n Δ t and the subject makes a preference "metastatement" in L^1 that he desires the value i_s for i at this instant. The actual value i(n) is determined by G_Θ which selects i_s with probability of Θ and i_m with probability of 1 - Θ. Hence, the subject's initial degree of control over i is 0 and his ultimate control over i, when he has learned the skill, is absolute. Table 2 compares the case of machine control of i, using G, of "unfettered subject" control and of an adaptive metasystem, in terms of the value of T. (See Table 2. over page).

2.7. <u>Stability</u>

Equations 15a and 15b describe the behaviour of the coupled two subskill system. Notice that the general m subskill interference term is c/m-1 = C when m = 2. Now consider this system in the region after the initial transients have died out, i.e. the individual subskills have <u>all</u> reached equilibrium. Then in the nomenclature of 2.3. we are dealing with the region $t_1 < t < t_2$.

in which the individual subsystems are in equilibrium (approximately, since the subsystem time constants are very much less than the overall system time constant) and we can apply equations 15.

Differentiating equations 15a and 15b and putting

$$\frac{dI_1}{dt} = X_1 \quad \text{and} \quad \frac{dI_2}{dt} = X_2$$

we get

Fig. 14.

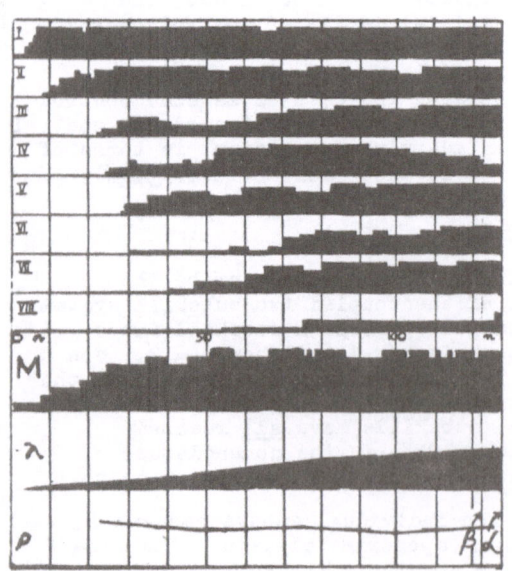

Fig. 15.

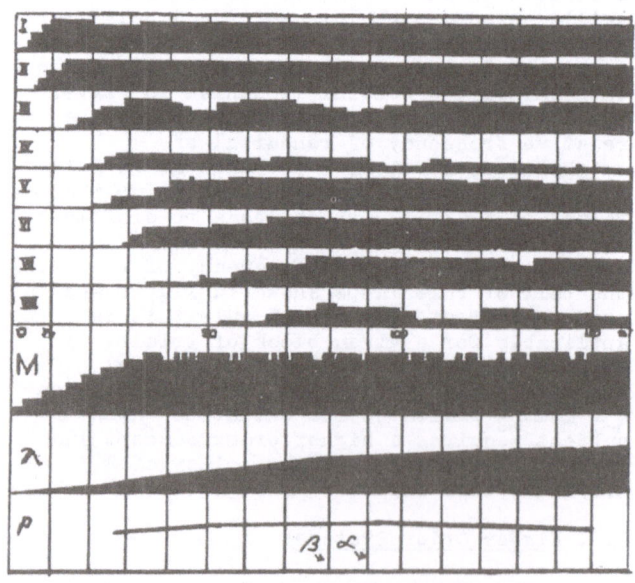

Fig. 16. Characteristics for perceptual discrimination skill from different subjects. I through VII = $\mu_{max} - \mu_I^{(n)}$ through $\mu_{max} - \mu_{VII}^{(n)}$.

84

$$\frac{dX_1}{dt} = \frac{b}{1-c^2} \; X_2 - \frac{ac}{1-c^2} \; X_1 \quad \ldots\ldots\ldots\ldots (16a)$$

$$\frac{dX_2}{dt} = \frac{a}{1-c^2} \; X_1 - \frac{bc}{1-c^2} \cdot X_2 \quad \ldots\ldots\ldots\ldots (16b)$$

or

$$\begin{bmatrix} \frac{dX_1}{dt} \\[2mm] \frac{dX_2}{dt} \end{bmatrix} = \frac{1}{1-c^2} \begin{bmatrix} -ac & b \\ a & -bc \end{bmatrix} \begin{bmatrix} X_1 \\ X_2 \end{bmatrix} \quad \ldots\ldots\ldots (17)$$

and the Hurwitz criterion for the stability of this system is $C > 1$.

Now the steady state solution of equ (16a) is

$$X_1 = \frac{b}{ac} \; X_2$$

and (16b) is

$$X_2 = \frac{a}{bc} \; X_1$$

whence in the steady state $C = 1$ and so $X_1 = \frac{b}{a} X_2$ in consonance with the optimal control policy devised in 2.3.

We interpret the requirement $C > 1$ as the condition for stable interaction while the subject is acquiring the higher order Type 1 concept, and the condition $C = 1$ as the condition for stable learning when the Type 1 concept has been acquired and this is also the condition under which the machine control policy developed in 2.3. is optimal. Note that both conditions satisfy the von Foerster criterion $\frac{dR}{dt} > 0$ where $R = 1 - \frac{H_o}{I}$ that the whole system can be self organising since in either case $\frac{dI_i}{dt} > 0$ on average.

We comment that $1 - R$ has the characteristics of a Lyapunov function defining the system stability conditions, but shall not discuss the implications of this remark.

2.8. Stable Systems

The argument for stability of a real system is supported by the data already provided. However, the point can be more elegantly demonstrated in the case of a structured skill with 8 different subskills. The structured skill concerned is the perceptual discrimination of 8 classes of trajectory against a disturbing visual background signal (the amplitude of which, in the case of the ith class of trajectories is interpreted as η_i). The display and the response board for this skill are shown in Fig.13. (the small squares in Fig. 13. are backilluminated to define the disturbing background and the trajectory. The subject discriminates a trajectory by pressing the response buttons).

Typical response curves are shown in Fig. 14., Fig. 15. and Fig. 16. which apply to our argument after the value of M, the number of alternative subskills, is stable at 7 in Fig.14. and 8 in Fig.15. and Fig.16. For the present purpose the initial transient in which M rises to its stable value is irrelevant since the control procedure entailed in this part of the response is not described in the present paper.

TABLE 2
Number of blocks of six trials to achieve criterion performance, for coupled systems stabilised under 3 different conditions.

T_A	T_B	T_C
83	50	46
93	63	49
105	72	54
106	72	57
108	74	58
108	78	61
110	78	63
134	80	64
134	92	68
147	103	81

Condition A: Value of $i(n)$ determined by subject.
Condition B: Value of $i(n)$ determined by machine.
Condition C: Adaptive metasystem.

Using Jonckheere's Trend Test $s/\sigma = 4.6$ The Null Hypothesis is rejected in favour of the predicted trend $T_A > T_B > T_C$ at the .01% level of significance ($p < .0001$).

References.

1. G. Pask. Tech. Note under U.S.A.F. contract AF61(052)402. ASTIA, 1962.
2. G. Pask. In K. Austwick: TEACHING MACHINES, Pergamon, 1964.
3. G. Pask. In S. Beer et Al. ADVANCES IN CYBERNETICS. Academic Press, 1965.
4. G. Pask. Proc. 4th Con. Int. Assoc. Cybernetics, Namur, 1964.
5. B.N. Lewis & G. Pask. In R. Glaser (ed.): TEACHING MACHINES AND PROGRAMMED LEARNING Nat. Ed. Assoc., 1965.
6. G. Pask. 2nd Con. Int. Assoc. Cybernetics, Namur, 1961.
7. B.N. Lewis & G. Pask. PERCEPTUAL AND MOTOR SKILLS, 14, 1962.
8. B.N. Lewis. In M. Goldsmith: MECHANISA-TION IN THE CLASSROOM. Souvenir Press, 1963.
9. G. Pask. NEW SCIENTIST, June 10, 1961.
10. G. Pask. DATA AND CONTROL, February 1964.
11. G. Pask. In M. Frank: KYBERNETISCHE MASCHINEN. Fisher-Verlag, 1964.
12. G. Pask & B.N. Lewis. Annual Reports under U.S.A.F. contract AF61(052)402. 1961-1965, Washington.
13. L.S. Vygotsky. THOUGHT AND LANGUAGE. J. Wiley, 1962.
14. A. Luria. THE ROLE OF SPEECH IN THE REGULATION OF BEHAVIOUR, Pergamon, 1961.
15. G. Pask. In Wiener & Schade: PROGRESS IN BIOCYBERNETICS, Elsevier, 1964.
16. G. Pask et Al. Psychophysical Experiment and Demonstration, London Conf. on Psychology, 1964.
17. G. Pask. TRANS. SOC. INSTR. TECHNOL. June, 1960.
18. G. Pask. CONTROL ENGINEERING, Nov. 1959.
19. G. Pask. 10th Int. Con. Electronics, Rome, 1963.
20. G. Pask. CONTROL, Jan-April, 1965.
21. R.M. Martin. INTENSION AND DECISION Prentice Hall, 1964.
22. S. Gorn. PROC. SYMPOS. PURE MATHEMATICS 5, 1962.
23. R. Tarjan. In Bollinger et Al: OPTIMIZ-ING AND ADAPTIVE CONTROL. Washington 1963.
24. W. Ross Ashby. INTRODUCTION TO CYBER-NETICS, Chapman, 1957.
25. G. Pask. DIALECTICA, 66/67, 1963.
26. G. Pask. 3rd Con. Int. Assoc. Cybernetics Namur, 1961.
27. G. Pask. AN APPROACH TO CYBERNETICS. Hutchinsons, 1961.
28. G. Pask. DOCTORAL THESIS, London, 1964.
29. G. Pask. Br. Jnl. Stat. Psych. (forthcoming).
30. J. Piaget. LOGIC AND PSYCHOLOGY, Basic Books, 1957.
31. G.A. Miller et Al. PLANS AND THE STRUC-TURE OF BEHAVIOUR. Holt. 1960.

32. G.A. Miller & N. Chomsky. In Luce et Al.. HANDBOOK OF MATHEMATICAL PSYCHOLOGY. McGraw Hill, 1964.
33. L. Pun. Int. Fed. Aut. Con. Rome 1963.
34. A. Newell. In C.M. Popplewell: INFORMATION PROCESSING '62. Nth. Holland, 1963.
35. A. Koestler. THE ACT OF CREATION. Hutchinson, 1964.
36. J.A. Deutsch. THE STRUCTURAL BASIS OF BEHAVIOUR. Cambridge, 1960.
37. C. Hull. PRINCIPLES OF BEHAVIOUR. Appleton Century, 1943.
38. E.C. Tolman. In S. Koch: PSYCHOLOGY, A STUDY OF A SCIENCE, I, 2, McGraw Hill, 1959.
39. W.K. Estes. In S. Koch, op.cit. 1959.
40. C. Osgood. EXPERIMENTAL PSYCHOLOGY. Chicago, 1954.
41. E.R. Guthrie. PSYCHOLOGY OF LEARNING. Holt, 1951.
42. J.Z. Young. A MODEL OF THE BRAIN. Oxford, 1964.
43. H. Harlow. In S. Koch. op. cit. 1959.
44. M. Minsky. Proc. I.R.E. 49. 1961.
45. A. Newell et Al. PSYCHOL REV.65,1958.
46. G. Pask. Proc. Berlin Conf. on Cyber-netics, 1964.
47. G. Pask. BUL. MATH. BIOPHYSICS (In press).
48. A. W. Burke. In Yovits et Al. SELF ORGANISING SYSTEMS, Pergamon, 1960.
49. A. Culbertson. THE MINDS OF ROBOTS Chicago, 1963.
50. O. Selfridge. In: THE MECHANISATION OF THOUGHT PROCESSES, H.M.S.O. London 1959.
51. G. Pask. Proc. IFAC Conf. Basle, 1963. Butterworth, 1965.
52. G. Pask. In M. Rubinoff: ADVANCES IN COMPUTERS, Vol.5. Academic Press, 1964.
53. H. von Foerster. In Yovits et Al., op. cit. 1960.
54. H. von Foerster. In Bernard et Al. BIOLOGICAL PROTOTYPES AND SYNTHETIC SYSTEMS. Plenum, 1962.
55. G. Pask. In Yovits et Al. op. cit. 1960.
56. B.N. Lewis & G. Pask. PROGRAMMED LEARNING. July 1964.
57. B.N. Lewis. In C. Jones et Al. Conf. on DESIGN METHODS. Pergamon, 1963.
58. G. Pask. In W.G. Greene: NATO CONF. ON COMMAND AND CONTROL SYSTEMS. (In press).

GENERAL DISCUSSION OF PAPERS ON ADAPTIVE CONTROL
IN BIOLOGICAL SYSTEMS

Dr. S.R. Caplan (Israel) summarised his work on the self adaptive control of energy conversion and its relation to the mechanical behaviour of muscle.

He demonstrated that the force-velocity relation for muscle, as discovered experimentally by A.V. Hill (ref. 1 and 2), could be derived by considering the properties of an energy conversion system in which two coupled irreversible thermodynamic processes take place. The treatment was facilitated by the use of the 'degree of coupling' (q) between the input and output of the system, the analogy being coupling between electrical networks (ref. 3). The parameter q completely determines both maximum efficiency of the system and the efficiency when the output is maximal. If the value of q is close to unity, the power output may reach a maximum at extremely different values of the load, depending on the source impedance of the input signal. Therefore, to obtain high output power over a wide range of loads a regulator is required which adjusts the input in response to the load.

In the case of muscle, the energy convertor (q now less unity) has the form in which the regulator can assess the load resistance through the converter itself, i.e. a converter subject to self-adaptive control. In order to make the analogy complete, the energy converter must be linear. The regulator programme may then be identified with an arbitrary function of load resistance. If this function is constant, the information in the programme is minimal and includes only the q parameter of the converter and its operational limits. With these restrictions, the operation of the energy converter is analogous to that of muscle, where limitations are put on the force generated and on the velocity of shortening. The output characteristic is then identical with the classical force-velocity equation of muscle contraction.

REFERENCES

1. Hill, A.V., Proc. Roy. Soc. B, <u>126</u>, 136 (1938)

2. Hill, A.V., Proc. Roy. Soc. B, <u>159</u>, 297 (1964)

3. Kedem, O. and Caplan, S.R., Trans. Faraday Soc., <u>61</u>, 1897 (1965).

PARAMETER ESTIMATION AND MODELLING TECHNIQUES

An Open-Loop Procedure for Process Parameter
Estimation Using a Hybrid Computer

by Paul Alper
Division of Automatic Control,
The Technical University of Norway.
Trondheim, Norway

INTRODUCTION

In the field of estimation and identification one principle not stressed often enough is that the experimenter's approach to the problem should vary according to and take into account what is known "a priori" about the problem. A method which works when nothing is known about the system is most likely to be grossly inefficient when much is known about the system; a method which works quite well (i.e., fast, precise, stable) for a system where much is known should not necessarily be expected to be applicable in a wider scope. More picturesquely as the black box becomes grayer the experimenter ought to be able to utilize this knowledge to his advantage.

In the situation where the dynamics of a system are completely known except for the coefficients of the differential equation then it behoves us to seek a method which incorporates as much of this a priori knowledge as possible rather than, let us say, a simple hill climber which blindly and slowly eventually inches its way to success; at the same time, one would like a method which is robust enough to still work should some of the assumptions be inaccurate.

When the configuration of the system is assumed given, model matching methods have been used extensively in the past for estimating parameters. One of the main drawbacks, however, when using a model matching technique is that once the feedback loops which adjust the estimates of the parameters are closed it becomes exceedingly difficult to perform even a rudimentary analysis. Statements concerning the gain of the feedback loop, the type of input signal desired, or the influence of other parameters usually must be qualitative rather than quantitative. An open-loop procedure, since there is no non-linear feedback, would be expected to obviate these difficulties to some extent.

This paper discusses some of the implications of an open-loop method of Meissinger (1) for determining process parameters and it is shown that this approach in addition to being systematic and easily mechanized is capable of analytic treatment such that good use is made of the "a priori" knowledge, while still possessing a certain degree of robustness.

OPEN-LOOP TECHNIQUE

Meissinger's procedure as originally formulated assumed that the error, e; the difference between the output of the model, z, and the output of the process, y, is approximated by a Taylor series expansion

$$e = \sum_{1}^{N} \Delta\phi_i u_i \qquad (1)$$

where the parameter influence coefficient u_i is

$$u_i = \frac{\partial e}{\partial \phi_i} = \frac{\partial z}{\partial \phi_i} \qquad (2)$$

and ϕ_i is the estimate of the process parameter a_i and

$$\Delta\phi_i = a_i - \phi_i \qquad (3)$$

and by alternating between measurement and adjustment the error can be reduced to zero with ϕ_i assuming the value a_i. The block diagram is shown in Figure 1.

The open-loop nature of the method implies certain benefits in terms of ease of analytic treatment; measuring the parameter influence coefficients while leaving open the feedback loop which adjusts ϕ_i results in the exact determination of the partial derivatives. This is not the case when the measurements and adjustments are done simultaneously (as is usually the situation in ordinary continuous or closed-loop steepest descent approaches) leading to possible instability (2). Furthermore, using an open-loop technique, it is easily seen how the choice of the input testing signal influences the interaction of the adjustments.

This comes about due to the fact that equation (1) may be written as

$$\overline{E} = A\ \overline{\Delta\phi} \qquad (4)$$

where

$$E_i = \langle eu_i \rangle \qquad (5)$$

$$A_{ij} = \langle u_i u_j \rangle \qquad (6)$$

"Bracketted numbers in the text refer to similarly numbered references at the end of this paper"

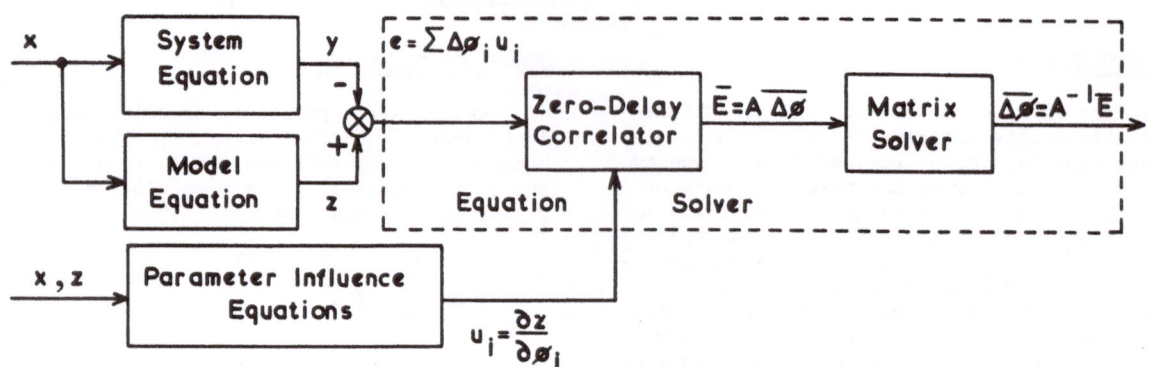

Fig. I Block Diagram

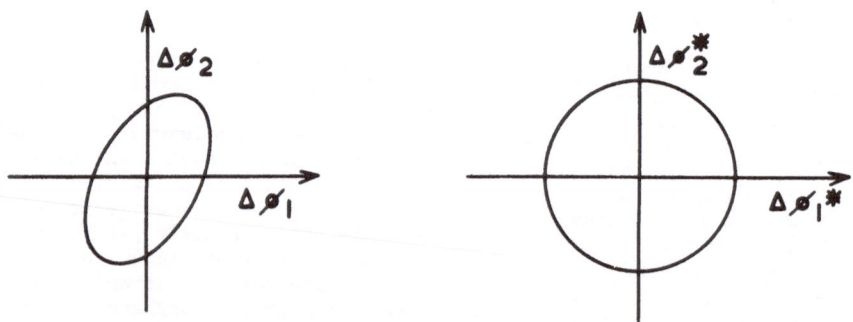

Fig. 2a. Original Coordinates.   Fig 2b. Canonical Coordinates.

$$\overline{\Delta \phi} = \begin{bmatrix} \Delta \phi_1 \\ \vdots \\ \Delta \phi_i \\ \vdots \\ \Delta \phi_N \end{bmatrix} \qquad (7)$$

and thus the elements of the sensitivity matrix A can be determined in a relatively straight forward manner using power spectral analysis because each element is merely the integral of the cross spectra of the relevant parameter influence coefficients. The effect of using too short a measuring time is reflected by the statistical variations in the element of the sensitivity matrix and therefore it is possible to choose a measuring time consistent with small statistical variations and reasonable time of convergence.

In addition, the open-loop approach implies that, employing an orthogonal transformation, it is possible to deal with canonical estimates such that the old estimates can be more accurately (computationally) determined by first calculating $\overline{\phi}^*$, the canonical variables, and then transformed back using

$$\overline{\Delta \phi} = Q \overline{\Delta \phi^*} \qquad (8)$$

where Q^{-1} AQ diagonalizes the sensitivity matrix and then a simple transformation reduces the spread to zero. Pictorially, in two dimensions, it becomes analytically possible to change the error contours from that shown in Figure 2a, a tilted ellipse to that shown in Figure 2b, a circle.

Moreover, if the input signal has a known spectrum then it may be possible to calculate all the various A_{ij}'s and thus reduce $\overline{\Delta \phi}$ to zero in one step. For example, if the system and the model are given respectively by

$$\frac{dy}{dt} + ay = x \qquad (9)$$

$$\frac{dz}{dt} + \phi z = x$$

then if x is white noise

$$\Delta \phi = \phi - \frac{4\phi^3}{(\phi + a)^2} \qquad (10)$$

from which a can easily be obtained. Should the input signal be of unknown power spectrum one could still make a reasonable assumption as to its nature and determine a suitable gain which will insure stability when operating with adjustment alternating with measurement. The experimentor thus, depending on his degree of confidence or ability to perform all the cross correlations, can choose his manner of operation.

Obviously, it would be difficult for a pure analog computer to perform all these calculations but such a procedure lends itself readily to hybrid computation and the differential equations of the process, of the model and of the parameter influence coefficients are easily programmed with the analog computer performing merely the necessary integration called for and the digital computer (in this instance, GIER) doing all the rest such as the necessary logic, multiplication, function generation and switching of modes.

In Figure 3 is shown a typical parameter estimation run for a non-linear process having fixed parameters with ϕ_1 the estimate of the inverse of the time constant of the process and ϕ_2 the estimate of the coefficient of the non-linearity. In Figure 4 is shown a typical parameter estimation run for a non-linear system with time-varying parameters.

Some caution is in order because the delay introduced by the digital processing can cause a self-excited oscillation especially when a configuration such as shown in Figure 5 is used for generating terms such as

$$\frac{1}{s + \phi}$$

Some improvement can be achieved by making use of the parameter influence coefficient of the delay. The problem can be greatly lessened if the analog does more than just providing the necessary integrations but this would in turn decrease the accuracy of the results.

CONCLUSION

Meissinger's open-loop method is seen to have several analytic benefits due to the fact that the parameter influence coefficients are well defined quantities such that the experimentor may take advantage of the a priori knowledge he possesses. The more he is able to calculate in terms of the cross correlations the better he can use these calculations to achieve results more quickly and more accurately. If his a priori calculations are somewhat in error the technique is robust enough to overcome this error because of the iteration procedure of measurement alternating with adjustment.

REFERENCES

(1) Meissinger, H.F., Parameter Optimization by an Automatic Open-Loop Computing Method, 4th International Conference on Analog Computation AICA, Paper No. 69, Brighton, England.

(2) Margolis, M., and Leondes, C.T., A Parameter Tracking Servo for Adaptive Control Systems, IRE Wescon Conv. Record, p. 104, August 1959

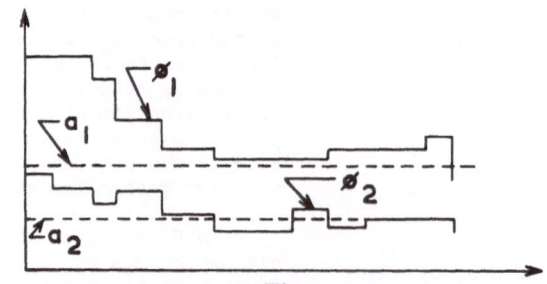

Time

Fig. 3. Parameter Estimation for a Nonlinear System.

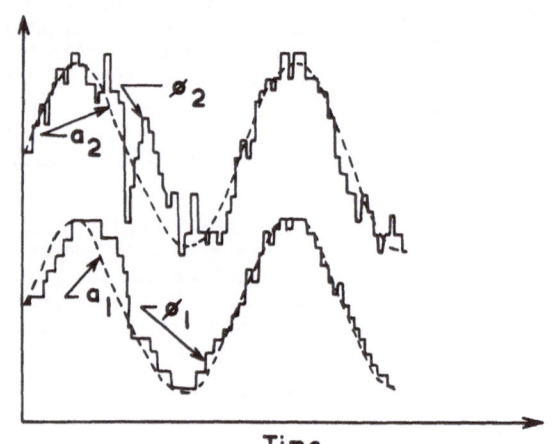

Time

Fig. 4. Parameter Estimation for a Linear System.

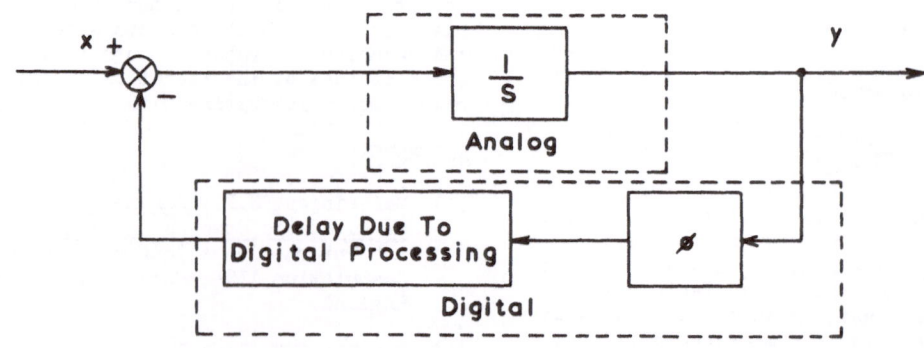

Fig. 5   Generation of $1/(s+\phi)$

94

DISCUSSION

Professor G.D.S. MacLellan (U.K.) asked what types of system non-linearity had been tried in the example of Figure 3?

Dr. Alper replied that only a very simple system was investigated, a first order system with a square-law characteristic in the feed back path. In his opinion, the system is not really critical and the extra effort to change the form of system is entirely due to re-programming the computer.

Professor L. von Hamos (Sweden) asked what are the time scales on the adaptation steps of Figures 3 and 4?

Dr. Alper replied that the adaptation stepping times must be chosen to be consistent with two requirements:

(a) We want convergence to be as fast as possible.

(b) When the system is stochastic, we must correlate over a minimum time to get a reasonable estimation, and yet not so long as to be inefficient.

In general, 2 or 3 times the time constant of the longest mode of the differential equation governing the system is sufficient.

P. Eykhoff (Netherlands) asked in the case of Figure 1, how closely must the model approximate to the actual system in order that a good estimate of the system be obtained? Further, is there a region of convergence around the system for which a solution can be obtained?

Dr. Alper replied that although only a limited range of models have been tried, his experience was that one can have a large discrepancy and still get an answer, although in this case it may be necessary to measure and adjust several times. The region of convergence could be extended by taking to higher order the Taylor's expansion of equation (1), but this has not been done due to limited computer storage space and time.

Mr. T. Horrocks (U.K.) commented in writing on the "open-loop" procedure used by Dr. Alper as follows:

"It seems to me that the moment Dr. Alper says 'I measure and adjust, measure and adjust ...', his "open-loop" assumption falls down. What he has done, in effect, is to introduce another loop into the system. It is a sampled-data loop, with himself as the switch of the sample-hold device. If that is so then of course he can be mechanized out of the loop and we are back to an ordinary closed-loop system.

This seems to me to be an excellent illustration of Professor Truxal's contention that the question of the loop being open or closed is often a matter merely of the point of view of the observer. Certainly one ought to be extremely careful in talking about "open" and "closed" loops.

NUMERICAL IDENTIFICATION OF LINEAR DYNAMIC SYSTEMS

FROM NORMAL OPERATING RECORDS

by   Prof. Karl-Johan Åström *
     Dr. Torsten Bohlin
     IBM Nordic Laboratory
     Solna, Sweden

ABSTRACT

A technique for numerical identification of a discrete time system from input/output samples is described. The purpose of the identification is to design strategies for control of the system. The strategies are obtained using linear stochastic control theory.

The parameters of the system are estimated by Maximum Likelihood. An algorithm for solving the M.L. equations is given. The estimates are in general consistent, asymptotically normal and efficient for increasing sample lengths. These properties and also the parameter accuracy are determined by the information matrix. An estimate of this matrix is given.

The technique has been applied to simulated data and to plant data.

INTRODUCTION

In order to apply linear stochastic control theory, the process to be controlled should be described in terms of linear differential or difference equations driven by the input signals and disturbances in the form of weakly stationary stochastic processes with rational spectra. In this paper we describe a technique for determining such models from samples of the process inputs and outputs.

We restrict outselves to the case of

- difference equations
- single input and single output
- time invariant models

As will be discussed in section 7, these restrictions are not essential for the application of the technique.

If it is also assumed that the disturbances are normal there is a canonical form (2.1) containing a finite number of unknown parameters, for the class of models of interest. Each set of parameters together with the input sequence determine uniquely the distribution of the output sequence. Conversely, given a sample of the observed output we have a statistical parameter estimation problem. We will solve it using the method of maximum likelihood. For alternative approaches to the identification problem we refer

to [3], [4], [5], [8], [15], [18], [19], [21], [24].

* now at Lunds Institute of Technology, Sweden

The problem is stated and commented on in section 2. In section 3, an algorithm is given for maximizing the likelihood function, and in section 4 the statistical properties of the estimates are investigated. Section 5 contains an example. An alternative interpretation of the technique is given in section 6.

The algorithm has been tested on artificially generated data and has generally been able to identify the parameters with the theoretical accuracy. It has also been applied to design control strategies for regulating a paper machine [2]. A large amount of data has been analyzed. In general, it has in this application, been possible to fit the data with models of low order (first or second order with delay). As a rule, we encountered very few numerical difficulties with the standard algorithm. So far, our practical experience with the technique is, however, limited to systems of comparatively low order. The results of the practical experiments have been very satisfactory.

STATEMENT OF THE PROBLEM

2.1. Model of the system to be identified

Consider a discrete time single-input single-output dynamical system whose input/output relation can be described by the equation

$$A(z^{-1}) y(t) = B(z^{-1}) u(t) + \lambda C(z^{-1}) e(t) \qquad (2.1)$$

where $\{u(t)\}$ is the input $\{y(t)\}$ the output and $\{e(t)\}$ a sequence of independent normal $(0, 1)$ random variables. Furthermore, z denotes the shift operator [22].

$$z\, x(t) = x(t+1) \qquad (2.2)$$

and $A(z)$, $B(z)$ and $C(z)$ are polynomials

$$A(z) = 1 + a_1 z + \ldots + a_n z^n$$

$$B(z) = b_0 + b_1 z + \ldots + b_n z^n$$

$$C(z) = 1 + c_1 z + \ldots + c_n z^n \qquad (2.3)$$

We also introduce the row-vectors a, b and c whose components are a_i, b_i and c_i respectively.

The following assumptions are made

- the functions $A(z^{-1})$ and $C(z^{-1})$ have all their zeros inside the unit circle

- there are no factors common to all three polynomials $A(z)$, $B(z)$ and $C(z)$

"Numbers within brackets refer to similarly-numbered references at the end of this paper".

The assumption that the function $A(z^{-1})$ has all its zeros inside the unit circle implies that the homogeneous equation corresponding to (2.1) is asymptotically stable. The assumption that there are no factors common to $A(z)$, $B(z)$ and $C(z)$ implies that every state of the system (2.1) is controllable either from u or from e. [13] This is no loss in generality. Neither is there any loss in generality to assume that the leading coefficients of the polynomials $A(z)$ and $C(z)$ are unity we can, however, not make this assumption for the polynomial $B(z)$. Also, notice that in (2.1) the degrees of all the polynomials $A(z)$, $B(z)$ and $C(z)$ formally are the same. If this is not desired we can put some of the coefficients equal to zero.

The system represented by the equation (2.1) is in fact the general, single-input single-output linear discrete time dynamical system, with normal disturbances having rational power spectra. Notice in particular that systems with time delays also can be represented by the model (2.1) if the time delay is an integer multiple of the sampling interval.

The system model (2.1) contains $4n + 2$ parameters, the $3n + 2$ coefficients of the equation (2.1) $a_1, a_2, \ldots, a_n, b_0, b_1, b_2, \ldots, b_n, c_1, c_2, \ldots, c_n, \lambda$ and n initial conditions for the equation (2.1). The initial conditions add little to the problem and are assumed zero. In practice, it is often necessary to include a constant level as an additional parameter. This adds nothing of interest to the identification problem and is therefore neglected. The complete problem including the initial conditions and the constant level is considered in [1].

2.2. Problem statement

We now formulate the identification problem as follows

PROBLEM

Given the input $\{u(t), t = 1, 2, \ldots, N\}$ and observations of the output $\{y(t), t = 1, 2, \ldots, N\}$ find an estimate of the parameters of the model (2.1).

Special cases of this problem are well-known

1. $n = 0$, regression analysis [6].

2. $b_0 = b_1 = \ldots = b_n = c_1 = c_2 = \ldots = c_n = 0$, estimation of parameters in autoregressive processes [10], [16].

3. $b_0 = b_1 = \ldots = b_n = a_1 = a_2 = \ldots = a_n = 0$, estimation of parameters in a moving average [10], [23], [26].

4. $b_0 = b_1 = \ldots = b_n = 0$, parametric estimation of rational power spectra [2], [7].

5. $c_1 = c_2 = \ldots = c_n = 0$, least squares modelbuilding [12].

6. $a_i = c_i$, $i = 1, 2, \ldots, n$, identification of noisefree process with measurement errors [15].

The general case has been considered by Galtieri [9].

2.3. Minimum variance prediction and control algorithms

Before proceeding to the solution of the stated problem we will demonstrate that once a model of type (2.1) is obtained it is a very simple matter to derive the minimum mean square control algorithm. This will be discussed in detail elsewhere, let it therefore suffice to give an example. For the case of simplicity, we assume that $b_0 = 0$ and $b_1 \neq 0$. Consider the situation at time t. The data $\ldots y(t - 1)$, $y(t), \ldots, u(t - 1)$ have been observed. The crucial step in the derivation of minimum mean square control algorithms is to find the minimum mean square prediction. It is well-known that this problem is solved if we express $y(t + 1)$ as a function of the data $\ldots y(t - 1)$, $y(t), \ldots, u(t - 1)$, $u(t)$ and a residual which is independent of the data. From the equation (2.1) we can immediately obtain such an expression. Solving (2.1) in terms of $y(t + 1)$ we get

$$y(t + 1) = \lambda e(t + 1) + A^{-1}(z^{-1}) B(z^{-1}) u(t + 1)$$
$$+ A^{-1}(z^{-1}) [C(z^{-1}) - A(z^{-1})] \lambda e(t + 1) \qquad (2.4)$$

Eliminating $e(t + 1)$ using (2.1) we get

$$y(t + 1) = \lambda e(t + 1) + C^{-1}(z^{-1}) B(z^{-1}) zu(t)$$
$$+ C^{-1}(z^{-1}) [C(z^{-1}) - A(z^{-1})] zy(t) \qquad (2.5)$$

Due to the assumptions made we find that the series expansion in powers of z^{-1} of the operators $C^{-1}(z^{-1}) B(z^{-1})$ and $C^{-1}(z^{-1}) [C(z^{-1}) - A(z^{-1})]$ have no constant terms. The right member of (2.5) only depends on the data $y(t)$, $y(t - 1), \ldots, u(t), u(t - 1), \ldots$ and $e(t + 1)$. As $e(t + 1)$ is independent of the other terms of the right member we have obtained the desired expression. The last two terms of the equation (2.5) can thus be interpreted as the minimum mean square prediction of $y(t + 1)$ based on the data $y(t)$, $y(t - 1), \ldots, u(t), u(t - 1), \ldots$. The prediction error is $\lambda e(t + 1)$. As $e(t)$ is normal $(0, 1)$ the number λ has physical interpretation as the standard deviation of the prediction error.

Having obtained the minimum mean square predictor we will now derive the minimum mean square control law. We observe that

$$Ey^2(t + 1) \geq \lambda^2 \tag{2.6}$$

where equality holds if

$$u(t) = -B^{-1}(z^{-1})[C(z^{-1}) - A(z^{-1})]y(t) \tag{2.7}$$

The equation (2.7) is thus the minimum mean square control law. As $b_o = 0$ and $b_1 \neq 0$ the series expansion of the operator of the right member does only contain non-negative powers of z^{-1} and (2.7) is thus a physically realizable control law.

Thus we have demonstrated that under the particular assumptions $b_1 \neq 0$, $b_o = 0$ the minimum mean square control algorithm is easily obtained from the model (2.1). Hence once the identification problem is solved we have in fact also a solution to the minimum mean square control problem, and the statement made in the introduction is proven.

SOLUTION

The problem as stated in section 2 is a statistical parameter estimation problem. We will solve it by the method of maximum likelihood. We first give an algorithm for the maximum likelihood estimator and we will later show that the estimates have desirable properties as the number of observations increase.

3.1. The likelihood function

Let $p[\{y(t)\} \mid \{u(t)\}, a, b, c, \lambda]$ be the probability density function of the outputs $\{y(t)\}$ given the inputs $\{u(t)\}$ and the parameters a, b, c, λ. The likelihood function is defined as the function p regarded as a function of the parameters and with the observed values $\{y(t)\}$ and $\{u(t)\}$ inserted [6], [14], [27]. The function is thus a stochastic variable. We will now derive an expression for the likelihood function. It follows from (2.1) that the numbers $\epsilon(t)$ defined by

$$C(z^{-1}) \epsilon(t) = A(z^{-1}) y(t) - B(z^{-1}) u(t) \tag{2.8}$$

are independent and normal $(0, \lambda)$. The logarithm of the probability density function of $\{\epsilon(t)\}$ now becomes

$$L = \frac{1}{2\lambda^2} \sum_{t=1}^{N} \epsilon^2(t) - N \log \lambda + const. \tag{2.9}$$

Since $\{y(t)\}$ is a one-to-one transformation of $\{\epsilon(t)\}$ and the Jacobian is 1, L is also the logarithm of the likelihood function. The logarithm of the likelihood function is thus obtained from

(2.9) where the "errors" ϵ are computed from the input $\{u(t)\}$ and the output $\{y(t)\}$ by (2.8). The likelihood function is thus a function of the parameters a, b, c, λ and of n initial conditions of (2.8). For simplicity we will here assume that the initial conditions of (2.8) are zero. This is not essential. For an analysis of the complete case we refer to [1].

3.2. Maximizing the likelihood function

We observe that the function L can be maximized with respect to the parameters a, b and c separately. To do this we introduce the function $V(\theta)$ defined by

$$V(\theta) = \frac{1}{2} \sum_{t=1}^{N} \epsilon^2(t) \tag{2.10}$$

where $\theta = col (a, b, c)$. Maximizing L is equivalent to minimizing the loss function V. When we have found $\hat{\theta}$ such that $V(\hat{\theta})$ is minimal we get the maximum likelihood estimate of λ from

$$\hat{\lambda}^2 = \frac{2}{N} V(\hat{\theta}) \tag{2.11}$$

and all parameters are estimated. We observe that $V(\theta)$ is a quadratic function of a and b but that the dependence on c is more complex. Thus we cannot obtain an analytical solution.

3.3. Numerical algorithm

To maximize the likelihood function minimize the loss function $V(\theta)$ we use the following Newton-Raphson algorithm

$$\theta^{k+1} = \theta^k - [V_{\theta\theta}(\theta^k)]^{-1} V_\theta(\theta^k) \tag{2.12}$$

where V_θ denotes the gradient and $V_{\theta\theta}$ the matrix of second partial derivatives of $V(\theta)$.

For a discussion of the algorithm (2.12) and related ones see [1], [11].

The partial derivatives of the loss function are obtained by straightforward differentiation. We get

$$\frac{\partial V}{\partial \theta_i} = \sum_{t=1}^{N} \epsilon(t) \frac{\partial \epsilon(t)}{\partial \theta_i} \tag{2.13}$$

$$\frac{\partial^2 V}{\partial \theta_i \partial \theta_j} = \sum_{t=1}^{N} \frac{\partial \epsilon(t)}{\partial \theta_i} \cdot \frac{\partial \epsilon(t)}{\partial \theta_j} + \sum_{t=1}^{N} \epsilon(t) \frac{\partial^2 \epsilon(t)}{\partial \theta_i \partial \theta_j}$$
$$\tag{2.14}$$

The derivatives of $\epsilon(t)$ are obtained by differentiating the difference equation (2.8).

$$C(z^{-1}) \frac{\partial \epsilon(t)}{\partial a_i} = z^{-i} y(t)$$

$$C(z^{-1}) \frac{\partial \epsilon(t)}{\partial b_i} = -z^{-i} u(t) \qquad (2.15)$$

$$C(z^{-1}) \frac{\partial \epsilon(t)}{\partial c_i} = -z^{-i} \epsilon(t)$$

$$C(z^{-1}) \frac{\partial^2 \epsilon(t)}{\partial a_i \partial c_j} = -z^{-i-j} \frac{\partial \epsilon(t)}{\partial a_1}$$

$$C(z^{-1}) \frac{\partial^2 \epsilon(t)}{\partial b_i \partial c_j} = -z^{-i-j} \frac{\partial \epsilon(t)}{\partial b_1} \qquad (2.16)$$

$$C(z^{-1}) \frac{\partial^2 \epsilon(t)}{\partial c_i \partial c_j} = -2z^{-i-j} \frac{\partial \epsilon(t)}{\partial c_1}$$

The initial values for the difference equations (2.15) and (2.16) are zero. Notice that the second order partial derivatives of $\epsilon(t)$ that do not involve the coefficients c_i are all identically zero. Notice that the derivatives of the residuals are equivalent to Meissingers [20] sensitivity coefficients.

The equations (2.8), (2.10), (2.15) and (2.16) immediately suggest a recursive scheme for computing the loss-function $V(\theta)$ and its partial derivatives. Alternatively these functions can be obtained as outputs of linear dynamical systems. Notice that in (2.15) and (2.16) the derivatives with respect to different parameters in the same group (a, b or c) can be obtained by shifting. We get e.g. from (2.15)

$$\frac{\partial \epsilon(t)}{\partial a_i} = \frac{\partial \epsilon(t-i+1)}{\partial a_1} \qquad i \le t+1 \qquad (2.17)$$

This leads to considerable simplifications of the computations as it is only necessary to solve the equations (2.15), (2.16) for $i = j = 1$.

Also notice that by utilizing (2.17) it is also possible to simplify the computation of the first term of the right member of (2.14). We have e.g.

$$\sum_{t=1}^{N} \frac{\partial \epsilon(t)}{\partial a_i} \cdot \frac{\partial \epsilon(t)}{\partial b_j} = \sum_{t=1}^{N} \frac{\partial \epsilon(t-i+1)}{\partial a_1} \cdot \frac{\partial \epsilon(t-j+1)}{\partial b_1}$$

$$= \sum_{t'=1}^{N-j} \frac{\partial \epsilon(t'+j-i)}{\partial a_1} \cdot \frac{\partial \epsilon(t)}{\partial b_1} \qquad j \le i \qquad (2.18)$$

Similar formulas also hold for the other derivatives.

The algorithm for maximizing the likelihood function is thus

1. Put $\theta^k = \theta^0$ (starting value of θ)

2. Evaluate $V_\theta(\theta^k)$ and $V_{\theta\theta}(\theta^k)$ using (2.13), (2.14), (2.15) and (2.16)

3. Calculate θ^{k+1} from (2.12) and repeat from 2.

3.4. Starting value

The algorithm (2.12) requires a starting value. We observe that if the parameters c are given, then $V(\theta)$ is a quadratic function of a and b and the second partial derivatives of $\epsilon(t)$ are all zero. The iteration (2.12) then converges in one step from any initial value for the parameters a and b. In particular if we put $c_i = 0$, $i > 0$ we obtain in one step with the approximative second partial derivatives, the least squares estimate a^0 and b^0 of a and b. The initial value for the iteration (2.12) is then taken as $\theta^0 = \text{col}(a^0, b^0, 0)$.

LARGE SAMPLE PROPERTIES OF THE ESTIMATES

When the identification scheme is applied there are many problems which naturally arise. Typical examples are the following:

- What happens to the estimate as the number of observations (N) increases?

- How accurate is the estimate?

- Are there "better" ways of estimating the parameters?

- What systems are possible to identify?

- In cases we can choose the input signal, how should it be chosen?

- What order should be chosen for the model?

- Does the solution obtained correspond to the absolute maximum?

In this section we develop some means for dealing with such questions. Many of these are essentially answered by an investigation of the large sample properties of the estimates. It means that asymptotic expressions can be utilized. The problem of several local maxima, however, cannot be solved by the results of this section. In the sequel, we disregard it, and hence assume that $\hat{\theta}^N$ for all sufficiently large N is the parameter value corresponding to the absolute maximum of the likelihood function.

The investigation of the large sample properties is a purely statistical problem, and we will use concepts and methods from mathematical statistics to find them. The complete investigation is uncomfortably involved and detailed, and we have chosen to omit it from this paper. However, the results are presented below in the form of mathematical theorems, which are commented on with regard to their application to some of the above problems. The proofs are published in an IBM report [1].

To faciliate the reading we restate and define more closely some of the problems and introduce a few statistical concepts.

- Consistency, i.e. $\hat{\theta}^N$ converges to θ_o when N increases.

- Asymptotic normality, i.e. convergence in distribution of the quantity $\sqrt{N} (\hat{\theta}^N - \theta_o)$ to a normal variable.

- Asymptotic efficiency, i.e. equality of the covariance matrix of the limiting distribution to the Cramér-Rao lower bound for regular estimators.

For further definitions of terms see [27].

We want to have conditions that guarantee these properties. In the present case the properties depend on the input and the parameters θ_o.

Since θ_o is not known, the conditions should preferably be expressed in terms of the input u and the output y which are both known quantities. If this is possible we are then able to resolve whether a certain sequence of estimates $\{\hat{\theta}^N\}$ actually has one of the desired properties in each case. It is evident that this can be done with certainty only if the sample is infinitely long, and we will confine ourselves to this case when defining the required conditions. This means that we will not consider here the problem of constructing statistical tests for finite sample lengths N.

The consistency conditions can be used to resolve the following important problem: To be able to design an experiment in order to estimate θ_o we must know what class of input sequences u that are able to excite the system sufficiently enough to yield consistent estimates of the system parameter θ_o. Since at this stage we do not know θ_o (or y) we can utilize only u such that the estimates $\{\hat{\theta}^N\}$ are consistent irrespective of the value θ_o. We are interested in such (smaller) classes of u.

The results of this section solve the following problems:

1. What set in the parameter space does $\hat{\theta}^N$ converge into (Theorem 1 + Lemma 3).

2. In cases where $\{\hat{\theta}^N\}$ is not consistent, find (singular) function of $\hat{\theta}^N$ that is consistent (Theorem 2).

3. Find a class of inputs u and of system parameters θ_o such that $\{\hat{\theta}^N\}$ is consistent (Theorem 3).

4. Find conditions on u and y for $\{\hat{\theta}^N\}$ to be consistent (Corrolary, theorem 2).

5. Find conditions such that $\{\hat{\theta}^N\}$ is asymptotically normal and asymptotically efficient (Theorem 4).

6. When $\{\hat{\theta}^N\}$ is asymptotically normal and efficient find an estimate of the covariance matrix of the limiting distribution (Lemma 4).

The theorems are in a sense ergodic theorems, since they all deal with asymptotic properties of functions of a single sample y. The ergodic property establishes that a single realization of the process output can be used in place of an ensemble of realizations.

The following is a general regularity condition on the input sequence $\{u(t)\}$

Condition A

u(t) and the crossproducts u(t) u(t+T) be bounded and Cesaro summable, i.e. the following limits

$$\lim_{N \to \infty} \frac{1}{N} \sum_{t=1}^{N} u(t)$$

$$\lim_{N \to \infty} \frac{1}{N} \sum_{t=1}^{N} u(t) u(t+T)$$

exist for all finite T.

In the sequel it is necessary to distinguish between θ = an arbitrary point in the parameter space and θ_o = the true parameter point, i.e. the parameter point defining the observed output according to (2.1).

Denote the logarithm of the likelihood function by $L^N(y|\theta)$, its gradient vector $L_\theta^N(y|\theta)$, and its second derivative matrix $L_{\theta\theta}^N(y|\theta)$. Further, denote the vector of length N with components y(i) by y. Analogously for u.

4.1. Consistency

Lemma 1

Let R be a region in 3n+2 dimensional Euclidian space E^{3n+2} defined by

$$R = \{\theta \mid \lambda > 0, \text{ and all zeros of } A(z^{-1}) \text{ and } C(z^{-1}) \text{ lie strictly inside the unit circle}\}$$

Assume that u satisfies the condition A. Then

$$\lim_{N\to\infty} \frac{1}{N} L^N(y \mid \theta) = \lim_{N\to\infty} \frac{1}{N} E L^N(y \mid \theta) = L(\theta, \theta_o)$$

with probability one if $\theta \in R$ and $\theta_o \in R$.

Lemma 2

Let $R' \subseteq R$ be a closed set.

Assume u satisfies condition A.

Then $L(\theta, \theta_o)$ is an analytic function in R', and we have

$$\lim_{N\to\infty} \frac{1}{N} \operatorname{grad}_\theta L^N(y \mid \theta) = \operatorname{grad}_\theta \lim_{N\to\infty} \frac{L^N(y \mid \theta)}{N} = L_\theta(\theta, \theta_o)$$

with probability one. The relation also holds for higher derivatives.

The lemmas establish that the time average of the residuals $\epsilon^2(t)$ i.e. $\frac{1}{N} L^N(y \mid \theta)$ converges to its ensemble average, which is a differentiable function in the parameters θ. The conditions are mild and natural, i.e. the system, the model, and the optimal predictor of the noise component $C(z^{-1})e$ be asymptotically stable.

The lemmas are fundamental for

Theorem 1

Let S_o be a set in E^{3n+1} defined by

$$S_o = \{\theta \mid L(\theta, \theta_o) = L(\theta_o, \theta_o)\}$$

Assume u satisfies condition A, and that for all sufficiently large N, $\hat{\theta}^N \in R'$, where $R' \subseteq R$ is a closed set. Then

$$\lim_{N\to\infty} \| \hat{\theta}^N - P\hat{\theta}^N \| = 0$$

with probability one, where $P\theta$ is the projection on $S_o \cap R'$, i.e. the nearest point $\in S_o \cap R'$.

This theorem replaces a consistency theorem. It asserts that the estimates $\hat{\theta}^N$ converge into the set S_o, though not necessarily to a point.

It gives the consistency conditions in terms of conditions on (u, θ_o) through the

Corrolary

If the set S_o contains only one point (θ_o), $\{\hat{\theta}^N\}$ is strongly consistent.

The set S_o can be interpreted as the set of parameters θ that are equivalent to θ_o in the sense that any model with $\theta \in S_o$ generates realizations that for long samples have the same statistical behavior (same likelihood function) as the system output y. The condition is then natural since there is no way to judge from the output only which of the parameters $\theta \in S_o$ that generated the observed output.

The purpose of the following theorems is to characterize S_o and find conditions for S_o to contain only the point $\theta = \theta_o$.

Lemma 3

The set S_o, as defined in theorem 1 has the following property

$$S_o = R \cap S_o'$$

where S_o' is a linear set.

Hence, we know that in all cases $\hat{\theta}^N$ will at least converge into a hyperplane S_o. This suggests that components of $\hat{\theta}^N$ orthogonal to this hyperplane will be consistent. We want to be able to calculate such components. This can be done with the aid of

Theorem 2

Let $\Lambda^N(y \mid \hat{\theta}^N)$ be the diagonal matrix of eigenvalues of $\frac{1}{N} L^N(y \mid \hat{\theta}^N)$ and let $P^N(y \mid \hat{\theta}^N)$ be a matrix of corresponding (orthogonal) eigenvectors. Then

$$\lim_{N\to\infty} \| \Lambda^N(y \mid \hat{\theta}^N) P^{N^T}(y \mid \hat{\theta}^N)(\hat{\theta}^N - \theta_o) \| = 0$$

with probability one.

The rather complicated form of this theorem is due to our desire to express the projections in computable terms. The main difficulty arises from the fact that it cannot be shown that

$$\frac{1}{N} L^N_{\theta\theta}(y \mid \hat{\theta}^N) \text{ converges. Even if this is the}$$

case, the limit may have no unique set of eigenvectors so that it is difficult or impossible to define a convergent sequence of orthogonal transformations $\{P^N(y|\hat{\theta}^N)\}$. This is the case particularly if the limit is singular with at least two eigenvalues zero. The singular case is of practical interest, since it arises from choosing a too high order model or from the fact that the system is degenerate (not controllable) or not excited (see theorem 3). In such cases we can then use theorem 2 to find a more reasonable model by taking the projections orthogonal to S_o' as new parameters. The estimates of these new parameters are then consistent.

From theorem 2 we immediately obtain a solution of the problem 4) posed above through the

Corrolary

If $[\frac{1}{N} L_{\theta\theta}^N(y|\hat{\theta}^N)]^{-1}$ is bounded then $\hat{\theta}^N$ is consistent.

It is not shown that the converse is necessarily true. When calculating $\hat{\theta}^N$ the matrix $L_{\theta\theta}^N(y|\hat{\theta}^N)$ is actually computed. Each estimate is thus accompanied by a quantity which can be used to judge its significance. A complete characterization of S_o in terms of (u, θ_o) is given in the proof of theorem 3 [1].

4.2. Identifiability

From the equations characterizing S_o it is possible, at least in principle, to resolve whether they possess a unique solution $\theta = \theta_o$ in which case $\hat{\theta}^N$ is consistent. We can also construct conditions for this being the case. These conditions are complicated and impractical for applications. It is desirable to find simpler conditions, possibly more restrictive, that are sufficient to ensure that S_o contains only one point. The corresponding theorems will be called identifiability theorems, because they give conditions for the system to be identifiable from the input/output record, in the sense that the parameter estimates are consistent. It is attractive to think of the conditions in the following terms:

A system at rest is defined by the parameters θ_o according to (2.1), $u = 0$ and regarded as a black box containing the unknown parameters θ_o. In order to draw conclusions about the contents of this black box it is necessary to excite the system by applying some $u \neq 0$ and observe the response. The input must excite all components of θ_o and must be sufficiently persistent, since the response is obscured by noise. It is evident that parameter components that cannot

be reached by the input (or the noise) can never be estimated. Hence, some controllability requirements are also needed.

Definition

A process is said to be completely identifiable if the maximum likelihood estimates of θ_o are consistent.

Definition

A bounded signal u is said to be persistently exciting of order m if the limits

$$\bar{u} = \lim_{N\to\infty} \frac{1}{N} \sum_{t=1}^N u(t) \quad \text{and} \quad r_u(T) = \lim_{N\to\infty} \frac{1}{N} \sum_{t=1}^N u(t)u(t+T)$$

exist and if the matrix

$$R_u = \{ r_u(i-j) \mid i, j = 1, \ldots, m+1 \}$$

is positive definite.

We can now state the main result.

Theorem 3

The process (2.1) is completely identifiable if the input is persistently exciting of order 2n and every state of the process is controllable either from u or e.

The first condition is easy to verify in practice, since u is known. The second condition is of less importance in a practical case, we can never verify it before the experiment and we can always verify it after by means of theorem 2. It is, however, useful as a means of diagnosis: Why are the estimates inconsistent?

4.3. Asymptotic normality

Theorem 4

Assume that S_o contains only the point θ_o so that $\hat{\theta}^N$ is consistent. Then the stochastic variable $L_{\theta\theta}(\theta_o, \theta_o)\sqrt{N}(\hat{\theta}^N - \theta_o)$ is asymptotically normal $(0, -L_{\theta\theta})$.

If in addition $L_{\theta\theta}(\theta_o, \theta_o)$ is nonsingular, then $\hat{\theta}^N$ is asymptotically normal $(\theta_o, -\frac{1}{N} L_{\theta\theta}^{-1})$.

Since $NL_{\theta\theta}(\theta_o, \theta_o) \sim L_{\theta\theta}^N(y|\theta_o)$ = the information matrix, the estimates are also asymptotically efficient.

The fact that the estimates are asymptotically efficient means in practice that we cannot expect to find an estimator with a greater accuracy for long samples.

The asymptotic normality implies that the distribution of $\hat{\theta}^N$ is completely known that confidence regions for the parameters can be determined and that approximate significance tests can be performed. To perform the tests an estimate of the covariance matrix is required. This is obtained from

Lemma 4

Assume that lemma 2 and theorem 1 hold. Then

$$\| \frac{1}{N} L_{\theta\theta}^N(y|\hat{\theta}^N) - L_{\theta\theta}(P\hat{\theta}^N, \theta_o)\| \to 0$$

with probability one.

Besides solving the consistency problem, the matrix $\lim_{N\to\infty} \frac{1}{N} L_{\theta\theta}^N(y|\hat{\theta}^N)$ then also yields an estimate of the accuracy of the estimated parameters in cases where the parameter estimates are consistent.

In practice we do not have an infinite sample and then the result means that whenever we can invert the matrix $\frac{1}{N} L_{\theta\theta}^N(y|\hat{\theta}^N)$ without difficulty, we may consider the obtained estimate $\hat{\theta}^N$ as consistent. The accuracy of the estimate is given approximately by the inverse of the matrix. If the matrix is nearly singular then the inverse will always contain some very large diagonal elements, and we may either say that the corresponding component of $\hat{\theta}^N$ is not consistent, or it is consistent but with a very large standard deviation. The practical result is the same, namely the conclusion that we have included too many parameters.

4.4. Tests of the order of model

According to Theorem 2, the second derivative matrix $L_{\theta\theta}^N(y|\hat{\theta}^N)$ can be used as an indication that there are redundant parameters in the model and also to determine a new, non-redundant set of parameters. Hence, we may in practice determine the order of the model by repeating the identification with increasing order using some measure of singularity of the matrix as a test figure. However, the following alternative may sometimes be preferable:

We observe that if the model order is not less than the order of the system then the residuals $\{\epsilon(t), t = 1, \ldots, N\}$ form a series of independent normal variables. Obviously the converse is also true, and we have then another test on the order:

If the residuals form a sequence of independent variables then the order of the model is equal to or greater than the system order. A simple test of independence is to compute the covariance function

$$\frac{1}{N-T} \sum_{t=1}^{N-T} \epsilon(t)\, \epsilon(t+\tau) \text{ for a few delays.}$$

$$\tau = 1, 2, 3, \ldots .$$

A quick method is to count the sign changes, the number of which should be $\approx \frac{1}{2} N$ for a sequence of independent variables.

EXAMPLE

As an example we will consider the identification of the following system

$$y(t) = \frac{z^{-1} + 0.5 z^{-2}}{1 - 1.5 z^{-1} + 0.7 z^{-2}} u(t)$$

$$+ \lambda \frac{1 - z^{-1} + 0.2 z^{-2}}{1 - 1.5 z^{-1} + 0.7 z^{-2}} e(t) \qquad (5.1)$$

Three cases are considered

1. $\lambda = 0.4$
2. $\lambda = 1.8$
3. $\lambda = 7.2$

In the experiment 240 pairs of input/output variables (u, y) were generated using the equation (5.1). The random numbers $\{e(t)\}$ were obtained as suitably scaled sums of twelve rectangularly distributed pseudorandom numbers obtained from a modified Fibonacci series. The same sequence of pseudorandom numbers were used in all three cases. In Figure 1 we show the chosen input and the output y in the three cases. As a reference we have also in each case shown the output for $\lambda = 0$.

The identification scheme described in this paper was applied to the generated data. The estimates of the model parameters obtained are given in Table 1. In this table we also give the standard deviations of the estimates which are computed from the estimate of the covariances given by the matrix $\hat{\lambda}^2 V_{\theta\theta}^{-1}$.

From this table we find that the estimates of the coefficients b_1 and b_2 are getting increasingly inaccurate in the experiments while the accuracy of the coefficients a and c are unaffected. This is very natural since the response of the deterministic part of the model is corrupted by an increasing amount of noise. The estimates of the parameters c_1 and c_2 should in general not be expected to depend on the noise amplitude.

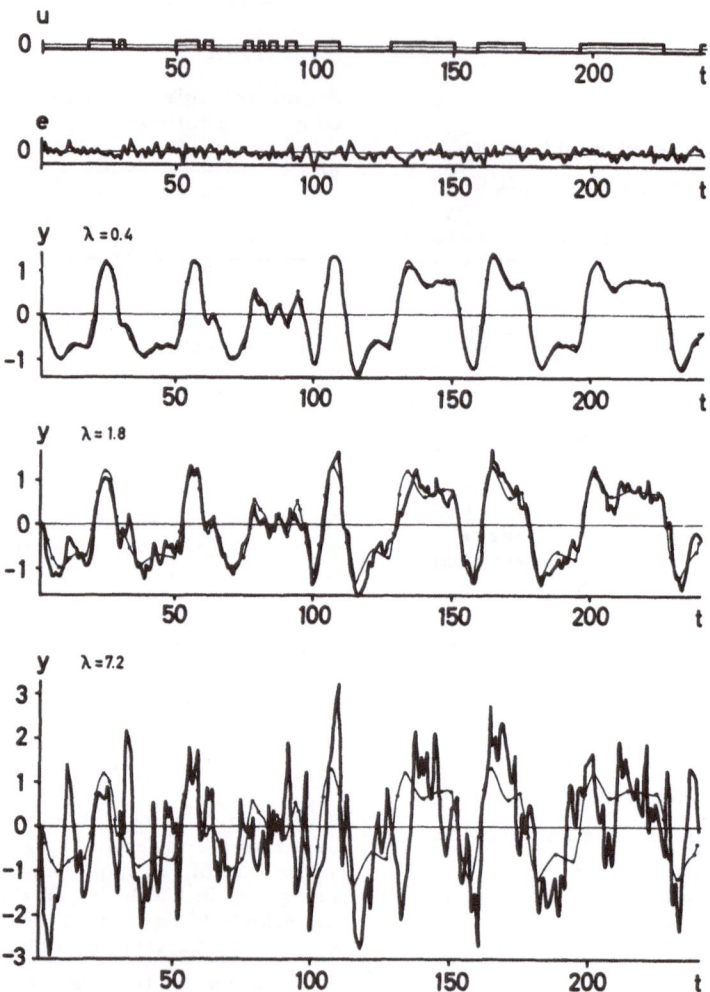

Figure 1   Input sequence, sequence of random numbers, and
            output sequences for test example.

TABLE 1

Estimated and true parameter values for cases 1, 2 and 3

Parameter	CASE 1	CASE 2	CASE 3	TRUE
$-a_1$	1.512 ± 0.008	1.544 ± 0.03	1.586 ± 0.06	1.500
a_2	0.705 ± 0.005	0.720 ± 0.02	0.722 ± 0.06	0.700
b_1	1.025 ± 0.04	1.161 ± 0.16	1.338 ± 0.6	1.000
b_2	0.413 ± 0.05	0.076 ± 0.2	-0.313 ± 0.6	0.500
$-c_1$	0.978 ± 0.06	1.015 ± 0.07	1.039 ± 0.10	1.000
c_2	0.158 ± 0.06	0.151 ± 0.07	0.143 ± 0.07	0.200
λ	0.419 ± 0.019	1.880 ± 0.08	7.572 ± 0.3	

TABLE II

Successive estimates of the parameters

STEP	c_1	c_2	b_1	b_2	$a_1 - c_1$	$a_2 - c_2$	LOSS FUNCTION
0	0.000000	0.000000	0.000000	0.000000	0.000000	0.000000	17794.97
1	0.000000	0.000000	1.793699	1.215727	-0.669223	0.067462	7696.58
2	-0.953107	0.036294	1.939274	-1.258858	-0.658175	0.654775	7162.93
3	-0.992611	0.108536	1.370642	-0.282776	-0.558668	0.583354	6891.57
4	-1.038508	0.134742	1.389544	-0.396304	-0.549532	0.585416	6882.01
5	-1.035053	0.142406	1.332628	-0.294744	-0.548205	0.578284	6880.34
6	-1.038635	0.142974	1.337403	-0.313218	-0.547242	0.578912	6880.12
7	-1.038668	0.143086	1.337638	-0.313263	-0.547204	0.578810	6880.12
8	-1.038671	0.143088	1.337638	-0.313265	-0.547203	0.578809	

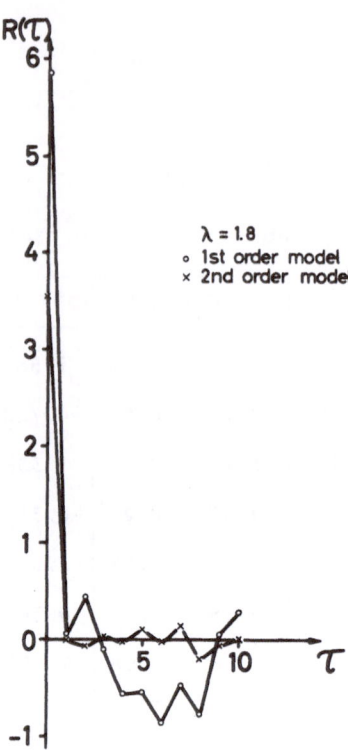

Figure 2   Covariance functions for residuals of first and
second order models of second order system.

TABLE III

First order partial derivatives of the loss function

STEP	$\dfrac{\partial V}{\partial c_1}$	$\dfrac{\partial V}{\partial c_2}$	$\dfrac{\partial V}{\partial b_1}$	$\dfrac{\partial V}{\partial b_2}$	$\dfrac{\partial V}{\partial (a_1 - c_1)}$	$\dfrac{\partial V}{\partial (a_2 - c_2)}$
0	0.0000	0.0000	1339.5956	-1450.3118	—	—
1	197.1321	-704.2079	0.0001	0.0000	-0.0001	-0.0049
2	9352.8641	8950.7504	532.4016	513.5850	—	—
3	-479.7225	-960.5074	78.6591	97.8004	1007.8257	1459.4972
4	-167.4281	-163.6832	-20.8965	-27.5368	241.4333	300.0271
5	8.0683	-8.4731	14.5029	15.7974	-107.2980	-124.7181
6	3.0052	2.3351	-0.6982	-0.6224	8.6651	9.0144
7	0.0131	-0.0039	-0.0010	0.0000	0.0030	0.0011

To demonstrate the convergence of the algorithm we give in table II the successive iterates for Case 3 and in table III the gradients of the loss function in the various iteration steps. Notice in table II the large difference between the least squares estimate (step 1) of a_1, a_2, b_1 and b_2 and the maximum likelihood estimate.

To illustrate the test of the order of the system as discussed in section 4, the data of Case 2 was also identified as a first order system. The covariances of the residuals for the first and second order models are graphed in Figure 2.

COMPARISON WITH MODEL ADJUSTMENT TECHNIQUES

There are many ways to approach the identification problem. Two approaches are represented by

- statistical parameter estimation
- model adjustment

In the first case the problem is put in a probabilistic framework and sufficiently many assumptions are made in order to insure that the methods of mathematical statistics can be applied. In the model adjustment technique [4], [17], [25] a model characterized by some parameters and a criterion are postulated. The problem is then to adjust the model parameters in such a way that the criterion is satisfied. The model adjustment technique is very general in the sense that complicated models and criteria can be used. The result of the model adjustment procedure is a set of parameter values. When the problem is approached as a parameter estimation problem many more assumptions must be made, instead the results are much more far reaching. The parameters as well as their confidence intervals are obtained, questions related to significance of the estimates can be answered. In many situations it is rewarding to consider a particular problem from both points of view. Statistical considerations may suggest a suitable criterion for the model adjustment procedure. In a situation where the assumptions required by the statistical approach are not fulfilled we can still obtain a solution to a model adjustment problem. A typical case may be the situation when the assumption on normality of the residuals is not fulfilled.

Classically, the two techniques have been developed in parallel. Let is suffice to mention least squares fitting of linear models.

So far in this paper the identification problem has been discussed entirely from the point of view of statistical parameter estimation. We will now interpret our procedure as a model adjustment technique.

Consider the equation (2.4). We recall that the last two terms of the right member of this equation can be interpreted as the prediction of $y(t+1)$ based on $y(t), y(t-1), \ldots$ and $u(t), u(t-1), \ldots$ and that the quantity $\lambda \epsilon(t+1)$ has physical interpretation as the error of the one-step ahead prediction of $y(t)$. Now consider the equation (2.8) which we rewrite as

$$\epsilon(t+1) = y(t+1) - \{C^{-1}(z^{-1})[C(z^{-1}) - A(z^{-1})]y(t+1)$$
$$+ C^{-1}(z^{-1}) B(z^{-1})u(t+1)\} \qquad (6.1)$$

A comparison with the equation (2.4) now shows that the last two terms of the right member can be interpreted as the prediction of $y(t+1)$ based on $y(t), y(t-1), \ldots$ and $u(t), u(t-1), \ldots$. In the algorithm (2.8) the number $\epsilon(t+1)$ can thus be interpreted as the difference between $y(t+1)$ and its one step ahead prediction based on $y(t), y(t-1), \ldots$, and $u(t), u(t-1), \ldots$.

Now consider the one step ahead predictor

$$\hat{y}(t) = C^{-1}(z^{-1}) B(z^{-1})zu(t-1)$$
$$+ C^{-1}(z^{-1})[C(z^{-1}) - A(z^{-1})]zy(t-1) \qquad (6.2)$$

as the model, and let the criterion be the sum of the squares of the prediction errors i.e.

$$V = \frac{1}{2} \sum_{t=1}^{N} [y(t) - \hat{y}(t)]^2 = \frac{1}{2} \sum \epsilon^2(t) \qquad (6.3)$$

Compare with equations (2.5) and (2.10). Notice that in the ordinary model reference techniques the model is a deterministic input/output relation while in our case the model is a predictor.

As we intend to use the results of the identification procedure to calculate the minimum variance control strategy i.e. a control strategy such that the one-step ahead prediction of the output is zero we find that even with the model adjustment interpretation our identification procedure has the required properties. Also notice that the identification algorithm solves the prediction problem for a stationary process with unknown, but rational power spectrum.

It is also of interest to compare the algorithm (2.12) with the algorithms currently used in model reference techniques. Blandhol [4] only evaluates the function $V(\theta)$. Judging from our experience it is very difficult to get a reasonable convergence rate by probing techniques using the values of $V(\theta)$. Blandhol also confirms this.

Notice that $V(\theta) / V(\theta_o)$ is asymptotically independent of N. This implies that the loss function does not get "sharper" with an increasing number of observations and that the "sharpness" of the minimum <u>alone</u> does not determine the accuracy of the estimates.

The gradient $V_\theta(\theta)$ is evaluated in some model reference techniques that are implemented in adaptive systems e.g. [17], [20]. In these cases the parameter adjustment routine is chosen as

$$\theta^{k+1} = \theta^k - \alpha V_\theta(\theta^k)$$

Notice that a more effective algorithm is obtained with very little extra computational effort, using an approximate second partial derivative i.e. neglecting the last term of (2.14). We then conclude that it appears worthwhile to consider this modification in model reference adaptive systems currently in use. By including this feature we would also obtain an estimate of the information matrix and thus also of the accuracy of the estimated parameters.

EXTENSIONS

There are many ways in which the problem can be generalized. The identification scheme can be immediately generalized to continuous time. The convergence proofs are, however, more difficult in this case and some modifications might be necessary.

The extension to multiple inputs is trivial. Both the algorithm and the convergence proofs generalize immediately. The extension to multiple outputs is more difficult. The crucial problem is to find a suitable structure. Once the structure is given, the generalization is immediate.

The algorithm can also be extended directly to non-linear and/or time variable systems with known structure. Consider for example, the following system:

$$x(t + 1) = g(x(t), u(t), t)$$

where $g(x, u, t)$ is a function which contains some unknown parameters. Let u be the input(s) of the system and let the output y be given by

$$y(t) = x_1(t) + z_1(t) + c_o e(t) + \varkappa$$

where the vector $z(t)$ is given by

$$z(t + 1) = Fz(t) + Ge(t)$$

The system described by these equations is an arbitrary non-linear system with a single output, with a random disturbance in the output that is stationary and has spectrum of order 2m. The problem is to identify the unknown parameters of the function $g(x, u, t)$, the constants c_o and $\varkappa$, and the elements of the matrices F and G. This identification problem can be solved immediately using the technique described in the report. To obtain the likelihood function we first write $e(t)$ as a function of the inputs and the observations. We get

$$x(t + 1) = g(x(t), u(t), t)$$

$$z(t + 1) = Fz(t) + \frac{1}{c_o} G[y(t) - x_1(t) - z_1(t) - \varkappa]$$

$$\varepsilon(t) = c_o e(t) = y(t) - x_1(t) - z_1(t) - \varkappa$$

and the logarithm of the likelihood function is

$$-L(y; \theta) = \frac{1}{2c_o^2} \sum_{t=1}^{N} \varepsilon^2(t) + N \log c_o + \frac{N}{2} \log 2\pi$$

We can now proceed in exactly the same way as was done in section 3 to obtain an algorithm to maximize $L(y | \theta)$.

The results can also be generalized in a different direction. So far. we have assumed that the estimate should be calculated from a complete record of inputs and observations. Such a situation is referred to as <u>off-line</u> estimation. In certain applications, particularly in connection with adaptive control, the problem is different because the inputs and outputs are obtained recursively in time. This situation is referred to as <u>on-line</u> identification. Due to the recursive structure of the computation scheme only minor modifications are required to obtain an on-line identification. Some preliminary numerical experiments with very encouraging results have been performed.

The choice of model structure has not been discussed in this paper. There is, however, one point we would like to comment upon. The important feature of the model (2.1) is that it contains only one noise source $e(t)$. This is essential, for the reason that it enables us to solve (2.1) for $e(t)$ in terms of $y(t)$ and $u(t)$. According to the representation theorem for stationary random processes it is always possible to find a representation such as (2.1). However, in many cases a different representation would appear more natural. Consider for example the case when there are independent measurement errors. In such a case we would obtain a model of the type

$$A(z^{-1})\, x(t) = B(z^{-1})\, u(t) + \lambda C(z^{-1})\, e(t)$$

$$y(t) = x(t) + \mu v(t) \qquad\qquad (7.1)$$

where $\{e(t)\}$ and $\{v(t)\}$ are sequences of independent equally distributed $(0,1)$ random variables. The disturbances $e(t)$ and $v(t)$ represent process disturbances and measurement errors. In the model (7.1) we thus have two noise sources and one output. To solve the identification problem for the model (7.1) we can proceed by the method at maximum likelihood. The negative logarithm of the likelihood function is

$$-\log L = \frac{1}{2\lambda^2} \sum_{t=1}^{N} \epsilon^2(t) + \frac{1}{2\mu^2} \sum_{t=1}^{N} [y(t) - x(t)]$$

$$-\frac{1}{2} N \log \lambda\mu + \text{const.}$$

where

$$C(z^{-1})\, \epsilon(t) = A(z^{-1})\, x(t) - B(z^{-1})\, u(t)$$

Analyzing the details, we find that the problem of maximizing the likelihood leads to a two point boundary value problem for the equation (7.1) and its adjoint. The computational aspects of this have been investigated and tried. We have found that the computations are much more involved and time consuming than the corresponding computations for the model (2.1). It is also much easier to solve the minimum variance control problem for the model (2.1). The reason for this is that the necessary spectral factorization has already been carried out.

ACKNOWLEDGEMENT

The authors would like to thank Mr. S. Wensmark of the IBM Nordic Laboratory for programming and directing tests on several identification programs, and Mr. R.W. Koepcke of IBM Research for many stimulating discussions.

BIBLIOGRAPHY

[1] Åström, K.J., Bohlin, T. & Wensmark, S. Automatic construction of linear stochastic dynamic models for stationary industrial processes with random disturbances using operating records. IBM Nordic Laboratory, Sweden. June 1, 1965 (Report TP 18.150).

[2] Åström, K.J., Control Problems in Paper Making. IBM Scientific Computing Symposium: Control Theory and Applications, New York, Oct. 20, 1964.

[3] Bellman, R.E. & Kalaba, R.E., Quasilinearization and nonlinear boundary-value problems. New York, Elsevier, 1965.

[4] Blandhol, E., On the use of adjustable models for determination of system dynamics. Institutt for Reguleringsteknikk, NTH. Trondheim, Norway. March 1962. (Technical report no. 62-5-D).

[5] Box, G.E.P., Fitting empirical data. Annals New York Academy of Sciences. 86 (1960): p. 792-816.

[6] Cramér, H., Mathematical methods of statistics. Princeton Univ. Press, (1946) 1958.

[7] Durbin, J., Efficient fitting of linear models for continuous stationary time series from discrete data. Bulletin de l'Institut International de Statistique. Tokyo. 38 (1960): 4, p. 273-282.

[8] Eykhoff, P., Process parameter estimation. Progress in control engineering. Ed. R.H. MacMillan. Vol. 2. London, Heywood, 1964, p. 161-208.

[9] Galtieri, C.A., Problems of estimation in discrete-time processes. IBM Research Laboratory. San Jose, Calif. Aug. 26, 1964. (Report RJ-315).

[10] Grenander, U., Stochastic processes and statistical inference. Arkiv för Matematik. 1 (1950): 17, p. 195-277.

[11] Kale, B.K., On the solution of likelihood equations by iteration processes. The multiparametric case. Biometrica. 49 (1962): p. 479-486.

[12] Kalman, R.E., Design of a self-optimizing control system. ASME Trans. 80 (1958): 2, p. 468-478.

[13] Kalman, R.E., Mathematical description of linear dynamical systems. J. on Control. 1 (1963): 2, p. 152-192.

[14] Kendall, M.G. & Stuart, A., The advanced theory of statistics. Vol. 2. London, Griffin, 1961.

[15] Levin, M.J., Estimation of the characteristics of linear systems in the presence of noise. D.Sc. thesis. Columbia Univ., New York, 1959.

[16] Mann, H.B. & Wald, A., On the statistical treatment of linear stochastic difference equations. Econometrica. 11 (1943): 3 & 4, p. 173-220.

[17] Margolis, M. & Leondes, C.T., On the theory of adaptive control systems: The learning model approach. 1st International Congress IFAC. Automatic and remote control. Moscow. 1960. Proceedings, Vo. 2, p. 556-563.

[18] Maslov, E.P., Application of the Theory of Statistical Decisions to the Estimation of Object Parameters , Automation and Remote Control, 24 (1964) 1214-1226.

[19] Mayne, D.Q., Estimation of system parameters. Symposium on optimal control. Imperial College of Science and Technology, London. 1964. Paper No. 9.

[20] Meissinger, H.F., Parameter influence coefficients and weighting functions applied to perturbation analysis of dynamic systems. 3rd International analogue computation meetings. Opatija, Yugoslavia. 1961. Proceedings, p. 207-216.

[21] Mishkin, E. & Brown, L. (editors) Adaptive Control Systems, McGraw Hill, New York, 1961.

[22] Ragazzini, J.R. & Franklin, G.F., Sampled data control systems, New York, McGraw Hill, 1958.

[23] Walker, A.M., Large-sample estimation of parameters for moving average models. Biometrica. 48 (1961): p. 343-357.

[24] Westcott, J.H., The parameter estimation problem. 1st International Congress IFAC. Automatic and remote control. Moscow. 1960. Proceedings. Vol.2. p. 779-787.

[25] Whitaker, H.P. et al., Design of a model reference adaptive control system for aircraft. MIT Instrumentation Laboratory. Boston. 1958. (Report R-164).

[26] Whittle, P., Estimation and information in stationary time series. Arkiv för Matematik. 2 (1952): 23, p. 423-434.

[27] Wilks, S.S., Mathematical statistics. New York, Wiley, 1962.

Discussion

P. Eykhoff (Netherlands) asked what was the reason for the choice of the maximum likelihood procedure among the many possible methods of parameter estimation available, such as least squares, maximum likelihood, or minimum risk estimates.

Prof. Åström said that this was a question he and his co-author were forced to answer in their search for a practical method which was quick yet gave good results. They began by trying least squares, as it is the simplest method, but found it unsatisfactory because the estimates were biased if the noise was high. The maximum likelihood was tried next, and was found to have the desired consistency and asymptotically efficient properties discussed in the text, which is the best which can be hoped for. If the data is quite clean (i.e. there is little noise involved) the least squares gives the same answer, and indeed, any method will give a good answer, but if the data is noisy, then the maximum likelihood method is much better.

Dr. F. R. Himsworth (U.K.) asked how necessary is the assumption of normal errors, that is, how sensitive is the procedure to the form of the disturbance?

Prof. Åström replied that the assumption of normality is very necessary for the theory, in order to get an analytic form for the likelihood function, but in practice it is not too important. With simulated data, the authors have used rectangularly distributed errors, and the method has evaluated out the value of the coefficients very well.

Mr. D. W. Mayne (U.K.) considered that the method proposed for system identification is excellent in that it is very simple compared to other methods of estimating the parameters of state vector models. He asked whether these results can be extended to multi-input, multi-output systems and thought that this would depend on whether a suitable form of the model could be obtained.

Prof. Åström stated that the procedure can be extended immediately to multi-input systems by adding appropriate terms to equation 2.1. However, for multi-output systems, the problem is more difficult as there is no general structure of the form of equation 2.1, and no general results can be obtained. If such a structure can be proposed, however, which contains a certain number of parameters, then the method can be applied.

Written contributions to the discussion

Mr. R. J. High (U.K.)

Further to Professor Åström's comments about the bias he obtained when using least squares analysis of recorded plant data, I would like to point out that the use of filters in reference below does substantially reduce the high frequency components of the noise and appears to allow adequate characterization of plant dynamics and I would be glad of his comments upon this.

Author's reply

There are two reasons why we do not use prefiltering

1. There is always a large amount of arbitrariness in the choice of filters.

2. Our reason for solving the identification problem is that we want to obtain strategies for controlling a process. If prefiltering is used we unnecessarily introduce arbitrarily chosen dynamic elements into the control law.

Reference

Bray, J.W., High, R.J., McCann, A.D. and Jemmeson, H. "On-line model making for a chemical plant", Transactions of the S.I.T., Vol. 17, No.3. pp.65 - 75, September 1965

See also Mr. R.J. High's discussion of Mr. H.A. Barker's paper in this volume.

MULTIVARIABLE MODELS FOR CONTROL SYSTEMS

by H. Anthony Barker
Lecturer in Electrical Engineering
The University of Glasgow
Glasgow, Scotland

ABSTRACT

The determination of a model for a multivariable system is an essential part of the control problem. This paper describes a method by which such a model may be realised and made to self-adjust in sympathy with variations of the system parameters. Special criteria are given for the synthesis of linear models, and a method for using these models in optimization schemes is suggested.

INTRODUCTION

Most formulations of the control problem contain implicitly the problem of obtaining a model of a multivariable system. Such a model, whether mathematical or physical, is obtained either through the laws governing a known system structure or through the responses of a system to known external stimuli. In the latter case, the structure of the model is not necessarily restricted to resemble that of the system and the modelling problem becomes that of determining a structure which, when subjected to the same external stimuli as the system, produces the same responses.

MODELLING THEORY

The models described here are applicable to systems in which external stimuli u_1, u_2, ..., ..., u_ℓ of a common type produce responses z_1, z_2, ..., z_m of the same type. The external stimuli constitute a column set of system input signals U_ℓ, and the responses constitute a column set of system output signals Z_m. U_ℓ is also the set of input signals for the model, in which it is expanded by fixed linear or nonlinear operators to form a set of model basis signals X_n. A set of model output signals V_m is formed from X_n by means of the linear transformation.

$$V_m = C_{mn} X_n \qquad (1)$$

Fig. 1 shows the general scheme.

Each signal of V_m is required to resemble the corresponding signal of Z_m as closely as possible; the differences define the set of error signals E_m, through

$$E_m = Z_m - V_m \qquad (2)$$

The model is optimal when the total energy (for transient signals) or the average power (for random or periodic signals) of each signal of E_m is least. This condition obtains when

$$\frac{\partial \overline{e_i^2}}{\partial c_{ij}} = -2\overline{e_i x_j} = 0$$
$$\text{for } \begin{array}{l} i = 1,2,\ldots\ldots,m \\ j = 1,2,\ldots\ldots,n \end{array} \qquad (3)$$

$$\frac{\partial^2 \overline{e_i^2}}{\partial c_{ij}^2} = 2\,\overline{x_j^2} > 0$$

where $\overline{}$ denotes the operation which, when applied to the square of a signal, defines either its total energy or its average power, as appropriate. This operation also defines a Hilbert space[1] in which the scalar product of two signal vectors s_1 and s_2 is $\overline{s_1 s_2}$.

From (3) the model is optimal when

$$\overline{E_{m_{opt}} X_n'} = 0 \qquad (4)$$

where $_{opt}$ denotes the optimal value.

$E_{m_{opt}}$ is therefore orthogonal to the subspace $\mathcal{X}_n$ which is spanned by X_n, and for which X_n forms a basis if, as assumed, it is a linearly independent set. Since from (1) $V_{m_{opt}}$ is in $\mathcal{X}_n$, it follows from (2) that $V_{m_{opt}}$ is the projection of Z_m on to $\mathcal{X}_n$ and $E_{m_{opt}}$ is the projection of Z_m on to the complementary subspace $\mathcal{X}_n^\perp$.

From (4)

$$\overline{Z_m X_n'} - C_{mn_{opt}} \overline{X_n X_n'} = 0 \qquad (5)$$

"Superior numbers refer to similarly-numbered references at the end of this paper".

so that

$$C_{mn_{opt}} = \overline{Z_m X_n'} \left[\overline{X_n X_n'} \right]^{-1} \qquad (6)$$

COEFFICIENT ADJUSTMENT PROCEDURE

If X_n is prescribed, the modelling problem reduces to that of obtaining $C_{mn_{opt}}$. This may be accomplished by implementation of the equation

$$K^2 T \frac{dC_{mn}}{dt} = E_m X_n' \qquad (7)$$

that is

$$K^2 T \frac{dC_{mn}}{dt} + C_{mn} X_n X_n' = Z_m X_n' \qquad (8)$$

An approximate solution may be obtained for this equation by means of a decomposition technique. Each signal s is taken as the sum of a moving average $\overline{s}$ and a remainder $\underline{s}$ (Fig.2); the average $\underline{s}$ is zero, and for stationary signals the moving average $\overline{s}$ becomes the average s.

(8) may then be separated into two distinct equations

$$K^2 T \frac{d\overline{C}_{mn}}{dt} + \overline{C}_{mn} \overline{X_n X_n'} = \overline{Z_m X_n'} \qquad (9)$$

and

$$K^2 T \frac{d\underline{C}_{mn}}{dt} + \underline{C}_{mn} \overline{X_n X_n'} = Z_m X_n' - \overline{C}_{mn} \overline{X_n X_n'} \qquad (10)$$

provided that $\underline{C}_{mn} \underline{X_n X_n'}$ is negligible.

If X_n and Z_m are stationary, (9) becomes

$$K^2 T \frac{d\overline{C}_{mn}}{dt} + \overline{C}_{mn} \overline{X_n X_n'} = \overline{Z_m X_n'} \qquad (11)$$

with solution from initial conditions $\left[\overline{C}_{mn} \right]_0$ given by [2]

$$\overline{C}_{mn} = \overline{Z_m X_n'} \left[\overline{X_n X_n'} \right]^{-1}$$

$$+ \left[\left[\overline{C}_{mn} \right]_0 - \overline{Z_m X_n'} \left[\overline{X_n X_n'} \right]^{-1} \right] e^{-\frac{\overline{X_n X_n'}}{K^2} \frac{t}{T}} \qquad (12)$$

$$-\frac{\lambda_j}{K^2} \frac{t}{T}$$

$\overline{C}_{mn}$ contains natural modes e

corresponding to the eigenvalues $\lambda_1, \lambda_2, \ldots\ldots, \lambda_j, \ldots\ldots, \lambda_n$ of $\overline{X_n X_n'}$. Since these eigenvalues are real, positive and distinct [3], the natural modes are convergent, and $\overline{C}_{mn}$ converges to $C_{mn_{opt}}$ at a rate which depends principally on the ratio of the smallest eigenvalue to $K^2 T$.

In terms of the moving averages

$$\frac{\partial \overline{e_i^2}}{\partial c_{ij}} = -2\overline{e_i x_j} = -2K^2 T \frac{d\overline{c}_{ij}}{dt} \text{ for } \begin{array}{l} i=1,2,\ldots,m \\ j=1,2,\ldots,n \end{array} \qquad (13)$$

which shows that the trajectories of the moving averages $\overline{C}_{mn}$ on the m average error power surfaces are those of steepest descent. The trajectories of the actual coefficients C_{mn} are perturbed about these steepest descent trajectories by $\underline{C}_{mn}$. (10) shows that $\underline{C}_{mn}$ is produced by smoothing $Z_m X_n' - \overline{C}_{mn} \overline{X_n X_n'}$, and the smoothing depends on the value of $K^2 T$, the choice of which represents a compromise between the amount of smoothing required and the effect on rapidity of convergence of the natural modes.

In particular, after convergence of the natural modes, the coefficients C_{mn} are perturbed about their optimal values $C_{mn_{opt}}$ by signals produced by smoothing $E_{m_{opt}} X_n'$. If the performance of the model is good, and the average power of each optimal error signal of the set $E_{m_{opt}}$ is small, then the average power of each signal of the set $E_{m_{opt}} X_n'$ is small and the coefficients C_{mn} have only small perturbations about their optimal values $C_{mn_{opt}}$.

LINEAR MODELS

An important class of models is that in which the operators used to expand U_ℓ are linear. For models of this class, the principle of superposition shows that each pair of signals u_h and z_i may be considered separately. Each multivariable model may therefore be partitioned into lm simple models, each with a single input and output.

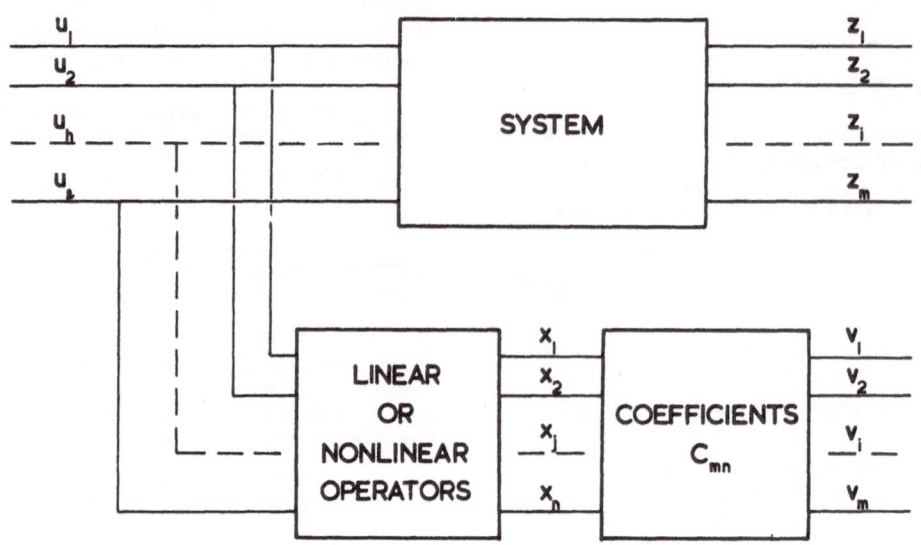

FIG. I MULTIVARIABLE SYSTEM AND MODEL

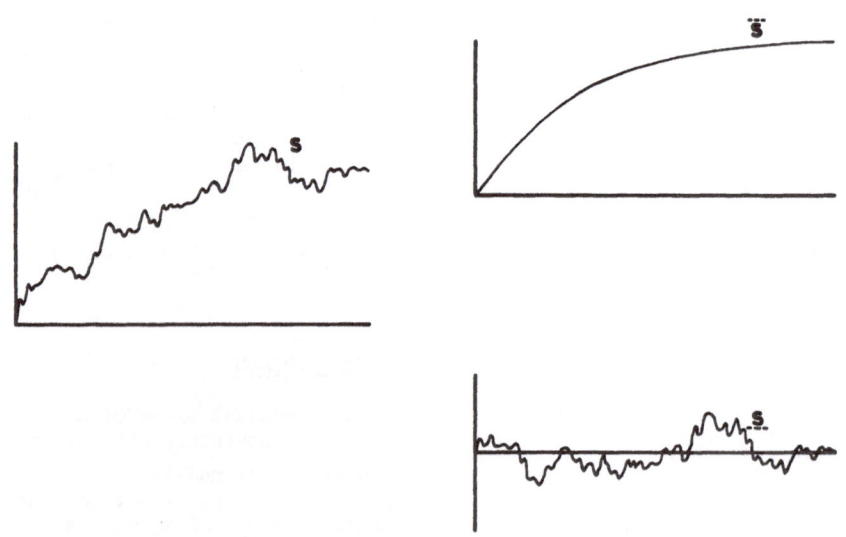

FIG. 2 DECOMPOSITION OF SIGNALS

A correspondingly simple expansion of u_h may be chosen to form the truncated set $X_{n'}$; this is defined by

$$x_j(p) = \frac{q_j}{p+p_j} u_h(p) \quad \text{for } j = 1,2,\ldots,n' \quad (14)$$

If the frequencies of the set $P_{n'}$ are permitted to assume any real or complex values, then models which use an expansion of this type are capable of modelling any linear lumped parameter system exactly, and any linear distributed parameter system approximately. In order to model the linear lumped parameter system exactly, it is necessary to choose the set of frequencies $P_{n'}$ to match a corresponding set of modal frequencies for the system. Since this involves either a priori knowledge of the system or the introduction of a separate procedure for adjusting the frequencies $P_{n'}$, an approximate model of the system is often preferable. In such a model the frequencies $P_{n'}$ may be chosen in any convenient manner, provided that the average power of the resulting optimal error signal $e_{i_{opt}}$ is within the required limits. An economical method is to choose these frequencies so that

$$\frac{p_j}{p_j+1} = r \quad \text{for } j = 1, 2, \ldots, n'-1 \quad (15)$$

where p may be real or complex, and r is a constant. In the resulting model, the errors in modelling systems with modal frequencies within the range p_1 to $p_{n'}$ may be closely controlled by choice of r.

STATIC GAIN MODELS

For the class of linear models under consideration, the optimal model for the i th part of the multivariable system is given by

$$v_{i_{opt}}(p) = \sum_{j=1}^{n'} \frac{q_j}{p+p_j} c_{ij_{opt}} u_h(p) \quad (16)$$

The static gain $g_{ih_{opt}}$ of this optimal model is obtained by letting $p \to 0$ in (16). If q_j is equal to p_j this is given by

$$g_{ih_{opt}} = \frac{v_{i_{opt}}(0)}{u_h(0)} = \sum_{j=1}^{n'} c_{ij_{opt}} \quad (17)$$

Hence the static gain of each part of the multivariable system is approximated simply by the sum of a number of optimal coefficients. The model is therefore suitable for measuring gradients in extremal control schemes, and provides an alternative to methods which require integration of correlation functions[5].

CONCLUSIONS

Models may be obtained for multivariable systems by expansion of the input signals. The optimisation of the expansion coefficients, which has a simple geometrical interpretation, may be accomplished by a self-adjusting procedure in which the choice of parameters represents a compromise between rapidity and accuracy of coefficient setting.

In the linear case, partitioning of the model structure allows considerable simplification, and a correspondingly simple expansion of each input signal in which the modal frequencies progress geometrically is satisfactory. The static gain of each part of the linear model is obtained by simple summation of coefficients, and may be used as a measure of gradient in extremal control schemes.

NOMENCLATURE

U_ℓ	set of input signals
X_n	set of basis signals
Z_m	set of system output signals
V_m	set of model output signals
E_m	set of error signals
C_{mn}	set of constants
$P_{n'}$	set of frequencies
χ_n	subspace with basis X_n
T	time constant
t	time
s	signal
p	Laplace variable
g	static gain
d	total differential
∂	partial differential
λ	eigenvalue
K,q,r,	constants
'	transpose
-1	reciprocal

$\perp$	complement
	average
----	moving average
	signal with zero average
$\sum$	sum
opt	optimal
h,i,j, ,m,n,n'	suffices

ACKNOWLEDGEMENT

The research described in this paper is supported by the Science Research Council of Great Britain.

REFERENCES

(1) Halmos, P.R., _Introduction to Hilbert Space and the Theory of Spectral Multiplicity_, Chelsea, New York, 1951.

(2) Bellman, R., _Stability Theory of Differential Equations_, McGraw-Hill, New York, 1953.

(3) Courant, R. and Hilbert D., _Methods of Mathematical Physics_, Interscience, New York, 1953.

(4) Barker, H.A., _The Application of Systems of Orthogonal Signals to the Solution of the Optimization Problem_, Ph.D. Thesis, Cambridge University, 1965.

(5) van der Grinten, P.M.E.M., _The Application of Random Test Signals in Process Optimization_, Proceedings Second International Congress, I.F.A.C., Basle, 1963.

Discussion

Dr. A. Straszak (Poland) asked three questions. Firstly how linear or non-linear operators are chosen for a given system? Secondly whether there exists a relationship between the dimensionality of the system and the number of basis signals used? Thirdly, as the number of parameters to be adjusted varies as the square of the dimensionality of the input or output vector, what method is suggested for use with large scale systems?

Dr. Barker replied, on the first question, that no general method existed for the choice. He had indicated towards the end of the paper how a given number of linear operators could be chosen in such a way as to minimise the error. He commented that the unifying concept behind his method was to be found in projection operator theory and that any future advance in the method of choice is likely to stem from this theory. Replying to the second question Dr. Barker considered that it should be possible to model well using exactly the same number of basis signals as the state space dimensionality. He had, however, found that, using fixed operators, it is more convenient to use one more basis signal, and thus to allow a degree of freedom to cope with a change in dynamics. He is presently looking at the problem of how the optimal control of the model found by using three basis signals compares with that using two basis signals, for a system with two state variables. On the third question Dr. Barker commented that any method for dealing with large systems is bound to be difficult, and decomposition techniques will probably have to be used, dividing the system into hierarchies or series elements. Although the method could, in theory, be applied to this model it would probably only be necessary to examine a small part of the system where continuous monitoring of the dynamics is taking place.

Prof. R.J.A. Paul (UK) noted that the choice of linear operators had been limited to real poles. He asked whether this choice would be adequate to model a system which had significant complex conjugage poles.

Dr. Barker explained that in the paper logarithmically spaced real poles were used to model a system where the poles were moving along the real axis, and good results were obtained. He stated that this model will also describe systems which have similarly spaced poles not far off the real axis. However the model can be extended directly to have complex conjugate poles, so that it will cover a wider range of systems.

An application of the method to determine a model for a visco-elastometer had required the use of complex conjugate poles. In this case the pole-zero representation was inadequate.

Prof. J.L. Douce (UK) asked how time delay parameters known to be present in the system are identified. He also asked the author to compare the estimation times as between his method and that of Van der Grinten.

Dr. Barker replied that using the method of this paper, the impulse response can only be represented by the appropriate set of complex or real poles and zeros. Hence only a smoothed approximation can be obtained. The accuracy of estimation will depend on the frequency content of the input signal, and white noise will not give a good answer as the change of phase at high frequencies will not be followed by the lumped parameters used in the model. In an open loop method Dr. Barker had used pseudo random codes (which can easily be delayed) as a test signal, putting the delayed version into the usual type of model.

Commenting on the comparison with the method of Van derGrinten, Dr. Barker considered that his method would take longer for a very accurate answer to be obtained. However the criterion in adaptive systems is to obtain a reasonable estimate very quickly, and in this case, his method is comparable in estimation time with that of Van derGrinten.

[Editors note. Both Dr. Barker and Prof. Douce mentioned two system time constants as an estimation time for which the method described in the paper and that of Van derGrinten gave reasonable estimates for one parameter.]

WRITTEN CONTRIBUTIONS TO THE DISCUSSION

Mr. R. J. High (UK)

In a paper published recently (ref. 1) we have represented a small chemical plant by six first order lags grouped into two sets of three. The gain of each of the resulting third order functions being fitted by a continuously updated regression.

We have recently been varying the time constants of the six poles and studying the effect that this has upon the residual mean square error of the regression. We have found that over a wide range of operation conditions and of input signal frequencies the residual mean squared error varies very little. Thus, in our case at least, the positioning of the poles is not critical and I wondered if Dr. H. A. Barker had any similar experience.

Dr. Barker in reply.

My experience is that in representing systems by models which consist of sets of first order real poles and sets of second order complex poles the positioning of the poles is not critical, provided that their dynamics span the dynamics of the system. It is this insensitivity to pole position which enables the type of model described in the paper to operate successfully by gain changes only.

Dr. W. D. T. Davies (UK)

I was extremely interested in the results Dr. Barker presented for a single input - single output system using, I think, five model paths. These were produced, if I understand him properly, by choosing the basic approximating functions to contain between them a spread of five basic poles, and then he assumes that the 'unknown process' transfer function has its poles contained in this region.

I would thus like to ask Dr. Barker whether he has considered optimising the model poles as well as the corresponding gains.

This would entail choosing arbitrary poles in each model path and adjusting these poles to approximate the system poles, rather than choosing a spread of model poles to cover the system poles. Using this approach, it should be possible (in theory at least) to simulate complex poles as well as real ones.

Dr. Barker in reply.

I have constructed models in which the poles as well as the gains are optimized; this is easily accomplished by a simple extension of the method described here based on equating to zero the partial derivatives of $\overline{e_i^2}$ with respect to the model frequencies p_j. Such models require almost exactly twice as much computing equipment as those described here, and in return for this possess the capability of reducing the error signals to zero. In situations where the small residual errors of models with fixed poles are an embarrassment, models with adjustable poles may be preferred.

An interesting feature of models with adjustable poles is that for systems of first or second order the model parameters converge almost to those of the system in the presence of considerable amounts of noise, while for higher order systems a relatively small amount of noise is sufficient to permit almost any reasonable combination of model parameters to be obtained. Anyone who has attempted the related problem of fitting exponentials to a third order transient response will appreciate this difficulty.

Reference

1. Bray, J.W., High, R.J., McCann, A.D. and Jemmeson, H. On-line model making for a chemical plant, Trans. Soc. Instrument Technology, September, 1965, pp.65 - 75

PROCESS PARAMETER ESTIMATION AND SELF ADAPTIVE CONTROL

by Peter C. Young
Research Student*
Loughborough College of
Advanced Technology
Loughborough, Leicestershire, England

ABSTRACT

The paper describes a method of automatic process parameter estimation which has been mechanised with the aid of hybrid (analogue-digital) equipment. The technique is characterised by it's simplicity, and differs from earlier schemes of this type in not requiring direct measurement of the input and output time derivatives of the process. The performance of the system is discussed, with particular reference to the effect of uncertainty on the sampled data caused by spurious noise contamination. Finally, a simple identification-adaptive control system for a non-stationary process is described, which utilises the hybrid parameter estimation techniques.

INTRODUCTION

One type of automatic process identification scheme which has received a great deal of attention in recent years is the 'Process Parameter Estimation' system. The basis of such systems is an a priori assumption on the form of the mathematical relationships which provide a reasonable description of the process dynamic performance characteristics (Ref.1). The paper describes a technique for estimating the parameters of a differential equation or Laplace transform transfer function description of the process, using hybrid (analogue-digital) equipment. These techniques are important not only in their straight forward role in the experimental evaluation of dynamic processes (The Analysis of Dynamic Experiments), but also because of their possible application to the problem of self adaptive control. This latter application is discussed and a simple identification-adaptive control system for a non-stationary process is described, which utilises the hybrid parameter estimation techniques.

1.1 THE TECHNIQUES OF PARAMETER ESTIMATION

The principal object of process parameter estimation is to obtain a realistic mathematical description of the process from a knowledge of it's normal operating characteristics, and making use of any a priori information which is available on the nature of the process. Under this definition it is possible to divide the various approaches to the problem into two broad classes (Ref.1).

(A) "Using a Physical Model", in which the characteristics of a physical model of some form (usually an electrical network) are manipulated in such a way that they converge towards the characteristics of the process itself, in some pre-defined sense.

(B) "Using an Explicit Mathematical Relation", in which the process is characterised by certain numerical values, which are chosen in order to satisfy a pre-determined and applicable mathematical relationship. This is possibly, a rather larger category than class (A), because the term 'numerical values', which appears in the definition, is capable of quite wide interpretation.

The parameter estimation schemes dealt with in this paper come within this class (B) category. Fundamentally, they are concerned with the description of a physical process in terms of a general mathematical expression of the form,

$$\sum_{r=o}^{r=k} a_r y_r = f \qquad \ldots\ldots\ldots\ldots(1)$$

where, $f = f(t)$ can be considered an arbitrary input or forcing function to the process
$y_r = y_r(t)$ are time dependent variables occurring in the process
$a_r = a_r(t)$ are slowly variable or fixed coefficients or parameters

Equation (1) can be written in matrix form,

$$Y^T A = f \qquad \ldots\ldots\ldots\ldots(2)$$

or $Y^T A - f = 0 \qquad \ldots\ldots\ldots\ldots(3)$

where, $Y^T = [y_0, y_1, y_2 \ldots y_k]$

$A^T = [a_0, a_1, a_2 \ldots a_k]$

let $\hat{A}^T = [a_{0c}, a_{1c}, a_{2c} \ldots a_{kc}]$ be the estimated value of the matrix A^T.

In general, it is possible to define an error signal E, where,

$$E = Y^T \hat{A} - f \qquad \ldots\ldots\ldots\ldots(4)$$

or,

$$E = Y^T(\hat{A}-A) \qquad \text{(from Eqn.(2)} \ldots\ldots\ldots\ldots(5)$$

Here, $E = E(t)$, the error in satisfaction of the process equation (or more simply the 'Satisfaction Error') is seen to be a function of the error between the actual and estimated values of A.

It is possible to arrive at an estimate of which minimises E or some function of E, in a number of ways.
(A) By systematic adjustment of the parameters

*now with Dept. of Eng., University of Cambridge.

"Superior numbers refer to similarly-numbered references at the end of this paper"

118

a_{ro} using the methods of steepest descent, it is possible to continuously minimise certain functions of the satisfaction error by means of either analogue or iterative digital techniques. This approach has been investigated and discussed in(Ref.2.)

(2) By sampling either the variables y_r and f or certain functions of these variables at S instants of time (where $S \geq k+1$) and so generating $k+1$ linear simultaneous algebraic equations in $k+1$ unknowns, a_{ro} ($r=o \rightarrow r=k$). When $S>k+1$ the samples can be used to perform a 'least squares' estimation.

(3) Alternatively, it can be shown that by special treatment of the variables y_r and f it is possible to generate the required algebraic equations from samples taken at only one instant of time.

These latter two approaches to the problem have been dealt with briefly in(Refs.3,4 and 5.) These investigations have been concerned with a special case in which the process can be described by the equation,

$$\sum_{n=o}^{N} a_n \frac{d^n y}{dt^n} = \sum_{m=o}^{M} b_m \frac{d^m f}{dt^m} \qquad \cdots\cdots\cdots (6)$$

so that an error function, $E_s = E_s(t)$, may be defined by

$$E_s = \sum_{n=o}^{N} a_{no} \frac{d^n y}{dt^n} - \sum_{m=o}^{M} b_{mo} \frac{d^m f}{dt^m} \qquad \cdots\cdots\cdots (7)$$

where, without loss of generality $b_{oo} = 1.0$

The analysis shown in Appendix 1 illustrates how, by performing a Laplace transformation on equation (6) and by then operating on the resultant relationship by a modulating function of the form $c/(s+c)^L$ (where s is the Laplace operator and c is a constant), it is possible to express the relationship between a_n and b_m in the form,

$$\sum_{n=o}^{N} a_n \left[\left(\frac{d^n y}{dt^n} \right) \right]_L = \sum_{m=o}^{M} b_m \left[\left(\frac{d^m f}{dt^m} \right) \right]_L \qquad \cdots\cdots (8)$$

where the brackets $[(\ )]$ indicate that the function enclosed has been physically filtered by L low pass filters each having the transfer function $c/s+c$ (see Fig.14).

The modified satisfaction error, E_m can now be defined, where

$$E_m = \sum_{n=o}^{N} a_{no} \left[\left(\frac{d^n y}{dt^n} \right) \right]_L - \sum_{m=o}^{M} b_{mo} \left[\left(\frac{d^m f}{dt^m} \right) \right]_L \cdots (9)$$

This is a particular case of equation (4) of the form,

$$E_m = Y_L^T \hat{A} - f_L \qquad \cdots\cdots\cdots\cdots (10)$$

in which,

$$\hat{A}^T = [a_{oo}, \ldots, a_{Nc}, -b_{1c}, \ldots, -b_{Mc}]$$

$$Y_L^T = \left[[(y_o)]_L, \ldots, [(y_N)]_L, [(f_1)]_L, \ldots, [(f_M)]_L \right]$$

and $f_L = [(f)]_L = [(f_o)]_L$

here, $y_N = \frac{d^N y}{dt^N}$; $f_M = \frac{d^M y}{dt^M}$

Appendix 1 shows that the variables $[(y_n)]_L$ (where $n=1 \rightarrow N$) and $[(f_m)]_L$ (where $m=1 \rightarrow M$) may be generated by the simple algebraic summation of the variables appearing at various stages in the chain of low pass filters shown in Fig.14 (i.e. such signals as $[(f_o)]_1, \ldots, [(f_o)]_L$ and $[(y_o)]_1, \ldots, [(y_o)]_L$; $L \geq M+N+1$).

In effectively removing the need for direct sampling of the higher order input and output derivatives, which can be difficult or even impossible in practice and can intensify the problem of spurious noise contamination, this 'Method of Multiple Filters' becomes an essential prerequisite to the practical implementation of the type of parameter estimation scheme described in this paper.

If the required samples are taken at S discrete instants of time, it is possible to generate S equations of the form,

$$E_{mj} = \sum_{n=o}^{N} a_{no} \ _j[(y_n)]_L - \sum_{m=o}^{M} b_{mo} \ _j[(f_m)]_L \cdots\cdots (11)$$

where $j = 1,2, \ldots S$ and $b_{oo} = 1.0$

Now, let the function Γ be defined, where

$$\Gamma = \sum_{j=1}^{S} E_{mj}^2 \qquad \cdots\cdots\cdots\cdots (12)$$

and $S \geq M+N+1$.

The necessary condition for a minimum of the function Γ is the equality to zero of all its first order partial derivatives with respect to a_{no} ($n=o \rightarrow N$) and b_{mo} ($m=1 \rightarrow M$). Thus,

$$\frac{\partial \Gamma}{\partial a_{no}} = 2 \sum_{j=1}^{S} E_{mj} \cdot \frac{\partial E_{mj}}{\partial a_{no}} = 0 \quad (n=o \rightarrow N)$$

$$\qquad \cdots\cdots (13)$$

$$\frac{\partial \Gamma}{\partial b_{mo}} = 2 \sum_{j=1}^{S} E_{mj} \cdot \frac{\partial E_{mj}}{\partial b_{mo}} = 0 \quad (m=1 \rightarrow M)$$

These M+N+1 linear simultaneous algebraic equations can be solved by normal methods to supply the required 'least squares' estimates of the M+N+1 parameters ($a_{oo}, a_{1o}, \ldots, a_{Nc}, b_{1o}, b_{2o}, \ldots, b_{Mo}$).

In deciding on the quantity of data to be used for each separate computation, it is necessary to strike a compromise between two rather contradictory requirements. On the one hand there is the need to accumulate and process enough information to provide a reasonable statistical estimate of the unknown parameters. On the other hand there is the need to restrict sampling time so that any non-stationary effects which the

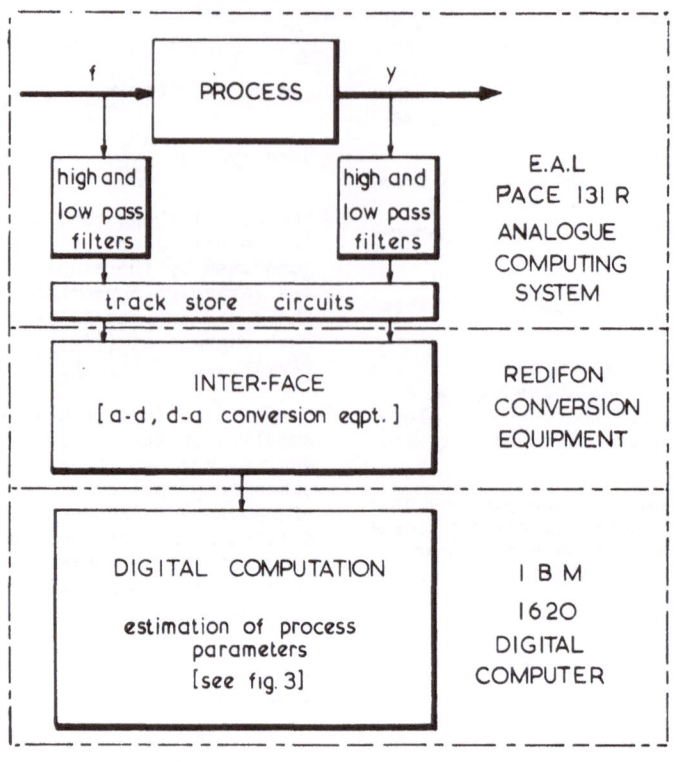

fig.1. mechanisation of hybrid P.D.C.
using general purpose hybrid
equipment

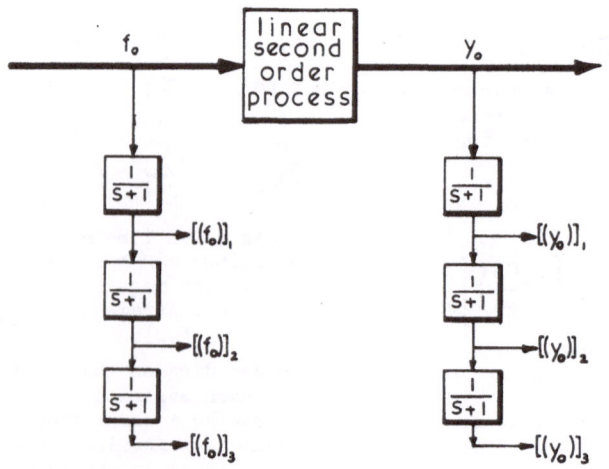

fig.2. multiple filters applied to a second order
process [c=1, high pass filters not shown]

process may have do not result in the use of obsolete data and consequent error in the computation of the parameters.

1.2 THE MECHANISATION OF THE PARAMETER ESTIMATION SCHEME

An automatic Parameter Determination Computer (P.D.C.) utilising the techniques of parameter estimation discussed in the previous section can be mechanised with the aid of analogue and digital elements. Fig.1 illustrates how such a 'Hybrid P.D.C.' has been simulated on the general purpose hybrid computational facility of Loughborough College of Technology (Refs.4 and 5).

It will be noticed that, in addition to the low pass filters which are necessary for the Method of Multiple Filters, high pass filters are incorporated in the system. By 'd.c. blocking' the signals in this manner, the detrimental effect of incorrect reference levels, or bias errors, is avoided and the need for scaling circuitry in the hybrid system is reduced ((Ref.4) and section 1.3).

Consider the application of this system to a simple second order process which can be reasonably described by the differential equation,

$$a_0 y + a_1 \frac{dy}{dt} + a_2 \frac{d^2 y}{dt^2} = f \qquad \ldots\ldots\ldots\ldots(14)$$

or $a_0 y_0 + a_1 y_1 + a_2 y_2 = f_0 \qquad \ldots\ldots\ldots\ldots(15)$

Choosing L=3 and c=1 rad/sec the algebraic relationships derived in Appendix 1 can be used to obtain the following equations for the filtered state variables,

$$\left. \begin{array}{l} [(y_2)]_3 = [(y_0)]_3 - 2[(y_0)]_2 + [(y_0)]_1 \\ \\ [(y_1)]_3 = -[(y_0)]_3 + [(y_0)]_2 \end{array} \right\} \ldots\ldots\ldots(16)$$

Now, if S is chosen to be 6, the signals $[(y_0)]_1$, $[(y_0)]_2$, $[(y_0)]_3$ and $[(f_0)]_3$ (Fig.2) can be fed to the track store circuits and sampled at six discrete instants of time. They are then passed via the conversion equipment to the digital computer. The digital computer is programmed to perform the algebraic summations detailed above and then insert the results into equation (13), (Fig.3). These three linear algebraic simultaneous equations in three unknowns (M+N+1=3) are then solved by normal methods to provide a least squares estimate of the process parameters. These estimates can then be either printed out, or processed and sent back to the analogue side of the equipment in the form of adaptive controller parameters (see section 2.1).

Up to the present time, a large amount of the work carried out on the Hybrid P.D.C. system has been concerned with second order processes of this type. This restriction was initially imposed for two principal reasons.

Firstly, the second order mathematical model provides a simple example to test the

feasibility of the identification scheme, and yet provides a reasonably realistic description of a number of physical processes.

Secondly, practical limitations inherent in the general purpose hybrid equipment used in the simulation prevented more sophisticated mechanisation.

Ideally, the sampling frequency and sample size per computation should be capable of wide variation in order that well conditioned simultaneous equations may be generated and well defined parameter estimates obtained. At the time the investigations described in this paper were carried out, it was only possible to sample a set of four variables every two seconds. In addition, the number of samples taken per computation was maintained at a fixed value to ease the problem of digital programming. These restrictions imposed a practical limit on the type of process which could be investigated; only relatively slow variation in the process parameters could be tolerated, undamped natural periods needed to be less than 15 seconds and noise levels had to be reasonably low in order that the variance in the estimates was not to great.

Similarly, in a truly practical system, it should be possible to modify the time constants of the high and low pass filters, in order that the P.D.C. can be 'matched' to the process under investigation (see section 1.3). In the simple case described here, they were maintained at a fixed value of one second in order to further simplify the mechanisation and the digital programming.

In addition to it's use in the identification of second order analogue processes the equipment described here has been applied to the investigation of a d.c. position control servo system. The servo had a number of mildly non-linear characteristics and presented a reasonably realistic problem in process identification. Fig.4 shows a comparison of the actual frequency response of the servo and that calculated from the estimated parameters. The fact that this estimation was achieved with a sample size maintained at 6 sets per computation (S=6 equation (12)), illustrates the potentialities of the system even in it's present elementary form.

1.3 THE USE OF FILTERS IN THE PARAMETER ESTIMATION SCHEME

The statistical least squares approach to parameter estimation of the Hybrid P.D.C. system is further aided by the low pass filters inherent in the Method of Multiple Filters. In effect, the filters introduce a degree of memory into the system by weighting each sample with past data (Ref.6). The fact that different low pass filters are able to weight the samples in different ways, also suggests that a more sophisticated sampling technique may be evolved and so form the basis of future equipment (preliminary work indicates that this should be possible, see Appendix 1 and section 1.5)

Both the high and low pass filter time

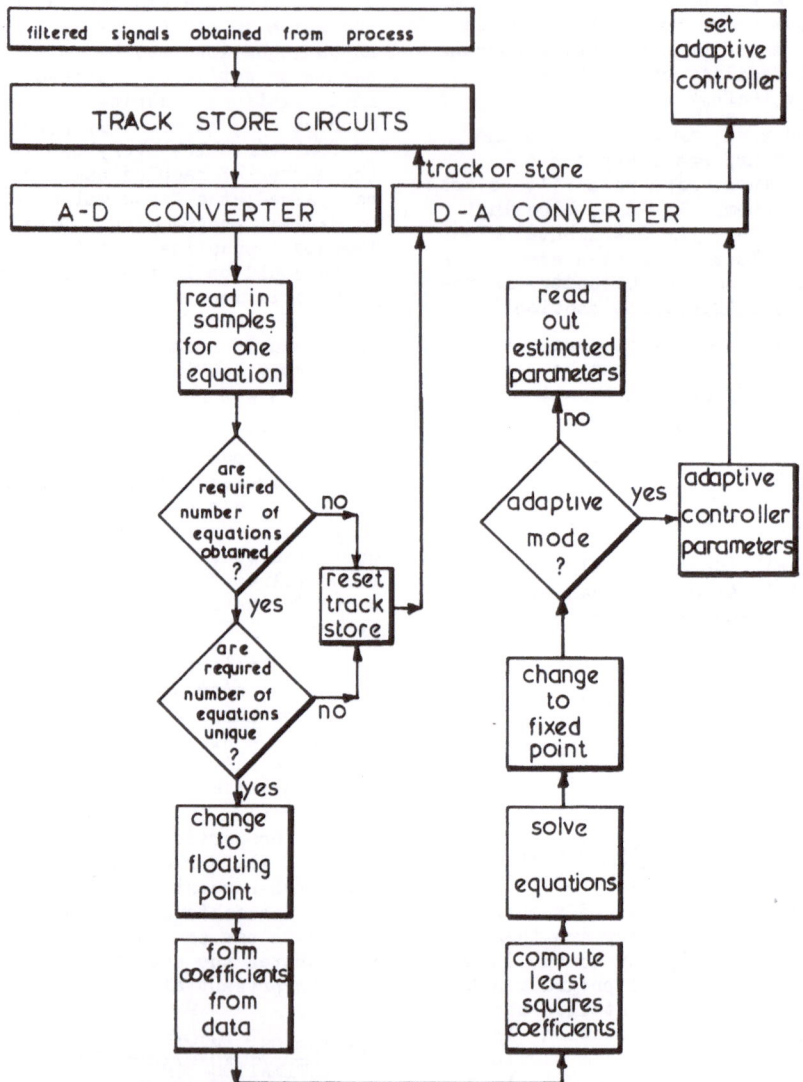

filtered signals obtained from process

TRACK STORE CIRCUITS

A-D CONVERTER

D-A CONVERTER

track or store

set adaptive controller

read in samples for one equation

are required number of equations obtained ?

no

reset track store

yes

are required number of equations unique ?

no

yes

change to floating point

form coefficients from data

read out estimated parameters

no

adaptive mode ?

yes

adaptive controller parameters

change to fixed point

solve equations

compute least squares coefficients

fig. 3.   schematic block diagram of hybrid p.d.c. digital programme

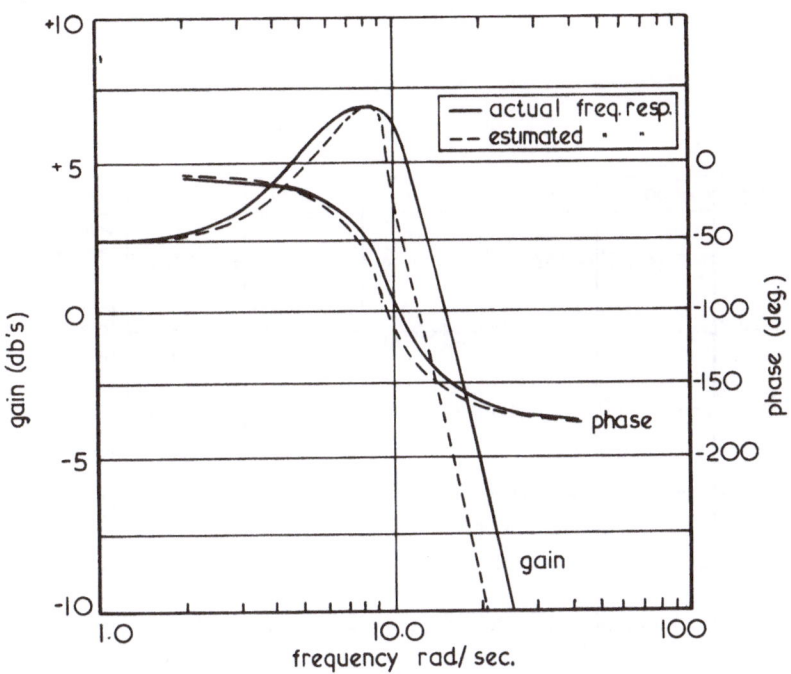

fig. 4.   comparison of actual and estimated
frequency response of the d.c. servo

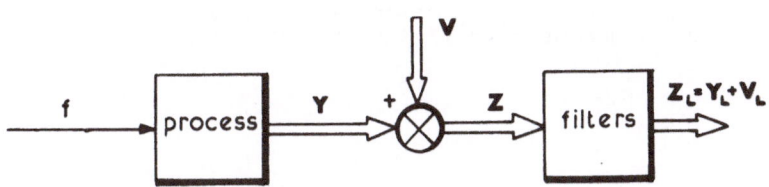

fig. 5 noise contamination of process
signals

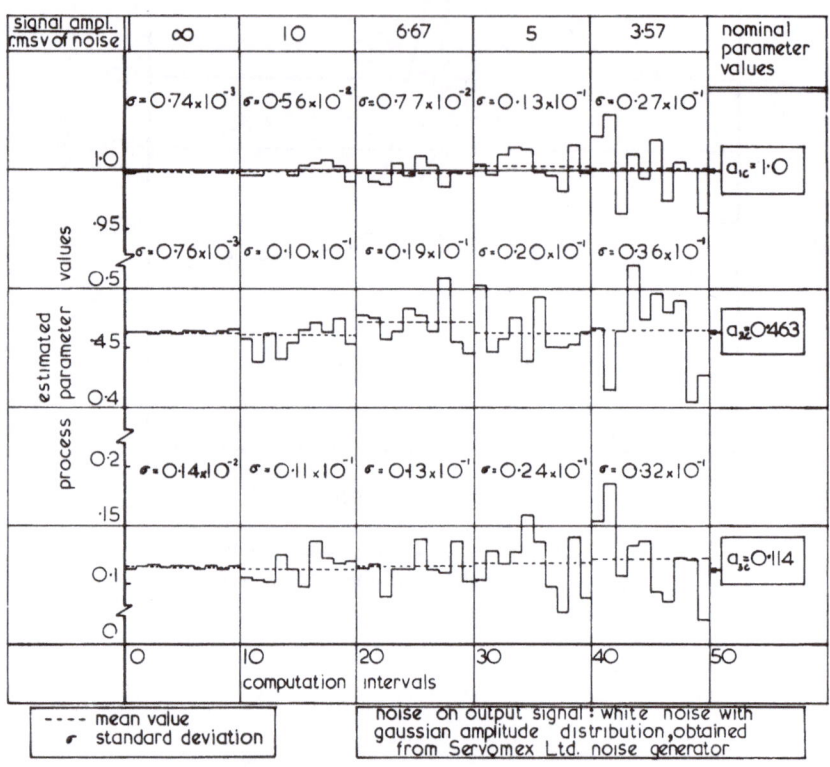

fig.6. typical example showing the effect of noise contamination of the process o/p signal on parameter estimation accuracy
[second order process activated by ±25 volt amplitude step signals]

constants can be considered as design parameters in systems of this type:

The time constant of the low pass filters determines how much importance is attached to older data (Ref.6). If the time constant is chosen to be large then the samples are quite heavily 'weighted' with past data. Consequently, even if the process comes to rest the parameter estimation is not affected for some time because the P.D.C. 'remembers' the effect of previous activation. Alternatively, if the process is non-stationary, and the parameters are changing fairly rapidly, then the time constant should be chosen much smaller in order that obsolete data is not utilised.

The time constant of the high pass filters controls the attenuation of the low frequency components of the signals. If possible it should be selected to 'block' any effects such as bias errors or slow drifts about the equilibrium point whilst at the same time passing any useful low frequency components of the signals.

In addition to these points, the low and high pass filters have considerable effect on the transient response characteristics of the sampled signals, a point which must be considered when deciding what sampling frequency is desirable.

The choice of filter characteristics for any particular practical application of the Hybrid P.D.C. has to be guided by consideration of all the above points. With the present fixed component system, the designer must take into account the possible range of parameter variation, the maximum rate of variation which might be expected, and the types of input excitation which are most likely to occur. He should then choose those filter characteristics which provide the best compromise between any conflicting requirements, and so produce the most satisfactory overall system performance.

1.4 THE EFFECT OF NOISE CONTAMINATION OF THE PROCESS SIGNALS

Equation (11) can be written in the matrix form,

$$E_{mj} = {}_jY_L^T\hat{A} - {}_jf_L \qquad \ldots\ldots\ldots\ldots(17)$$

where,

$$_jY_L^T = \left[{}_j[(y_0)]_L, \ldots, {}_j[(y_N)]_L, {}_j[(f_1)]_L, \ldots \right.$$
$$\left. \ldots, {}_j[(f_M)]_L \right] \qquad \ldots\ldots(18)$$

$$_jf_L = {}_j[(f_0)]_L$$

and $j = 1, 2, \ldots\ldots S$

conditions for a minimum of Γ are,

$$\frac{\partial \Gamma}{\partial \hat{A}} = 0$$

i.e. $$\sum_{j=1}^{S} {}_jY_L \cdot {}_jY_L^T\hat{A} - {}_jY_L \cdot {}_jf_L = 0$$

This may be written,

$$\sum_{j=1}^{S} {}_jY_L \cdot {}_jY_L^T(\hat{A} - A) = 0 \qquad \ldots\ldots\ldots\ldots(19)$$

Now if S is large and a common factor $^1/S$ is introduced this equation can be expressed in the form,

$$\langle Y_L \cdot Y_L^T \rangle (\hat{A} - A) = 0 \qquad \ldots\ldots\ldots\ldots(20)$$

where the angular brackets denote the expectation operator.

In practice the process signals are filtered by a high pass filter in addition to the low pass filters inherent in the Method of Multiple Filters. In effect, these high pass (or d.c. blocking) filters extract the mean value of each of the elements of Y_L (as defined by (10)). Thus $\langle Y_L \cdot Y_L^T \rangle$ can be considered as the covariance matrix of Y_L, cov. $[Y_L, Y_L]$.

$$\therefore \text{cov. } [Y_L, Y_L](\hat{A} - A) = 0 \qquad \ldots\ldots\ldots\ldots(21)$$

One factor of great importance to the practical implementation of any parameter estimation scheme is it's sensitivity to uncertainty on the sampled signals caused by such effects as measurement noise. Consider then the matrix block diagram Fig.5, in which the vector, $Y = Y(t)$, is contaminated by additive noise $V = V(t)$, where

$$V^T = [v_0, v_1, \ldots, v_{M+N+1}]$$

The observed vector is now Z_L and,

$$Z_L = (Y+V)_L = Y_L + V_L \quad \text{(by the principle of superposition)}$$

The elements of V_L are dependent upon the nature of the noise and on the characteristics of the filters (high and low pass) used in the mechanisation.

Now, in this noisy case, E_{mj} will be defined as,

$$E_{mj} = {}_jZ_L^T\hat{A} - {}_jf_L \qquad (j=1, 2, \ldots, S)$$

and the equation equivalent to (21) becomes,

$$\text{cov. } [Z_L, Z_L]\hat{A} - \text{cov.}[Z_L, Y_L]A = 0 \quad \ldots\ldots(22)$$

This equation can be expanded by use of the relationship, $Z_L = Y_L + V_L$. However, in the case where the input to the process contains no components of the noise V,

$$\text{cov.}[V_L, Y_L] = 0$$

Thus equation (22) becomes,

$$\{\text{cov.}[Y_L, Y_L] + \text{cov.}[V_L, V_L]\}\hat{A} - \text{cov.}[Y_L, Y_L]A = 0$$

so that,

$$\hat{A} - A = -\text{cov.}[Y_L, Y_L]^{-1} \cdot \text{cov.}[V_L, V_L]\hat{A} \ldots\ldots\ldots(23)$$

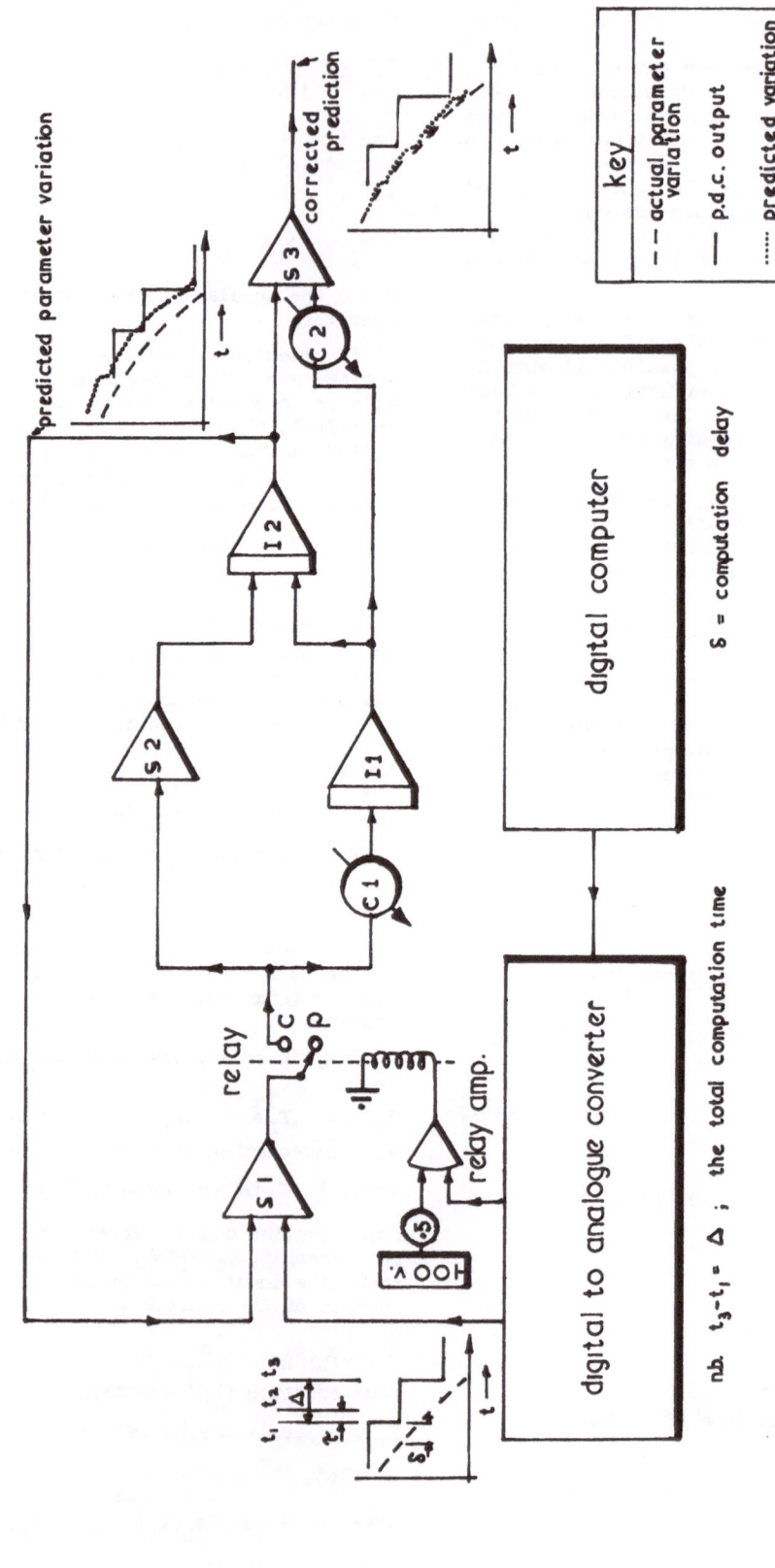

fig.7   first order   prediction   circuit

nb. $t_3 - t_1 = \Delta$; the total computation time

δ = computation delay

The effect of the noise or 'uncertainty' on the observed signals is apparent if this equation is compared with equation (21). In the long term the parametric estimate error, $\hat{A}-A$, is biased by a factor dependent upon the statistical properties of the noise and noise free process signals.

Fig.6 demonstrates the effect of various levels of additive white noise on the output signal y(t) of a simple second order process. As might be expected from the theoretical consideration just discussed quite high levels of uncertainty can be tolerated. In the case of white noise, the low pass filters used in the Method of Multiple Filters can help to reduce the magnitude of the elements of the covariance matrix, cov.$[V_L, V_L]$ (equation (23)),as compared with the elements of the straight noise covariance matrix,cov.$[V, V]$. Consequently, the parametric estimate error, $\hat{A}-A$, is correspondingly less in any system using the Method of Multiple Filters, as compared with a similar system based on straight observation of the state variables.

The limitations of the present mechanisation prevented large sets of observation being taken. As a result, it has only been possible to compute short term statistical properties, and equation (23) has not been verified experimentally. However, V.S. Levadi (Ref.7) has derived a similar expression for a continuous analogue system using direct observation of the state variables, and has shown that excellent correlation exists between theory and experiment.

1.5 THE PROBLEM OF THE TOTAL COMPUTATION TIME DELAY

In common with many other parameter estimation schemes, the Hybrid P.D.C. is limited to the investigation of processes having fixed or slowly variable parameters. Although in some measure this is the result of restrictions imposed by the mathematical assumptions which are the basis of the techniques, it arises principally from more practical considerations. The computer requires a finite time to sample data and to perform its computations, and as a result parameter estimates can only be supplied at discrete intervals of time. During the sampling and computation period the variation of the parameters must be within bounds otherwise outdated estimates are obtained which do not provide a satisfactory description of the actual process characteristics.

It is possible that the total computation time delay (i.e. including sampling, A-D and D-A transfer times) could be reduced by the design of a special purpose hybrid computer. This could also reduce physical size and make the equipment a practical proposition for use in on-line adaptive control applications and as a tool in the analysis of dynamic experiments. In addition, an initial feasibility study has indicated that a further stage of the hybrid equipment may be developed. The operation of this equipment would be based on method 3 mentioned in Appendix 1 and Ref.3. By utilising at least M+N+1 different sets of low pass filters (Fig.15) it appears that all the information required to determine the M+N+1 parameters could be extracted by taking samples at only one instant of time. In any practical system of this type it would probably be best to reach a compromise, incorporating only the minimum M+N+1 sets of filters, but at the same time sampling at more than one instant of time in order that sufficient data is available for use in the statistical least squares procedure.

In the present system all of the S samples required for each computation are taken at one sampling interval. A possible improvement could be obtained by taking only S_1 samples, where $S_1 < S$ and could be made equal to unity. This would be accompanied by the removal of the S_1 oldest samples taken during previous sampling intervals. This approach has been attempted in the case of a non-stationary process, but the slow operation time of the present hybrid equipment has resulted in inferior performance.

Reduction of the total computation time delay by means of the methods indicated above cannot provide the complete answer to this problem. The need to accumulate a large amount of information on the performance of the process, in order that a reasonable parameter estimate can be obtained, necessarily results in unavoidable computation delays. One simple approach to the problem is being developed at the present time. It concerns the introduction of first order analogue prediction elements based on the past rate of change of the parameters. The circuit diagram of such an element is shown in Fig.7. The relay is closed to C during the period t_1-t_2. In this condition the circuit has critically damped second order characteristics with rapid response time. Consequently, the output of the integrator I2 is quickly corrected to the level of the input to S1 from the P.D.C. (i.e. the new parameter estimate). Activation of the relay to p at time t_2 results in zero output from S2 and an output from I1 which is proportional to the difference between the last two parameter estimates. Thus the output of I2 represents the predicted variation in the parameter, and the output of S3 is this predicted variation corrected for the computation delay. The operation of the circuit is described in more detail in (Ref.8). Fig.8 illustrates the effect of introducing the predictive element into a P.D.C. which is tracking a ramp change in a process parameter. First order prediction of this type would not be so effective in dealing with other than ramp changes in the parameters, but it is possible that higher order elements may be developed which overcome this difficulty.

1.6 FUTURE DEVELOPMENT OF THE PARAMETER ESTIMATION TECHNIQUES

Uncertainty caused by additive noise contamination of the sampled signals can be a very real problem with any system relying on measured data. The 'bias' effect of additive noise has been

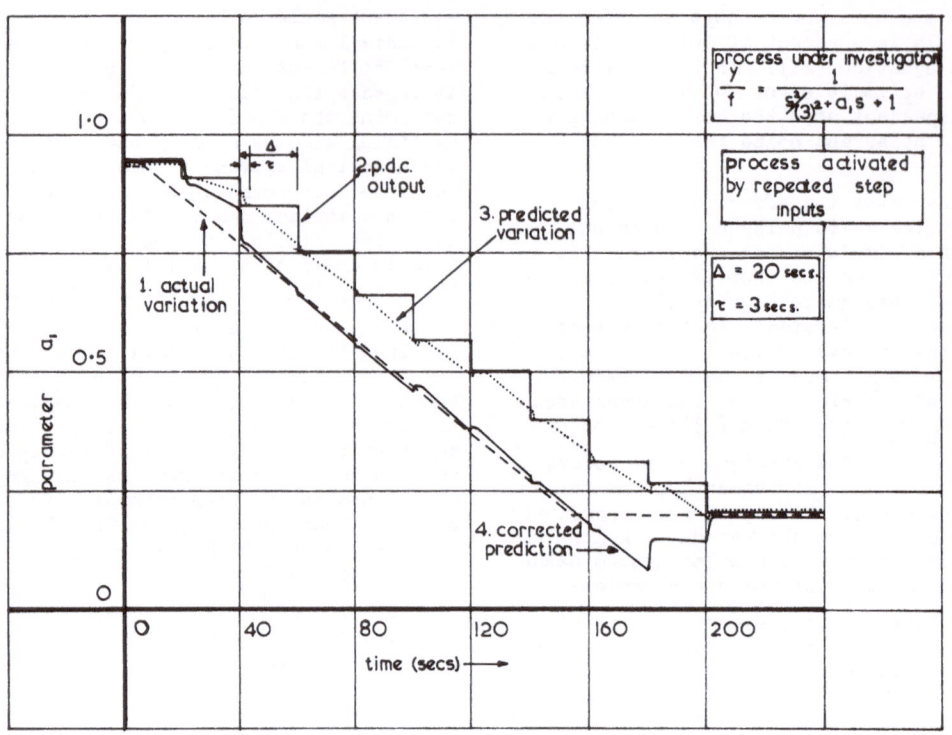

fig.8.   hybrid pdc II with first order analogue prediction elements
investigating a non stationary second order process

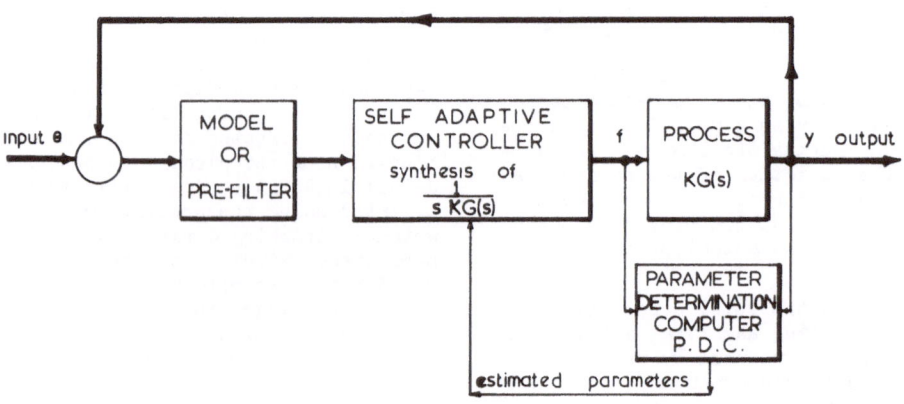

fig.9.   schematic diagram showing principal components of the SESAC system

discussed in section 1.4. One possible approach to the problem is mentioned in Appendix 2. It entails the use of an auxiliary model constructed from the parameter estimates. The technique would be a straight development of work carried out on a similar analogue parameter estimation scheme, by Victor S. Levadi (Ref.7).

An extension of the hybrid techniques to higher order and non-linear processes needs to be investigated. Although no extensive difficulties are envisaged as far as higher order linear processes are concerned (except for the necessary increase in sampling and computation time), the problem of identifying certain types of non-linear process should pose a number of difficult questions. Recent investigations have been concerned with the estimation of the second order equivalent parameters of higher order processes. Results suggest that the estimated parameters provide good agreement in the time domain (a second order model constructed with the aid of the estimated parameters and subjected to the same input as the higher order process under investigation, produces an output response which agrees very well with the output response of the process itself).

2.1 APPLICATION OF AUTOMATIC PARAMETER ESTIMATION TO THE PROBLEM OF SELF ADAPTIVE CONTROL

In 1958 R.E. Kalman (Ref.9) discussed a generalised concept of control in which he compared the role of the self adaptive controller with that of the control systems designer. In accordance with this concept, it is suggested that the controller should perform three major functions. These are :

1. Estimation of the dynamic characteristics of the process during normal operation.

2. Determination of the characteristics of the control element of reference to this information.

3. Construction or modification of the control element in accordance with the details computed in 2. above.

A simple self adaptive control system, designed in accordance with these basic principles, and using the techniques of parameter estimation discussed in the previous sections, has been described in (Ref.4.) The fundamental idea behind this Hybrid Satisfaction Error Self Adaptive Control, S.E.S.A.C., system is illustrated in Fig.9. In a more advanced form this system might be used in the self adaptive auto-stabilisation of aircraft.

Theoretically the self adaptive controller in the S.E.S.A.C. system is intended to continuously synthesise a function of the inverse transfer function of the process to be controlled. However, in practical terms the structure of the controller will often be limited to a second order form. As a result, it is more realistic to regard it as a variable parameter proportional-integral-derivative element (three term controller) which shapes the input demand to the process in order to produce a desired output response.

The system achieves adaptive control by using the information obtained from the P.D.C. in a certain organised manner. However, this same information might well be used in the design of systems having completely different adaption algorithms to that of S.E.S.A.C. So far, no attempt has been made to improve, or modify in any way the basic system. However, it is intended to investigate the S.E.S.A.C. principle in more detail subsequent to the realisation of a completely acceptable method of parameter estimation.

Despite it's limitations, the present S.E.S.A.C system does provide an excellent framework in which to assess the on-line operation of the Parameter Determination Computer. Fig.10 illustrates the control of a non-stationary second order process. In this case step changes in the process parameters are detected quickly and accurately by the P.D.C. and it is possible to maintain tight control over the transient response characteristics. This can be compared with the more realistic case shown in Fig.11. Here the system has been applied to the d.c. servo system mentioned earlier (section 1.2). Although the control action is not as good as in the previous example it is quite satisfactory considering the elementary nature of the present P.D.C. The effect of the control system on the step response of the d.c. servo is shown in Fig.12.

Finally, the improvement in control obtained by incorporating analogue prediction circuits into the P.D.C. (Appendix 2) is demonstrated in Fig.13. Here the system is controlling an analogue second order process having a ramp variation in one of it's parameters. The reduction in response error, ϵ, obtained by the use of a single prediction element on the variable parameter channel is apparent from this diagram.

2.2. IMPROVING THE PERFORMANCE OF THE S.E.S.A.C. SYSTEM

The S.E.S.A.C. approach to control is attractive because of it's simplicity. However, it is just this simplicity which provides some of the major limitations and disadvantages of the system.

The basic control principle must be further developed so that it may be applied in any practical situation. At the moment, it would appear to be limited to low order linear processes. Further investigation must be aimed at finding out whether the system can be used with non-linear or higher order processes. As mentioned earlier, initial attempts at applying the present system

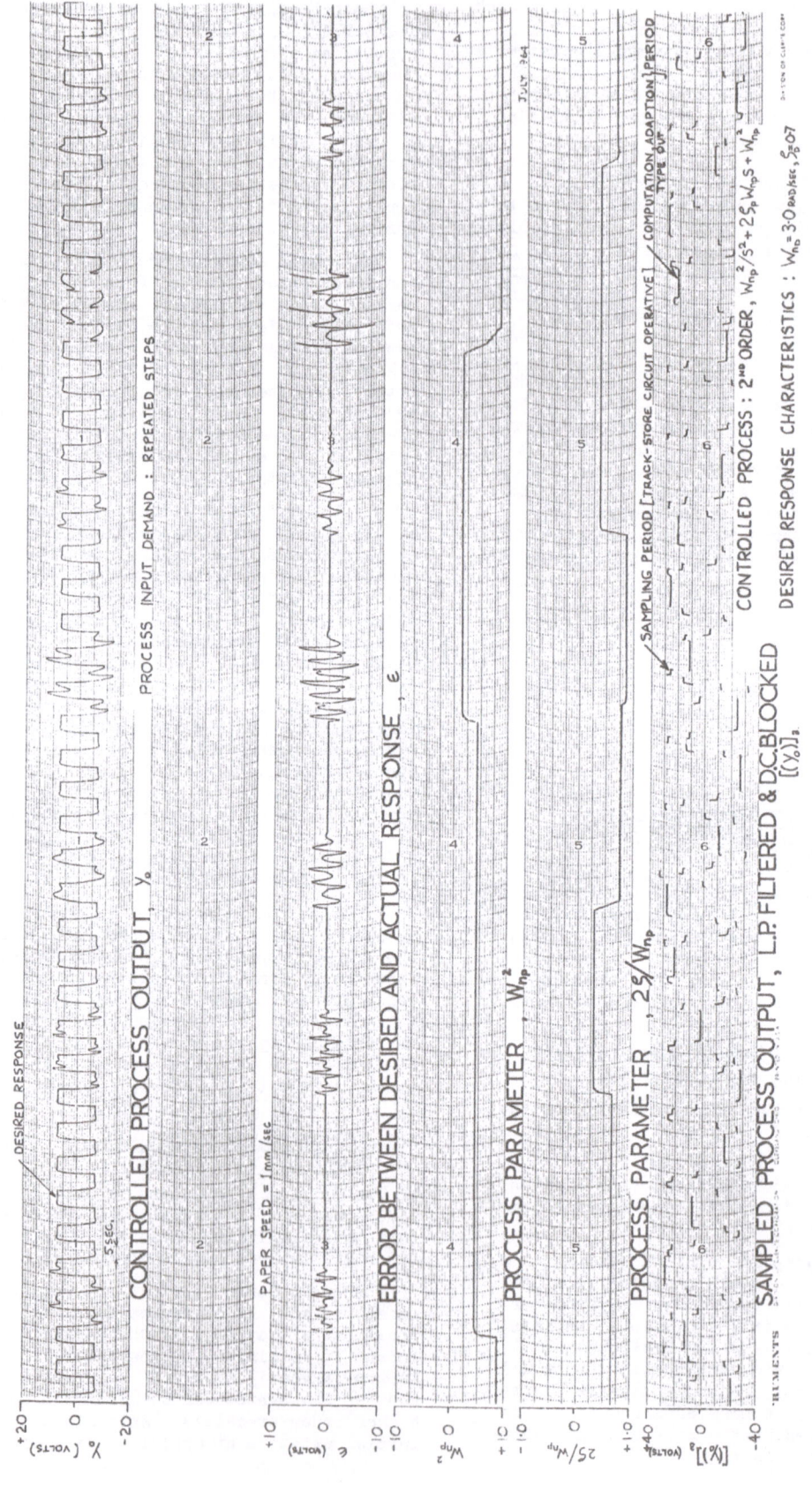

fig.10. hybrid SESAC system applied to a 2nd order analogue process

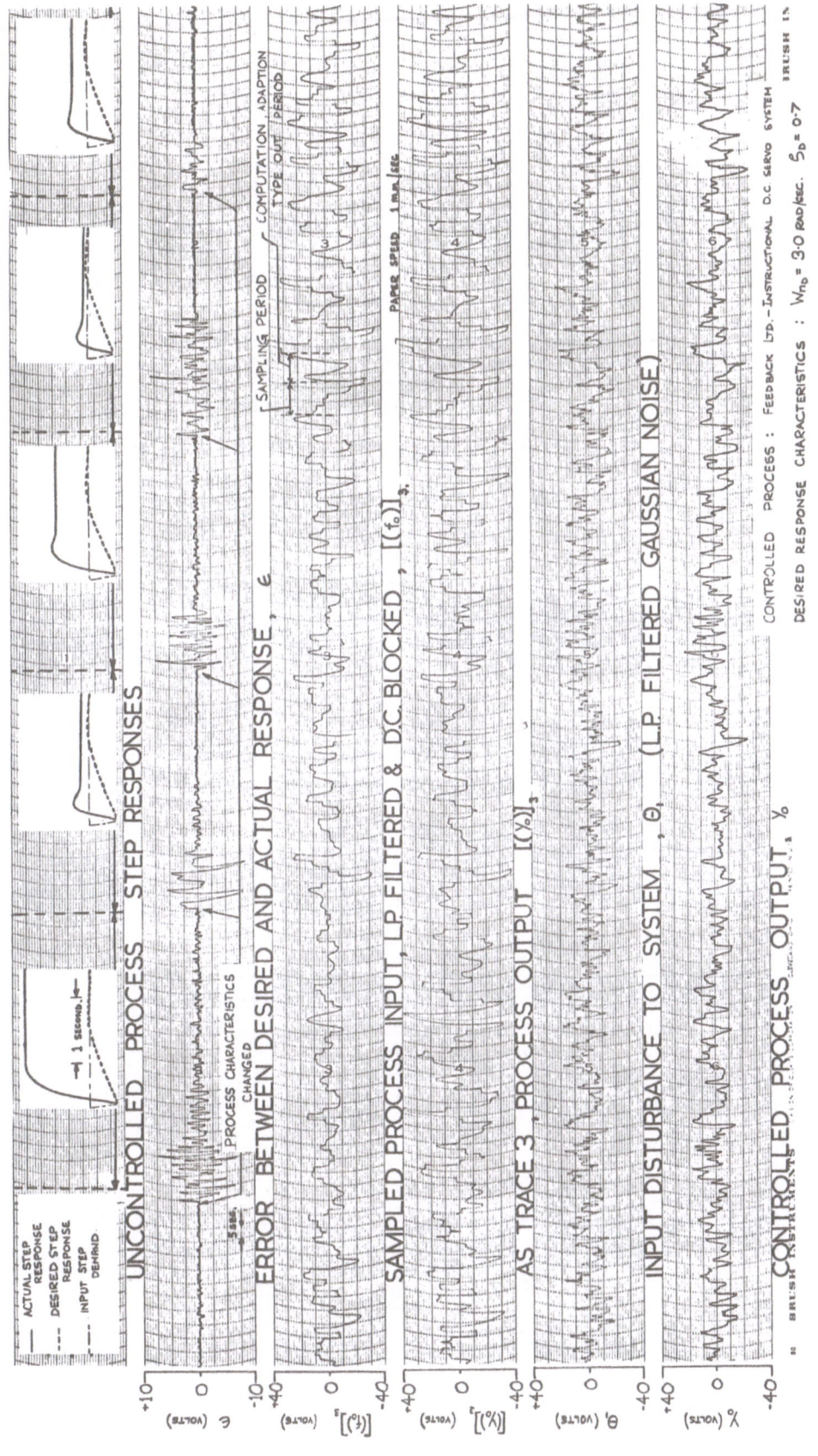

fig.ll. hybrid SESAC system applied to a d.c. position control servomechanism

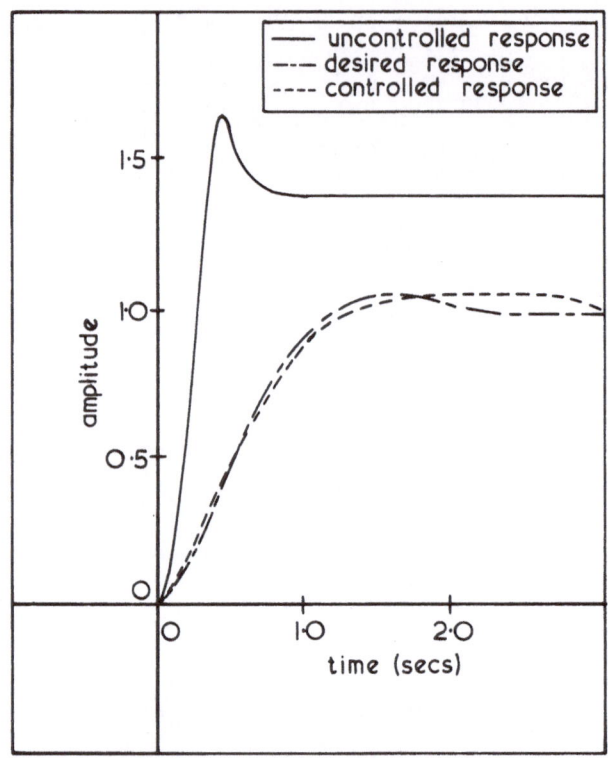

fig.12. hybrid S.E.S.A.C. system applied to a dc position
control servo. : step response for one particular
servo condition

to a d.c. position control servo have provided encouraging results, and it is possible that even in it's basic form the system may have wider application than at first envisaged.

As a next step in the development of this approach to adaptive control, the information obtained from the P.D.C. could be used to further improve the overall performance of the system. A knowledge of the process parameters could provide the basis for optimisation calculations (to be performed in the digital side of the equipment), which would be aimed at minimising or maximising certain performance criteria. An example which immediately springs to mind is the minimisation of gust effects in an aircraft control system. Dynamic Programming and/or variational techniques (including Pontryagins Maximum Principle) would provide the basis for such computations. Another possibility is the incorporation of extra feedback loops around the process to provide some improvement in performance and so make more simple the task of the adaptive portion of the controller.

Circulatory noise within the closed loop puts a limit on the final accuracy of the present parameter estimation scheme. It's effect will have to be evaluated in more detail and investigations should proceed to discover whether any method of easing the problem can be found. The methods suggested in Appendix 2 would be most effective when no circulatory noise exists in the system. However, if they are successful they should help to aleviate the difficulties to some extent.

The open loop nature of the parameter estimation methods means that the stability of the adaptive loop itself is not in question. Nevertheless, the stability of the whole S.E.S.A.C. system is dependent upon the speed of identification (Ref.10). The use of more sophisticated special purpose hybrid equipment and simple prediction elements provides a reasonable approach to this problem.

The basic S.E.S.A.C. system described in this paper has no inherent check on whether the adaptive system is maintaining satisfactory control. This will have to be remedied, possibly by the use of 'model reference' trimming adjustments.

Finally the memory store inherent in the digital side of the hybrid equipment provides the future possibility of 'second generation' self adaptive control. It is envisaged that the store would have the secondary purpose of imparting a rudimentary 'learning' capacity to the control system by supplying a long term memory for the calculated process parameters. In the case of a self adaptive aircraft autostabilisation system, these solutions could be used again should similar environmental conditions occur in the future. Control parameters could be set up on the basis of this information, so reducing adaption time; the P.D.C. would merely act as a check on these values and 'update' them if necessary.

3.1 CONCLUSIONS

A method of automatic process parameter estimation has been described which is easily mechanised from hybrid (analogue-digital) equipment. The estimation is based on the normal operation input signals to the process and does not require the injection of special test disturbances. A technique is introduced which eliminates the need for direct measurement of the derivatives of the process input and output signals which has been the major disadvantage of earlier systems of this type. It is shown that uncertainty on the sampled signals, caused by such effects as measurement noise, tends to bias the parameter estimates. One possible approach to this problem is suggested.

One application of the automatic parameter estimation scheme is in the field of self adaptive control systems. This is discussed and the simulation of a simple identification-adaptive system is described. In addition to it's use with linear second order analogue processes this system has been used to control a d.c. position control servomechanism.

APPENDIX 1.

The Method of Multiple Filters

Consider the general differential equation,

$$\sum_{n=0}^{N} a_n \frac{d^n y(t)}{dt^n} = \sum_{m=0}^{M} b_m \frac{d^m f(t)}{dt^m} \qquad \dots\dots\dots(1.1)$$

Performing a Laplace transformation on this equation and then multiplying by a modulating function of the form $\dfrac{c^L}{(s+c)^L}$ one obtains,

$$\frac{c^L}{(s+c)^L}\left[\sum_{n=0}^{N} a_n\left[s^n y(s) - \sum_{p=1}^{n} s^{n-p}\left(\frac{d^{p-1} y(t)}{dt^{p-1}}\right)_{0+}\right]\right]$$

$$= \frac{c^L}{(s+c)^L}\left[\sum_{m=0}^{M} b_m\left[s^m f(s) - \sum_{p=1}^{m} s^{m-p}\left(\frac{d^{p-1} f(t)}{dt^{p-1}}\right)_{0+}\right]\right]$$

$$\dots\dots\dots(1.2)$$

where the brackets $(\ )_{0+}$ indicates the initial conditions of the enclosed variables.

Considering initially the left hand side of equation (1.2), it is possible to expand binomially in terms of $(s+c)$ those terms which do not depend upon the intial conditions of $y(t)$ and its derivatives. Thus,

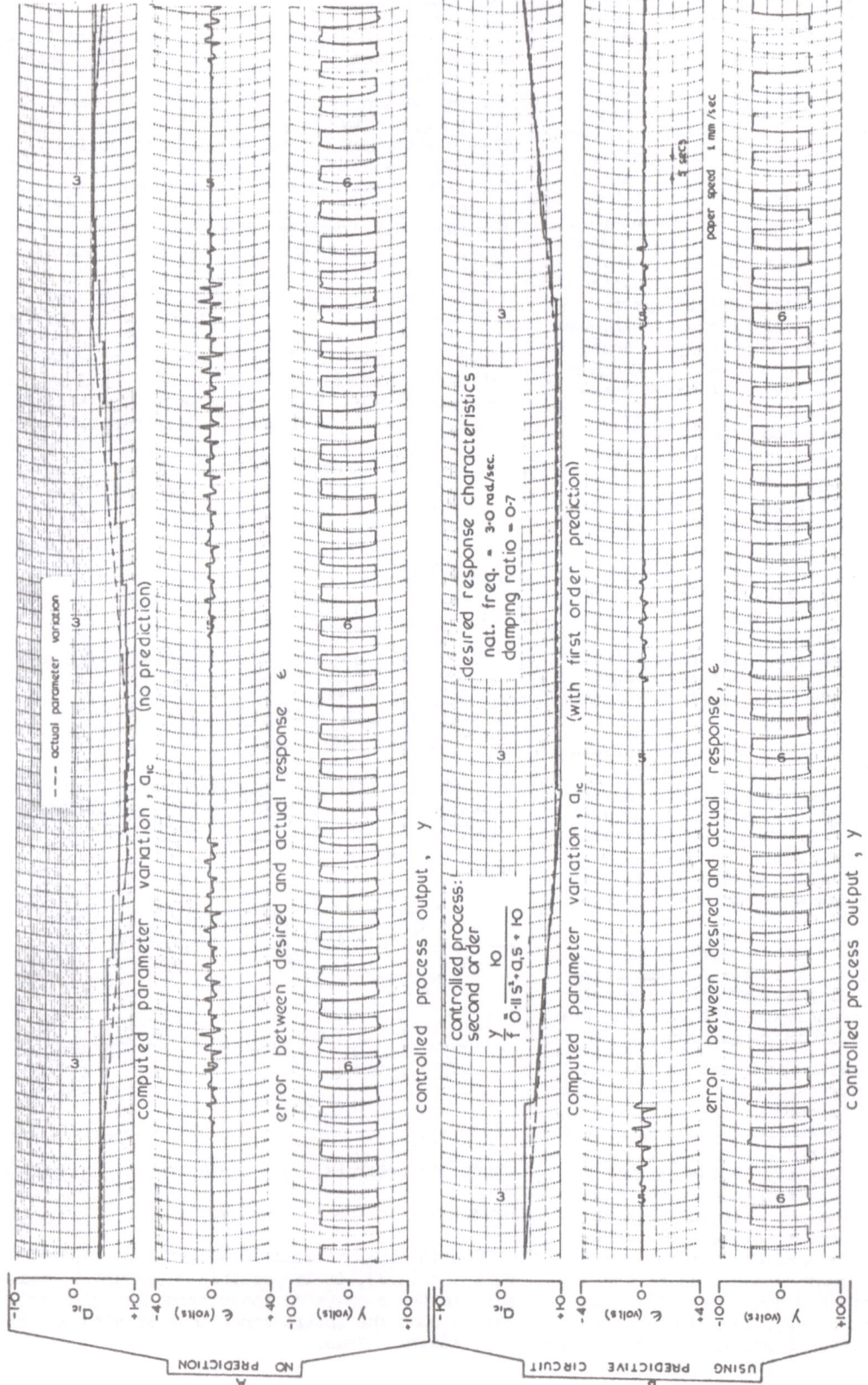

fig.13. improvement in adaptive control by the introduction of first order prediction element into the p.d.c

(A) ramp variation in parameter a_1 – no prediction

(B) similar to (A) but using predictive circuit

134

$$\frac{c^L}{(s+c)^L} s^n y(s) = (-1)^n c^n \left\{ \frac{c^L}{(s+c)^L} y(s) \right.$$

$$- \frac{nc^{L-1}}{(s+c)^{L-1}} y(s) + \frac{n(n-1)}{2!} \frac{c^{L-2}}{(s+c)^{L-2}} y(s)$$

$$\left. - \frac{n(n-1)(n-2)}{3!} \frac{c^{L-3}}{(s+c)^{L-3}} y(s) + \dots \right\} \quad \dots\dots(1.3)$$

Taking the inverse Laplace transformation of this equation, one can obtain an expression of the form,

$$[(y_n)]_L = (-1)^n c^n \left\{ [(y_0)]_L - n[(y_0)]_{L-1} \right.$$

$$\left. + \frac{n(n-1)}{2!} [(y_0)]_{L-2} - \dots \right\} \quad \dots\dots(1.4)$$

where, $y_n = \frac{d^n y(t)}{dt^n}$ $(n=o \to N)$

and, $[(y_0)]_L$; $[(y_0)]_{L-1}$; $[(y_0)]_{L-2}$ etc.,

may be considered physically as the outputs of a series of identical low pass filters as shown in Fig.14.

Inspection of equation (1.2) will show that the inverse Laplace transformation of the remaining terms on the left hand side of the equation, that is the terms involving initial conditions of the dependent variable y(t) and it's derivatives, will produce functions which die out very rapidly as time progresses, provided the filter time constant 1/c is reasonably small.

The right hand side of equation (1.2) may be treated in a similar manner to the left hand side and the various filtered input derivatives obtained by an expression of the form,

$$[(f_m)]_L = (-1)^m c^m \left\{ [(f_0)]_L - m[(f_0)]_{L-1} \right.$$

$$\left. + \frac{m(m-1)}{2!} [(f_0)]_{L-2} - \dots \right\} \quad \dots\dots(1.5)$$

where, $f_m = \frac{d^m f(t)}{dt^m}$ $(m=o \to M)$

and, $[(f_0)]_L$; $[(f_0)]_{L-1}$; $[(f_0)]_{L-2}$ etc.,

may be considered physically in a similar manner to the analogous functions in y(t)   (see eqn. (1.4) and Fig.14).

The foregoing analysis suggests that the continuous filtration of the various terms appearing in equation (1.1) by a filter of the form

$\frac{c^L}{(s+c)^L}$ enables the parameters a_n and b_m to be related by equation (1.6),

$$\sum_{n=0}^{N} a_n [(y_n)]_L = \sum_{m=0}^{M} b_m [(f_m)]_L \quad \dots\dots(1.6)$$

where the brackets $[( )]_L$ indicate that the enclosed function has been filtered L times by a filter of the above form, and

$y_n = \frac{d^n y(t)}{dt^n}$ (similarly for f_m).

Equation (1.6) may be considered true for all time $t > \varepsilon$, where ε is a small interval of time immediately following the initiation of the filtration process and whose magnitude is dependent upon the filter time constant, 1/c. Equations (1.4) and (1.5) allow the functions $[(y_n)]_L$ and $[(f_m)]_L$ (n=o $\to$ N; m=o $\to$ M) to be obtained by the summation of signals originating from the outputs of the various low pass filters, Fig.14.

This simple procedure, provides a method of generating the relationship shown in equation (1.6) which does not require direct measurement of the derivatives of the input and output signals. This fact is important because these derivatives can be difficult or even impossible to obtain directly in practice.

A similar, although more complex, analysis might be used to indicate how non-identical filters could be used for the same purpose.

Equations of the form of equation (1.6) may be solved for the parameters a_n and b_m in a number of ways.

1. Functions of the modified satisfaction error, E_m, may be continuously minimised. Where,

$$E_m = \sum_{n=0}^{N} a_{nc} [(y_n)]_L - \sum_{m=0}^{M} b_{mc} [(f_m)]_L \quad \dots\dots(1.7)$$

this method is similar to that discussed in Ref.2. The minimisation could be accomplished for fairly low order systems by using the steepest descent of the particular error parameter hypersurface selected.

2. b_0 may be assumed unity without loss of generality. Thus samples of the various signals required may be taken at a minimum of M+N+1 discrete instants of time to provide M+N+1 linear algebraic equations in M+N+1 unknowns ($a_0 \to a_N$; $b_1 \to b_M$). When samples are taken at more than M+N+1 instants of time they can be used in a least squares estimation (section 1.1 of the paper).

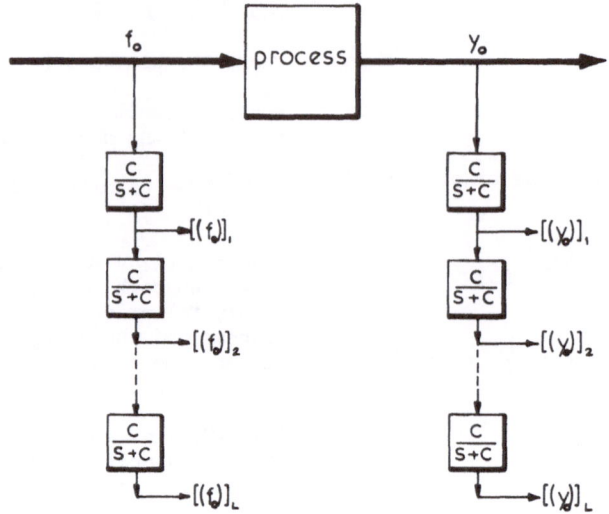

fig.14. multiple filters I

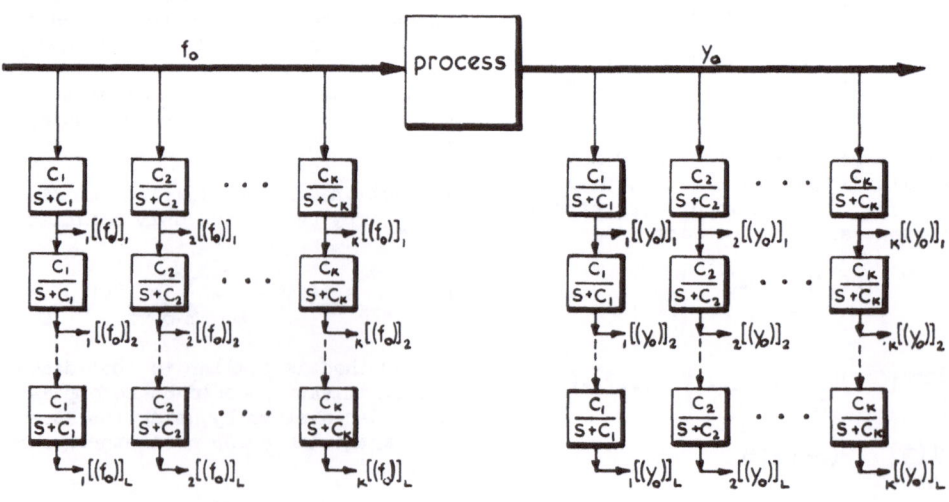

fig.15. multiple filters II   [note K⩾M+N+1]

3. Alternatively, at least M+N+1 sets of different filters may be used (see Fig.15) in order that M+N+1 equations in M+N+1 unknowns may be generated by the use of samples taken at only one instant of time.

APPENDIX 2.

Elimination of Noise Effects

One disadvantage of the simple least squares approach to parameter estimation is the presence of a bias on the parametric estimate if the samples signals are contaminated by noise. This appendix deals with a particular approach to the problem which has been discussed by V.S. Levadi (Ref.7).

A model of the process is constructed with the aid of the estimated parameters $\hat{A}$. This model is subjected to the same input as the process and let it's output be denoted by y_{oc} (i.e. the estimated value of y_o). If this signal is filtered and sampled in a similar way to the actual process output (Fig.16), it is possible to estimate a vector $\hat{Y}_L$, where,

$$\hat{Y}_L^T = \left[[(y_{oc})]_L, \ldots\ldots, [(y_{Nc})]_L, [(f_{1c})]_L, \ldots \right.$$
$$\left. \ldots, [(f_{Mc})]_L \right]$$

using the method of multiple filters.

E_{mj} can be defined as in section 1.4 by,

$$E_{mj} = {_j}Z_L^T \hat{A} - {_j}f_L \qquad (j=1,2,\ldots.S)$$

but let the least squares equations for the function, $\Gamma = \sum_{j=1}^{S} E_{mj}^2$, be replaced by,

$$\sum_{j=1}^{S} {_j}\hat{Y}_L \cdot {_j}Z_L^T \hat{A} - {_j}\hat{Y}_L \cdot {_j}f_L = 0$$

where, as before, the prefix j indicates that the function is sampled and is the jth sample of a set.

Thus, in place of equation (22) of section 1.4 one obtains,

$$\text{cov.}[\hat{Y}_L, Z_L]\hat{A} - \text{cov.}[\hat{Y}_L, Y_L]A = 0 \quad \ldots\ldots(2.1)$$

If the input to the process and the model, f, is uncorrelated with the additive noise, v, then,

$$\text{cov.}[\hat{Y}_L, Z_L] = \text{cov.}[\hat{Y}_L, Y_L]$$

and,

$$\text{cov.}[\hat{Y}_L, Y_L]\{\hat{A} - A\} = 0 \qquad \ldots\ldots(2.2)$$

In other words, if the least squares equations are generated in the special way detailed above it should be possible, in the long term, to remove the bias effect of the additive noise contamination. Levadi has used this approach in a continuous parameter estimation scheme and has shown that it achieves the desired effect.

This appendix has dealt with one possible approach to the additive noise problem encountered in a parameter estimation scheme such as that described in the present paper. The problem is a classical one in statistical parameter estimation theory and the method suggested is certainly not the only one available. Other techniques might be based on Kalman - Bucy filter theory (Ref.11) or Maximum Likelihood Estimates (Refs.12 and 13).

NOMENCLATURE

Note: Boldface upper case letters denote a matrix quantity. A superscript T used with such a boldface letter indicates a transposed matrix When shown in more detail the elements of the matrix are enclosed by square brackets.

t	time
f	(f(t)) arbitrary input or forcing function to a process
y_r	($y_r(t)$) time dependent variables of the process. (r=o→k)
a_r	($a_r(t)$) slowly variable or fixed coefficients or parameters which appear in the mathematical model of the process. (r=o→k)
a_{rc}	estimated values of the parameters a_r.
f_m	$\dfrac{d^m f}{dt^m}$, arbitrary input and it's derivatives in general linear differential equations of process. (m=o→M)
y_n	$\dfrac{d^n y}{dt^n}$, output dependent variable and it's derivatives in general linear differential equation of process. (n=o→N)
b_m	coefficients of input and it's derivatives in general linear differential equation of process. (m=o→M)
a_n	coefficients of dependent variable and it's derivatives in general linear differential equation of process. (n=o→N)
$\left. \begin{array}{c} b_{mc} \\ a_{nc} \end{array} \right\}$	estimated values of b_m and a_n
s	Laplace operator.
S	integer denoting sample size.
${_j}[(\ )]_L$	brackets enclosing a time variable function indicates that the function enclosed has been filtered L times by a low pass filter having a Laplace transform transfer function $^c/s{+}c$ (where c is a constant). If j is present it indicates that the total filtered quantity is sampled and is the jth sample of a set. (j=1→S)

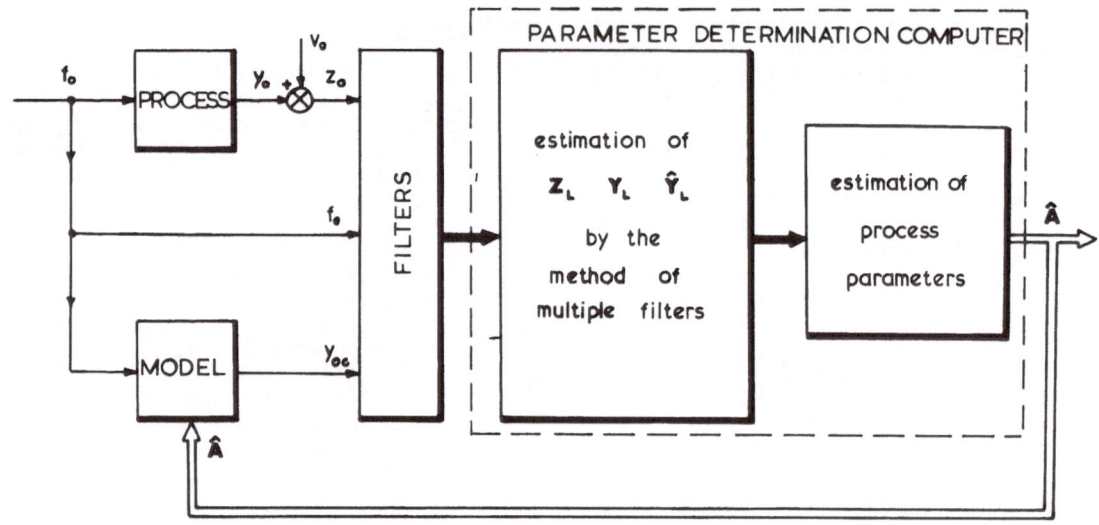

fig.16.   the auxiliary model approach to the noise problem

$_j^f\mathbf{Y}^T$ $\qquad$ $\left[\; y_0, y_1, \ldots, y_k \right]$ or $\left[y_0, y_1, \ldots, y_N, f_1, f_2, \ldots, f_M \right]$

$\mathbf{Y}_L^T$ $\qquad$ $\left[[(y_0)]_L, \ldots, [(y_N)]_L, [(f_1)]_L, \ldots, [(f_M)]_L \right]$

$_j\mathbf{Y}_L^T$ $\qquad$ $\left[_j[(y_0)]_L, \ldots, _j[(y_N)]_L, _j[(f_1)]_L, \ldots, _j[(f_M)]_L \right]$

$\mathbf{Y}_L^T$ $\qquad$ $\left[[(y_{0c})]_L, \ldots, [(y_{Nc})]_L, [(f_{1c})]_L, \ldots, [(f_{Mc})]_L \right]$

$\qquad$ estimated state variables obtained by way of an auxiliary model constructed from the parameter estimates .

$\mathbf{A}^T$ $\qquad$ $\left[a_0, a_1, \ldots, a_k \right]$ or $\left[a_0, a_1, \ldots, a_N, b_1, b_2, \ldots, b_M \right]$

$\hat{\mathbf{A}}^T$ $\qquad$ as for $\mathbf{A}^T$ but with the estimated values of the parameters for elements.

E $\qquad$ error in satisfaction of the process equation — 'satisfaction error'.

E_{mj} $\qquad$ modified satisfaction error,

$$E_{mj} = \sum_{n=0}^{N} a_{nc}\, _j[(y_n)]_L - \sum_{m=0}^{M} b_{mc}\, _j[(f_m)]_L$$

$(j = 1 \to S)$

Γ $\qquad$ $\displaystyle\sum_{j=1}^{S} E_{mj}^2$

$\langle \; \rangle$ $\qquad$ denotes expectation operator (ensemble average).

cov. $\qquad$ covariance matrix.

v_k $\qquad$ additive noise contamination $(k = 0 \to M+N+1)$

$\mathbf{V}^T$ $\qquad$ $\left[v_0, v_1, v_2, \ldots, v_{M+N+1} \right]$

$\mathbf{V}_L$ $\qquad$ column matrix of noise components. The elements depend upon the elements of and on the nature of the filters used in the estimation scheme.

z_k $\qquad$ observable process variables $(k = 0 \to M+N+1)$

$\mathbf{Z}_L$ $\qquad$ $\left[\mathbf{Y} + \mathbf{V} \right]_L = \mathbf{Y}_L + \mathbf{V}_L$

ACKNOWLEDGMENT

The investigations described in this paper were carried out under a research scholarship in the Department of Aeronautical and Automobile Engineering, Loughborough College of Advanced Technology. The paper is published with the permission of the Head of Department, Professor K.L.C. Legg. The author wishes to thank Mr. C.D. Dwyer for his collaboration on the various problems of hybrid instrumentation and digital programming. Thanks are also due to Dr. J.D. Roberts of the University of Cambridge for his helpful discussion on the paper.

REFERENCES

(1) Eykhoff, P., "Some Fundamental Aspects of Process Parameter Estimation", Trans. I.E.E.E. on Automatic Control, Vol.AC-8, No.4, October 1963, pp.347-57.

(2) Young, P.C., "The Determination of the Parameters of a Dynamic Process", The Radio and Electronic Engineer, Vol.29, No.6, June 1965, pp.345-61.

(3) Young, P.C., "In-Flight Dynamic Checkout: a discussion", Trans. I.E.E.E. on Aerospace, Vol.AS-2, No.3, July 1964, pp.1106-11.

(4) Young, P.C., and Dwyer, C.D., "A Hybrid Self Adaptive Control System", recipient of P.A.C.E. Prize 1964, awarded by Electronic Associates Ltd.

(5) Dwyer, C.D., "Hybrid Computation", Masters Thesis submitted to University of Leicester.

(6) Truxal, J.G., "Control System Synthesis", McGraw-Hill Book Company Inc., New York, 1955, p.55 et seq.

(7) Levadi, V.S., "Parameter Estimation of Linear Systems in the Presence of Noise", Paper presented at the International Conference on Microwaves, Circuit Theory and Information Theory, Tokyo, 7th-11th September, 1964.

(8) Young, P.C., "A Simple Analogue Predictive Circuit", internal technical note Department of Aeronautical and Automobile Engineering, Loughborough C.A.T.

(9) Kalman, R.E., "Design of a Self Optimising Control System", Trans. A.S.M.E., Vol.80, No.2, February 1953, pp.852-62.

(10) Young, P.C., "Parameter Estimation and Self Adaptive Control", A.L.C. Thesis, Loughborough College of Advanced Technology, 1965.

(11) Kalman, R.E. and Bucy, R.S., "New Results in Linear Filtering and Prediction Theory", Trans. A.S.M.E., Vol.83, 1961, pp.95-108.

(12) Linnik, Y.V., "Method of Least Squares and Principles of the Theory of Observation", Pergamon Press, London, 1961.

(13) Aström, K.J. and Bohlin, T., "Numerical Identification of Linear Dynamic Systems from Normal Operating Records", paper presented at the I.F.A.C. (Teddington) Symposium 1965, on "The Theory of Self Adaptive Control Systems", Teddington, England, 14th-17th September 1965.

Discussion

The following remarks were based on
Mr. Young's verbal contribution to the Symposium
before the text of his paper was available.
Mr. Eykhoff (Netherlands) interpreted Mr. Young
as implying that each filter provided a coefficient of the differential operators of the
differential equation describing the generalized
model - a point in fact possible only with pure
differentiators.

Mr. Rowe (U.K.) remarked on the similarity
of the method and that described by Mishkin and
Braun for generating Laguerre polynomial
coefficients. Mr. Young stated in clarification
that the "Method of Multiple Filters" generated
certain functions of the derivative terms which
could then be used, in the manner described in
his paper, to arrive at a least squares estimate
of the unknown parameters. Mr. Young considered
the problem of removing noise induced bias and
saw promise in a technique similar to that suggested by V. S. Levadi. By the incorporation
of an auxiliary model of the process based on
the estimated parameters, it should be possible
to discriminate against any noise present on the
sampled signals and so reduce the bias effect.
Again, an adaptive filter of the Kalman-Bucy type
might be utilised to obtain optimal unbiased
estimates, whilst at the same time providing an
answer to the optimal regulator problem.
Mr. Young stated that the method described in the
paper was not related to the method of Laguerre
polynomials dealt with by Mishkin and Braun.
However, if Mr. Rowe was interested in an approach
to identification along those lines he should be
referred to the paper, "The Synthesis of Dynamical
Models of Plants and Processes" by W. D. T. Davies,
which was available in the Proceedings of the
U.K.A.C. Convention on Advances in Automatic
Control, Nottingham, 1965.

Mr. Duckenfield (U.K.) suggested that feedforward of observed noise might prove helpful.

Dr. J. H. Andreae (U.K.)

Nearly 4 years ago in the field of Artificial Intelligence a learning machine called STeLLA was conceived. After three years of computer simulation of the machine in a variety of problem environments, the design is greatly improved, although the main features have survived. Since the machine is being simulated with a vehicle-steering control problem which is second order, non-linear and stochastic it would appear that STeLLA now qualifies for the title of "adaptive controller".

Currently we are developing hardware, based on a statistical digital adaptive circuit called the ADDIE, for building parallel-processing computers like STeLLA. Some of the remarks made at this Symposium seem to be relevant to our learning machine/parallel computer/adaptive controller.

When Professor Fu was discussing his learning controller without external supervision, he suggested that "the controller correlates each measurement V with all possible control situations". In a machine with a limited information store and with a reasonable number of input channels, such a procedure would be too time-consuming, if not impossible. Professor Fu goes on to say that this "controller uses its own decision to direct the learning process". In our machine an internal measure of performance is generated from the (incomplete and intermittent) externally-provided performance feedback and this measure is used to provide reinforcement which does, in a sense, direct the learning process. It is not clear to me that Professor Fu intends any more than this.

Dr. Kwakernaak's definition of admissible policies makes me wonder whether it would not be more appropriate to talk about "admissible problems" for a general-purpose adaptive controller or learning machine. If our learning machine can become an admissible controller for a number of problems, then we can talk about these problems as being admissible problems. In many cases, however, we shall be happy if our adaptive controller converges on a stable control policy, be it admissible or not.

Dr. Åström describes a predictive controller for linear systems which he says can be extended to multiple input systems. In STeLLA a Bayesian predictor with adaptive logic is used for the forward prediction of trajectories, but it suffers severely from the necessary restriction on information storage. The main control policy of the machine is generated by a path-learning procedure which could be described as a kind of on-line, incomplete dynamic programming. The total strategy of the machine can be viewed as the forward prediction of trajectories into a policy space which has been learned as a set of generalized paths into the goal or demand region.

Professor Li's remark about the open and closed loop behaviour of the human operator can be applied to our learning machine. The closed loop character of the learning stage gradually turns over to an efficient open loop procedure as useful experience is filtered out into a definite control policy. A change in the problem environment calls in the closed loop behaviour and efficiency again gives way to learning.

In the field of artificial intelligence it was natural to use a state description for the machine and to present it with non-linear and stochastic problems. On the other hand, the advantages of linearization and the need for optimization were neglected. We were more concerned with such questions as (a) How much can we afford to let the machine remember? (b) How much should we tell the machine at the start? (c) How can we help the machine in its task? (d) How can we tell the machine what we want it to do and then change our minds as we see the consequences of our requests? (e) How can we communicate with the machine so that we can get it to do that part of the job which we cannot manage without giving it the whole job?

Finally, there are many problems common to the fields of artificial intelligence and automatic control, but most of these will remain 'in the air' if the lack of suitable hardware continues to prevent the proper testing and realization of postulated adaptive control schemes.

Prof. K. J. Åström in reply.

It was interesting to hear that Mr. Andreae had storage difficulties using a Bayesian predictor. We have noticed the same thing and therefore discarded the Bayesian approach to the identification problem. The identification

algorithm described in our paper requires very
little storage space.

Dr. W. D. Ray (U.K.)

It was most interesting to note at a symposium devoted to control theory the use being
made in the papers by Fu, Kwakernaack, Åström
and Bohlin of such statistical extravaganza as
the Bayes, minimax and maximum likelihood
theories of inference in conjunction with
statistical decision theory and sequential
analysis. However, a surprising omission was
that no mention was made of the work of Box and
Jenkins (J. R. Statist. Soc. B, 24, 2, 1962),
which was concerned with precisely the same
problems as those dealt with by the above authors
namely adaptive control and optimisation. Box
and Jenkins produced predictor or controller
equations which automatically adapt to any polynomial trends in the input signals, i.e. they
track the non-stationary statistics automatically
without the requirement, incidentally of their
estimation (see also Ray and Wyld. J.R. Statist.
Soc. B, 27, 1, 1965). This approach is a lot
more embracing than Wiener's theory of control
for example, and takes care of many time-varying
parameter situations implicitly, not simply a
first-order Markov Process as mentioned by
Mr. Kwakernaack.

One small point in the paper by Åström and
Bohlin. Although their paper is an elegant use
of the maximum likelihood principle for the
estimation of the parameters in the control
equations, it often happens in practice that the
efficiency of control is not very sensitive to
the values of the parameters. In other words the
likelihood surface will probably be very flat near
the stationary point and hence no great loss will
accrue if near-optimum estimates are used. Very
often straight evaluation of the likelihood
function for a specimen set of values of the
parameters will be enough.

Prof. K. J. Åström in reply

Dr. Ray's comment is very relevant. We
have, however, also found cases where the likelihood function has sharp ridges. In these
situations it is very difficult to find the
maximum using straight forward probing. Let me
also emphasize that at present we have very poor
knowledge of the shapes of likelihood surfaces.

MODEL REFERENCE AND ADAPTIVE CONTROLLERS

SYNTHESIS OF MODEL REFERENCE ADAPTIVE SYSTEMS BY LIAPUNOV'S SECOND METHOD

by B. Shackcloth and R. L. Butchart
Department of Aeronautics and Astronautics
The University
Southampton, England

ABSTRACT

In this paper, model reference adaptive systems having linear processes and models of the same form and order are considered. The The model parameters have fixed values which represent an ideal or desired form of the process. Liapunov's second method is applied to synthesize the adaptive loop guaranteeing the stability under certain conditions. The results are verified experimentally using an analogue computer. A further investigation of the stability problem uses two theorems due to Kats and Krasovskii.

INTRODUCTION

Model reference systems have always been prominent in the self adaptive control field. Some of the early work was carried out by Whitaker[1] at M.I.T. and later more fully developed with the help of Osborn and Kezer[2]. Leondes and Donalson[3] have recast previous work on parameter tracking systems and they investigate the stability problem using Liapunov theorems due to La Salle[4]. Liapunov's second method has also been used by Grayson[5] to synthesize stable model reference systems. Fishwick and Davies[6], working on system identification, have produced a system very similar to that of Whitaker.

The principle of the model reference system is that they operate by forcing the output of a controlled plant to follow that of a model which represents some ideal or desired form of the plant. The basic concept is illustrated in Fig.1. A process with transfer function p(s) may have some of its parameters varying. Associated with the process is a controller. In the figure it is shown as a gain K_c but it will usually be more complex. The combination of process and controller will be called the plant. The input to the plant, R, is also fed to a model with transfer function m(s). The outputs from plant and model, θ_s and θ_m, are compared to give an error signal, E, and this is used in an adaptive loop to adjust the controller to compensate for process variations. In this paper it will be assumed that the process transfer function is known and linear and that the model is of the same form and order.

With the form of controller shown in Fig.1, it is assumed the process variations can be adequately compensated by changing the controller gain K_c. The system developed at M.I.T.[2] attempts to minimise $\int E^2 dt$ and gives an adaptive loop of the form:

$$\dot{K}_c = B E \theta_m , \qquad (1)$$

where B is the adaptive loop gain. Except by making simplifying assumptions, a stability analysis of this system is not easy. As an example, showing that stability problems exist, a system will be considered where the plant and model are governed by second order equations:-

$$(1 + T_1 D + T_2 D^2) \theta_m = KR , \qquad (2)$$

$$(1 + T_1 D + T_2 D^2) \theta_s = K_v K_c R . \qquad (3)$$

T_1 and T_2 are constants and K_v is the process gain which may be varying. The adaptive loop must vary K_c to keep the plant gain $K_c K_v$ equal to the model gain K. Substracting Eq.(3) from Eq.(2) gives:

$$(1 + T_1 D + T_2 D^2)(\theta_m - \theta_s) = (K - K_v K_c) R . \qquad (4)$$

If an adaptive loop of the form (1) is used and if K_v, θ_m and R are constant, then Eq.(4) and Eq.(1) can be reduced to the equation:

$$(T_2 D^3 + T_1 D^2 + D + B K_v \theta_m R) \theta_s = B K_v \theta_m R \theta_m . \qquad (5)$$

Eq.(5) is a third order system which could be unstable if the product, $B K_v \theta_m R$, is large enough. This is an undesirable feature and it could be a severe restriction on the adaptive loop gain B. If so, the result would be a sluggish adaptive response.

Our aim in this investigation was to apply Liapunov's second method to such model reference systems, and initially it was attempted to obtain stability bounds on simple systems which used the M.I.T. adaptive loop form. However, little success was achieved and consequently our attention was diverted to a synthesis technique. To apply this a positive definite function is assumed for V and then conditions

Superior numbers refer to similarly-numbered references at the end of this paper.

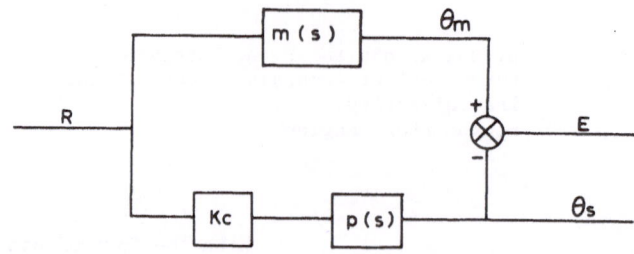

FIGURE 1. MODEL REFERENCE SYSTEM WITH GAIN CONTROL

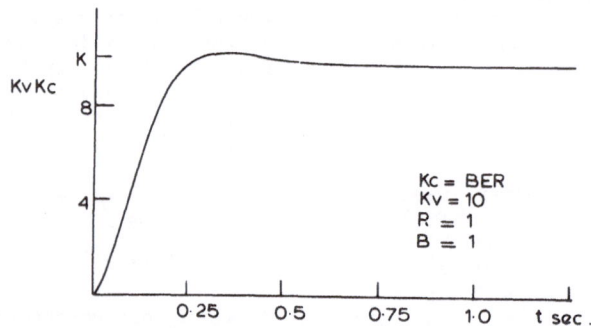

FIGURE 2. ADAPTIVE STEP RESPONSE.

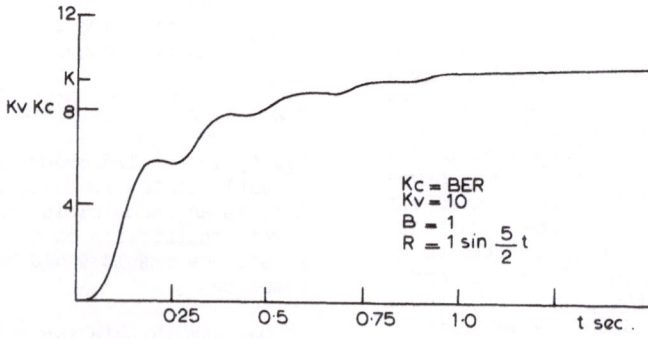

FIGURE 3 RESPONSE WITH R SINUSOIDAL.

imposed on the form of the adaptive loops, so that $\dot{V}$ is made at least negative semi-definite. The stability of the system is thus guaranteed if it is possible to adopt the requisite adaptive loops.

FIRST ORDER PLANT WITH CONTROLLABLE GAIN

As an example of how the technique may be applied consider the model and plant to be governed by the equations:

$$(1 + T D) \; \theta_m = KR , \qquad (6)$$

$$(1 + T D) \; \theta_s = K_v K_c R \qquad (7)$$

The time constant T and the gain K are fixed and known but the process gain K_v is unknown and possibly time varying. The aim of the design is thus to determine a suitable adaptive loop to control K_c so that the plant and model gains are made the same. Now since $E = \theta_m - \theta_s$ the plant and model equations may be combined to form an error equation:

$$(1 + T D) \; E = (K - K_v K_c)R \qquad (8)$$

Writing $x_1 = K - K_v K_c$ and considering initially K_v to be a constant, another system equation applies:

$$\dot{x}_1 = -K_v \dot{K}_c , \qquad (9)$$

One of the major difficulties in applying Liapunov's second method is to determine a suitable choice for a V function; however, for this system a fairly simple one may be used:

$$V = B K_v T E^2 + x_1^2 \quad \text{where B is a constant} \quad (10)$$

$$\therefore \; \dot{V} = -2E^2 K_v B + 2x_1 K_v (BER - \dot{K}_c). \qquad (11)$$

Thus if the adaptive loop is:

$$\dot{K}_c = BER, \qquad (12)$$

where B is the adaptive loop gain, then:

$$\dot{V} = -2 \; E^2 \; K_v B . \qquad (13)$$

Now if B , K_v and T are all positive, V is positive definite and $\dot{V}$ is negative semi-definite, and so according to Liapunov's theorems if the adaptive loop of eq. 12 is used the system must be stable. If the further condition that R does not always remain zero may be imposed then asymptotic stability may be guaranteed,

i.e. as $t \rightarrow \infty$ $E \rightarrow 0$ $x_1 \rightarrow 0$

At this juncture it is convenient to define a term which we have originated, and which may be used as a measure of the goodness of a particular model reference system as regards stability. This term is an adaptive step response, and in this case is defined as the response K_c where R is a constant and K_v is altered by a step. For the previous example

where use of the M.I.T. adaptive loop led to a third order system equation, such a response may be unstable. It would seem unlikely that the stability would be improved if K or R were time varying. Thus a stable adaptive step response might be regarded as the first requisite of a model reference system.

If the first order plant and model above are considered, then it can be seen from the V function analysis that if the adaptive loop of eq.12 is used then this first requirement of the system is satisfied. However, the V function analysis goes further than this as it is unaltered if R is time varying in a continuous bounded manner. With such an input, stability will still be guaranteed and if R is non zero for all time asymptotic stability is present.

To verify the theoretical conclusions about this system, it was simulated on a PACE TR - 48 analogue computer. Low frequency imputs were used (< 10 c.p.s.) and a small value of time constant (T = .05) chosen. As can be seen from fig.2,3 whether R is a constant or sinusoidal the plant gain $K_v K_c$ tends eventually to settle at the value K of the model gain as predicted.

Synthesis technique with K_v time varying.

Many model reference systems have been designed assuming that the process parameters are constant or only varying slowly, and then analogue computer studies are made to see how the system behaviour is affected if these variations become rapid. In this study we attempted to investigate theoretically the effect of rapid process variations.

The previous model reference system with first order plant and model may again be considered and with K_v time varying the system equations become:

$$T \; \dot{E} = -E + x_1 R , \qquad (14)$$

$$\text{and} \quad \dot{x}_1 = -K_v \dot{K}_c - K_c \dot{K}_v \qquad (15)$$

Two methods of approach to the problem are presented in this paper: the first being on extension of the synthesis technique, and the second utilises other concepts of stability for such systems.

The synthesis method may be applied if K_v, $\dot{K}_v$ and R are all continuous bounded functions, and then by choosing a V function of the form:

$$V = x_1^2 + BTE^2 , \qquad (16)$$

$$\dot{V} = -2x_1 K_c \dot{K}_v - 2 BE^2 , \qquad (17)$$

if the adaptive loop is $K_v \dot{K}_c = BER \qquad (18)$

Now since $\dot{V}$ is not always negative the system stability cannot be guaranteed; however, since for B, T positive V is positive definite,

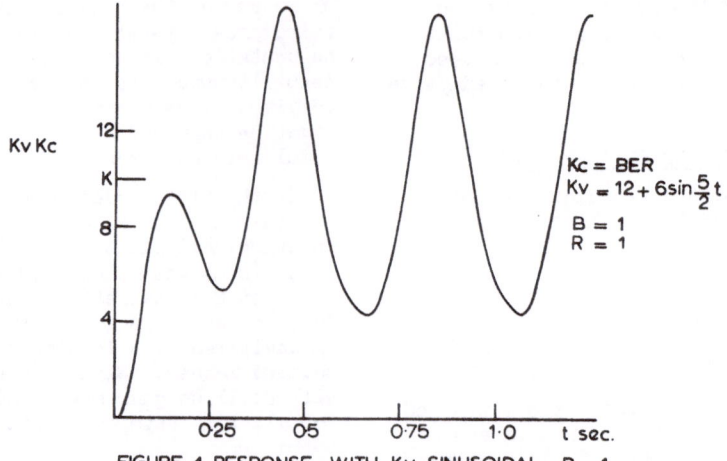

Kv Kc
12
K
8
4
0·25
0·5
0·75
1·0
t sec.
Kc = BER
Kv = 12 + 6sin 5/2 t
B = 1
R = 1

FIGURE 4 RESPONSE WITH Kv SINUSOIDAL B = 1

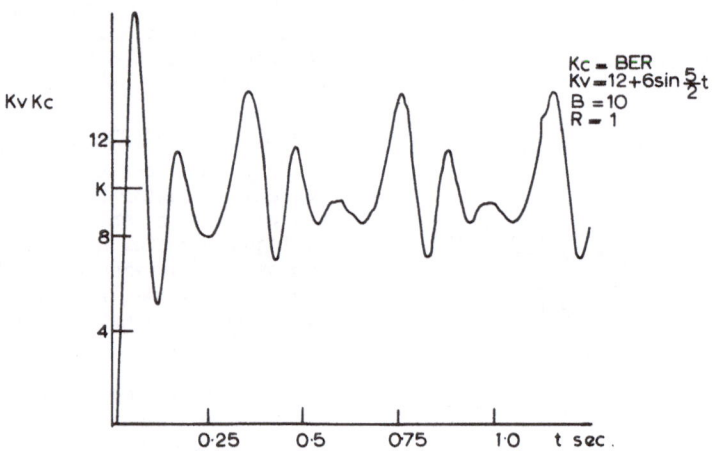

Kv Kc
12
K
8
4
0·25
0·5
0·75
1·0
t sec.
Kc = BER
Kv = 12+6sin 5/2 t
B = 10
R = 1

FIGURE 5. RESPONSE WITH Kv SINUSOIDAL B = 10.

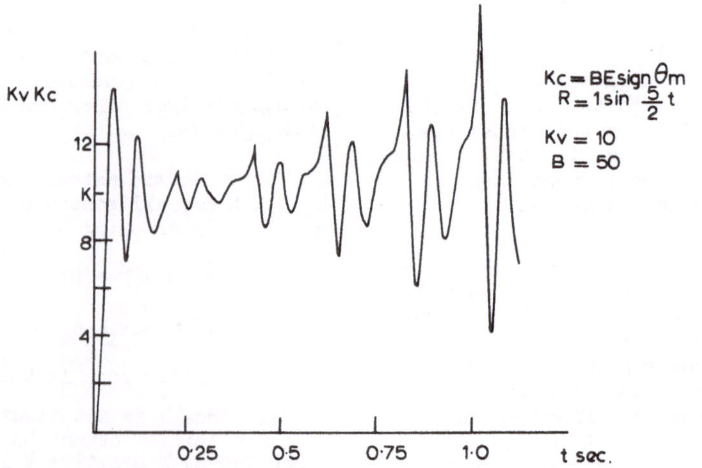

Kv Kc
12
K
8
4
0·25
0·5
0·75
1·0
t sec.
Kc = BEsign θm
R = 1sin 5/2 t
Kv = 10
B = 50

FIGURE 6 UNSTABLE RESPONSE WITH R SINUSOIDAL.

at each instant the sign of $\dot{V}$ determines whether the motion tends to, or moves away from, the origin $x_1 = 0$ $E = 0$, although it can never remain at this origin because of the disturbance term $K_c \dot{K}_v$. As $\dot{V}$ is composed of a negative semi-definite term and a sign indefinite term the system motion must be around the origin in some manner. Also increasing B would seem to have no adverse effect on the system stability (from the form of $\dot{V}$) and from eq. 15 by doing this the effect of the disturbance term on the system behaviour would be reduced. Thus it would appear that the motions x_1 and E about zero would have smaller amplitudes, i.e. the plant and model gains and outputs would become more nearly the same, the larger the gain B was.

These conclusions are assuming that the adaptive loop is of the form of eq.18, but in practice this is not realisable as the variations of K_v are unknown; consequently it would seem that if K_v were always positive the same adaptive loop as for K_v constant i.e. $\dot{K}_c = BER$ would lead to similar results.

When a positive sinusoidal variation is assumed for K_v in an analogue computer simulation the results that are indicated theoretically are verified, and in figs. 4, 5 it can be seen that the motion of $K_v K_c$ is about the value K and that by increasing the adaptive loop gain the amplitude of this motion may be reduced.

Application of the Krasovskii Theorems.

When K_v is varying and R is constant, two theorems due to Kats and Krasovskii[7] allow some analysis of the stability problem. The theorems concern a linear stationary system:

$$\dot{x} = \underline{A}(y)\,\underline{x}\,, \qquad (19)$$

whose parameters are functions of a Markov process y. In the first theorem (ref.7: Theorem 6.1 p.1239), if the system of eq.(19) is asymptotically stable in the mean, then a positive definite Liapunov function exists whose derivative is negative definite. The second theorem deals with the problem when a Gaussian process is added to the right hand side of eq.(19) If the linear system is asymptotically stable then the addition of the Gaussian process will make the disturbed system stable in probability under constantly acting disturbances.

When the model and system are governed by first order equations, the two theorems can be applied if K_v alone is varying. First a new variable is chosen:

$$Z = K_c - K/K_v\,.$$

The system equations become:

$$T\dot{E} = -E - K_v Z R\,,$$
$$\dot{Z} = BER - \frac{d}{dt}(K/K_v)\,. \qquad (20)$$

Here the synthesized adaptive loop is being used. To satisfy the theorems K_v must be a Markov process and the simplest form it can take is when only two values a_1 and a_2 are allowed. The probability of a transfer from a_1 to a_2 in time δt is α_{12} and from a_2 to a_1, α_{21}. Eq.(20) is first considered without the disturbance term $\frac{d}{dt}(K/K_v)$. A sufficient condition for asymptotically stability in the mean is:

$$(a_1 - a_2)^2 \; \alpha^2 \; < \; \frac{4a_1 \; a_2 \; (\, 2\alpha + \frac{1}{T}\,)}{T}$$

It can be seen that this condition is more likely to be met if $(a_1 - a_2)$, the step is small, α, the probability of a change is small and if T, the time constant of the process, is small. These are conditions that one would expect and the influence of T is of particular interest.

From the second theorem of Kats and Krasovskii that was mentioned, the system of eq.(20) will not be made unstable if the disturbance term $\frac{d}{dt}(K/K_v)$ is a Gaussian process. Unfortunately it is not Gaussian but it becomes a reasonable approximation to this as the number of values K_v is allowed to have increases. The method can be extended to deal with such cases but computation becomes a problem. The main interest of the results is the way that they display the influence that the system parameters have on stability.

Other adaptive control loops.

The synthesis technique has indicated a form of adaptive loop in which a multiplier is needed to obtain the product E x R; it might seem possible that R could be regarded as a weighting function and so the adaptive loop may then be written:

$$\dot{K}_c = BE \; \text{sign} \; R \quad \text{sign } R = 1, R > 0; \; \text{sign } R = -1,$$
$$R < 0; \; \text{sign } R = 0, \; R = 0\,. \qquad (21)$$

Although an attempt was made to use Liapunov's analysis on the resultant system a suitable function could not be determined. Thus to obtain some idea of the system stability, the system governing equation with K_v constant was studied further; for R either positive or negative this is:

$$\ddot{x}_1 + \frac{1}{T}\dot{x}_1 + \frac{K_v B}{T} R \; \text{sign } R.\, x_1 = 0\,. \qquad (22)$$

By making a Fourier analysis of the term R sign R when R is sinusoidal and considering solely the fundamental frequency of this expression, the governing equation becomes identical to the Mathieu equation[9] for a damped motion. The stability regions of the Mathieu equation are most complex (p.40, 96 ref (9)) but generally the more negative the coefficient of x_1 can become the more likelihood there is for instability to occur; but even when this coefficient is always positive as in the above equation, stability is not guaranteed. Thus it would seem that by simplifying the adaptive loop mechanisation in this manner the stability of the system might suffer.

A simplification of the mechanisation of the M.I.T. adaptive loop may also be made to $\dot{K}_c = BE$ sign θ_m, and for K_v constant and $\theta_m \neq 0$, a governing equation:

$$\ddot{x}_1 + \frac{1}{T} \dot{x}_1 + \frac{K_v B}{T} R \text{ sign } \theta_m x_1 = 0 , \qquad (23)$$

may be devised. Now although this is very similar to the previous equation the coefficient of x_1 may become negative for some time, and so it would appear that there would be more chance of instability than with using the adaptive loop $\dot{K}_c = B E$ sign R.

This was confirmed by analogue computer studies and although much careful searching was made no instability with R sinusoidal using $\dot{K}_c = BE$ sign R for various values of gain was experienced. However with $\dot{K}_c = BE$ sign θ_m more instability was evident and an example of this is shown in fig.6 where it can be clearly seen that the motion amplitude increases with time.

HIGHER ORDER SYSTEMS

If the process transfer function is:

$$p(s) = K_v / 1 + T_1 D + \ldots + T_n D^n, \qquad (24)$$

and the model transfer function:

$$m(s) = K / 1 + T_1 D + \ldots + T_n D^n, \qquad (25)$$

will there still be a loop which adjusts K_c, and is stable? Once again, in seeking an answer, the choice of the tentative Liapunov function is vital. First make:

$$y = K_c - K/K_v, \qquad (26)$$

and form a function:

$$V = T_n \underline{x}' \, \underline{H} \, \underline{x} + y^2 , \qquad (27)$$

where $\underline{x} = (E, \dot{E}, \ddot{E} \ldots E^{(n-1)}),$

and $\underline{H}$ is the Hermite matrix of the coefficients of the nth order differential equation for E. This matrix is described in Ref.(10). The function V is positive definite and its derivative negative semi-definite (K_v and R Constant) if:

$$\dot{K}_c = \dot{y} = B R (T_{n-1} E^{(n-1)} + T_{n-3} E^{(n-3)} \ldots) \quad (28)$$

A stable adaptive step response is guaranteed.

TWO CONTROLLABLE PARAMETERS

As well as designing adaptive loops so that the plant gain may be adjusted correctly, the synthesis technique may be extended to plants having more than one controllable parameter.

For example if the first order plant considered previously had both its gain and time constant adjustable then its equation may be written:

$$(1 + a_1 D) \theta_s = b_1 R , \qquad (29)$$

where a_1 and b_1 are plant parameters comprising both the uncontrollable parameter and the adjustable element. The model form is:

$$(1 + a D) \theta_m = b R , \qquad (30)$$

where a and b are constant, and again it is desired to make the plant and model the same. Again an error equation may be derived and this is:

$$(1 + a D) E = (b - b_1) R - (a - a_1) \dot{\theta}_s. \quad (31)$$

By choosing a V function of the form:

$$V = a E^2 + \frac{1}{B} (b - b_1)^2 + \frac{1}{C} (a - a_1)^2, \quad (32)$$

$$\dot{V} = - 2 E^2 , \qquad (33)$$

if the adaptive loops are $\dot{b}_1 = BER$ and

$$\dot{a}_1 = -CE\dot{\theta}_s \qquad (34)$$

Thus by using these adaptive loops V is a Liapunov function and stability may be guaranteed. Of course if for example:

$$b_1 = K_v K_c ,$$

then $\dot{b}_1 = \dot{K}_v K_c + \dot{K}_c K_v$, and it is necessary to use for practical examples $\dot{K}_c = BER(K_v > 0)$: for each plant parameter only the controllable element may be adjusted.

CONCLUSIONS

In conclusion it has been shown that to guarantee the stability of a model reference system, use of the M.I.T. adaptive loop might make it impossible to fulfil the system specifications. By designing model reference systems so that they are stable these problems may be avoided, and in this paper a method of synthesizing stable systems has been evolved by using Liapunov's second method.

ACKNOWLEDGEMENTS.

The work reported here is part of a general study of the use of Liapunov functions carried out under Mr.P.C.Parks in the Institute of Sound and Vibration Research at the University of Southampton. The work is supported by the D.S.I.R. contract No. 9634/09 and Ministry of Aviation contract No. 9634/07.

REFERENCES

(1) Whitaker, H.P. Proc.of the Self Adaptive Flight Control Systems symposium. WADC TR 59-49 March, 1959.

(2) Osburn, P.V., Whitaker, H.P. and Keezer, A. _New developments in the design of adaptive control systems._ IAS Paper No.61-39.

(3) Donalson, D.D. and Leondes, C.T. _A model referenced Parameter Tracking technique for Adaptive Control Systems._ IEEE Trans. on Applications and Industry. Sept.63, pp. 241-262.

(4) La Salle,J.P. and Rath, R.J. _A new concept of stability._ Paper 415 RIAS. Aug.1962.

(5) Grayson, L.P. _The design of nonlinear and adaptive systems via Liapunov's second method._ Polytechnic Institute of Brooklyn Report PIB MR1-937-61.

(6) Davies, W.D.T. and Fishwick, W. _Proc.UKAC Convention on "Recent Advances in Automatic Control"._ April 1965, Proc. I. Mech.Engrs.1965

(7) Kats, I. and Krasovskii, N.N. _On the stability of systems with random parameters._ PMM 1960. 24(No.5).

(8) Kalman, R.E. and Bertram, J.E. Trans. ASME. Series D, 1960, 82, No.2, June, 1960.

(9) McLachlan, N.N. _Theory and Applications of Mathieu functions._ Dover Publications 1947

(10) Parks, P.C. _Analytic methods for investigating stability - linear and non-linear systems._ A survey. I.Mech.E. Paper. May 1964. for discussion on "Stability of Systems".

Written contributions to the discussion

Mr. P.J. Burt (UK)

It is interesting to compare this Liapunov synthesis with the Whitaker synthesis. In two ways at least the former is an improvement. It uses no parameter influence filters thus reducing the amount of on-line computation. Also the decreasing function is a function of the parameter errors as well as the system error, which presumably lessens the parameter inter-action. It will be remembered that in the Whitaker synthesis the function minimised is a function of the system error only. The effect is that in the Whitaker synthesis the adaption of many parameters proceeds more slowly simul-taneously than individually. In one way the Liapunov synthesis is impractical in that the two parameter system requires the process out-put rate to be formed. What is the effect of using practical approximations to the often physically unrealisable rate such as high pass filtering instead of differentiation ?

In reply the authors comment:

The practical bias of this question may expose a weakness in our paper. We suspect that Mr. Burt would know more about the advantages and problems of using high pass filtering in place of a rate signal than we do. In fact, it is not something we have experience of. Without experimental results, it is difficult to comment on the problem of parameter interaction. This will certainly occur but whether the problem will take on the same importance as in the M.I.T. system we cannot say.

Mr. T. Horrocks (UK)

I think that the authors have made a very brave attempt to demonstrate the superiority of the Liapunov approach over the Whitaker method but perhaps they will forgive me if I am not convinced. It is true, as they have shown, that the characteristic equation of the system, including the adaptive loop, in a Whitaker analysis, is at least one order higher than the original equation of the system. Thus in their analysis the second order system has become a third order system. But although now an increase in the gain of the adaptive loop may make the Whitaker system unstable, their system if not unstable, may be made unacceptable, which is not much better. In any case this argument is limited to very low order systems.

What is more important is the main flow in the philosophy of their approach. Because they have started from a Liapunov function the system, they say, cannot be unstable. I will skip the usual argument that a convenient positive definite function with similar restrictions on its first derivative may not exist. I will limit myself to pointing out that the degree of stability is in no way determined by taking a Liapunov function. But the Liapunov function is based on a mathematical model of the system which is certainly imperfect. Thus there is a danger of finishing with an unstable system in spite of the theory.

Secondly, the Liapunov function chosen is of necessity an arbitrary one, usually based on quadratic forms. Who is to say that a system designed on this basis is not the worst possible whereas a choice of another Liapunov function would have produced a much better one?

Thirdly, their mechanisation involves them in obtaining a series of derivatives. The difficulties of such requirements in practice and the dangers inherent in differentiating in a noisy environment are too well known to require comment.

The fact remains that the Whitaker system is the only one so far which has proved fast, simple and reliable.

The authors replied:

The V function that we have used in our design is not so arbitrary as Mr. Horrocks makes out, as we are choosing adaptive loops to make V a Liapunov function, and in so doing ensure that when the process parameters are constant the plant and model forms become the same, and the error tends to zero. The system stability is not guaranteed when these parameters are time varying, however, analogue computer studies confirm that by increasing the adaptive loop gains, the plant and model may be made approximately the same and the error may be confined to a small region about zero. How such a system could be the 'worst possible' is difficult to envisage, although it seems likely that other V functions leading to possibly simpler adaptive loops may exist. If the mathematical model is not a true representation of a practical system then no system devised using this model would behave as predicted.

Mr. Horrocks' comment about the difficulties involved in obtaining signal derivatives in the presence of noise we would concede, but of course derivatives of θs would be needed in the M.I.T. adaptive loops if it were required to control plant parameters other than the gain, and so then the same problem would arise.

For high order systems use of the M.I.T. form of adaption would probably need small adaptive loop gains so that the overall system is stable, whereas using our form of control no theoretical limits are set on these gains. Consequently of the two schemes of adaption ours might well lead to a system response which was more acceptable.

SOME PROBLEMS OF ANALYTICAL SYNTHESIS
IN MODEL REFERENCE CONTROL SYSTEMS BY THE DIRECT METHOD OF LYAPUNOV

by S. D. Zemlyakov, USSR

USSR National Committee on Automatic Control
Moscow, USSR

In the design of self adaptive control systems, great attention has been paid to the principle of design without gradient search using reference models [1 to 4]. Such systems are described by non-linear differential equations with variable coefficients. In order to investigate such systems the direct method of Lyapunov is most frequently applied [5]. It can be used both for the analysis of the control system, for the synthesis of its structure and for the selection of controller parameters [2 to 6].

The direct method of Lyapunov permits one to predict sufficient conditions of stability of the system which, as a rule, are more rigid than the necessary conditions. The application of the sufficient stability conditions according to Lypunov for controller synthesis often requires obtaining a greater volume of information about the system than is necessary for satisfactory operation. However these sufficient conditions of system stability according to Lyapunov permit one to find a structure which is adequate to secure the effectiveness of the system; furthermore the investigation of the system either by more general methods than that of Lyapunov or by the methods of mathematical modelling allows for synthesis of systems with smaller amount of available information (absence of high orders of differentiation, etc.)

Frequently investigation into the stability of adjusting model-reference control systems is reduced to finding the zero solution for a differential equation of matrix form: e.g.

$$\dot{x} = Ax + y \qquad (1)$$

where

$$x = \begin{Vmatrix} x_1 \\ x_2 \\ \cdot \\ \cdot \\ \cdot \\ x_n \end{Vmatrix}, \quad A = \begin{Vmatrix} 0 & 1 & 0 & \cdots & 0 \\ 0 & 0 & 1 & \cdots & 0 \\ \cdot & \cdot & \cdot & \cdots & \cdot \\ 0 & 0 & 0 & \cdots & 1 \\ -b_{0,0} & -b_{1,0} & -b_{2,0} & \cdots & -b_{n-1,0} \end{Vmatrix}, \quad y = \begin{Vmatrix} 0 \\ 0 \\ \cdot \\ \cdot \\ \cdot \\ 0 \\ f \end{Vmatrix}$$

$b_{i,0}$ = const $(i = 0,1,\ldots,n-1)$; $f = f(t,x)$.

Assume that the following condition is met

$$f \equiv 0 \text{ at } x = 0 \qquad (2)$$

- -

"Numbers in square brackets refer to similarly numbered references at the end of this paper."

and matrix A is not singular for negative real parts of the characteristic equation

$$|A - \lambda E| = 0 . \qquad (3)$$

To investigate the stability of the zero solution for Equation (1) select a Lyapunov function of quadratic form

$$V(x) = x'Px , \qquad (4)$$

where P is non-singular symmetric matrix for which the condition of positive definite $V(x)$ is satisfied. Find the derivative of $V(x)$ which satisfies Equation (1)

$$\dot{V}(x) = \dot{x}'Px + x'P\dot{x} . \qquad (5)$$

Denote

$$A'P + PA = Q , \qquad (6)$$

$$W(x) = x'Qx . \qquad (7)$$

Then Equation (5) can be written as

$$\dot{V}(x) = x'Qx + 2x'Py . \qquad (8)$$

Applying the Lyapunov method the sufficient condition for the asymptotic stability of the zero solution of Equation (1) is that $\dot{V}(x)$ should be negative definite. Since A and P are non-singular matrices and the matrix A has negative real parts of the roots of the characteristic equation, then, as proved by Lyapunov in [5], the matrix Q can always be made to satisfy the condition of negative definiteness of the quadratic form of Equation (7). In this case the condition of asymptotic stability reduces to non-positiveness of the expression

$$x'Py . \qquad (9)$$

If the matrix P is written as

$$P = \begin{Vmatrix} P_{11} & \cdots & P_{1n} \\ \cdot & & \\ \cdot & & \\ \cdot & & \\ P_{n1} & \cdots & P_{nn} \end{Vmatrix} , \qquad (10)$$

153

then the following equation is true

$$x'Py = f(p_{1n}x_1 + p_{2n}x_2 + \ldots + p_{nn}x_n) \ . \quad (11)$$

For instance, if

$$f = -\rho(t) \, \text{sign} \, (\beta_1 x_1 + \beta_2 x_2 + \ldots + \beta_n x_n) \quad (12)$$

where

$$\rho(t) \geqslant 0 \ ; \quad \beta_j > 0 \ (j = 1,2,\ldots,n) \ ,$$

the condition of positiveness of (9) and therefore the sufficient condition of the asymptotic stability for the zero solution of Equation (1) reduces to the requirement that the relation below be valid

$$\frac{\beta_1}{p_{1n}} = \frac{\beta_2}{p_{2n}} = \ldots = \frac{\beta_n}{p_{nn}} = \text{const} > 0 \ . \quad (13)$$

Assume that the system discussed is characterized by a differential equation of the form

$$x^{(n)} + [b_{n-1}(t) + \kappa_{n-1}]x^{(n-1)} + \ldots$$

$$+ [b_0(t) + \kappa_0]x = \kappa_g \, g(t) \ , \quad (14)$$

where $b_0(t), b_1(t), \ldots, b_{n-1}(t)$ are time-variable parameters of the control system

$\kappa_g, \kappa_0, \kappa_1, \ldots, \kappa_{n-1}$ are the controller parameters and

$g(t)$ is the control signal.

Represent the parameters of the plant and the controller as

$$\left. \begin{aligned} b_i(t) &= \bar{b}_i + \Delta b_i(t) \ , \\ \kappa_i &= \bar{\kappa}_i + \Delta \kappa_i \ , \\ \kappa_g &= \bar{\kappa}_g + \Delta \kappa_g \ , \end{aligned} \right| \quad (i = 0,1,\ldots,n-1) \quad (15)$$

where $\bar{b}_i$ are some constant quantities

$\Delta b_i(t)$ are variable components of the plant parameters,

$\bar{\kappa}_g, \bar{\kappa}_i$ are constant,

$\Delta \kappa_g, \Delta \kappa_i$ variable components of the controller's parameters.

Denote

$$\bar{b}_i + \bar{\kappa}_i = b_{i,0} \ ; \quad \bar{\kappa}_g = \kappa_{g,0} \ .$$

As the reference model select a filter for which a differential equation of the following form applies

$$x_M^{(n)} + b_{n-1,0} \, x_M^{(n-1)} + \ldots + b_{0,0} x_M = \kappa_{g,0} \, g(t) \ . \quad (16)$$

Denote the difference between the dynamics of the actual plant and that of the reference model as ε

where $\qquad \varepsilon = x - x_M \ . \quad (17)$

From Equations (14), (16) and bearing in mind (15), (17) obtain the equation for ε

$$\varepsilon^{(n)} + \left[b_{n-1,0} + \Delta b_{n-1}(t) + \Delta \kappa_{n-1} \right] \varepsilon^{(n-1)} + \ldots$$

$$+ \left[b_{0,0} + \Delta b_0(t) + \Delta \kappa_0 \right] \varepsilon =$$

$$= - \left[(\Delta b_{n-1}(t) + \Delta \kappa_{n-1}) x_M^{(n-1)} + \ldots \right.$$

$$\left. + (\Delta b_0(t) + \Delta \kappa_0) x_M \right] + \Delta \kappa_g \, g(t) \ . \quad (18)$$

In the algorithms for readjustment of the controllers parameters consider three terms which are generally written in the form

$$\Delta \kappa_i = f_{1i}(\varepsilon, \dot{\varepsilon}, \ldots, \varepsilon^{(n-1)}, x, \dot{x}, \ldots, x^{(n-1)})$$

$$+ f_{2i}(\varepsilon, \dot{\varepsilon}, \ldots, \varepsilon^{(n-1)}, x, \dot{x}, \ldots, x^{(n-1)}) +$$

$$+ \int_{t_0}^{t} f_{3i}(\varepsilon, \dot{\varepsilon}, \ldots, \varepsilon^{(n-1)}, x, \dot{x}, \ldots, x^{(n-1)}) dt \ , \quad (19)$$

$$\Delta \kappa_g = f_g(\varepsilon, \dot{\varepsilon}, \ldots, \varepsilon^{(n-1)}, g) \ , \quad (i = 0,1,\ldots,n-1) \quad (20)$$

where f_{1i} are discontinuous functions of the variables

f_{2i}, f_{3i}, f_g are linear functions of the variables;

t_0 is the switching time for integral components;

t is the present instant.

Equation (18) will be written in some what

different form

$$\varepsilon^{(n)} + b_{n-1,0}\varepsilon^{(n-1)} + \ldots$$

$$+ b_{0,0}\varepsilon = -\left[(\Delta b_{n-1}(t) + \Delta \kappa_{n-1})x^{(n-1)} + \ldots \right.$$

$$\left. + (\Delta b_0(t) + \Delta \kappa_0)x\right] + \Delta \kappa_g\, g(t) . \qquad (21)$$

Denote

$$\Delta \kappa_g\, g(t) = \rho_1(t);$$

$$-\left[\Delta b_{n-1}(t)x^{(n-1)} + \ldots + \Delta b_0(t)x\right] = \rho_2(t);$$

$$\qquad\qquad (22)$$

$$-\left[\Delta \kappa_{n-1}x^{(n-1)} + \ldots + \Delta \kappa_0 x\right] = \rho_3(t);$$

$$f(t) = \rho_1(t) + \rho_2(t) + \rho_3(t) .$$

Then equation 21 will be written as

$$\varepsilon^{(n)} + b_{n-1,0}\,\varepsilon^{(n-1)} + \ldots + b_{0,0}\varepsilon = f(t) . \qquad (23)$$

Assume that generally

$$f(t) = f(t,\varepsilon,\dot{\varepsilon},\ldots,\varepsilon^{(n-1)}) . \qquad (24)$$

Next denote

$$\varepsilon^{(i)} = x_{i+1} . \qquad (25)$$

Then equation (23) can be represented by the matrix form discussed earlier

$$\dot{x} = Ax + y . \qquad (1)$$

As the variation laws for $\Delta \kappa_g$, $\Delta \kappa_0$, ..., $\Delta \kappa_{n-1}$ assume the expressions

$$\Delta \kappa_g = -\frac{\tau}{g(t)}\,\text{sign}\,\sigma - \Delta\bar{\kappa}_g\,\text{sign}\,g\,\text{sign}\,\sigma,$$

$$\qquad\qquad (26)$$

$$\Delta \kappa_i = \Delta\bar{\kappa}_i\,\text{sign}\,x_{i+1}\,\text{sign}\,\sigma \;(i = 0,1,\ldots,n-1)$$

where

$$\sigma = \beta_1 x_1 + \beta_2 x_2 + \ldots + \beta_n x_n;$$

$$\beta_j = \text{const} > 0 \;(j = 1,2,\ldots,n); \quad \tau = \text{const} > 0; \quad (27)$$

$$\Delta\bar{\kappa}_i = \text{const} > 0; \quad \Delta\bar{\kappa}_g = \text{const} > 0 .$$

Consider the condition

$$\Delta\bar{\kappa}_i \geqslant |\Delta b_i(t)| \quad (i = 0,1,\ldots,n-1) \qquad (28)$$

if this is true then it follows from (22) that

$$\rho_3(t) = -\bar{\rho}_3(t)\,\text{sign}\,\sigma , \qquad (29)$$

where

$$\bar{\rho}_3(t) = |\rho_3(t)| ,$$

while $\quad \bar{\rho}_3(t) \geqslant \rho_2(t) . \qquad (30)$

Consequently

$$\rho_2(t) + \rho_3(t) = -\left[\bar{\rho}_3(t) + \rho_2(t)\,\text{sign}\,\sigma\right]\text{sign}\,\sigma , \qquad (31)$$

where

$$\rho_0(t) = \bar{\rho}_3(t) + \rho_2(t)\,\text{sign}\,\sigma \geqslant 0 ,$$

and the expression for $f(t)$ is of the form

$$f(t) = -\left[\tau + \rho_0(t) + \mu(t)\right]\text{sign}\,\sigma , \qquad (32)$$

where

$$\mu(t) = \Delta\bar{\kappa}_g\,|g| \geqslant 0 .$$

Expression (32) is equivalent to (12) if

$$\rho(t) = \tau + \rho_0(t) + \mu(t) ; \quad \rho(t) > 0 . \qquad (33)$$

When we find

$$\text{sign}\,\sigma = \begin{cases} +1 & \text{at } \sigma > 0 , \\ 0 & \text{at } \sigma = 0 , \\ -1 & \text{at } \sigma < 0 , \end{cases} \qquad (34)$$

condition (2) is evidently also true. From the principle of the system structure the truth of the condition (3) follows.

Then for asymptotic stability of the actual system dynamics it is sufficient that relation (13) be valid, assuming operation of the model with the same discontinuous terms in the algorithms for variation of the controller parameters (varying by (26) if expression (28) is true). If the range of variation in the variable parameters of the plant is large correspondingly large amplitudes of discontinuous variation in the controller parameters are required.

Then, as experience shows, at a relatively high level of disturbance the parameter of the

sliding mode* deteriorate and practical implementation of the system is handicapped.

REFERENCES

1.   Adaptive control systems. Edited by
E. Mishkin and L. Braun. McGraw Hill, 1961

2.   Donaldson, D.D., Leondes, C.T. - "A model
reference parameter tracking technique for
adaptive control systems", IEEE Trans. Applic.
and Ind., No. 68, 1963.

3.   Evlanov, A.G. - A self adaptive control
system with gradient search using the method of
auxiliary operators. Journal. Acad. Sciences of
the USSR. Technical Cybernetics, No. 1, 1963.

4.   Krutova, I.N., Ruthkovskii, V.U. - A self
adaptive control system with a model. Technical
Cybernetics, Nos. 1 and 2, 1964.

5.   Lyapunov, A.M. - The general problem of
stability of motion. State Publishing House
USSR, 1950.

6.   Hiza, T.G., Li, C.C. - Analytical
synthesis of a class of model-reference time-
varying control systems, IEEE Trans. Applic. and
Ind., No. 69, 1963.

*EDITOR'S NOTE ON THE SLIDING MODE

The Russian term скользящий режим,
(sliding mode), has no simple technical equivalent
in English. It refers to a mode of operation of
on-off controlled systems characterised by the
presence of an oscillation about the switching
surface in the phase space.

Thus, the figure shows a switch line in two
dimensional phase space with co-ordinates x and
ẋ. The equation of this switch line may be of
the form

$$\mathrm{sgn}\left[\, x + f(\dot{x})\, \right] = 0$$

For sliding mode dynamics to exist any
trajectory which encounters the switch line gives
rise to subsequent motion described by trajectories
oscillating about this line.

REFERENCES

1.   Paper 5.5 in these proceedings by
S. V. Yemelyanov et al.

2.   Tsypkin Ya Z., Theory of relay systems of
automatic control, page 117, Moscow 1955
(in Russian)

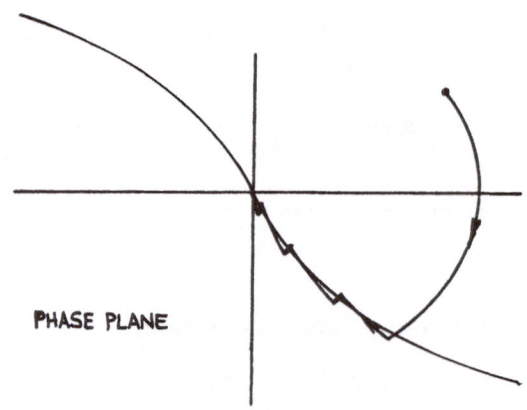

PHASE PLANE

BASIC PROPERTIES AND SOME DYNAMICS PROBLEMS OF ADAPTIVE MODEL REFERENCE SYSTEMS

by B. N. Petrov, V. Yu. Ruthkovsky and I. N. Krutova
USSR National Committee of Automatic Control
Institute of Automatics and Telemechanics
Moscow, USSR

We shall discuss the basic properties and some dynamics problems of adaptive model-reference systems which do not use extremum search procedures. The reference model represents some desired process and the adjustment of the controller's parameters depends upon differences between the coordinates of the plant and those of the model as well as between their derivatives [4]. As compared with other model-reference systems [2,3] those discussed here allow a significantly greater rate of plant parameter change and the adaptive transient process is less dependent on the form of the control signals.

Suppose that n parameters change in the equation which characterizes the dynamics of the plant and the **actuator**. To effect accurate adaptation it is necessary to readjust n parameters in the control law. However, in engineering practice it is normally sufficient to maintain the differences between the coordinates of the plant and those of the model within a certain range; only a small number of controller parameters then require adjustment. To decide which parameters of the controller are to be adjusted under given conditions it is necessary to study the basic properties aquired by the system when separate adaptive loops are in operation.

The initial problem in studying the dynamics of this class of system may be considered to be to find these properties. At the same time, to study the rules for calculating signals used in the adaptive loops and to formulate laws relating parameter changes to the stability and quality of the control processes.

The equations of the system dynamics are written in the form

$$\sum_{\alpha=0}^{\kappa} a_{\alpha}^{*}(t) x^{(\alpha)} = -b^{*}(t)(\mu+f) \qquad \text{(plant), (1)}$$

$$\sum_{\gamma=0}^{\tau} c_{\gamma} \mu^{(\gamma)} = \sigma , \quad \tau + \kappa = n \qquad \text{(actuator), (2)}$$

$$\sigma = \kappa_{b}\left(\sum_{i=0}^{n-1} \kappa_{i} x^{(i)} - \kappa_{g} g\right) \qquad \text{(control law), (3)}$$

$$\left. \begin{aligned} \kappa_{i} &= \bar{\kappa}_{i} + \kappa_{iu} Z_{i} + \sum_{j=i}^{n-1} \kappa_{ij}(x^{(j)} - x_{m}^{(j)})\, \text{sign}\, x^{(i)} , \\ Z_{i} &= \int_{o}^{t} (x^{(i)} - x_{m}^{(i)})\, \text{sign}\, x^{(i)}\, dt , \\ \kappa_{g} &= \bar{\kappa}_{g} - \kappa_{gu} Z_{g} - \sum_{j=0}^{n-1} \kappa_{gj}(x^{(j)} - x_{m}^{(j)})\, \text{sign}\, g , \\ Z_{g} &= \int_{o}^{t} (x - x_{m})\, \text{sign}\, g\, dt \end{aligned} \right\}$$

(Adaptive loops) (4)

$$\sum_{i=0}^{n} d_{i} x_{m}^{(i)} = g + F \qquad \text{(model). (5)}$$

Here x is the controlled plant coordinate,

x_{m} the model coordinate,

μ the coordinate of the actuator,

σ the control law,

$a^{*}(t)$, $b^{*}(t)$, c_{γ} the adjustable parameters,

κ_{iu}, κ_{gu}, κ_{ij}, κ_{gj} the constant paramters,

g, f are the control signal and the disturbance,

F a signal connected with disturbance f.

The structural diagram of the system is shown in Fig. 1.

If $\kappa_{gu} = 0$ is assumed equal to zero then for the system governed by equations 1 to 5 the following theorem is true.

Theorem If the total gain of the controller is given by

$$\kappa_{b} \cdot \frac{b^{*}(t)}{a_{\kappa}^{*}(t)\, c_{\tau}} = x_{0} = \text{const}$$

"Numbers in square brackets refer to similarly numbered references at the end of this paper"

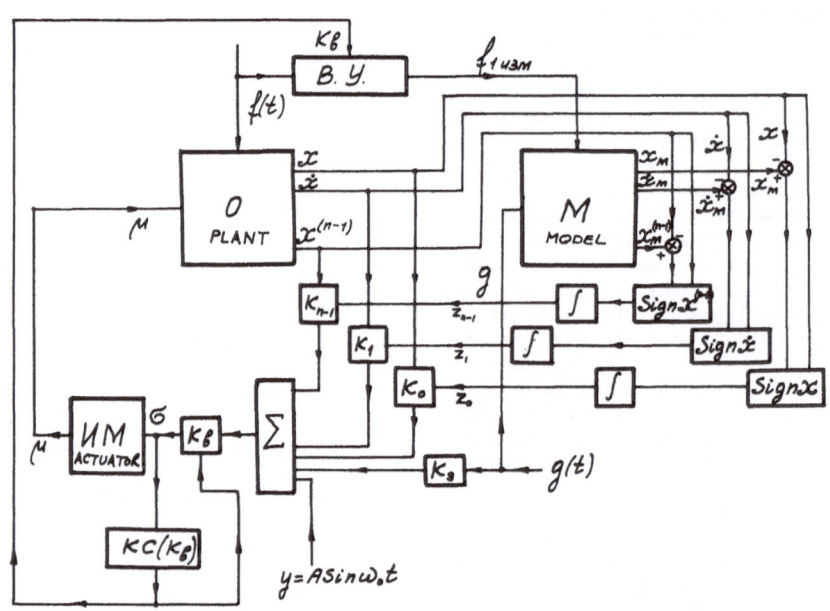

Figure 1

and if κ_i is given by (4), then set (1-5) is stable, and the plant's parameters, after having arbitrarily changed, remain constant starting from some time $t = t_0$ then at $g(t) \neq 0$ and $f(t) \equiv 0$ in the stabilized state the integrals z_i define unambiguously a_i, where a_i are the coefficients of the operator

$$A(p) = \sum_{i=0}^{n} a_i p^i = \frac{1}{a_\kappa^* c_z} \sum_{\alpha=0}^{\kappa} a_\alpha^* p^\alpha \sum_{\gamma=0}^{\tau} c_\gamma p^\gamma \left(p = \frac{d}{dt} \right),$$

while the stationary values

$$Z_{i,st} = \frac{x_0 \kappa_g \bar{d}_i - x_0 \bar{\kappa}_i - a_i}{x_0 \kappa_{iu}} . \qquad (6)$$

The parameter of the model would be selected as

$$d_n = \frac{1}{x_0 \kappa_g} . \qquad (7)$$

If $f(t) \neq 0$, then besides $g(t)$ the model should be fed with the signal

$$F(t) = -\frac{1}{\kappa_b \kappa_{go}} \sum_{\gamma=0}^{\tau} c_\gamma f^{(\gamma)} . \qquad (8)$$

Evidently, to calculate $F(t)$ it is necessary to measure the disturbance. In a number of cases $f(t)$ causes a deviation of x of which the modulus is far less at every moment than that of the deviation $x(t)$ caused by the control signal. In this case it is not necessary, to measure $f(t)$ in other words we assume that $F(t) \equiv 0$. Integrals Z_i will have errors whose moduli are not greater than the quantity

$$\Delta Z_i = \left| \frac{1}{\kappa_b \kappa_{iu} x_m^{(i)}} \sum_{\gamma=0}^{\tau} c_\gamma f^{(\gamma)} \right| . \qquad (9)$$

In cases where the errors ΔZ_i are large and it is not feasible to measure the disturbance the system discussed is not practicable.

Let us note some properties of the system which are of great practical importance. If the control signal is such that in stationary operation $x_m^{(i)} \equiv 0$, $i = \ell \leq n-1$, then equation (6) for $i = \ell$, $\ell + 1$, ..., $n-1$ cannot be valid and integrals Z_i increase to their maximum values. To prevent the erroneous increase of Z_i it is

necessary to introduce into the subintegral expressions the following non-linear function as a multiplier

$$\phi(x_m^{(i)}) = \begin{cases} 1 & \text{at } |x_m^{(i)}| > \Delta_i , \\ 0 & \text{at } |x_m^{(i)}| \leq \Delta_i , \end{cases} \qquad (10)$$

where Δ_i is the insensitivity zone,

$$i = 0,1,\ldots,n-1 .$$

In [5] it was proved that in the system discussed above the error $\varepsilon = x - x_m$ at $Z_i = Z_{i,st}$ is invariant in respect to signals g and f. Therefore the adaptive system which does not search for the extremum may be astatic of any order. In particular if only the parameter κ_0 is adjusted by the integral law, the system is astatic of the first order, if κ_0 and κ_1 are adjusted then it is astatic of the second order, etc. The property of adaptivity is caused by the integral components Z_i. However, if factors $a_i(t)$ change rapidly Z_i tracks a_i with a large delay. In view of this, the system should also be of low sensitivity to variation of its parameters.

If the operation does not require an accurate compensation of a_i in the stationary mode, then the desired quality of the adaptive process may be achieved by making the system insensitive to certain signals. There is then no need to introduce κ_i into the corresponding parameters. The insensitivity is obtained by introducing

$$\sum_{j=i}^{n-1} \kappa_{ij} (x^{(j)} - x_m^{(j)}) \text{ sign } x^{(i)} .$$

When studying the system's properties it is also important to discover the relationship between the number of adjustable parameters and the insensitivity of the system to changes of plant parameters. Analysis of system dynamics for any derivations of κ_i and κ_g makes it possible to solve the problem of the proper system design since adaptation and increase of insensitivity may be obtained through adjustment of either κ_i or κ_g.

All these properties and the related formulation laws for the adjustable parameters can best be found through analysis of stability. As an example consider a second order system

$$a_2 \ddot{\varepsilon} + a_1 \dot{\varepsilon} + \kappa_{11} \dot{\varepsilon}^2 \text{ sign } \dot{\varepsilon} + a_0 \varepsilon + \kappa_{ou} Z_0 \varepsilon +$$
$$+ \kappa_{oo} g_0 \varepsilon \text{ sign } (\varepsilon + g_0) + \kappa_{go} g_0 \varepsilon \text{ sign } g_0$$
$$= (\bar{\kappa}_g - a_0 - \kappa_{ou} Z_0) g_0 , \qquad (11)$$

159

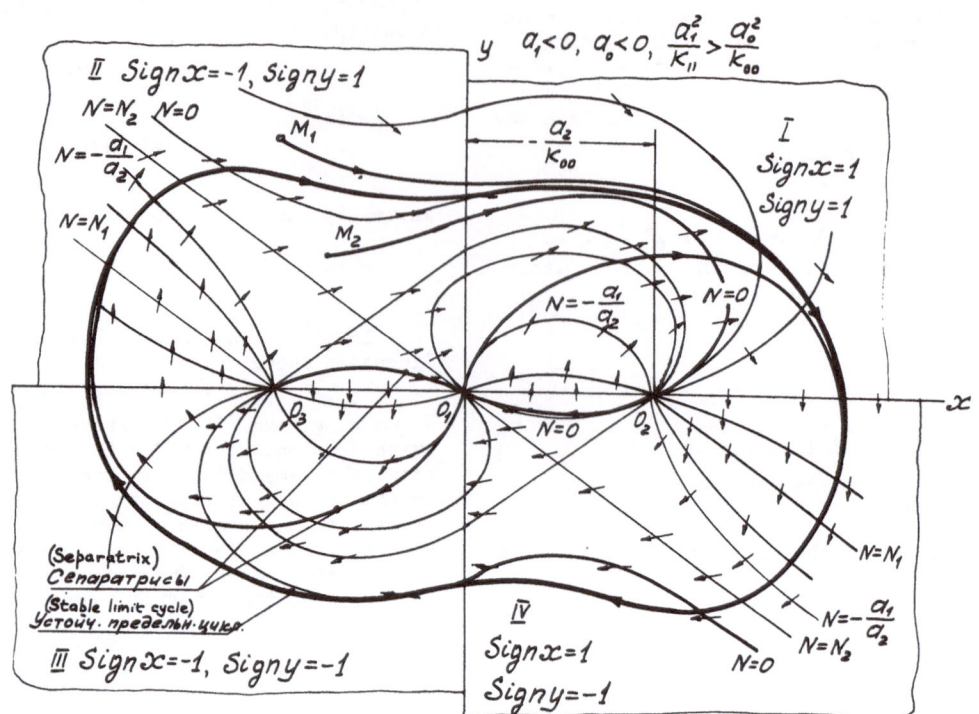

Figure 2

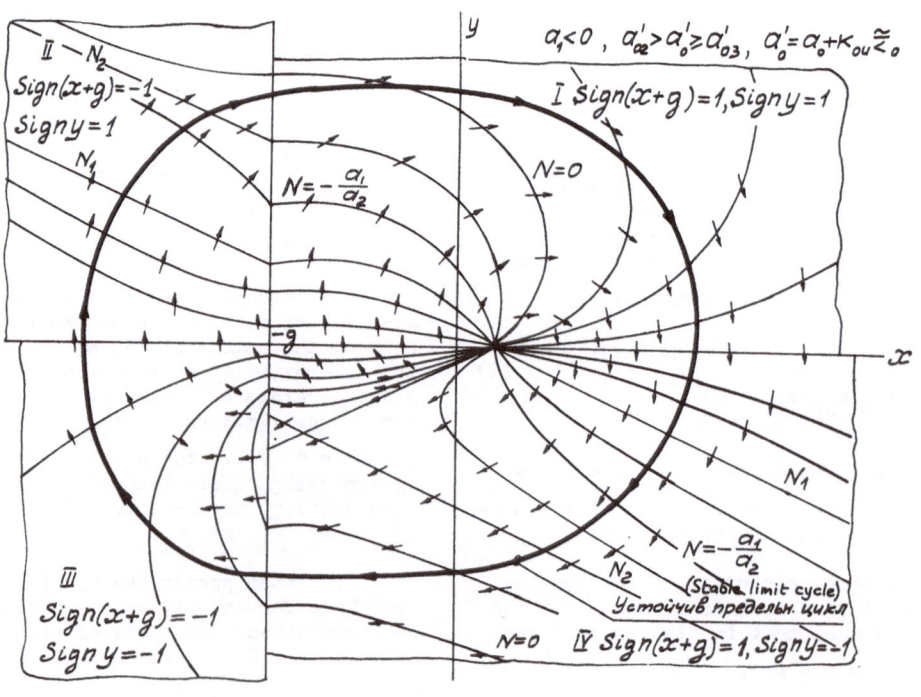

Figure 3

160

where

$$x - g_0 = \varepsilon \ , \qquad \int \varepsilon \ \text{sign} \ x \ dt = Z_0 \ , \qquad g_0 = \text{const} \ .$$

The dynamics of the system may be studied in the phase space ε, Z_0, $\dot{\varepsilon} = y$. The singular points on the plane x,y at $g = 0$ are $x = 0$, $x_{2,3}$ sign $x_{2,3} = -\frac{a_0}{\kappa_{11}}$; $y_{1,2,3} = 0$. The presence of adjustable parameters κ_0 and κ_1 yields a selfoscillating system even at $a_0 < 0$, $a_1 < 0$ [6]. In the latter case self-oscillations will occur and the phase portrait may exhibit either one limit cycle, which covers all the three singular points or two limit cycles near points x_2 and x_3 (Fig. 2). Analysing the system at $g = g_0 = \text{const}$ and $\kappa_{ou} = 0$ it can be seen that generally the dynamics remain as in the case of free oscillations. At $g(t) = \text{const}$ there is a singular point on the phase plane which can be stable at $a_1 > 0$ and unstable at $a_1 < 0$. In the latter case we have a limit cycle in the phase plane, see fig. 3. If a_0 is changed then the singular point moves along the axis ε to the right or to the left.

Adjustment of κ_1 by the law $\kappa_{11} (\dot{x} - \dot{x}_M)$ sign $\dot{x}$ increases the system's stability while adjustment of κ_0 makes the system stable at $a_0 < 0$. However, in the stationary state an error occurs which vanishes when an integral is introduced into κ_0.

REFERENCES

1. Krasovskii, A. A. - The Dynamics of continuously self-adjusting systems [Math. Physics. State Publishing House, USSR] 1963.

2. Whitaker, H. P. - An Adaptive System for Control of the Dynamic Performance of Aircraft and Spacecraft. JAS Paper, N 59-I00, June, 1959.

3. Donalson, D. D., Leondes, C. T. - Model Referenced Parameter Tracking Technique for Adaptive Control Systems, I. The Principle of Adaptation. IEEE. Trans. on Applic. and Industry, Sept., 1963.

4. Krutova, I. N., Ruthkovskii, V. U. - A self-adaptive control system with a model. Tech. Cybernetics Nos. 1,2, 1964.

5. Petrov, B. N., Ruthkovskii, V. U. - On the invariance of self-adjusting systems using a model. Proceedings of the Academy of Sciences of the USSR, vol. 161, No. 3, 1965.

6. Krutova, I. N., Ruthkovskii, V. U. - A study of a self-adjusting system of the second order using a model. Automat. i. Telemech., vol. XXVI, No. 1, 1965.

ADAPTIVE CONTROLLERS WITH PARAMETER INDENTIFICATION OR PREDICTION

A SELF ADAPTIVE SYSTEM EMPLOYING HIGH SPEED PARAMETER IDENTIFICATION

by Dr. J.F. Meredith,
Smiths Aviation Division,
Cheltenham,
Glos., England.

and A.J. Dymock,
Smiths Aviation Division,
Cheltenham,
Glos., England.

1. INTRODUCTION

To provide satisfactory handling characteristics with a modern high performance aeroplane an autostabiliser is a necessity. Since aerodynamic parameters vary widely over the flight regime it is not usually possible to obtain satisfactory augmentation of the natural aeroplane stability without some alteration in the controller parameters. It is desirable both from the point of view of the pilot and also for convenience in the design of automatic control facilities that the demand response of the aeroplane be made to have a specified, or model, form over the entire flight envelope. To do this the controller parameters must vary to the same extent as those for the aeroplane.

Considering the short period pitching motion of an aeroplane the transfer function relating the elevator input (η) and the pitch rate response ($\dot{\theta}$) is

$$\dot{\theta}/\eta \;\; = A = \frac{a(D + b)}{D^2 + cD + d}$$

For a modern supersonic aeroplane the parameters 'a' and 'd' may vary by a factor of 8 in one minute, 'b' and 'c' may vary by factors of 2-3 in the same period, under extreme conditions. The average value of d over the flight envelope varies between one aeroplane and another but a value of mean $\sqrt{d}$ = 5 radians/sec may be regarded as typical.

2. HILL CLIMBING USING CONTROLLER PARAMETERS

The very rapid rates of change of the parameters of the aeroplane mean that a hill climbing procedure carried out directly on the controller parameters and designed to minimise the response error of the system, is likely to be too slow[1,2] because of the number of steps required and the fact that each step requires a fresh T seconds of signal to allow it to be assessed. Figure 1 shows the standard deviation of the ratio of the time average of the square of a Gaussian random process to the probability average. This indicates the necessary slowness of any identi-

fication technique involving the accurate measurement of the average of signals specified only in a probability sense. Thus the time necessary to make a measurement, which belongs to a population having standard deviation of 20% of the mean value, is given by $\Omega T = 50$. So that for a system for which $\Omega = 5$, T = 10 seconds and each point on the hill will require that time for evaluation.

3. IDENTIFICATION TECHNIQUES HAVING NECESSARY SPEED

If the system is linear it is sufficient for the synthesis of an optimum controller that the parameters of the aeroplane transfer function be known. These may be determined by minimising the error between the response of the aeroplane and the response of a dummy aeroplane fed with the same input and having the same form of transfer function as the real aeroplane but with undetermined coefficients. Since the transfer function of the real aeroplane remains unaltered during such a hill climbing process, only a single T second sample of input and output signal need be employed. This, as will be discussed in the next section, enables the search procedure to be mechanised in such a way that the time it takes is arbitrarily small compared with the time T required to gather information.

A second advantage of this technique, which is rather difficult to assess quantitatively may be seen by reference to Figure 2. This shows the probability of determining the correct sign of the difference between the probability averages of two random variables from two measurements of time average, as a function of the averaging time (T) and the ratio of the probability averages. The error signals are assumed independent, Gaussian and to have autocorrelation function

$e^{-\Omega|t|}$ so that a multiple of the integrals of their squares have approximately a χ^2 distribution with $n(\approx \Omega T + \frac{1}{2})$ degrees of freedom[3].

Thus taking $\Omega = 5$, as before, the figure in-

Superior numbers refer to similarly-numbered references at the end of this paper

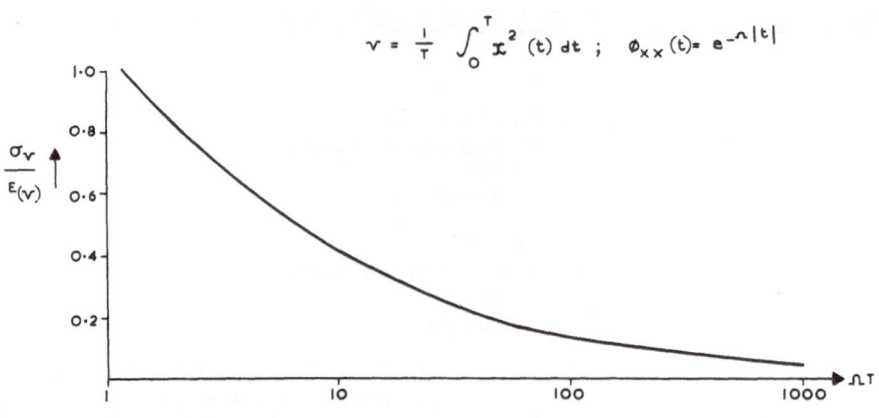

$$\gamma = \frac{1}{T} \int_0^T x^2(t)\,dt \quad ; \quad \phi_{xx}(t) = e^{-\Lambda|t|}$$

FIG 1

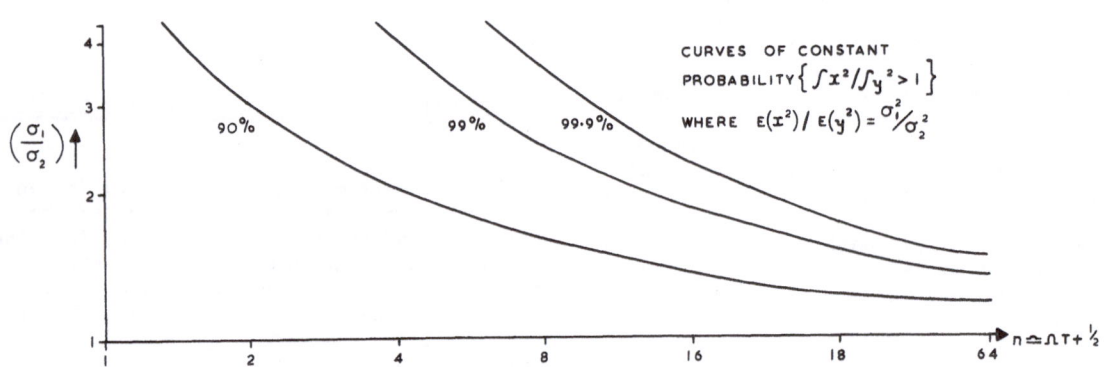

CURVES OF CONSTANT
PROBABILITY $\left\{ \int x^2 / \int y^2 > 1 \right\}$
WHERE $E(x^2) / E(y^2) = \sigma_1^2 / \sigma_2^2$

FIG 2

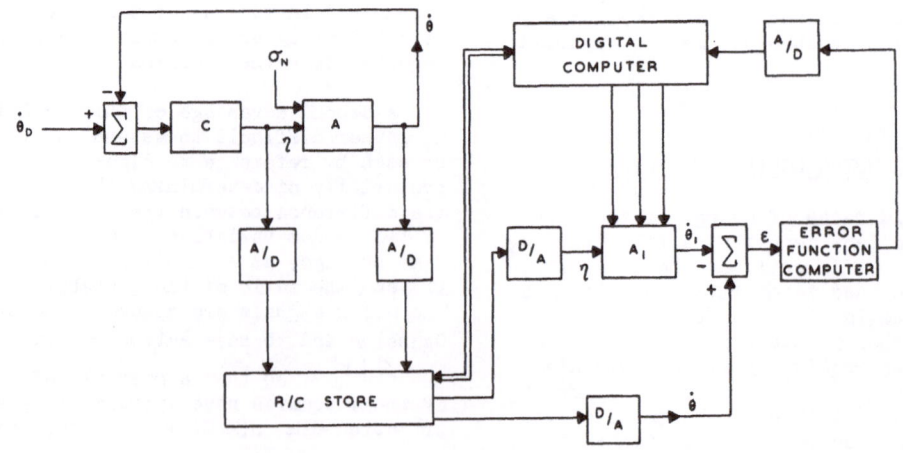

FIG 3

166

dicates that the 99% confidence limit is achieved for $\sigma_1/\sigma_2 \approx 1.5$ with a ten second sample of signal. However this is likely to be pessimistic when small differences in mean square error result from small differences in parameter values because in these circumstances the two signals will be highly correlated, since the input signal is the same in each case. Thus the probability of the sign of the difference being correctly determined over any finite time is consequently increased.

4. MECHANISING OF THE IDENTIFICATION PHASE

Although it has been indicated that the averaging of signals for ten seconds may suffice to define the error at one point of a hill climbing process with sufficient accuracy, it is still necessary that the entire identification phase be performed in as short a time as possible. This condition is satisfied by running the dummy aeroplane on an accelerated timescale using the arrangement indicated in Figure 3.

The system is disturbed continuously by an input $\dot{\theta}_D$, the elevator and pitch rate signals are sampled, converted to binary form and stored in the recycling (R/C) store. When this store is full, that is after ten seconds, the overall controller signals the R/C store to replay, an initial set of parameters having already been set into the dummy system by the digital computer. The replayed signals are converted to analogue form and the elevator signal feeds the dummy aeroplane (A_1) whose response $\dot{\theta}_1$ is compared with the $\dot{\theta}$ signal from the R/C store.

The modulus of the error $\dot{\theta} - \dot{\theta}_1$ is integrated and at the end of the replay cycle the output of the integrator is sampled, converted and fed into the digital computer. On the basis of this and the computer programme a new set of parameters are set into the dummy aeroplane and the R/C store again stimulated into its replay mode. This process is continued until an optimum point has been obtained after which the store is stimulated into its record mode again. The advantages of this method of searching are:

1. that the parameters of the hill being climbed remain constant throughout the process, and

2. that the time to perform the identification may be made as small as desired compared with the record cycle.

Thus in the system being investigated the playback time is 1/400 of the record time and the identification process takes about 1 second to perform.

5. METHODS OF SPEEDING THE IDENTIFICATION

As described the operation of the system consists of a record phase followed by an identification phase and then by a second record phase. This means that the controller parameters may be updated only once per cycle (10 + 1 seconds), so on average the parameter information is $11\frac{1}{2}$ seconds out of date (being 6 seconds when it is obtained and 17 seconds before it is replaced). This average may be improved by a factor of two for an additional storage capacity equal to approximately 10% of the capacity of the R/C store. In this mode recording and identification proceed continuously, the data used being updated after every completed identification.

6. EFFECT OF GUST DISTURBANCES

The error in identification, ϵ (Figure 3) is given by

$$\epsilon = \frac{C}{1 + AC} (A - A_1)\dot{\theta}_D + \frac{1 + A_1 C}{1 + AC} A\sigma_N$$

where σ_N is the input due to gusts.

So that if we define

$$\nu = \int_0^T |\dot{\theta}_D| \, dt \Big/ \int_0^T |A\sigma_N| \, dt \, , \text{ then}$$

as $\nu \to 0$ the point of minimum error tends to $(1 + A_1 C) = 0.$

Figure 4 shows histograms of the distribution of the parameter d for various signal/noise ratios ν and also the variation of the mean d of the distribution as a function of ν.

Alternative forms of dummy system exist for which $\dot{\theta}_1$ is not correlated with gust signals, but even in these cases the mean of the distribution will not necessarily remain constant because of asymmetry of the error surface. Thus it is not sufficient in the presence of gusts to improve the effective accuracy of identification by averaging successive values of identified parameters. It is necessary to find a method of improving the effective signal/noise ratio.

6.1. Stacking

The aeroplane is regarded as linear between elevator and pitch rate. If $\eta^{(i)}$ represents the elevator signal during the ith record phase and $\dot{\theta}^{(i)}$ represents the corresponding pitch rate response, then

$$\sum_{i=1}^{n} k_i \eta^{(i)} \text{ will produce a response } \sum_{i=1}^{n} k_i \dot{\theta}^{(i)}$$

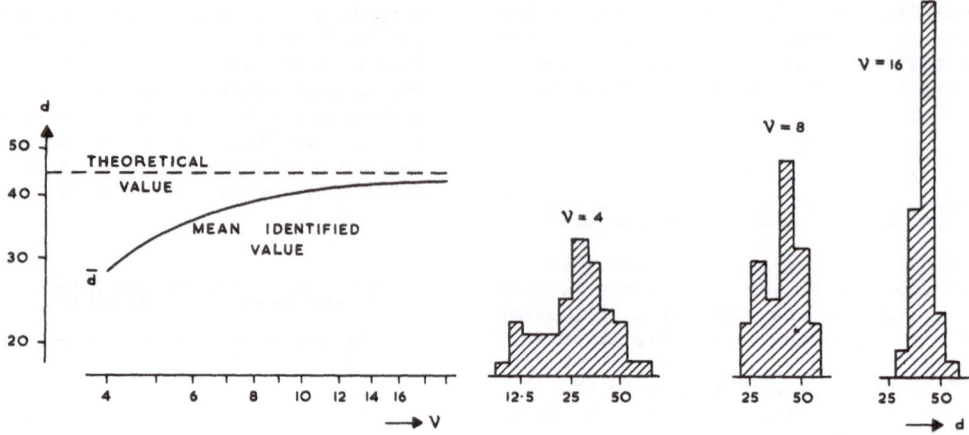

FIG 4

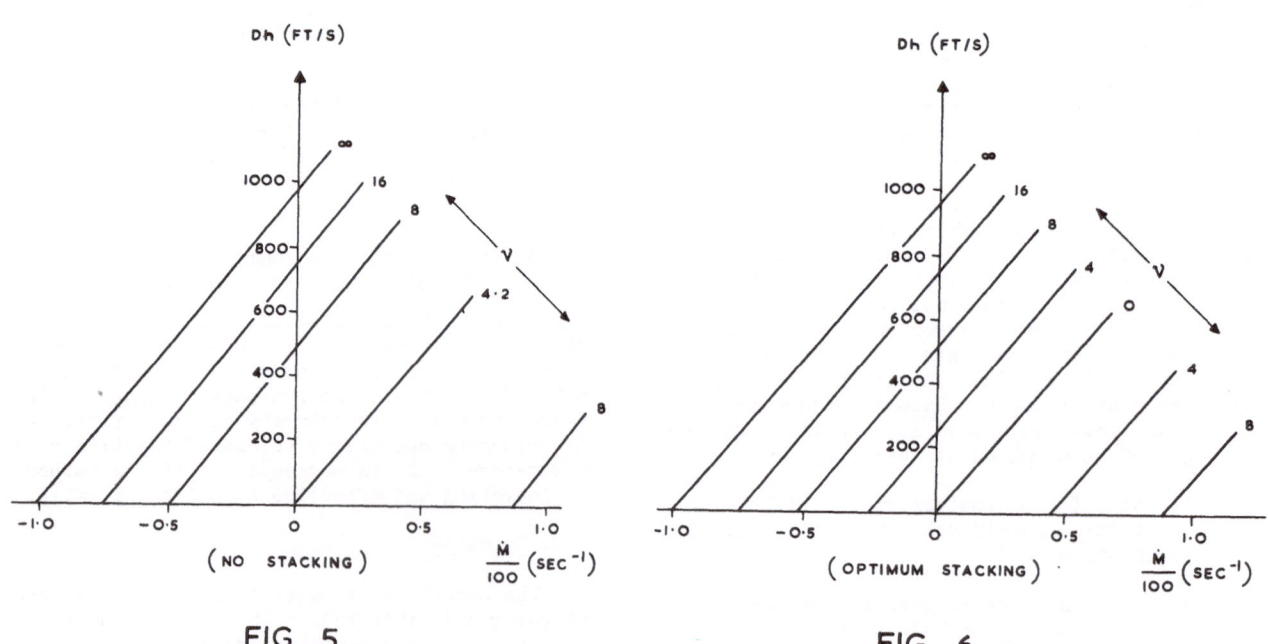

FIG 5

FIG 6

168

If σ_N is the gust signal during the ith record cycle, and if the parameters of A remain substantially unaltered over the period of the summation, we have (Figure 3)

$$\sum_{i=1}^{n} k_i \eta^{(i)} = \frac{C}{1+AC} \sum_{i=1}^{n} k_i \dot{\theta}_D^{(i)} - \frac{AC}{1+AC} \sum_{i=1}^{n} k_i \sigma_N^{(i)}$$

$$\sum_{i=1}^{n} k_i \dot{\theta}^{(i)} = \frac{AC}{1+AC} \sum_{i=1}^{n} k_i \dot{\theta}_D^{(i)} + \frac{A}{1+AC} \sum_{i=1}^{n} k_i \sigma_N^{(i)}$$

Now if the $\dot{\theta}_D^{(i)}$ signals are independent of i, that is the disturbance input is periodic with respect to the record cycle, and σ_N is a random function of time with zero mean value

$$\text{then} \qquad \frac{\displaystyle\sum_{i=1}^{n} k_i \eta^{(i)}}{\displaystyle\sum_{i=1}^{n} k_i} = \left(\frac{C}{1+AC}\right)\dot{\theta}_D - \left(\frac{AC}{1+AC}\right)\frac{\displaystyle\sum_{i=1}^{n} k_i \sigma_N^{(i)}}{\displaystyle\sum_{i=1}^{n} k_i}$$

and this latter term tends to zero as $n \to \infty$

$$\text{Similarly} \qquad \frac{\displaystyle\sum_{i=1}^{n} k_i \dot{\theta}^{(i)}}{\displaystyle\sum_{i=1}^{n} k_i} = \left(\frac{AC}{1+AC}\right)\dot{\theta}_D + \left(\frac{A}{1+AC}\right)\frac{\displaystyle\sum_{i=1}^{n} k_i \sigma_N^{(i)}}{\displaystyle\sum_{i=1}^{n} k_i}$$

This means that superimposing or stacking recordings allows the effect of gusts to be attenuated to any desired extent without requiring any increase in the capacity of the R/C store.

The effective time constant of stacking can be varied by choice of the factor k_i, with which previous information in the R/C store is weighted with respect to new information, as a function of the estimated signal/noise ratio. The input signal $\dot{\theta}_D$ may most conveniently be locked to the record cycle by making the computer generate $\dot{\theta}_D$ in the form of a chain code[4].

7. LIMITATIONS ON RATE OF FOLLOWING

Stacking enables improvement in signal/noise ratio to be obtained at the price of identification taken over longer time. Thus if the aeroplane parameters are changing with time, then there is an optimum for k_i which is a function of ν and the rate of change of parameters.

Figures 5 and 6 show the loci of the minimum signal/noise ratio, ν, which will lead to identification which, with 95% confidence, will not be in error by more than a factor of 2 at any time before the next identification is obtained.

These graphs are drawn for a particular aeroplane case, and with and without stacking, but do not include the effect of continuous identification.

The definition of signal/noise ratio given in section 6 was chosen because it is independent of the controller setting. It should be realised that this is not the only possible; a more practically useful definition is

$$\left[\frac{RMS \; \dot{\theta} \; \text{(GUSTS + PERTURBATION INPUT)}}{RMS \; \dot{\theta} \; \text{(GUSTS ONLY)}} - 1\right] \quad \text{which}$$

although a function of the controller setting is relatively invariant over the range of interest. Signal/noise as defined by this expression gives numerical values about 5 times smaller than those appearing in, for example, figures 4 and 5.

8. MAINTAINING THE OPTIMUM CONTROLLER

The system discussed enables the parameters of the transfer function of the aeroplane to be determined and thence the controller which will produce model demand response to be synthesised. However the system as such contains no feedback to indicate the degree of success achieved.

To evaluate for each controller setting an error criterion measuring the departure of the response from model response is to return almost to a direct hill climbing on the controller.

However occasional erratic identifications can be eliminated by predicting on the basis of previous parameter identifications the values to be expected in the current identification and accepting only if it satisfies some criterion which depends upon allowable rates of change of parameter values. This does not eliminate permanent faults (i.e. wrongly set gain) but this may be detected by other means.

Prediction is also useful in that it enables estimates of rate of change of parameter and signal/noise ratio values to be obtained as the gradient of, and the scatter of points about, the best fit straight line. These quantities are required to control the degree of stacking. Furthermore such estimates may also be used to alter the number of points in the prediction. In addition the inherent lag between true parameter values and those identified may be reduced by predicting the appropriate distance ahead. Work is currently concentrated on such problems.

9. DESIGN OF THE OPTIMAL CONTROLLER

To satisfy the requirement for a demand response which has a specified form (characterised by the transfer function M) it is sufficient, for the type of system illustrated in Figure 3, to provide a controller C of the form $\frac{M}{1-M} \cdot \frac{1}{A}$. If a more general control law is considered, in which the elevator angle is merely required to be a linear function of pitch rate and pitch

169

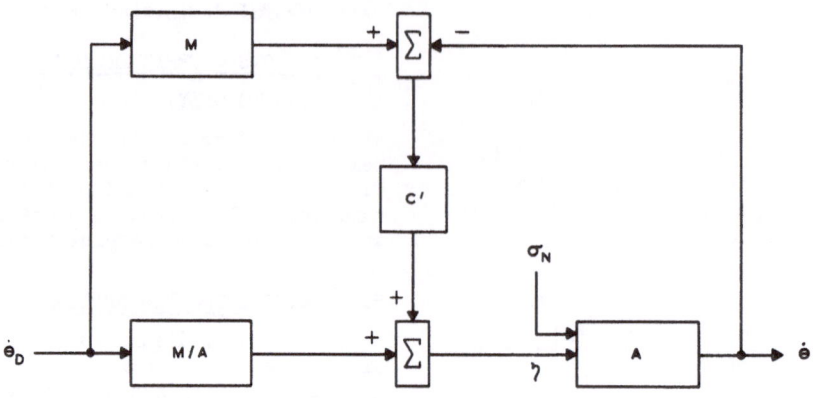

FIG 7

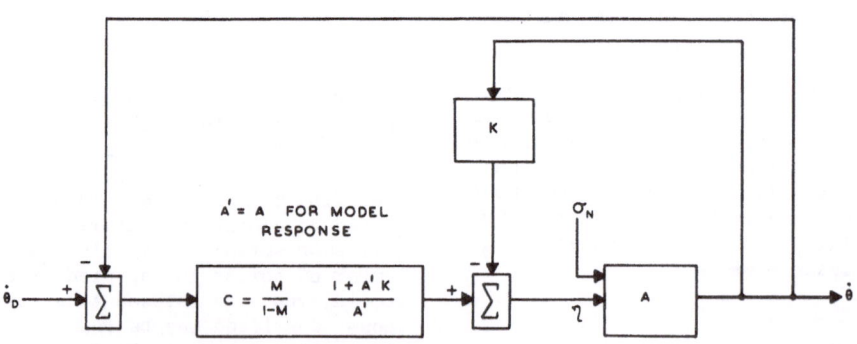

FIG 8

170

rate demand, then an infinite number of possibilities exist, all of which will give model response. The most general system of this form is illustrated in Figure 7 where C' is an arbitrary linear function.

This includes an open loop system (C' = 0), a high gain system (C' → ∞) and the system of figure 3 (C' = $\frac{1}{A} \cdot \frac{M}{1-M}$) as special cases. Since C' is arbitrary it may be chosen to satisfy some additional criterion. Possible criteria are:

(a) to minimise the disturbance of the aeroplane by gusts,

(b) to minimise the effect, on the demand response of the system, of errors in the determination of aeroplane parameters,

and for each of these it is necessary to consider the effect of additional lags and nonlinearities such as may occur in, for example, the elevator servo and actuator.

9.1. Supplementary Damping Loop

9.1.1. Effect of controller parameter errors

If the stability of the system of Figure 3 is examined for the case in which the aeroplane parameters (a, b, d) and the controller parameters (a', b', d') are unequal, it is found that instability results if $(\frac{d' - d}{d})$ is too large.

The most severe limitation occurs when aeroplane natural frequency is of the same order as model frequency, and the damping of the aircraft is low. The latter quantity may be artificially increased by adding an elevator signal proportional to pitch rate (Figure 8). Model response may still be obtained by setting in C the parameters corresponding to the damped aircraft $(\frac{A}{1 + AK})$.

If it is required that the system be at least neutrally stable for a 100% error in a' and d', then the minimum required gain of the damping loop may be obtained as a function of the aeroplane and controller parameters.

An upper limit which may be taken as a simple approximation is $K = \frac{\omega}{a'}M$.

9.1.2. Effect of lags, etc.

The lags which inevitably exist in the elevator servo and on the pitch rate signal can cause high frequency instability if the damping loop gain is too high. This is one reason which prevents a simple high gain system from being used, without some additional features to prevent, or control, this potential instability. Sub-

stitution of realistic values for a modern supersonic aircraft shows that the gains required by the preceding section are unlikely to lead to difficulties from this cause.

9.1.3. Effect of Gusts on Pitch Rate Response

With the system of Figure 8, the pitch rate due to gusts is given in the case of correct controller C by

$$\dot{\theta} = (1 - M)(\frac{1}{1 + AK})A\sigma_N$$

and the error between the system response and that of the ideal or model M is

$$\dot{\theta} - M\dot{\theta}_D \approx (\frac{A - A'}{A})(\frac{1}{1 + AK})(1 - M)M\dot{\theta}_D$$

It would appear therefore that since the factor $\frac{1}{1 + AK}$ appears in both expressions, improvement in gust response and reduction in the effect of parameter errors are likely to proceed together. A similar argument may also be applied to the generalised system of Figure 7.

10. OPTIMISATION OF GUST RESPONSE

A possible method of choosing C' (Figure 7) is to minimise some measure of the disturbance to the aircraft due to gusts. $\int (\eta^2 + \theta^2)$ was chosen as a suitable function. It seems reasonable that $\dot{\theta}$ will be small if both η and θ are. A difficulty which arises is that the optimum C' depends on the gust disturbance, which is unknown. Accordingly, some simplification is necessary.

10.1. Supplementary damping loop

The value of the constant K in Figure 8 which minimises $\int (\eta^2 + \theta^2)$ has been calculated for a sinusoidal disturbance, and the resultant optimum K plotted as a function of frequency. It is found that the average value over the range of frequencies of interest is of the same order as that obtained in paragraph 9.1.1., i.e. $K = \frac{\omega}{a}M$

10.2. Dynamic Programming

A further possibility is to minimise

$\int (\eta^2 + \theta^2)$ in response to arbitrary initial conditions on the state variables of the system. This is a standard example of dynamic programming, which in the case of a quadratic 'cost function' ($\eta^2 + \theta^2$), and a linear system, is soluble analytically[2]. Taking as a simple example the case b = 0, the optimum control law obtained is

$$\eta = G(1 + \tau D) \theta$$

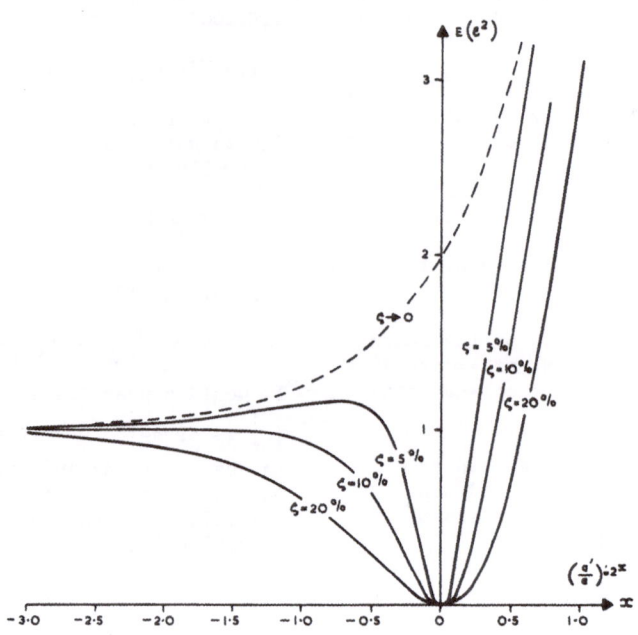

FIG 9 SECTIONS THROUGH ERROR SURFACE, SHOWING
VARIATION WITH AIRCRAFT DAMPING

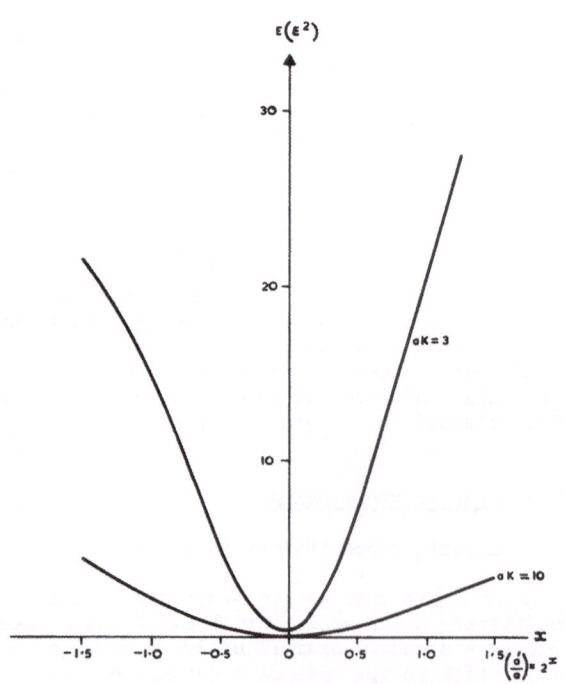

FIG 10 ERROR SURFACE FOR IDENTIFICATION OF
DAMPED AIRCRAFT

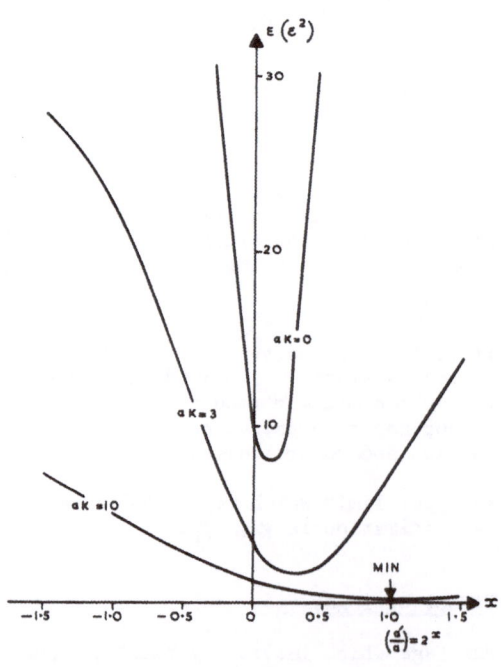

FIG 11 SHIFT IN POSITION OF MINIMUM DUE TO SMALL
LAG IN DUMMY SYSTEM

172

where $G = -\frac{1}{a}\left\{\sqrt{(d^2 + a^2)} - d\right\}$; $G^2\tau^2 = -\frac{2}{a}G$

If also $(-a) = d$ (which is approximately true for the aircraft on which most of the work has been performed), the optimum control is

$$\eta = 0.41\left\{1 + \frac{2.2}{\sqrt{|a|}} D\right\}\theta$$

and the controlled system has natural frequency $1.2\sqrt{d}$, and damping 38%.

11. MEASUREMENT OF AIRCRAFT PARAMETERS WITH DAMPING LOOP (Figure 8)

In attempting to identify the aircraft transfer function two alternatives present themselves:

1. The elevator signal (η) may be used as the input to the dummy system. The error criterion will then be a minimum when the parameters of the dummy system equal the corresponding parameters of the real aircraft.

2. Or the input to the dummy may be set equal to the output of the controller (block 'C' in the diagram). Since the transfer function, $\frac{A}{1 + AK}$ between this point and the pitch rate signal is of the same analytical form as the aircraft transfer function, A, the same form of dummy system may be used as in method (1). The optimum dummy system parameters will now, however, be a function both of the aircraft parameters and the damping loop gain.

11.1. Comparison of Error Surfaces

One means of comparing these two methods is by the shape of the two error surfaces. Figure 9 shows a section through the error surface for identification of the natural aircraft. The curves are drawn for different aircraft damping ratios.

Noteworthy features are the narrow valley containing the actual minimum, especially for low damping ratios, and the flat portion of the curve to the left of the minimum. Both of these features increase the difficulty of the identification procedure. In practice the values of the error criterion are initially examined for values of the dummy aircraft parameters (a')differing by powers of 2. We see that for the 5% damping case, which is a possible situation, this form of search could miss the true minimum completely. When the effect of sampling the elevator and pitch rate signals prior to recording them is included, a number of subsidiary minima appear, which again could lead to incorrect identification.

These unpleasant effects are only important for extremely lightly damped systems. However, the adaptive process must be able to cope with this situation, and so some modification to improve this behaviour is required. As shown in figure 9, a much more symmetrical curve is obtained for increasing damping ratios, and so it seems likely that identifying the artificially

damped aircraft would improve this aspect of the performance. This is indeed the case, as is shown in Figure 10.

Unfortunately, increasing the damping loop gain also reduces the sensitivity of the error criterion to changes in dummy system parameters. This is not surprising, since the damping loop was introduced in order to reduce the sensitivity of the pitch rate response to changes in the controller parameters, and these two effects are closely related.

A further reason for not using too high a damping loop gain is illustrated in figure 11, which shows the effect of the presence of a small lag (0.05 seconds) in the dummy system, but not in the real system. Such a difference could easily arise due to the presence of servo lags, for example, or merely due to the fact that the simplified transfer function used in the dummy system is not an accurate representation of the real aircraft dynamics over the whole frequency range. In the theoretical curves shown, a shift by a factor of 2 in the position of the minimum is illustrated (for aK = 10), and such a discrepancy has also been observed on the simulator.

To summarise, from the point of view of the shape of the error surface, it would appear advantageous, if not essential, in some cases, to identify the damped aircraft, provided that the damping loop gain is not too large. This argument will clearly apply whatever is the precise form of search procedure used.

11.2. Effects of Gusts and Backlash

Two further factors also influence the present discussion. These are the possible presence of backlash and gust disturbances.

The design of the control system took no account of the backlash which almost inevitably exists in the elevator servo. This was not unreasonable, since the object was to obtain a specified response to comparatively large pilot or autopilot inputs. For identification, however, it may be necessary to use very small disturbing signals in order not to cause discomfort to the passengers. The effect of backlash may now be significant. If the damped aircraft is identified, then any servo backlash will be included within the real system, and identification based on small input signals may give a totally false impression of the response of the real system to a large input; from this point of view identification of the natural aeroplane seems preferable.

Gust disturbances slow down the identification, as pointed out in section 6. It would be advantageous, however, if the average parameter identification remained constant. This implies that the input to the dummy system should be uncorrelated with the gusts. Due to the form of the various transfer functions, in the presence of a damping loop the output of the controller (C) is in general far less correlated with any gusts than the elevator signal. Thus from this point of view identification of the damped aircraft is preferable.

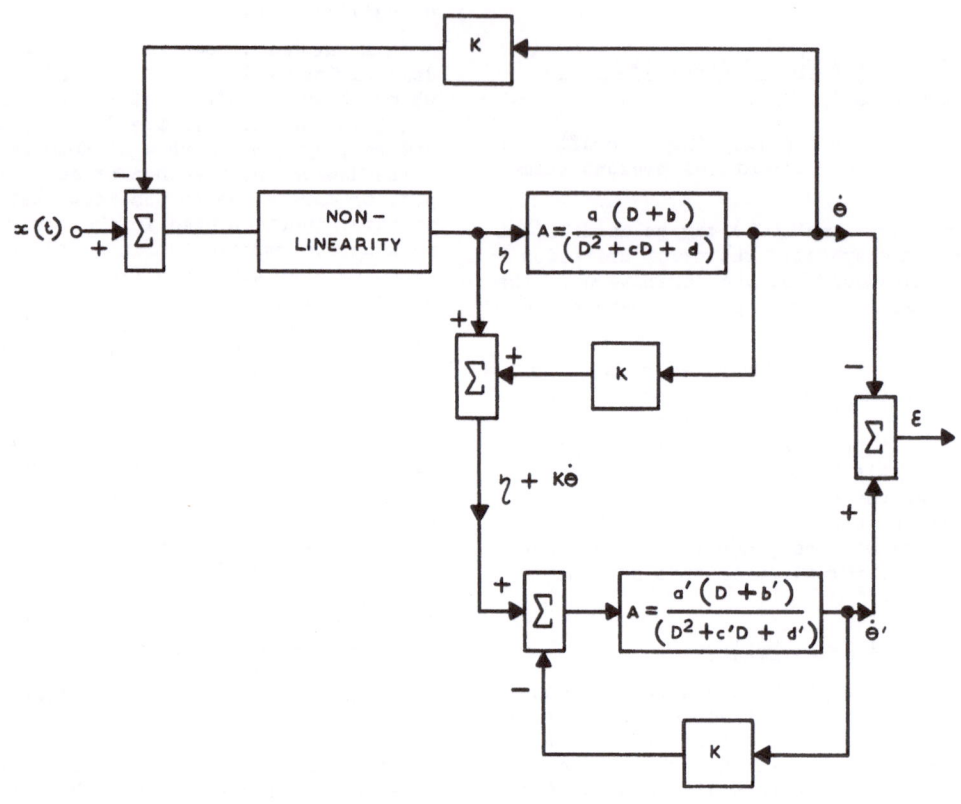

FIG 12 IDENTIFICATION OF DAMPED SYSTEM WITH
NON-LINEARITY

11.3. Compromise System

Figure 12 shows an attempt to combine the advantages of both methods, so far as this is possible. The dummy system is well-damped, so as to give a satisfactorily shaped error surface. The input to the dummy is a combination of pitch rate and true elevator position, so that in the absence of gusts correct identification will take place, irrespective of the presence of backlash. In the absence of backlash, the input to the dummy is the same as the controller output $x(t)$, so that the effect on the identified parameters of gust disturbances is minimised.

The situation in which both backlash and gust disturbances are present simultaneously has been examined on the simulator, using backlash sufficiently large to noticeably alter the elevator wave-form, and various levels of gust disturbances. Identifications obtained so far, using the configuration illustrated, have been as accurate as the corresponding identifications with no backlash.

12. CONCLUSION

A method has been described for the in-flight determination of the parameters of the aircraft short-period pitch transfer function. Consideration has been given to the optimum manner in which this information may be used to vary the control law parameters so as to obtain a fixed response to a pitch rate demand signal. Work on this project continues, but the results so far show reasonable promise that a system based on the principles described will be able to deal with the rapid changes in aircraft parameters experienced in practice.

13. ACKNOWLEDGEMENTS

The authors wish to thank the directors of Smiths Aviation Division for permission to publish this paper.

14. REFERENCES

1. HAMMOND, P.H., DUCKENFIELD, M.J. Automatic Optimisation by Continuous Perturbation of Parameters. Automatica 1963, 1, p.147.

2. WHITAKER, H.P., Use of Model Reference Adaptive Control to Achieve a Specified Performance. M.I.T. Report R.407, April 1963.

3. BLACKMAN, R.B., TUKEY, J.W. Measurement of Power Spectra from the Point of View of Communications Engineering. Dover 1959.

4. BRIGGS, P.A.N., HAMMOND, P.H., HUGHES, M.T.G. and PLUMB, G.O., Correlation Analysis of Process Dynamics Using Pseudo-Random Binary Test Perturbations. Proceedings of Convention on Advances in Automatic Control, April, 1965, Proc.Inst. Mech. Engrs. Vol. 179, part 3H, 1965

5. LEITMANN, G. Optimisation Techniques. Academic Press 1962.

Discussion

Mr. G. Belcher (UK) was not totally convinced that the stacking feature sufficiently attenuated the effect of wind gusts to allow operation in a signal/noise ratio environment lower than 4:1. The questioner argued that the large demand signals attendant with even moderate gusting would severely tax the pilot and called on figures 5 and 6 for confirmation. The question could not be fully resolved without justifying the details of the constraints imposed on the rate-of-change of the aircraft parameters.

Prof. J.G. Balchen (Norway) observed that in the absence of an external stimulus, the model would not match the aeroplane because the input to and output from Block A are highly correlated. Dr. Meredith proposed perturbing the control surfaces, in the absence of gusting, to assist identification. Prof. Balchen then proposed the additional complexity of including a complete controller with the reference model, although performance in the presence of gusts would not be improved.

Mr. M.J. Duckenfield (UK) suggested that feed-forward of gust disturbance (by the principle of invariance) would improve performance.

Written contributions to the discussion

P.C. Young (UK)

The authors present a very attractive method of automatically identifying the parameters which describe a dynamic process. As they point out rapid parameter estimation is essential to an identification-adaptive system such as that described in their paper. However, straight-forward reduction in identification time cannot provide a complete answer to the problems which arise when this type of control system is applied to a highly non-stationary process such as a modern high performance aircraft.

The need to accumulate a large amount of information, in order that a reasonable statistical parameter estimate can be obtained, necessarily results in unavoidable total computation time delays. This difficulty has been one of the subjects of investigations which have been carried out at Loughborough College during the past two years. An account of these investigations is presented in a paper which is published in the Proceedings of this Symposium (ref. 1). Other aspects of this work have been described in Refs. 2 and 3 in this paper.

The hybrid "Satisfaction Error Self Adaptive Control" (S.E.S.A.C.) system is basically very similar to the self adaptive system suggested by Meredith and Dymock. However, parameter identification is accomplished by a hybrid "Parameter Determination Computer" (P.D.C.) of rather different design to the "Learning Model" computer which they describe. The digital computation performed by the hybrid P.D.C. consists of the repeated solution of sets of linear algebraic simultaneous equations by a least squares approach. This provides an estimate of the coefficients of a Laplace transform transfer function description of the process.

The present system uses general purpose hybrid equipment, having comparatively slow A - D and D - A conversion and transfer rates. As a result, parameter estimates can only be supplied in a minimum of fifteen seconds. This total computation time delay is not only dependent upon the speed of operation of the hybrid equipment but also on the undamped natural period of the process under investigation as well as the effective signal/noise ratio of the sampled data obtained from the process. An extension to the present minimum of fifteen seconds appears necessary if the process has an undamped natural period greater than fifteen seconds, or if its signals are contaminated by high levels of additive noise.

Several possible methods of increasing the speed of estimation are available but they can only provide a limited solution to the basic problem. Whatever the reduction in computation time, the computer still requires a finite time to sample data and perform its computations. Consequently, the adaptive controller parameters can only be updated at discrete intervals of time. Any rapid variation in the controlled process parameters during this period can lead to poor control characteristics and possible instability.

Further improvement in performance can be obtained by means of a technique described in reference 1. It concerns the introduction of simple analogue elements which perform a first order prediction based on the past rate of change of the parameter. Although these prediction elements still do not supply a fully satisfactory solution, preliminary results have indicated that they can provide improved performance for any particular computation delay, whilst at the same time requiring only simple mechanisation from normal analogue equipment (see Fig. 7 in ref. 1).

Fig. 8 in ref. 1 illustrates the effect of introducing such a predictive element into a P.D.C. which is tracking a ramp change in a process parameter. First order prediction of this type would not be so effective in dealing with other than ramp changes in the parameters, but it is possible that higher order elements may be developed which overcome this difficulty.

This approach is still in its infancy but the encouraging results obtained so far suggest that techniques of this kind could help in the practical implementation of identification-adaptive schemes such as hybrid S.E.S.A.C., or that described by Meredith and Dymock.

Another point dealt with in Ref. 1 is the problem of obtaining a good parameter estimate in the presence of uncertainty on the sampled signals caused by such effects as measurement noise, or the random gust disturbance mentioned by Meredith and Dymock. The "Least squares" approach of the Hybrid P.D.C. system, coupled with the use of a technique known as the Method of Multiple Filters (see ref. 1), make it reasonably insensitive to uncertainty of this kind. However it is felt that modified techniques are required if the system is to be fully effective in a truly practical sense.

Two methods are suggested as being possible ways of dealing with the problem.

The first entails the use of an auxiliary model constructed from the parameter estimates. The technique would be a straight development of work carried out on a similar continuous analogue parameter estimation scheme by V. S. Levadi (see reference list in ref. 1). The second approach would be to filter the process signals by means of an adaptive "Kalman-Bucy" optimal filter (see reference list in ref. 1). This would have the additional purpose of providing an answer to the dual optimal regulator problem, which could be utilised in any adaptive control application of the P.D.C.

REFERENCE

1. "Process Parameter Estimation and Self Adaptive Control" by P. C. Young. Published in these proceedings.

REPLY BY DR. MEREDITH AND MR. DYMOCK

We agree that it should be possible to use some form of prediction to improve the accuracy of parameter identification. We have ourselves been experimenting with such techniques, both from the point of view of eliminating the time lag due to the use of out-of-date information, as described by Mr. Young, and also in order to detect any abnormal values which might be obtained, in our case, by exceptional gust conditions. In this manner erroneous values are prevented from reaching the adaptive controller.

The form of prediction used, however, requires careful consideration. As a simple example, suppose that we wished to predict the future value, x_3, of a parameter, from the previous two identifications x_1 and x_2. Linear prediction would lead to the value $2x_2-x_1$. In practice the identifications will not be precise, but might be regarded as having some probability distribution, with variance σ^2, say. Then the variance of the predicted value of x_3 will be $5\sigma^2$. Thus, dependent on the relative values of the rate of change of parameter, and accuracy of identification, the predicted value may be less accurate than even an out-of-date identification or the arithmetic average of several previous identifications.

It is true that, in theory, the effect of identification errors may be reduced by using more points in the prediction, but one must also take into account the fact that the greater the number of points used, the less likely it is that the parameter has been changing linearly over that interval of time.

One may visualise an 'optimal predictor' which is itself self-adaptive in nature, using constant prediction for low rates of change of the parameter, straight line prediction for larger but constant rates of change, and higher-order prediction for more complicated behaviour. Estimates of rate of change of the parameters, and standard deviation of the identifications, would need to be used to vary the mode of operation of the predictor, together with some a priori knowledge of the way in which the parameter might change.

Clearly there is considerable scope for further work here, and we shall be interested to learn of any results which have been obtained.

A NON-LINEAR SELF-ADJUSTING SYSTEM WITH LINEAR PREDICTION

by P. F. Klubnikin

USSR National Committee of Automatic Control
Moscow, USSR

INTRODUCTION

This paper is devoted to the analysis of a self-adjusting system of control using a digital computer. The parameters of the controlled plant, which is described by ordinary nonlinear differential equations, are not well known beforehand and change slowly with time.

The algorithm of the controller is constructed according to the principles described in reference 1. The prediction of system performance uses a linear model of the plant within a cycle of control signal computation. A study is made of the influence of the parameters which define the search and prediction processes; the effect on quality of control of differences between model and actual plant are also studied.

1. FORMATION OF PLANT MODEL

The structure of the control system is shown in fig. 1. In it are two self-adjusting loops. In the first of these a learning model is formed, in the second selection of parameters in the control algorithm takes place. In general the performance of the control system will depend on how well the model reflects the characteristics of the plant. In practice it is not possible to reach exact correspondence between model and plant, therefore the problem arises of optimal or rational selection of the model. A successful solution of this problem can be achieved for a specific plant by a study of its characteristics. The approach presented here is the simplest and most obvious one.

Suppose the plant can be described by system of equations as follows:

$$\dot{X}_j = \sum_{i=1}^{\lambda} B_{ji}(X_1, X_2, \ldots, t)X_i + A_j(X_1, X_2, \ldots, t)\delta_j \quad (1)$$

$$j = 1, 2, \ldots n .$$

where the δ_j control signals

and where

$$\left|\frac{\partial B_{ji}}{\partial t}\right| ; \quad \left|\frac{\partial A_j}{\partial t}\right| \ll |\dot{X}_j| . \quad (2)$$

(In what follows, for simplicity, we will consider only one control signal δ and one plant output X.)

Supposing $\delta = \delta_{oo} + \Delta\delta$ and $X_j = X_{oj} + \Delta X_j$ then from equation (1) we can formulate a linear model of the plant, for which we obtain:

$$\Delta\dot{X}_{jM} = \sum_{i=1}^{\lambda} b_{ji}(t)\,\Delta X_{iM} + a_j(t)\,\Delta\delta \quad (3)$$

where

$$\left|\frac{\partial a_j}{\partial t}\right| ; \quad \left|\frac{\partial b_{ji}}{\partial t}\right| \ll |\Delta\dot{X}_{jM}| .$$

Adjustment of model as given by equation (3) can be achieved according to the criterion

$$\Delta_{cp} = \frac{1}{T_0} \int_0^{T_0} (X - X_M)^2 dt \quad (4)$$

where X_M is the model variable corresponding to the controlled plant variable, or to the plant output variable (sign of Δ is always positive) or

$$\Delta_{cp}^* = \frac{1}{N_0} \sum_{v=0}^{N_0} [X(t-vT) - X_M(t-vT)]^2 \quad (5)$$

where T sampling time of computation.

Use of a model governed by equation 3 with sufficiently small Δ_{cp} - ensures stability of the control system.

Actually, setting equations (1) and (3) in the form of

$$X(t) = A\,\delta(t) \quad \text{and} \quad X_M(t) = A_M\,\delta(t)$$

where A and A_M are operators which for each function $\delta(t)$ belonging to set E, relate functions $X(t)$ and $X_M(t)$ also belonging to E. On the basis of equation (2) we have

$$||X(t)|| \leqslant |A| \cdot ||\delta(t)||$$

$$\text{and} \quad ||X_M(t)|| \leqslant |A_M| \cdot ||\delta(t)|| \quad (6)$$

Superior numbers refer to similarly-numbered references at the end of this paper.

Prior to editing this paper was translated from the Russian by a computer program devised by the Machine Translation Group, Autonomics Division, National Physical Laboratory

where
$$\|X(t)\| = \lim_{\lambda_0 \to \infty} \sqrt{\frac{1}{\lambda_0} \int_0^{\lambda_0} |X(t)|^2 \, dt}$$

and
$$|A| = \sup_{t \in 0, \infty} \frac{\|X(t)\|}{\|\delta(t)\|} \, .$$

$|A|$ is the conventional gain coefficient.

Then utilizing (6) and (4) we obtain

$$\Delta_{cp} \geqslant \|\delta(t)\|^2 \, (|A| - |A_M|)^2 \, . \qquad (7)$$

If the model loop is stable, which is ensured by appropriate selection of computer program parameters, then for sufficiently small Δ_{cp}, by virtue of (7), the system is also stable. This follows from the proximity of the values of $|A_M|$ and $|A|$, which, in the given case define stability (reference 2).

With sufficiently small Δ_{cp} the performance of the control system can be near to that of the prediction loop. Transforming equation (3) under the condition $b_{ji}, a_j = $ const to difference equations we obtain expressions for the digital computation of X_M as:

$$X_M(t) = \sum_{\lambda=1}^{\ell} c_\lambda X_M(t-\lambda T) + \sum_{\lambda=1}^{q} d_\lambda \, \delta(t-\lambda T) \qquad (8)$$

where c_λ and d_λ are constant coefficients independent of time, depending on a_j and b_{ji}.

2. ALGORITHM OF SELF-ADJUSTMENT

The general algorithm of self-adjustment and derivation of control signals is divided into two almost independent algorithms (see also ref.1). The first algorithm leads to the formation of the model, or the derivation of coefficients c_λ and d_λ in equation (8) by means of a search for the minimum of criterion (5) in space of c_λ, d_λ. In the case of a linear equation (1) and a well behaved function $\Delta_{cp} (c_\lambda, d_\lambda)$ a search carried out by the method of modified gradient allows us to find the minimum of Δ_{cp} comparatively quickly (reference 1). For a nonlinear plant the function $\Delta_{cp} (c_\lambda, d_\lambda)$ can have local sub-minima. In such a situation search by the gradient method does not guarantee against

acquiring a local minimum and does not always lead to valid results; for this reason it is expedient to use non-local methods of search. One such method is the method described in reference 3 which avoids secondary peaks and troughs of the function $\Delta_{cp} (c_\lambda, d_\lambda)$.

With regard to what has been said, we obtain the structure of main operations of first algorithm as follows:

(1) Input into the memory of the digital computer sequences of the variables

$$X(t), \, X(t-T), \, X(t-2T), \, \ldots.$$

and

$$X_m(t), \, X_m(t-T), \, X_m(t-2T), \, \ldots.$$

(2) Compute Δ_{cp}^* according to equation (5).

(3) Determine the partial derivatives:-

$$\frac{\partial \Delta_{cp}^*}{\partial c_\lambda} = \frac{\Delta_{cp}^* (c_\lambda + \Delta c, \ldots.) - \Delta_{cp}^* (c_\lambda, \ldots.)}{\Delta c}$$

$$\frac{\partial \Delta_{cp}^*}{\partial d_\lambda} = \frac{\Delta_{cp}^* (\ldots., d_\lambda + \Delta d) - \Delta_{cp}^* (\ldots., d_\lambda)}{\Delta d}$$

$$\lambda = 1, 2, \ldots \ell; \, q$$

and increments of coefficients:-

$$\Delta c_\lambda = - p_2 \frac{\partial \Delta_{cp}^*}{\partial c_\lambda} \, , \qquad \Delta d_\lambda = - p_2 \frac{\partial \Delta_{cp}^*}{\partial d_\lambda}$$

for hill climbing by the gradient method.

(4) Prepare the step of the hill climbing method of reference 3, that is define two points of local minima $O_1 (c_\lambda', d_\lambda')$ and $O_2 (c_\lambda'', d_\lambda'')$.

(5) Compute the hill climbing step

$$c_{\lambda 0} = c_\lambda + \frac{p_1 (c_\lambda'' - c_\lambda')}{\sqrt{Q}} \, , \qquad d_{\lambda 0} = d_\lambda + \frac{p_1 (d_\lambda'' - d_\lambda')}{\sqrt{Q}} \, .$$

$$Q = \sum_{\lambda=1}^{\ell} (c_\lambda'' - c_\lambda')^2 + \sum_{\lambda=1}^{q} (d_\lambda'' - d_\lambda') \, .$$

(6) Compute coefficients of model (equation 8)

$$c_{\lambda 1} = c_\lambda + \Delta c_\lambda \, , \qquad d_{\lambda 1} = d_\lambda + \Delta d_\lambda \, .$$

Note that, in realising the first algorithm only part of the operations 1-6 can be completed in

179

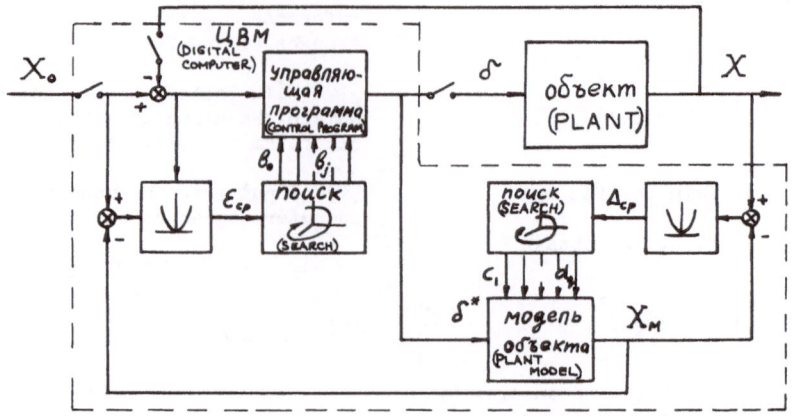

Fig. 1. Block diagram of system.

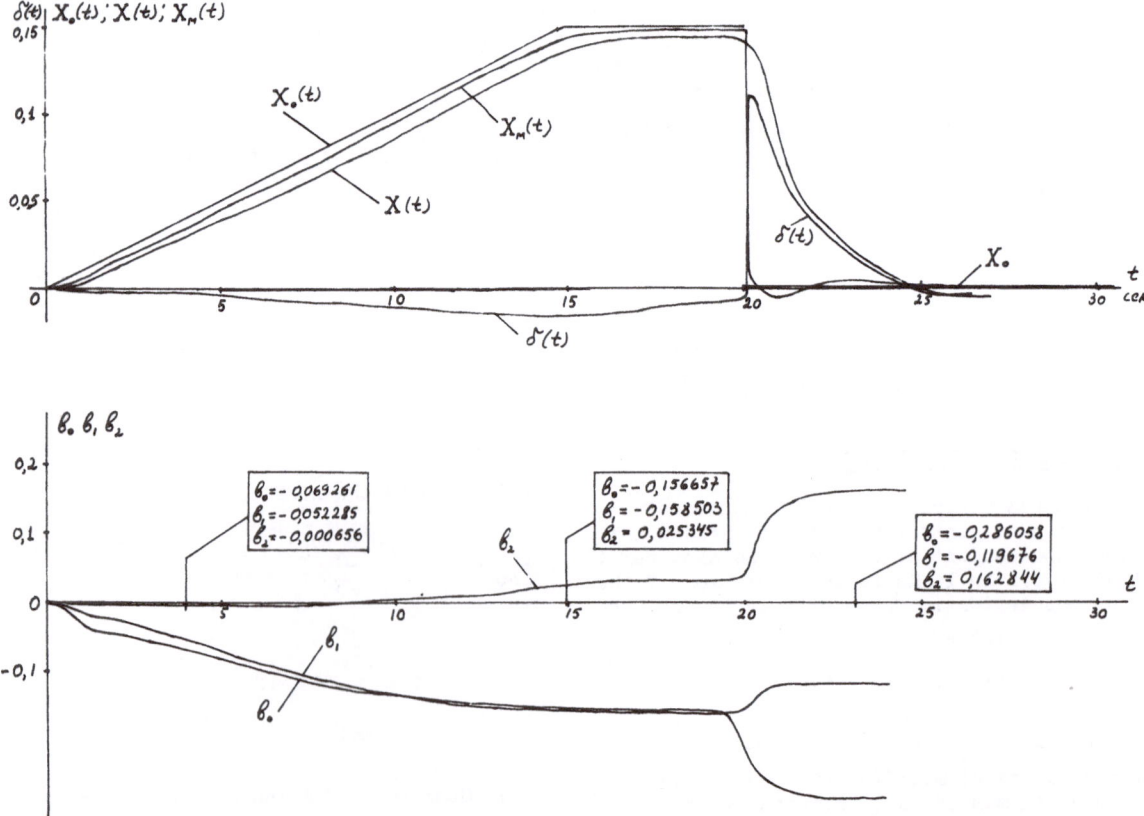

Fig. 2. System responses to ramp function input $X_0 = at$. (m = 4, K_0 = 8, b = 0.125, n = 3)

180

one cycle T; this allows us to decrease the amount of computation necessary. The second algorithm includes the following main operations.

(1) Input into the computer store the sequence of variables

$$X_0(t), \; X_0(t-T), \; X_0(t-2T), \; \dots \; .$$

(2) Compute X_{oe} by linear extrapolation;

$$X_{oe}(t + nT) = n[X_0(t) - X_0(t-T)] + X_0(t) \qquad (9)$$

$$n = 1,2,\dots$$

where n is the number of cycles of prediction.

By averaging the extrapolation we have:

$$X_{oe}(t+nT) = \frac{n}{m+1} \sum_{v=0}^{m} x_v [X_0(t-vT) - X_0(t-(v+1)T)] + X_0(t) \qquad (10)$$

$$0 < x_v < 1$$

x_v is the weight of previous differences.

(3) Compute $\delta(t)$

$$\delta(t) = b_0 \, \varepsilon(t) + b_1 \, \varepsilon(t-T) + \dots$$

$$-b_{i+1} \, \delta(t+T) - b_{i+2} \, \delta(t-2T) - \dots \qquad (11)$$

$$\varepsilon(t) = X_0(t) - X(t) \; .$$

(4) Compute $X_M(t+T)$ according to equation (8).

(5) Compute $\delta_e(t+T)$

$$\delta_e(t+T) = b_0 \, \varepsilon_e(t+T) + b_1 \, \varepsilon(t) + \dots$$

$$- b_{i+1} \, \delta(t) - b_{i+2} \, \delta(t-T) - \dots$$

$$\varepsilon_e(t+T) = X_{oe}(t+T) - X_M(t+T)$$

operations 4 and 5 are repeated n times, that is

$$X_M(t+nT) = \sum_{\lambda=1}^{\ell} c_\lambda X_M[t-(\lambda+n)T] + \sum_{\lambda=1}^{q} d_\lambda \delta[t-(\lambda-n)T] \qquad (12)$$

and

$$\delta_e(t+nT) = b_0 \, \varepsilon_e(t+nT) + b_1 \, \varepsilon_e[y-(1-n)T] + \dots$$

$$- b_{i+1} \, \delta_e[t-(1-n)T] - b_{i+2} \, \delta_e[t-(2-n)T] - \dots$$

(6) Compute mean error in prediction process

$$\varepsilon_{cp} = \frac{1}{n+1} \left\{ \varepsilon^2(t) + \sum_{v=1}^{n} \left[X_{oe}(t+vT) - X_M(t+vT) \right]^2 \right\} . \qquad (13)$$

(7) Compute partial derivatives of $\varepsilon_{op}(b_0,b_1,\dots)$ in equation (13)

$$\frac{\partial \varepsilon_{cp}}{\partial b_j} = \frac{\varepsilon_{cp}(\dots,b_j+\Delta b_j,\dots) - \varepsilon_{cp}(\dots,b_j,\dots)}{\Delta b}$$

$$j = 0,1,2 \dots$$

Δb is experimental increment in the coefficient.

(8) Compute increments of coefficients in equation (11) by hill climbing according to the method of gradient

$$\Delta b_j = - K_0 \, \frac{\partial \varepsilon_{cp}}{\partial b_j} \; .$$

(9) Determine current values of coefficients in equation (11)

$$b_{j1} = b_j + \Delta b_j$$

operations 3-9 are repeated m times.

(10) Compute control signal $\delta(t)$ using equation (11) in next cycle T. From expressions (10)-(13) it is evident that the prediction loop, including the plant model and the main control system loop, are mutually connected so far as information is concerned. Partial autonomy of prediction is inherent in the proposed method of self-adjustment. It allows, on the one hand, the successful achievement of coefficient adjustment in the control program by means of a model. On the other hand, it takes into account the true parameters and structure of the plant, and, to some degree, compensates for the difference between model and plant. The net effect is equivalent to partial self-adjustment of the main loop according to true parameters of the plant.

3.   EXPERIMENTAL WORK

In the analysis of this system major attention was devoted to the second loop.

The complexity of the computer algorithm due to the large number of parameters makes this analysis very difficult. For the second algorithm

n - number of prediction cycles

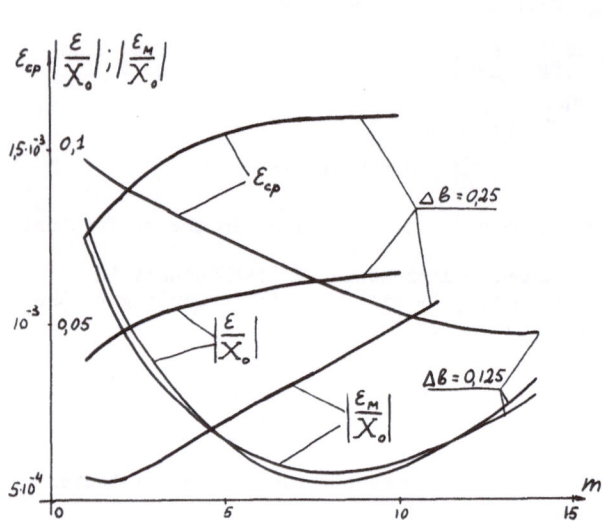

Fig. 3. Relationships with $K_0 = 4$, $n = 3$.

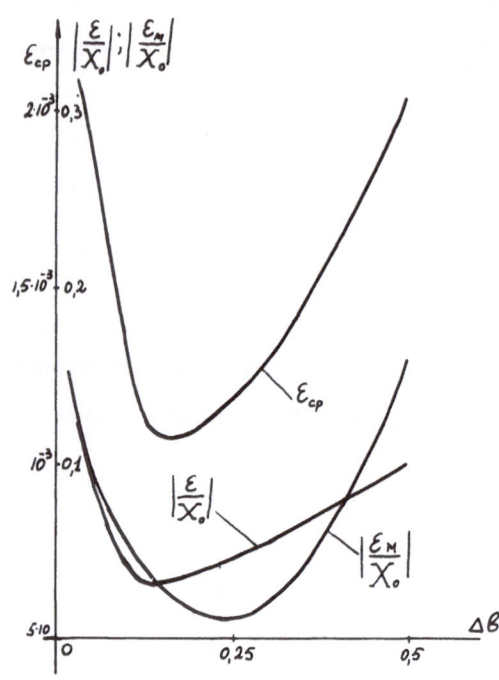

Fig. 4. Relationships with $K_0 = 2$, $m = 4$, $n = 3$.

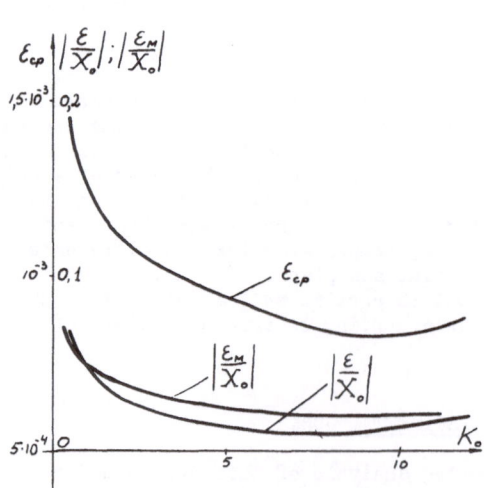

Fig. 5. Relationships with $m = 4$, $b = 0.125$, $n = 3$.

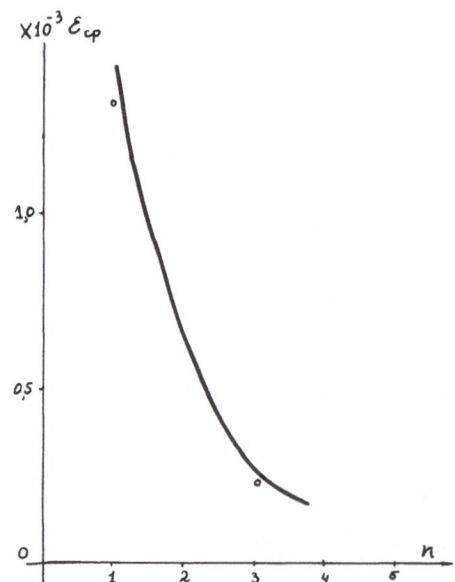

Fig. 6. Relationships with $K_0 = 8$, $m = 4$, $b = 0.125$.

182

m - number of steps

K_0 - magnitude of step in hill climbing according to gradient;

Δb - magnitude of experimental increment, number of cycles of averaging.

Experimental work was performed using the system of equation (1) of the form

$$\dot{X_1} = -B_{11}(X_4)X_1 + B_{12}(X_4)X_2 - B_{13}(X_4)X_3 + A_4(X_4,X_5)\delta ,$$

$$\dot{X_2} = X_1$$

$$\dot{X_3} = B_{31}(X_4)X_2 - B_{32}(X_4)X_3 , \qquad (14)$$

$$\dot{X_4} = B_{41}X_3 - B_{42}X_4$$

$$\dot{X_5} = B_{51}X_3X_4 ,$$

where $\qquad B_{ij} > 0 .$

Functions and coefficients B_{ji} were given arbitrarily but were chosen so that the plant was stable. Conditions (2) were implemented and slow modification of coefficients of first and second equations (14) was achieved

$$|A| = 0.9 \text{ to } 0.1 .$$

Control was produced by the variable $X_3 = X$. The model equation corresponded to equation (8) with $\ell = q = 3$. Control action $X_0(t)$ was given in the form of arbitrary, and also stepped functions. Random components were absent from $X_0(t)$. Studies were made of the processes of control and self-adjustment with modifications to the parameters K_0, m, n and Δb in the prediction loop. Analysis was done by the method of mathematical modelling of the control system. Initial values of coefficients b_j in equation (11) were set equal to zero, that is self-adjustment was begun under open loop conditions. After connection the system automatically found the sign of feedback and optimal values of the coefficients b_j (j = 0,1,2; i = 0 and j = 0,1,2,3,4). Initial values of the coefficients of model given by equation (8) were also estimated according to calculated data. By repeated experiments it was established that, independent of the form of $X_0(t)$, the process of self-adjustment is very stable and provides adequate quality of control. In fig. 2 are shown typical responses characterising the performance of the system. In fig. 3-6 are shown relationships between error ε_{cp}, $\varepsilon = X_0(t_1) - X(t_1)$ and $\varepsilon_M = X_0(t_1) - X_M(t_1)$ and the parameters m, Δb, K_0 and n with $X_0 = at$. Analysis of these relationships shows, that for m, Δb and K_0 there exist clearly expressed optimal values. From the description of the

second algorithm it is easy to conclude that the time spent on computation of control signals is mainly determined by the value of m, therefore it is desirable for m to be small. The optimal value of m with $\Delta b = 0.125$ proves to be m = 8 (fig. 3). With increase of Δb the optimal value of m is decreased, however this leads to increase of control errors (fig. 4). It is apparently expedient to have $\Delta bm \leqslant 1$. The optimal value of K_0 lies within the limits $K_0 = 5$ to $K_0 = 10$ (fig. 5). The control error depends strongly on the number of prediction cycles n (fig. 6) which should be not less than three. Special interest centres on the determination of the effect of differences between model and plant. To this end the self-adjustment loop was opened and experiments were performed with constant coefficients c_λ, d_λ (equation 8). In fig. 7 are shown graphs of errors ε, ε_M and $K_n = \frac{\delta(\infty)}{\varepsilon(\infty)}$ as functions of the ratio $|A_M|/|A|$. (Time constant of plant in variable X_3 was changed between the limits $T_{np} = 2$ seconds and $T_{np} = 10$ secs.) From the graphs it is evident, that control error ε remains small, even with significant differences between parameters of the model and plant; with large values of $|A_M|/|A|$ noticeable increase of error ε occurs. However the system remains efficient as is illustrated by the graphs in fig. 8, shown for $|A_M|/|A| = 2.56$.

These results confirm the earlier indication of better self-adjustment with partial autonomy of prediction. Modification of K_n (fig. 7) shows a tendency of the system to maintain constancy of gain $K_n|A|$ without leaving the stability zone. An automatic guarantee of sufficient stability is also a characteristic of this system, when operating in accordance with the principle of self-adjustment as described.

CONCLUSIONS

Control of a nonlinear plant described by a system of equations of form (1) can be successfully carried out by a computer which uses a self-adjusting control program as described. The algorithm employs prediction in an accelerated time-scale from a linear learning model of the plant. Analysis of this system shows high stability and efficiency of self-adjustment. Due to partial autonomy of the prediction loop the quality of control depends comparatively weakly on the accuracy of adjustment of plant model parameters. Relationships are derived between control error and the parameters of the prediction loop; optimal values of prediction loop parameters are thus established. However, the generality of these relationships are subject to further investigation.

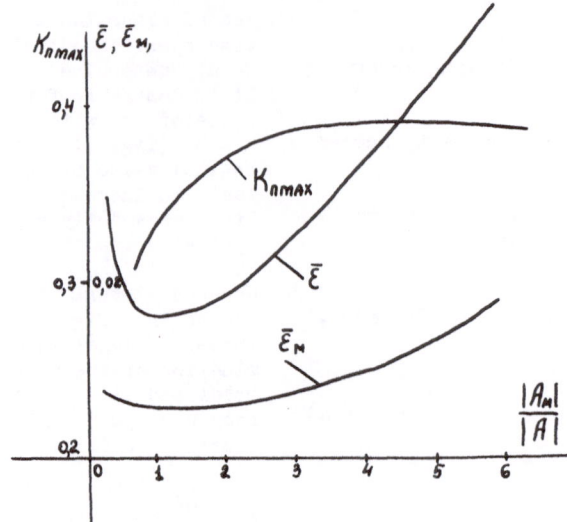

Fig. 7. Relationships with $K_0 = 8$, $m = 4$, $b = 0.125$,
n = 3.

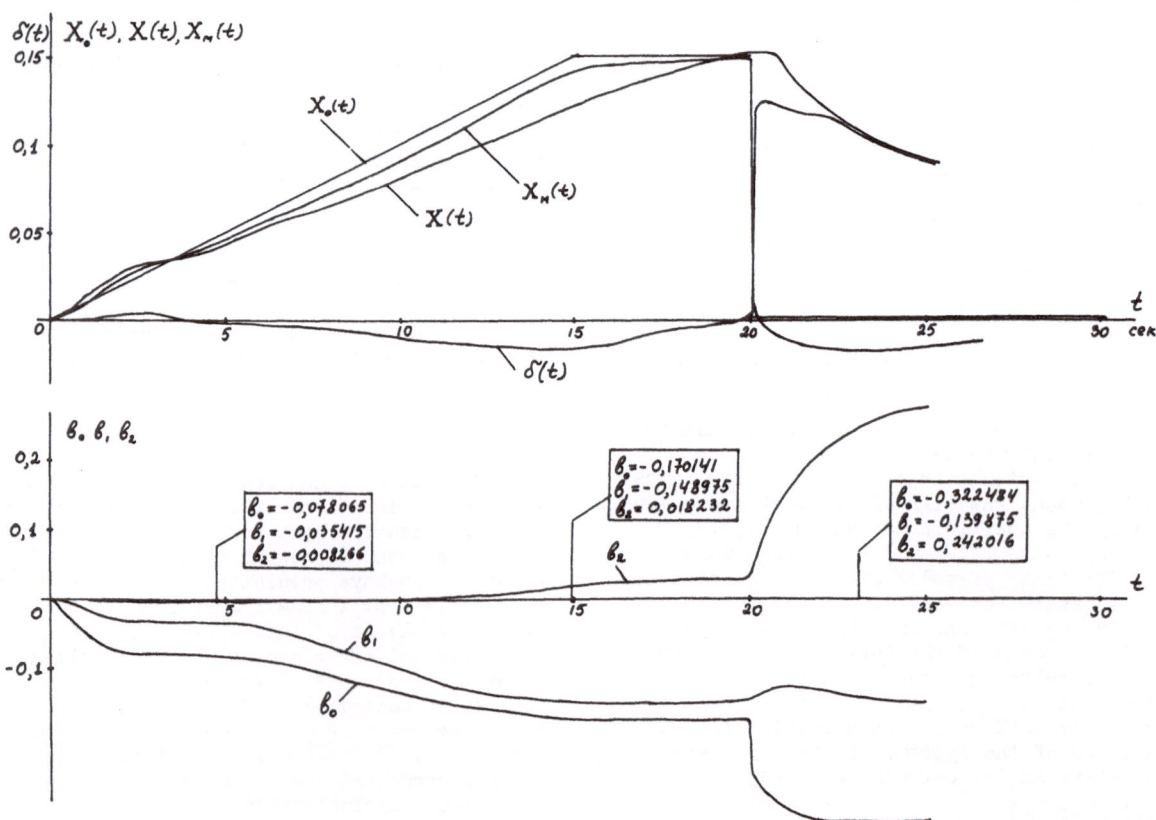

Fig. 8. Transient responses to ramp function input $X_0 = at$.
($K_0 = 8$, $m = 4$, $b = 0.125$, $n = 3$, $|A_M|$ / $|A|$ = 2.56)

REFERENCES

1.   P. F. Klubnikin, The realization of a self-adapting control program in a system with digital computer, Proceedings of the Second Congress of IFAC, Basle, 1963, Theory Section, p.481.

2.   A. Kudrevich, Stability of nonlinear systems with feedback, Automation and Remote Control, Vol. XXV, No. 8, 1964.

3.   I. M. Gelfend, M. N. Detlin, One some methods of control by complex systems, Progress of Mathematical Sciences, vol. XVII, issue, 1 (103), 1962.

METHODS OF SYNTHESIS OF STRUCTURES EQUIVALENT
TO SELF-ADJUSTING SYSTEMS FOR PLANTS WITH
VARIABLE PARAMETERS

by M. V. Meerov
USSR National Committee of Automatic Control
Moscow, USSR

ABSTRACT

Methods are devised to construct structures of systems without searching elements equivalent to self-adjusting systems for plants with time-variable parameters. Design of such structures reduces to finding the conditions for stability in the large of an equation with time variable free term. Sufficient conditions for stability in the large are obtained in the analytical form.

1. STATEMENT OF THE PROBLEM

The problem of self-adjustment arises in the case when it is known or when there is a sufficient degree of probability that the controlled plant has the extremum for some quality criterion selected by us and that if the optimum regime is achieved in any way in the course of the plant operation, it can be disturbed.

The causes of the disturbance of the optimum regime are as a rule of two kinds. In the first place a change in the characteristics and their displacements may be brought about by external noises; and in the second place, the plant parameters can be changed in the course of the plant operation that changes its characteristics. In both cases the change of the plant characteristics changes the regime and disturbs the optimality if it had been achieved before the characteristics were changed.

As a matter of fact the cause of change of characteristics of the controlled plant is not very important. What is essential is that the optimum regime is disturbed. Therefore it is very useful to consider the problems of self-adjustments for a plant with variable parameters, since a great number of very important plants can be reduced to the considered class.

For the purposes of the discussion it is important to divide plants into two classes: a) plants with complete information and b) plants with incomplete information. The first class includes plants whose characteristics are known beforehand; however, possible changes of the characteristics due to the causes stated above are not known beforehand. The second plants are those about which we know only that they have an extremum for the selected quality criterion but whose characteristics are unknown.

This situation called for appearance of the so-called self-adjusting systems. These systems are usually characterized with the presence of a searching element.

In the References[1,2] I stated the following problems.

1) Is the element for search of a compulsory component for any system with properties of self-adjustment?

2) Is it possible to find structures equivalent to self-adjusting systems but without a searching element?
In the References [1,2] it was shown that it is possible to design self-adjusting control systems without a searching element if we use the structures which admit an unlimited increase of gains without disturbing the stability[3]

This paper describes the results of further research into the problems stated above.

2. PLANTS WITH COMPLETE INFORMATION

We shall understand complete information about the plant in the above sense. As both external disturbances and changes of the plant's parameters change its characteristics, the problem discussed is closely connected with the problem of sensitivity. In this sense the following statement can be made: automatic control systems whose dynamic characteristics are low-sensitive to the change of the plant characteristics would be equivalent to self-adjusting systems and would require no searching element.

The latter statement is connected with the availability of information on the plant. Since we know all about the plant when disturbances are absent or the parameters are constant, the optimum operating conditions can be selected. In

Superior numbers refer to similarly-numbered references at the end of this paper.

186

the discussion that follows it is required that the properties of the control system (control with the plant) are not at all or little dependent on the plant characteristics. Then the operating conditions will remain optimal even when the plant characteristics have changed.

From the range of considered problems we shall choose one to which many other can be reduced, i.e. that of the plant with variable parameters. Assume that the plant is described by the equation with variable parameters

$$a_0(t)\frac{d^2 y(t)}{dt^2}+a_1(t)\frac{dy(t)}{dt}+a_2(t)y(t)=\varkappa(t)x(t) \quad (1)$$

We prefer a second order equation only for the sake of simplicity; further it will be made clear that the obtained results are true for any order of the equation.

Suppose that the optimum conditions of the plant operating is characterized by the value $y_{ref}(t)$ which is unknown. It is necessary to select such a control system that would realize the optimum $y_{ref}(t)$ operating conditions with any degree of accuracy irrespective of time varying plant parameters present.

Rules for variation of plant parameters have just one constraint. Any rules for variation of parameters will hold (in particular, even random), however, the moduli

$$|a_0(t)|,\,|a_0'(t)|,\,|a_1(t)|,|a_1'(t)|,\,|a_2(t)|$$

$$|a_2'(t)|,\,|\varkappa'(t)|\;\;u\;\;|\varkappa(t)|$$

must not exceed the given values. Let us try to solve this problem as we did in References [1,2,3]

In Fig.1 a block diagram is shown which is similar in design to the solution in references [1,2,3]

Processes in the system, the diagram of which is shown in Fig.1, are described by the following linear equation with variable coefficients

$$T\frac{a_0(t)}{\varkappa(t)}p^3y+\left\{T\cdot\frac{a_0'(t)\varkappa(t)-\varkappa'(t)a_0(t)}{[\varkappa(t)]^2}+T\frac{a_1(t)}{\varkappa(t)}+\right.$$

$$\left.+\varkappa_\delta\frac{a_0(t)}{\varkappa(t)}+\frac{a_0(t)}{\varkappa(t)}\right\}p^2y+\left\{T\frac{a_1'(t)\varkappa(t)-a_1(t)\varkappa'(t)}{[\varkappa(t)]^2}+\right. \quad (2)$$

$$+T\frac{a_2(t)}{\varkappa(t)}+\varkappa_\delta\frac{a_1(t)}{\varkappa(t)}+\frac{a_1(t)}{\varkappa(t)}+\varkappa_\delta^2 T\Big\}py+$$

$$+\left\{T\frac{a_2'(t)\varkappa(t)-\varkappa'(t)a_2(t)}{[\varkappa(t)]^2}+\frac{a_2(t)}{\varkappa(t)}+\varkappa_\delta\frac{a_2(t)}{\varkappa(t)}+\varkappa_\delta^2\right\}y=$$

$$=\varkappa_\delta^2 Tpy_{ЭT}+\varkappa_\delta^2 y_{ref}\ldots$$

Dividing Equation /2/ by $\varkappa_\delta^2$ and assuming $\frac{1}{\varkappa_\delta}=m$ after elementary calculations we obtain

$$m^2\left\{T\frac{a_0(t)}{\varkappa(t)}p^3y+\Big[T\frac{a_0'(t)\varkappa(t)-\varkappa'(t)a_0(t)}{[\varkappa(t)]^2}+\right.$$

$$+T\frac{a_1(t)}{\varkappa(t)}+\frac{a_2(t)}{\varkappa(t)}\Big]p^2y+\Big[T\frac{a_1'(t)\varkappa(t)-\varkappa'(t)a_1(t)}{[\varkappa(t)]^2}+$$

$$(3)$$

$$+T\frac{a_2(t)}{\varkappa(t)}+\frac{a_1(t)}{\varkappa(t)}\Big]py+\Big[T\frac{a_2'(t)\varkappa(t)-\varkappa'(t)a_2(t)}{[\varkappa(t)]^2}+$$

$$\left.+\frac{a_2(t)}{\varkappa(t)}\Big]y\right\}+m\left\{\frac{a_0(t)}{\varkappa(t)}p^2+\frac{a_1(t)}{\varkappa(t)}py+\frac{a_2(t)}{\varkappa(t)}y\right\}+$$

$$+Tpy+y=Tpy_{ref}+y_{ref}$$

Assume that $m\longrightarrow\infty$, which is equivalent to $\varkappa_\delta\longrightarrow\infty$. Equation /3/ shows that the equation degenerates into one with constant coefficients, and the degenerated system corresponds to the system with ideal reproduction of y_{ref}. As y_{ref} corresponds to the optimum operating conditions, the degenerated system obtained is equal to the optimum system. It follows from Equation /3/ that if $m\longrightarrow o$, then the obtained optimum operating do not depend on the variable parameters and in this sense we can consider the obtained system equivalent to a self-adjusting system.

However, these conclusions are true if we make sure that the eliminated members of Equation /3/ do not disturb its stability.

Let us consider this problem in more detail. Similar to the procedure in earlier papers where the equations with constant coefficients were considered (for example [3]), let us introduce the following transformation.

Replace the operator p by the ratio

$$\rho=\frac{q}{m}=\frac{1}{m}\frac{d}{dt} \quad (4)$$

187

$$a_0(t)\frac{d^2y}{dt} + a_1(t)\frac{dy}{dt} + a_2(t)\,y = K(t)\,x$$

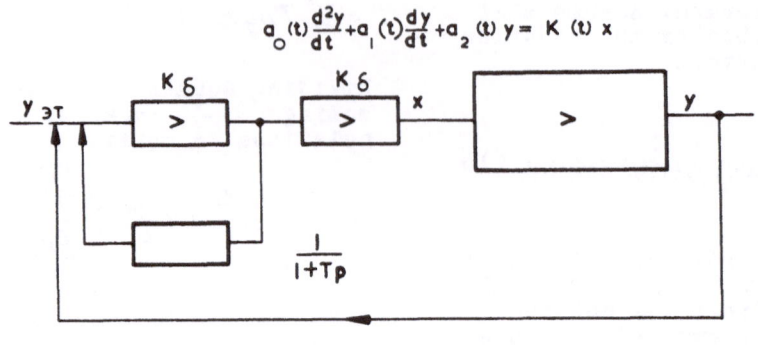

FIG. 1

$$a_0\frac{d^3y}{dy^3} + a_1\frac{d^2y}{dt^2} + a_2\frac{dy}{dt} + a_3\,\text{Sin}\,ty = a\,\text{Sin}\,tx$$

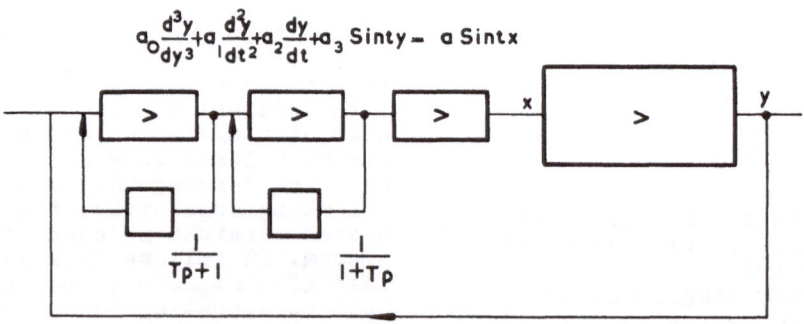

FIG. 2

By dividing Equation (3) by m^2 and substituting $\rho \frac{q}{m}$ we obtain, after simple transformations, the equation

$$T \frac{a_0(t)}{K(t)} \frac{q^3}{m^3} y + \left[T \frac{a_0'(t)K(t) - K'(t)a_0(t)}{[K(t)]^2} + \frac{a_1(t)}{K(t)} + \right.$$

$$\left. + \frac{a_2(t)}{K(t)} \right] \frac{q^2}{m^2} y + \left[T \frac{a_1'(t)K(t) - K'(t)a_1(t)}{[K(t)]^2} + T \frac{a_2(t)}{K(t)} + \right.$$

$$\left. + \frac{a_1(t)}{K(t)} \right] \frac{q}{m} qy + \left[T \frac{a_2'(t)K(t) - K'(t)a_2(t)}{[K(t)]^2} + \frac{a_2(t)}{K(t)} \right] y +$$

$$\frac{a_0(t)}{K(t)} \frac{q^2}{m^3} y + \frac{a_1(t)}{K(t)} \cdot \frac{q}{m^2} y + \frac{a_0(t)}{K(t)} \frac{1}{m} y +$$

$$+ T \frac{q}{m^3} y + \frac{1}{m^2} y = T \frac{q}{m^3} y_{ref} + \frac{1}{m^2} y_{ref} \tag{5}$$

Multiply now Equation (5) by m^3 and assume $m = 0$; then we have

$$T \frac{a_0(t)}{K(t)} q^3 y + \frac{a_0(t)}{K(t)} \cdot q^2 y + T q y = T q y_{ref}$$

or

$$T q^2 y + q \cdot y + T \frac{K(t)}{a_0(t)} y = T \frac{K(t)}{a_0(t)} y_{ref} \tag{6}$$

Now we have the auxiliary equation with variable parameters. If this equation corresponds to the convergent process, the initial system will be completely defined by the degenerate equation.

$$T p y + y = T p y_{ref} + y_{ref}$$

or

$$y = y_{ref} \tag{7}$$

First of all let us note the very important circumstance that Equation /6/ depends only on two variable parameters, that is, on the major coefficient derivative of the plant equation and on the plant gain. This considerably simplifies the research. It is essential to stress that if auxiliary Equation (6) satisfies the conditions of stability, then the control process is defined only by the degenerate equation. In this sense the realization of the system equivalent to a self-adjusting system reduces to the problem of stability in the large.
Consider some examples.

a) Assume that the coefficient $\gamma(t) = \frac{K(t)}{a_0(t)}$ varies in time much slower than the time of the transient response in the system, so that for the given process $\gamma(t)$ can be considered to be constant. Then Equation /6/ converts into an ordinary equation with constant coefficients and in order to ensure stability it is necessary and sufficient that the coefficients of /6/ should satisfy the Hurwitz condition. From this a very important property of $\gamma(t)$, even for this simplest example, follows.

Indeed, it is clear from the Hurwitz condition that the condition $\gamma(t) > 0$ must be met, i.e. $\gamma(t)$ should not be negative at any values of t, otherwise the system becomes unstable.

If we assume that $\gamma(t)$ can be negative, then the stability of the auxiliary equation can be ensured for the given case by introduction of constant component β so that

$$\beta - |\gamma(t)| > 0 \tag{8}$$

The effect of this can be illustrated by the following example. Assume that for the plant described by a third order equation the gain varies according to the sinusoidal law.
The equation for the plant is written as

$$a_0 \frac{d^3 y}{dt^3} + a_1 \frac{d^2 y}{dt^2} + a_2 \frac{dy}{dt} + a_3 \sin t \, y = a \sin t x$$

As it is shown in[2] for the given case the elimination of the gain effect changing by the sinusoidal law is realized by the circuit shown in Fig.2.
Fig.3 presents the oscillogram for the case when $\omega = 3.14$. It can be seen that the output quantity remains constant despite the changing gain: here $K = 5 - 4 \sin \omega t$ i.e. $K > 0$ for all t.

b) Now assume that the rate of changing of $\gamma(t)$ is commensurable with the rate of the transient response. In this case the problem is reduced to defining the stability conditions in variable parameters systems. An important result of this method is that the free term of the auxiliary equation of the first order is obtained as a variable parameter. This makes it possible to use in this case the theorem by V.M.Popov[4] for studying stability. The following section is devoted to this problem.
Note the following property of the obtained solution.

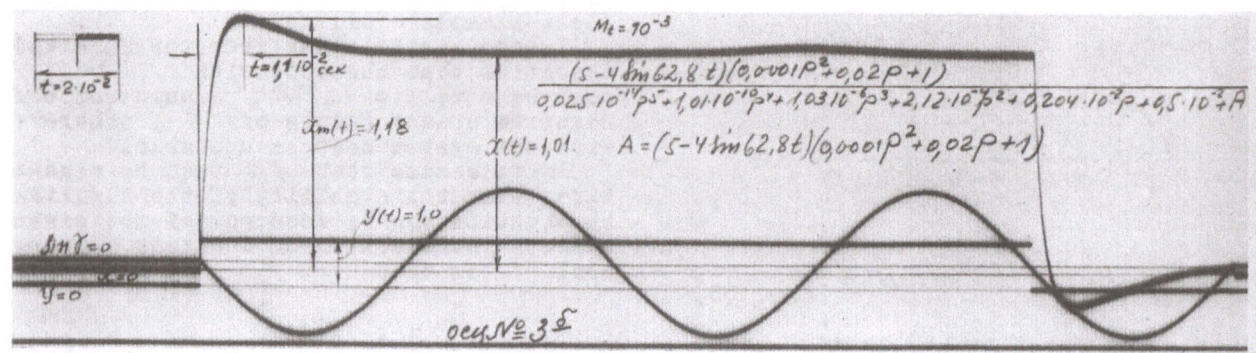

FIG. 3

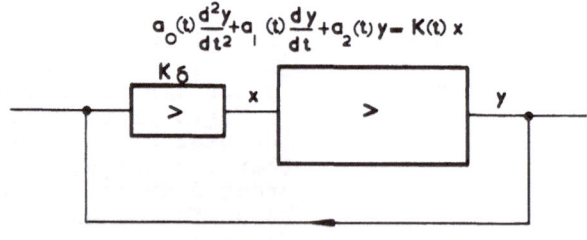

FIG. 4

190

We have assumed that all the parameters of the plant including the gain are time functions. However, thanks to control circuit adopted only the coefficient of senior derivative $a_0(t)$ and the plant gain $K(t)$ affect the stability. As for $a_1(t)$ and $a_2(t)$ their effect is suppressed by the gains K_δ so that they cannot affect the process or stability. This result is a matter of paramount importance because it allows one to make conclusions on the structural properties of the system. Indeed at first sight it may seem that elimination of the influence of variable parameters can also be achieved by means of a simpler circuit. Fig.4 represents the circuit with one high-gain amplifier.

If the gain is high enough, then in this case the influence of variable parameters can be eliminated. However, in this case the auxiliary equation will depend not only on $a_0(t)$ and $K(t)$ but on $a_1(t)$ as well.

Indeed in this case the equation will look like

$$a_0(t)\frac{d^2y}{dt^2} + a_1(t)\frac{dy}{dt} + a_2(t)y_k(t)K_\delta(y_{ref}-y) \qquad (9)$$

or after dividing this equation by K_δ we have

$$m\left[a_0(t)\frac{d^2y}{dt^2} - a_1(t)\frac{dy}{dt} + a_2(t)y\right] + K(t)y = K(t)y_{ref} \quad (10)$$

In this case if we assume that the coefficients change relatively slowly, for the system to be stable it is necessary (3) that

$$\frac{a_1(t)}{a_0(t)} > 0 \cdots \qquad (11)$$

but this is not enough; the degenerate part is represented by the equation

$$K(t)y = 0, \qquad (12)$$

whose stability requires special consideration. Hence, in this case the degenerate and auxiliary equations depend on variable parameters.

Conclusions resulting from this study are extended to the general case of a plant described by the equation of the n-th order with variable parameters. They consist in the following.

Assume that the plant is described by an equation of the n-th order with variable parameters including variable gain. In this case the number of variable parameters will equal $n+2$

The control system will consist of n amplifiers with sufficiently high gains ($n-1$) of which will have feedback with transfer functions $\frac{1}{1+T_ip}$ where $p=\frac{d}{dt}$

Similar to the procedure which was used for the case, when the plant was described by an equation of the second order one can prove that for $m_i = \frac{1}{K_{\delta_i}} \to 0$

$$\lim y = y_{ref}$$

$$K_\delta \longrightarrow \infty$$

For stability it is necessary and sufficient that the degenerate and auxiliary equations, each taken separately meet the conditions of stability. Here the degenerate equation will be a linear equation with constant coefficients and the auxiliary equation will have a variable parameter.

$$\gamma(t) = \frac{K(t)}{a_0(t)}$$

We shall discuss the methods of system design for plants with complete information; as for the plants with incomplete information, evidently, to find $y_{ref\,op}$ in this case it is necessary to introduce at least an episodical search and then to use the model of the plant with output value $y_{ref\,op}$ to maintain $y_{ref\,op}$ However this problem is outside the scope of this paper.

3. ANALYTICAL CONDITIONS OF STABILITY IN THE LARGE

It was shown in the previous chapter that the problem of design of a system, where the influence of variable parameters of the plant on the dynamic properties of the system is eliminated, or in other words, where what we call the property equivalent to self-adjusting is obtained, is reduced to the stability in the large of the system for the time variable free term of the auxiliary equation.

It is obvious that the theorem by V.M.Popov can be directly used for the case under consideration. Indeed, according to the theorem by V.M.Popov, the nonlinear characteristic has the constraint

$$0 \le \frac{\varphi(\sigma)}{\sigma} \le K$$

where $\mathcal{K}$ is a positive value, defining the inclination of the straight line, $\gamma(\sigma)$ is the output value of a nonlinear element and σ is its input.

In this case the connection between input and output of the element with the characteristic $\gamma(t)$ will be expressed in the following way: denote the input by x, and the output by $\gamma(t)x$

$$\frac{\gamma(t)x}{x} = \gamma(t); \cdots \qquad (13)$$

therefore it is necessary, that

$$0 < \gamma(t) < \mathcal{K} \qquad (14)$$

Thus to find the conditions of stability in the large we can use the graphical method which realizes the conditions of the theorem by V.M.Popov.

But it follows from the results obtained in the previous chapter, the graphical methods are not always convenient. Moreover, analytical conditions in this case could be very efficient because they can define necessary values of the constants T_i in transfer function of feedbacks and solve a number of problems.

Determination of analytical conditions of stability in the large is of principal importance.

As is known, the solution of a number of very important problems not only of the automatic control theory but of the circuit theory as well is reduced to finding positivity of real functions.

In fact the conditions of the theorem by V.M.Popov are reduced to finding the conditions of positivity of the real function.

According to[4] the sufficient conditions of stability in the large are expressed by the condition

$$Re(1+jg\omega)W(j\omega)+\frac{1}{\mathcal{K}} > 0 \qquad (15)$$

where g is a finite real number. $W(j\omega)$ is the amplitude phase characteristic of the open-loop linear part of the system.

As for the circuit theory, the conditions of positivity of real function define the realizability of the circuit. From this one can understand the great theoretical and practical importance of analytical conditions of stability in the large.

Let us consider this problem in case where the real function is positive.

As it is known the function $Z(p)$ is named positive, if its real part is positive for positive real parts of p

$$Re Z(p) \geqslant 0 \quad \text{at} \quad Re\, p \geqslant 0 \qquad (16)$$

If the function is real at real values of the argument then it is called the real positive function.

From Equation (16) it follows that ω for real values of the condition of the theorem by V.P.Popov[5] is reduced to finding the condition of positivity of the real function.

From the properties of the real positive function follows that the conditions of their realization will be found if we find the conditions of stability of the following equation

$$Q(p)+\mathcal{K}R(p)=0 \qquad (17)$$

for all values of $\mathcal{K}$

The problem can be formulated in the following manner: find conditions under which the stability region in the plane of complex parameter $\bar{\mathcal{K}}$ will be the whole positive real axis.

First of all it is clear that the polynomials $Q(p)$ and $R(p)$ must be the Hurwitzian polynomials due to requirements of stability of the equation (17) for $\mathcal{K}=0$ and $\mathcal{K}=\infty$ respectively[x]. For the existence of the segment of real axis, belonging to the stable region and originating in infinity it is necessary and sufficient that the system structure corresponding to Equation (17) belong to the class of stable systems for $\mathcal{K} \longrightarrow \infty$ [3] and that the necessary and sufficient conditions, formulated in [3] are fulfilled. Suppose that on the real axis „$\mathcal{K}$" of the plane $\bar{\mathcal{K}}$ there is a segment originating in the infinity and belonging to the stable region.

On the basis of the theorem of continuous dependence of roots on the coefficients of equation we can state that, besides the part of the stable region originating in the infinity, there is a segment of the real axis „$\mathcal{K}$" of the plane $\bar{\mathcal{K}}$ beginning in the origin of coordinates and also belonging to the stable region. Under these conditions the following statement can be formulated: if $Q(p)$ and $R(p)$ are Hurwitzian polynomials and there is a segment of real axis $Q(p)$ of the plane $R(p)$ originating in the infinity and belonging to

_____
[x] $Q(p)$ may be one zero root.

the stable region then the whole real axis $\mathcal{K}$ of the plane $\bar{\mathcal{K}}$ will belong to the stable region, if the imaginary part of the expression has no real positive roots. This statement is true due to the properties of the curve of D-decomposition. Thus we must define only the conditions under which in the structures, stable when $\mathcal{K}$ increases infinitely, the curve of D-decomposition will never cross the real axis „$\mathcal{K}$" of the plane $\mathcal{K}$

From (17) the equation of D-decomposition can be written as

$$\bar{\mathcal{K}} = \frac{Q(j\omega)}{R(j\omega)} = -\frac{Q_1(\omega) + jQ_2(\omega)}{R_1(\omega) + jR_2(\omega)} \qquad (18)$$

The imaginary part of Equation (18) can be written as

$$\frac{Q_1(\omega)R_2(\omega) - R_1(\omega)Q_2(\omega)}{R_1^2(\omega) + R_2^2(\omega)} \qquad (19)$$

Consequently, in order that the curve of D-decomposition should not cross the axis   , it is necessary and sufficient that the equation

$$Q_1(\omega)R_2(\omega) - R_1(\omega)Q_2(\omega) = 0 \qquad (20)$$

have no real roots.
In a general case if the zero root is omitted, Equation (20) can be written in the following form

$$\sum_{i=0}^{n} a_i \omega^{2i} = 0 \qquad (21)$$

From the discussion above it follows that the conditions of reality and positivity of the function $\frac{\mathcal{K}R(\rho)}{Q(\rho)}$ are reduced to conditions under which these polinomials $Q(\rho)$ and $R(\rho)$ satisfy the Hurwitz condition and to the absence of real roots in the polynomial, containing only even degrees ω and representing the imaginary part of the curve of D-decomposition over K.

Another approach to this problem is also legitimate. One may require that the real part of $\frac{\mathcal{K}R(j\omega)}{Q(j\omega)}$ for all values of $\omega > 0$ should be greater than zero. Under these conditions the function will be real and positive; in this case the numerator of the real part of the function which is the polynomial with even degrees should not have real roots.

Consider polynomial (21). We have

$$\sum_{i=0}^{n} a_i \omega^{2i} \qquad (22)$$

The power of polynomial (22) is equal to 2n.

Consider an auxiliary polynomial of power $n + 1$.

$$f(x) = b_{n+1} x^{n+1} + b_n x^n + \cdots B_0 \qquad (23)$$

Let us require that the roots of Equation (23) be real and different. For this purpose the coefficients of (23) should satisfy the conditions of aperiodical stability [6].

If the roots of (23) are real and different, the roots $f'(x)$ will be real and different also and will alternate with the roots of (23). The roots $f''(x)$ will be also real and will alternate with the roots of $f'(x)$

A theorem [7] is well known that if the roots of the functions $f(x), \psi(x)$ are real, different and alternating, then the equation

$$f'(x)\psi(x) - f(x)\psi'(x) = 0 \qquad (24)$$

has no real roots.
If we set $\psi(x) = f'(t)$ then the equation

$$[f'(x)]^2 - f(x)f''(x) = 0 \qquad (25)$$

has no real roots.
It follows from (23) that the power of (25) will be 2n. Make Equation (22) equal to (25)

$$\sum a_i \omega^{2i} = [f'(x)]^2 - f(x)f''(x) \qquad (26)$$

From (26) we obtain the system of recurrent relations necessary in defining the coefficients of the polynomial $f(x)$ by means of the coefficients of the polynomial (22).

Denote the obtained coefficients of the polynomial $f(x)$ as $A_{n+1}, A_n, \ldots A_0$ (some of these coefficients may be equal to zero) and compose the determinant expressing the conditions of reality for roots [6] of the   polynomial.

Now we have

$$\begin{vmatrix} (n+1)A_{n+1} & \cdots & A_{n1} & 0 & \cdots & 0 & 0 \\ n\,A_n & \cdots & A_{n1}(n+1)A_{n+1} & \cdots & 0 & 0 \\ \cdot & \cdot & \cdot & \cdot & \cdot & \cdot & \cdot & \cdot \\ \cdot & \cdot & \cdot & \cdot & \cdot & \cdot & \cdot & \cdot \\ 0 & 0 & \cdots & \cdot & \cdot & \cdots & (n+1)A_{n+1}A_n \end{vmatrix}$$

Thus, in order to make the system remain stable in the large, under the conditions previously adopted, it is sufficient that angular minors of the determinant in (27) exceed zero.

REFERENCES

(1) Meerov M.V., Synthesis of Systems with the Fixed Characteristics of Equivalent Self-Adjusting Systems. Proc. 2nd Congress of the IFAC, Butterworths, London-Munich, 1965

(2) Meerov, M.V., Structural aspect of a solution of the sensitivity problem. Proceedings of the IFAC/ETAN Symposium on Sensitivity Analysis, Dubrovnik, 1964.

(3) Meerov, M.V., Structural synthesis of high accuracy automatic control systems. State Publishing House of Physics & Mathematics of the USSR, Moscow, 1959 (Published in English by Pergamon Press)

(4) Aizerman, M.A. and Gantmaker, F.G., Absolute Stability of automatic control systems. State Publishing House of Physics & Mathematics of the USSR, Moscow, 1964.

(5) Naumov, B.N., Tsypkin Ya.Z. A Frequency Criterion for absolute process stability in nonlinear automatic control. Automation and Remote Control, Vol. 25, No. 6, 1964.

(6) Meerov, M.V., A criterion of aperiodicity. Izvestya OTN, No. 12, 1945.

(7) Grave, D.A., Bases of Higher Algebra. Kharkov, 1914.

Discussion

Discussion centred on the relation of the dynamics of the feedback loop to the plant dynamics. Mr. K. Norkin (USSR) was assured by the author that correct application of the technique elaborated in his book (ref. 3 of the paper) assured stability and freedom from oscillation. He elaborated also on his application of Popov's theorem, citing the written form of his paper for details. Prof. Leonard (West Germany) pointed out that work remained to be done on the possibility of saturation of the high gain amplifiers by noise when connected by delayed feedback as lead amplifiers, especially when connected in cascade. Dr. Meerov thought that an additional measurement of noise and further investigation were required.
Dr. Kwakernaak (Netherlands) wondered if the method outlined by Dr. Meerov might be applicable to the problem posed by Meredith and Dymock in the same session.

ON A METHOD OF SYNTHESISING
SYSTEMS WHICH ARE SELF ADAPTIVE TO VARYING PARAMETERS

by K. T. Tsaturyan
USSR National Committee of Automatic Control
Moscow, USSR

This paper relates to some aspects of synthesis of systems self adapting to variable parameters. Let the system be divided into two parts: (1) a plant with characteristics which are functions of the parameters of the environment, and (2) a regulator whose characteristics must be varied during normal functioning so that the system satisfies given technical requirements.

The behaviour of the plant is assumed to be described by means of a linear differential equation with variable coefficients α_i. These coefficients specified, not in the form of functions of time, but rather as varying in a predetermined manner inside some region γ (Fig. 1), which is the region of determining of plant coefficients. In such a case both the values of a coefficient itself and of its time derivatives are limited

$$\alpha_{i0} \leqslant \alpha_i \leqslant \alpha_{im} \qquad \ldots (1)$$

$$|\dot{\alpha}_i| \leqslant |\dot{\alpha}_{im}| . \qquad \ldots (2)$$

Further, the behaviour of the regulator is also described by means of a linear differential equation, which involves a number of adjustable coefficients λ_j, the values of each coefficient and its time derivatives being also limited. The set of equations, describing the behaviour of the plant and the regulator may be reduced to one differential equation of a high order. The coefficients A_K of this equation will depend on the variables α_i, λ_j and their derivatives. If there are slow variations of the coefficients α_i and λ_j, it is not difficult to find differential inequalities. In satisfying these inequalities the coefficients A_K are functions only of α_i and λ_j, that is

$$A_K = A_K (\alpha_i, \lambda_j) . \qquad \ldots (3)$$

The form of the equations (3) is defined wholly by the structure of the system. The coefficients α_i are functions of the environmental parameters, therefore the response of the system to a particular input signal depends on the environmental conditions.

To minimize this relationship and to satisfy the system requirements it is necessary to limit the region of variation of certain coefficients of the differential equation of the system:

$$A_{K0} \leqslant A_K \leqslant A_{Km} . \qquad \ldots (4)$$

Such inequalities as (4) in the space of system coefficients (Fig. 2) define the region G_*, the inside points of which correspond to the desired values of the system's coefficients. Equations (3) map the region γ (in the space of the coefficients α_i) into a region $G(\lambda_j)$ (in the space of the coefficients A_K). If, for the constant values of λ_j the region G wholly projects into region G_*, the necessity for self-adapting of the regulator's parameters is absent. Otherwise selfadaption is necessary.

Thus the problem of selfadaptive control is reduced to the following.

Given: (1) region of determining plant's coefficients from (1) and (2);

     (2) structure of the system from (3);

     (3) system requirements from (4).

It is required to define the relationships between the regulator's adjustable coefficients and the plant's variable coefficients

$$\lambda_j = \lambda_{j*} (\alpha_i) \qquad \ldots (5)$$

which satisfy system's requirements.

The use of computers allows the solution of the problem of synthesis in the following manner. Region γ is divided into discrete points. Desired values λ_{j*} for each point are defined by means of simulation, so as to satisfy system's requirements. When required, discrete relationships may be approximated in a form of continuous functions and represented as (4).

Consider a problem of choice of variation rate of the regulator coefficients using as an example a plant having two variable coefficients. In view of inequality (2) one may state that translation of the system (Fig. 1) from point $(\alpha_{11}, \alpha_{21})$ to point $(\alpha_{12}, \alpha_{22})$ can not take place in less time than

$$T_{12} = \frac{(\alpha_{12}-\alpha_{11})^2 + (\alpha_{22}-\alpha_{21})^2}{\dot{\alpha}_{1m}(\alpha_{12}-\alpha_{11}) + \dot{\alpha}_{2m}(\alpha_{22}-\alpha_{21})} . \qquad \ldots (6)$$

"Superior numbers refer to similarly-numbered references at the end of this paper"

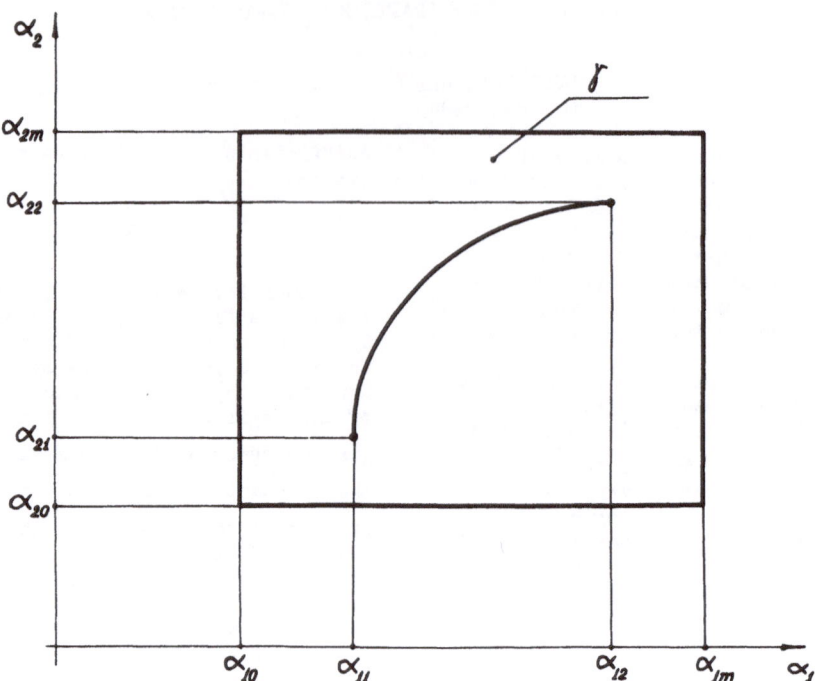

Fig. 1   Region of variation plant's coefficients

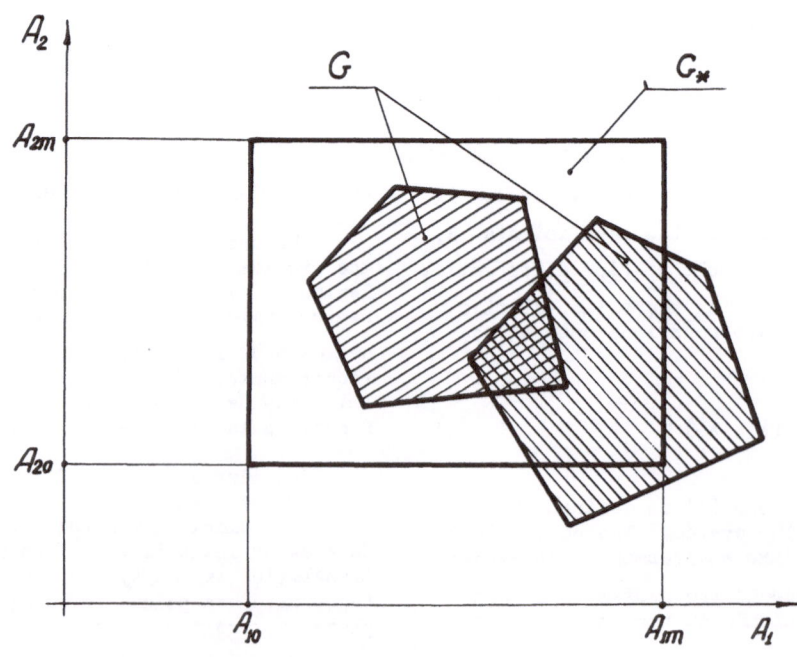

Fig. 2   Region of variation system's coefficients

196

Let the desired value of some adjustable coefficient λ_j for point 1 be λ_{j1} and for
point 2 be λ_{j2}. Then one should choose the
maximum rate of coefficient adjustment as:-

$$\dot{\lambda}_{jm} > \frac{\lambda_{j2} - \lambda_{j1}}{T_{12}} . \qquad \ldots (7)$$

To realise relationships (4) during normal
functioning of system it is necessary to measure
the plant coefficients by means of instruments.
However instead of measuring coefficients themselves it is possible to measure an equivalent
set of independent functions of these coefficients.
A method of measuring plant coefficients by
means of harmonic test signals has been described
in reference 1.

Thus the proposed method of synthesis
reduces virtually to the following. The desired
relationships between adjustable coefficients of
the regulator and the variable plant coefficients
are determined by simulation or by computer
solution, taking into account all possibilities.
The regulator includes a device for measuring
variable plant coefficients or certain independent functions of these coefficients.

The results of measurement using equations
obtained during simulation or design are used
to form the desired relationships.

The approach described is dictated by
practical considerations and is based on the
treatment of control systems with constant parameters. The method is applicable in the presence
of nonlinearities in the control circuit.

A system based on measuring plant coefficients will have higher speed of response,
because the longest operation of synthesis (optimization of regulator coefficients) is carried
out by separate computing outside the control
system.

The number of computing operations required
to design the system described is less than that
required to design the system minimizing any
figure of merit. To show the usefulness of the
chosen figure of merit it is necessary to demonstrate that when we reach the minimum of the
figure of merit the system requirements are
satisfied. Moreover it is necessary to show
that there is a single and sufficiently low
minimum of figure of merit in a given region.

Therefore one may suppose that there is no
more economical method than trying all versions
in designing various selfadapting systems.

REFERENCE

1.   Yerman, V. L., Sabolev, O. K. and
Tsaturyan, K. T.; Some Problems in the Theory of
Self-Adaptive Control Systems. Engineering
Cybernetics 1963, No. 3, p.60. [English
translation of Russian Journal.]

AN ADAPTIVE LAGGING SYSTEM

by Dr A.P. Roberts
Head of Department of
Engineering Mathematics
The Queen's University of Belfast
N. Ireland

INTRODUCTION

A system is considered in which the output should follow a randomly varying signal. A fixed time τ elapses between reception of the signal and the need to follow it. Thus if the received signal is $r(t)$ and the system output is $c(t)$, the error in following at time t is

$$e(t) = r(t - \tau) - c(t).$$

The random signal is assumed to be a continuously varying Gaussian process. In the initial design of the system, it is also assumed that the signal is stationary in the statistical sense. However the system is then made to adapt itself to slow or infrequent changes to the average magnitude of the signal. It is further assumed that the form of the spectral density of the signal is known.

The criterion of optimum design is that the mean square error in following should be minimum subject to a constraint on the mean square value of the input to the controlled object. The problem is similar to the design of an optimum lagging filter which was discussed by Wiener[1]. The same approach can be used here but the method is not completely satisfactory in that a suitable rational operator must be selected to approximate $e^{-p\tau}$ before a necessary integration can be performed. In this paper the problem is approached in a completely different way using the method of dynamic programming[2].

METHOD

The cost functional to be minimized is defined as

$$\int_0^T \{e^2(t) + \lambda u^2(t)\}\, dt,$$

where u is the control signal, i.e. the input to the controlled object. Without loss of generality, $t = 0$ represents the present instant of time. The future interval T is taken large enough for the integral to give a reasonable estimate of average value. λ is a Lagrangian multiplier which is eventually adjusted automatically so that the desired constraint is imposed on the mean square value of u.

The problem now is to find the function $u(t)$, $0 \leqslant t \leqslant T$, which will minimize the cost functional and in particular give $u(0)$, the optimum control signal at the present instant. Unfortunately, the integrand of the cost functional contains the desired output which is given for $0 \leqslant t \leqslant \tau$ but is an unknown random quantity thereafter. Hence the cost functional should really be written as the expected value of the original one. However, since the integrand is quadratic, r is Gaussian and no finite bounds are imposed on any variable, the optimum value of $u(0)$ is obtained if the original cost functional is retained but with the desired output replaced by its expected value for $\tau < t \leqslant T$ [3].

The dynamic programming formulation results in a first order partial differential equation with

$$z = \min_u \int_t^T \{e^2(s) + \lambda u^2(s)\}\, ds, \quad 0 \leqslant t \leqslant T,$$

as the dependent variable, and t and the system state variables forming the independent variables. The optimum control signal is a function of one of the partial derivatives so it is only necessary to solve for that derivative. The equation could be solved by the method of characteristics but as this would involve two-point boundary values, the solution would take too long. In this case it is better to use Merriam's series solution method[4]. A solution can then be found by one computation performed backwards in time from $s = T$ to $s = 0$.

Suppose that one computation takes a time $\Delta\tau$. During the interval of time $0 \leqslant t \leqslant \Delta\tau$, the optimum control signal $u(t)$ can be found for $\Delta\tau \leqslant t \leqslant 2\Delta\tau$. Repeated computation then provides the optimum control signal in successive time sections each being of duration $\Delta\tau$. Finally the value of λ used in the computation may be adjusted automatically[5] in order to achieve the desired mean square level for the control signal u.

Example

Suppose that the signal to be followed is

a continuous Markov process described by the equation

$$\dot{r} = -ar + \xi , \qquad (1)$$

where a is a positive constant and ξ is white noise such that

$$E\left[\xi\right] = 0 \quad \text{and}$$
$$E\left[\xi(t_1)\xi(t_2)\right] = k^2\delta(t_1 - t_2) \qquad (2)$$

Here E means expected value and $\delta(x)$ is the Dirac delta function. Now $r(t)$ is known for $0 \leqslant t \leqslant \tau$ but thereafter we shall use its expected value. Define the function

$$r_p(t) = r(t), \quad 0 \leqslant t \leqslant \tau$$
$$= r(\tau)e^{-a(t-\tau)}, \quad \tau \leqslant t. \qquad (3)$$

Let the controlled object be described by the differential equation

$$\ddot{c} = u, \qquad (4)$$

where c is the output. We shall write

$$c = x_1 \quad \text{and} \quad \dot{x}_1 = x_2. \qquad (5)$$

It is shown in the appendix, that the optimum control signal at the present instant $t = 0$ is $u^*(0)$, where $u^*(t)$ gives the minimum functional

$$z(x_1,x_2;t) = \min_u \left[\int_t^T \{(r_p - x_1)^2 + \lambda u^2\} \, ds \right]$$

$$= \min_u \left[\int_t^{t+\delta t} \{(r_p - x_1)^2 + \lambda u^2\} ds \right.$$

$$\left. + z(x_1 + \delta x_1, x_2 + \delta x_2; t + \delta t) \right]$$

$$= \min_u \lim_{\delta t \to o} \left[\{(r_p - x_1)^2 + \lambda u^2\}\delta t \right.$$

$$+ z + \frac{\partial z}{\partial x_1} x_2\delta t + \frac{\partial z}{\partial x_2} u\delta t$$

$$\left. + \frac{\partial z}{\partial t} \delta t + 0(\delta t^2) \right]. \qquad (6)$$

On subtracting z and dividing by δt, this equation becomes

$$0 = \min_u \left[(r_p - x_1)^2 + \lambda u^2 + x_2 \frac{\partial z}{\partial x_1} \right.$$

$$\left. + u \frac{\partial z}{\partial x_2} + \frac{\partial z}{\partial t} \right]. \qquad (7)$$

As no theoretical bounds are imposed on it,

the optimum u is

$$u^* = -\frac{1}{2\lambda} \frac{\partial z}{\partial x_2} \qquad (8)$$

Back substitution results in the following partial differential equation to solve for $\frac{\partial z}{\partial x_2}$,

$$(r_p - x_1)^2 + x_2 \frac{\partial z}{\partial x_1} - \frac{1}{4\lambda}\left(\frac{\partial z}{\partial x_2}\right)^2 + \frac{\partial z}{\partial t} = 0 \qquad (9)$$

The function z can be written as

$$z = z_o + z_1 x_1 + z_2 x_2 + z_{11}x_1^2 + z_{12}x_1 x_2 + z_{22}x_2^2 ,$$
$$\cdots \quad (10)$$

where $z_o, z_1, \ldots, z_{22}$ are functions of t only. Equation (9) holds for all values of x_1 and x_2 so when (10) is substituted into (9), we may equate the coefficient of each power and product of x_1 and x_2 to zero. This produces the following equations

$$-\dot{z}_o = r_p^2 - \frac{z_2^2}{4\lambda} , \qquad (a)$$

$$-\dot{z}_1 = -2r_p - \frac{z_2 z_{12}}{2\lambda} , \qquad (b)$$

$$-\dot{z}_2 = z_1 - \frac{z_2 z_{22}}{\lambda} , \qquad (c) \qquad (11)$$

$$-\dot{z}_{11} = 1 - \frac{z_{12}^2}{4\lambda} , \qquad (d)$$

$$-\dot{z}_{12} = 2z_{11} - \frac{z_{12}z_{22}}{\lambda} , \qquad (e)$$

$$-\dot{z}_{22} = z_{12} - \frac{z_{22}^2}{\lambda} . \qquad (f)$$

Since z is an integral from t to T, it must be zero when $t = T$ regardless of the values of x_1 and x_2. This means that $z_o, z_1, \ldots, z_{22}$ are all zero at $t = T$ and these must be the boundary conditions for equations (11).

SYNTHESIS OF OPTIMAL CONTROL

In order to find the optimum value of the control signal at the present instant, it is necessary to evaluate

$$\frac{\partial z}{\partial x_2} = z_2 + z_{12}x_1 + 2z_{22}x_2 \qquad (12)$$

at $t = 0$. x_1 and x_2 are the present values of the system output position and velocity which we assume can be measured. z_{12} and z_{22} can be found as functions of λ by solving equations (d), (e) and (f) of (11). As $T \to \infty$, z_{12} and z_{22} are asymptotic to $\frac{1}{\mu^2}$ and $\frac{1}{2\mu^3}$, respectively, where $\mu = 2^{-\frac{1}{2}}\lambda^{-\frac{1}{4}}$

  The evaluation of z_2 involves the simultaneous solution of all except (a) of equations (11). The presence of the arbitrary function r_p prevents solution by analysis and the value of z_2 must be computed. This can be done by computing backwards in time from $t = T$ when the boundary conditions on $z_1, z_2, \ldots, z_{22}$ are all zero. With sufficient technical ingenuity it is undoubtedly possible to perform the computation very quickly by analogue computational methods. However, the computation can only be performed repetitively and each solution will take a finite time, say $\Delta\tau$. Thus u^* can only be provided in time sections of duration $\Delta\tau$ which must be fed successively into the system. Control improves with decrease in $\Delta\tau$ and the consequent use of more recent information.

  As usual, much simplification results if T is large. z_{12} and z_{22} become simple functions of μ as stated above. Furthermore, it is shown in the appendix that z_2 may then be evaluated by solving just the two equations (b) and (c) of (11) backwards in time from $t = \tau$ to $t = 0$. During this interval z_{12} and z_{22} have their constant asymptotic values and the initial values of z_1 and z_2 are

$$z_1(\tau) = -\frac{2}{a}\{1 - f(\mu)\}\, r(\tau) \quad \text{and}$$

$$z_2(\tau) = -\frac{f(\mu)}{\mu^2}\, r(\tau),$$

where

$$f(\mu) = \frac{2\mu^2}{2\mu(\mu + 1) + a^2} \qquad (13)$$

SELF-ADAPTATION

  The Lagrangian multiplier λ was included in the cost functional in order to constrain the mean square value of the control signal. To achieve the desired level of constraint the value of μ in the computation of z_2, z_{12} and z_{22} must be adjusted to the appropriate value. A simple way of adjusting μ has already been demonstrated by the author[5] in another application. The signal u^* which is fed into the system is also squared and passed through a low pass filter with long time constant. The resulting signal is then compared with the desired mean square value of u^* and the error actuates the adjustment of μ.

REFERENCES

1. Wiener, N., 'Interpolation, extrapolation and smoothing of stationary time series', Wiley.

2. Bellman, R., 'Dynamic Programming', Princeton University Press.

3. Roberts, A.P., 'Application of Pontryagin's maximum principle to the design of control systems with randomly varying inputs', 1963, Trans. S.I.T., 15, 155.

4. Merriam, C.W., 'Use of a mathematical error criterion in the design of adaptive control systems', A.I.E.E., 78, Part 2, 506.

5. Roberts, A.P., 'Self-optimizing control systems for a certain class of randomly varying inputs', 1959, Trans. S.I.T., 11, 193.

APPENDIX

  We shall now examine the example discussed in the text to prove that the method does in fact give the optimum control. The problem is to choose the optimum $u(0)$ to minimise the cost functional

$$C_1 = E\left[\int_0^T \{(r - x_1)^2 + \lambda u^2\}dt\right] \qquad (A.1)$$

Since $r(t)$ is given for $0 \leqslant t \leqslant \tau$, C_1 can be written as

$$C_1 = \int_0^\tau \{(r - x_1)^2 + \lambda u^2\}dt$$
$$+ E\left[\int_\tau^T \{(r - x_1)^2 + \lambda u^2\}dt\right] \qquad (A.2)$$

It is necessary to show that the $u(0)$ obtained when minimizing C_1 is the same as the $u(0)$ obtained when minimizing

$$C_2 = \int_0^\tau \{(r - x_1)^2 + \lambda u^2\}dt$$
$$+ \int_\tau^T \{(r_p - x_1)^2 + \lambda u^2\}dt \qquad (A.3)$$

Let
$$z'(x_1, x_2, r; \tau) = \min_u E\left[\int_\tau^T \{(r - x_1)^2 + \lambda u^2\}dt\right] \qquad (A.4)$$

which is the minimum value of the second term in C_1. We shall now use the continuous time dynamic programming formulation given in references such as 6 and 7. For $\tau \leqslant t \leqslant T$,

$$z'(x_1,x_2,r;t) = \min_u E\left[\int_t^T \{(r - x_1)^2 + \lambda u^2\}ds\right]$$

$$= \min_u E\left[\int_t^{t+\delta t} \{(r - x_1)^2 + \lambda u^2\}ds\right.$$

$$\left. + z'(x_1 + \delta x_1, x_2 + \delta x_2, r + \delta r; t + \delta t)\right]$$

$$= \min_u \lim_{\delta t \to o} E\left[\{(r - x_1)^2 + \lambda u^2\}\delta t\right.$$

$$+ z' + \frac{\partial z'}{\partial x_1}\delta x_1 + \frac{\partial z'}{\partial x_2}\delta x_2 + \frac{\partial z'}{\partial r}\delta r$$

$$\left. + \frac{\partial z'}{\partial t}\delta t + \frac{1}{2}\frac{\partial^2 z'}{\partial r^2}\delta r^2 + \ldots\right]$$

$$= \min_u \lim_{\delta t \to o}\left[\{(r - x_1)^2 + \lambda u^2\}\delta t + z'\right.$$

$$+ \frac{\partial z'}{\partial x_1}x_2\delta t + \frac{\partial z'}{\partial x_2}u\delta t + \frac{\partial z'}{\partial r}(-ar\delta t)$$

$$\left. + \frac{\partial z'}{\partial t}\delta t + \frac{1}{2}\frac{\partial^2 z'}{\partial r^2}k^2\delta t + O(\delta t^2)\right] \qquad (A.5)$$

Thus

$$0 = \min_u\left[(r - x_1)^2 + \lambda u^2 + \frac{\partial z'}{\partial x_1}x_2 + \frac{\partial z'}{\partial x_2}u\right.$$

$$\left. - \frac{\partial z'}{\partial r}ar + \frac{\partial z'}{\partial t} + \frac{1}{2}\frac{\partial^2 z'}{\partial t^2}k^2\right] \qquad (A.6)$$

Then $u^* = -\frac{1}{2\lambda}\frac{\partial z'}{\partial x_2}$ and

$$0 = (r - x_1)^2 + x_2\frac{\partial z'}{\partial x_1} - \frac{1}{4\lambda}\left(\frac{\partial z'}{\partial x_2}\right)^2 - ar\frac{\partial z'}{\partial r}$$

$$+ \frac{1}{2}k^2\frac{\partial^2 z'}{\partial r^2} + \frac{\partial z'}{\partial t} \qquad (A.7)$$

Assume a solution of the form

$$z' = z_o' + z_1'x_1 + z_2'x_2 + z_3'r + z_{11}'x_1^2 + z_{22}'x_2^2$$

$$+ z_{33}'r^2 + z_{12}'x_1x_2 + z_{13}'x_1r + z_{23}'x_2r, \qquad (A.8)$$

where $z_o', z_1', \ldots, z_{23}'$ are functions of t only and all are zero at $t = T$ since z' is zero then regardless of the values of x_1, x_2 and r. Equation (A.7) holds for all values of x_1, x_2 and r so on substituting (A.8), the coefficient of each power and product of x_1, x_2 and r can be equated to zero. This results in the following set of simultaneous differential equations.

$$-\dot{z}_o' = -\frac{1}{4\lambda}z_2'^2 + k^2z_{33}', \qquad (a)$$

$$-\dot{z}_1' = -\frac{1}{2\lambda}z_2'z_{12}' \qquad , \qquad (b)$$

$$-\dot{z}_2' = z_1' - \frac{1}{\lambda}z_2'z_{22}' \qquad , \qquad (c)$$

$$-\dot{z}_3' = -\frac{1}{2\lambda}z_2'z_{23}' - az_3', \qquad (d)$$

$$-\dot{z}_{11}' = 1 - \frac{1}{4\lambda}z_{12}'^2 \qquad , \qquad (e)$$

$$\qquad\qquad\qquad\qquad\qquad\qquad (A.9)$$

$$-\dot{z}_{22}' = z_{12}' - \frac{1}{\lambda}z_{22}'^2 \qquad , \qquad (f)$$

$$-\dot{z}_{33}' = 1 - \frac{1}{4\lambda}z_{23}'^2 - 2az_{33}' , \qquad (g)$$

$$-\dot{z}_{12}' = 2z_{11}' - \frac{1}{\lambda}z_{22}'z_{12}' , \qquad (h)$$

$$-\dot{z}_{13}' = -2 - \frac{1}{2\lambda}z_{12}'z_{23}' - az_{13}', (i)$$

$$-\dot{z}_{23}' = z_{13}' - \frac{1}{\lambda}z_{22}'z_{23}' - az_{23}'. \qquad (j)$$

Notice that z_o' does not appear in (b), (c) or (d); also, z_o', z_1', z_2' and z_3' do not appear in (e), (f), (g), (h), (i) or (j). Furthermore, if the series (A.8) had been extended to higher powers, the corresponding differential equations of coefficients would not include the coefficients of (A.9), nor would they include any separate constant terms. Consequently the extra coefficients would remain equal to their boundary values of zero. This is the justification for taking the series (A.8) no further than the quadratic terms. Equation (10) in the text can be justified by similar reasoning.

Next, if we examine equations (b), (c) and (d) of (A.9), it is seen that each term on the right hand side has either z_1', z_2' or z_3' as a factor. Since z_1', z_2' and z_3' are all zero at $t = T$, they must therefore remain zero for other values of t. This leaves us with seven equations (A.9) compared with the six equations (11) for minimizing the second term in C_2 given in (A.3).

Let us now examine the series (10) and (A.8) and the equations (11) and (A.9) side by side to investigate the relationship between them.

$$z = z_o + z_1 x_1 + z_2 x_2 + z_{11} x_1^2$$
$$+ z_{12} x_1 x_2 + z_{22} x_2^2. \qquad (10)$$

$$z' = z'_o + z'_{11} x_1^2 + z'_{22} x_2^2 + z'_{33} r^2$$
$$+ z'_{12} x_1 x_2 + z'_{13} x_1 r + z'_{23} x_2 r. \qquad (A.8)$$

$$-\dot{z}_o = r_p^2 - \frac{1}{4\lambda} z_2^2 \qquad , \qquad (a)$$

$$-\dot{z}_1 = -2r_p - \frac{1}{2\lambda} z_2 z_{12} \qquad , \qquad (b)$$

$$-\dot{z}_2 = z_1 - \frac{1}{\lambda} z_2 z_{22} \qquad , \qquad (c)$$

$$-\dot{z}_{11} = 1 - \frac{1}{4\lambda} z_{12}^2 \qquad , \qquad (d) \qquad (11)$$

$$-\dot{z}_{12} = 2z_{11} - \frac{1}{\lambda} z_{12} z_{22} \qquad , \qquad (e)$$

$$-\dot{z}_{22} = z_{12} - \frac{1}{\lambda} z_{22}^2 \qquad . \qquad (f)$$

$$-\dot{z}'_{33} = 1 - \frac{1}{4\lambda} z_{23}^{'2} - 2a z'_{33} \qquad , \qquad (g)$$

$$-\dot{z}'_{13} = -2 - \frac{1}{2\lambda} z'_{12} z'_{23} - a z'_{13} \qquad , \qquad (i)$$

$$-\dot{z}'_{23} = z'_{13} - \frac{1}{\lambda} z'_{22} z'_{23} - a z'_{23} \qquad , \qquad (j) \qquad (A.9)$$

$$-\dot{z}'_{11} = 1 - \frac{1}{4\lambda} z_{12}^{'2} \qquad , \qquad (e)$$

$$-\dot{z}'_{12} = 2z'_{11} - \frac{1}{\lambda} z'_{12} z'_{22} \qquad , \qquad (h)$$

$$-\dot{z}'_{22} = z'_{12} - \frac{1}{\lambda} z_{22}^{'2} \qquad , \qquad (f)$$

$$-\dot{z}'_o = -\frac{1}{4\lambda} z_2^{'2} + k^2 z'_{33} \qquad . \qquad (a)$$

Firstly compare (d), (e) and (f) of (11) with (e), (h) and (f) of (A.9). The two sets of equations are identical. Also each set is self-sufficient in that it does not involve any variables from any other equations. Therefore the solutions for the two sets must be equal, i.e. $z_{11} = z'_{11}$, $z_{12} = z'_{12}$ and $z_{22} = z'_{22}$, and the corresponding terms in z and z' are equal.

Next examine equations (b) and (c) of (11). Now $r_p = r(\tau)e^{-a(t-\tau)}$ and $\dot{r}_p = -a r(\tau)e^{-a(t-\tau)} = -a r_p$. Guess that $z_1 = r_p z'_{13}$ and $z_2 = z'_{23}$ and substitute in equations (b) and (c) of (11) so that the latter become

$$-r_p \dot{z}'_{13} + a r_p z'_{13} = -2r_p - \frac{1}{2\lambda} r_p z'_{23} z'_{12}, \text{ i.e.}$$

$$-\dot{z}'_{13} = -2 - \frac{1}{2\lambda} z'_{12} z'_{23} - a z'_{13} \qquad \text{and}$$

$$-r_p \dot{z}'_{23} + a r_p z'_{23} = r_p z'_{13} - \frac{1}{\lambda} r_p z'_{23} z'_{22}, \text{ i.e.}$$

$$-\dot{z}'_{23} = z'_{13} - \frac{1}{\lambda} z'_{22} z'_{23} - a z'_{23}.$$

These equations are (i) and (j) of (A.9) so the original guess was correct. Corresponding terms in z and z' are therefore $r_p z'_{13} x_1 + r_p z'_{23} x_2$ and $z'_{13} x_1 r + z'_{23} x_2 r$, respectively. These are equal at $t = \tau$ when $r_p(\tau) = r(\tau)$.

Finally examine equation (a) of (11). Guess that $z_o = r_p^2 z'_{33}$ so that (a) of (11) becomes

$$-r_p^2 \dot{z}'_{33} + 2a r_p^2 z'_{33} = r_p^2 - \frac{1}{4\lambda} r_p^2 z_{23}^{'2}, \text{ i.e.}$$

$$-\dot{z}'_{33} = 1 - \frac{1}{4\lambda} z_{23}^{'2} - 2a z'_{33}.$$

This is (g) of (A.9) so again the original guess was correct. The corresponding terms in z and z' are $r_p^2 z'_{33}$ and $z'_{33} r^2$. Again these are equal at $t = \tau$.

The only term which has not been accounted for is z'_o. Thus

$$z'(\tau) - z(\tau) = z'_o(\tau), \qquad (A.10)$$

which is independent of the system state variables and of the signal r. For given k^2 and λ, $z'_o(\tau) = \text{constant} = K$, say.

Referring back to equations (A.2) and (A.3) and minimizing with respect to u, we have

$$\min_u C_1 = \min_u \left[\int_o^\tau \{(r - x_1)^2 + \lambda u^2\} dt \right.$$
$$\left. + z'(x_1, x_2, r; \tau) \right]$$
$$= \min_u \left[\int_o^\tau \{(r - x_1)^2 + \lambda u^2\} dt \right.$$
$$\left. + z(x_1, x_2, r; \tau) + K \right]$$
$$= \min_u \left[\int_o^\tau \{(r - x_1)^2 + \lambda u^2\} dt \right.$$
$$\left. + z(x_1, x_2, r; \tau) \right] + K = \min_u C_2 + K, (A.11)$$

since K is unaffected by choice of u. Therefore $u*(0)$ for minimum C_1 is the same as $u*(0)$ for minimum C_2, which is what we set out to prove.

When $(T - t)$ is very large, all the rates of change of coefficients which are given on the left hand side of equations (A.9) are zero except for $\dot{z}_0'$. Equations (b), ..., (j) of (A.9) are then algebraic equations which may be solved to give in particular

$$z_{12}' = 2\lambda^{\frac{1}{2}} = \frac{1}{\mu^2},$$

$$z_{22}' = 2^{\frac{1}{2}}\lambda^{\frac{3}{4}} = \frac{1}{2\mu^3},$$

$$z_{13}' = -\frac{2}{a}\left(1 - \frac{1}{1 + a^2\lambda^{\frac{1}{2}} + 2^{\frac{1}{2}}\lambda^{\frac{1}{4}}}\right)$$

$$= -\frac{2}{a}\{1 - f(\mu)\}, \qquad (A.12)$$

and $\quad z_{23}' = -\frac{2\lambda^{\frac{1}{2}}}{1 + a^2\lambda^{\frac{1}{2}} + 2^{\frac{1}{2}}\lambda^{\frac{1}{4}}} = -\frac{1}{\mu^2}f(\mu),$

where $\mu = 2^{-\frac{1}{2}}\lambda^{-\frac{1}{4}}$ and $f(\mu) = \frac{2\mu^2}{2\mu(\mu + 1) + a^2}$.

At $t = \tau$; $z_{12} = z_{12}'$, $z_{22} = z_{22}'$, $z_1 = r(\tau)z_{13}'$ and $z_2 = r(\tau)z_{23}'$. Consequently, from equations (8), (12) and (A.12), for $0 \leqslant t \leqslant \tau$, we have

$$u*(t) = -2\mu^4 \frac{\partial z}{\partial x_2} = -2\mu^4(z_2 + z_{12}x_1 + 2z_{22}x_2)$$

$$= -2(\mu^4 z_2 + \mu^2 x_1 + \mu x_2), \qquad (A.13)$$

where z_2 is given by the simultaneous differential equations

$$-\dot{z}_1 = -2r_p - \frac{1}{2\lambda}z_2 z_{12} = -2(r_p + \mu^2 z_2) \quad \text{and}$$

$$-\dot{z}_2 = z_1 - \frac{1}{\lambda}z_2 z_{22} = z_1 - 2\mu z_2,$$

with boundary conditions $\qquad\qquad (A.14)$

$$z_1(\tau) = -\frac{2}{a}\{1 - f(\mu)\}\, r(\tau) \quad \text{and}$$

$$z_2(\tau) = -\frac{1}{\mu^2}f(\mu)\, r(\tau).$$

REFERENCES

6. Florentin, J.J., 'Optimal control of continuous time, Markov stochastic systems', J. Electron. Contr., 1961, 10, 473.

7. Roberts, A.P., 'Optimum regulation', Trans. S.I.T., 1965, 17, 25.

Discussion

Dr. O.L.R. Jacobs (UK) saw a multilevel structure in Dr. Roberts' paper. He interpreted Dr. Roberts' use of the phrase "desired mean square value of control $u*$" as being different from most formulations that seek the minimum of mean square control. From this viewpoint, it seemed that a higher level controller was necessary to determine the Lagrangian multiplier λ.

Dr. Roberts observed in reply that, in general, his cost formulation will lead to a demand for as much control signal u as is available (bang-bang). The multiplier λ then minimizes the mean square value by imposing a minimum constraint automatically. It is this minimum that is desired and no prior value is assigned to u.

Some Problems in the Development of Adaptive Systems
using the Sensitivity Operator

by Juraj V. Medanić and Petar V. Kokotović,
Institute "Mihailo Pupin",
Belgrade,
Yugoslavia.

INTRODUCTION

Adaptive systems have been defined in a variety of ways but it is generally agreed that some degree of ignorance in the characteristics of the plant and/or the input signals must be involved in order to justify the adaptive approach. An adaptive system would be one having the feature of intentional changes of the system characteristics in order to achieve a specified performance in the presence of changing characteristics of its environment.

The main problem in adaptation is to obtain the desired performance so that the existing state of the system is not essential and the plant is usually identified only as an aid to the process of adaptation. The model-reference adaptive systems belong to a class of parameter optimization adaptive systems in which the desired performance of the system is specified implicitly through a reference model.[1],[2]. The application of the model-reference idea reduces the need for an explicit identification of the system. If there exists a way of driving the system parameters towards the optimum point, there is no real need of identifying the state of the system since the state will be changed any way if it is not the optimum state. If it is the optimum state, the model will be a good approximation of the system characteristics if a compatible controller is designed. Thus the identification of the only important state has been achieved in advance.

This of course, raises the problems of specifying a quality index for the system performance,[2],[5],[13],[21],[10], the choice of a compatible controller, [9],[10], and a method of finding the optimum state of the system.[2],[5],[8],[9],[16].

Finally, if the above problems have been solved, the resulting system will be time-varying and non-linear and a stability analysis has to be performed. Contrary to the variety of solutions proposed in connection with most of the above stated problems, the stability analysis of the system has generally been attempted by the application of the direct method of Liapunov. Unfortunately, most papers treat the stability analysis only of the simplest adaptive control systems and the question of stability has yet to be answered, even for systems already proposed in the references.[2],[15],[16],[12].

This paper presents some theoretical and experimental results obtained in the search for solutions of the above problems. The body of the paper consists of three sections. The first section reviews some important but neglected results in sensitivity analysis obtained recently which lead to the sensitivity points method. In the second section a simple and very appealing structure for self-optimization applicable to the synthesis of control systems is demonstrated. The third section contains some new experimental results obtained in fast adaptation of systems with the adaptive structures already proposed for slow adaptation and self-optimization.

1. The Sensitivity Operator Approach to the Problem of Generating a Gradient

The basic problem of optimization is the evaluation of the present state of the system and the assertion that there exists at least one state nearer the optimum than the present one. A quality index is used to measure the nearness of the existing state to the desirable state of the system, and a convex function of the error between the desired and actual output is used. The squared error or the integral of the squared error are usually employed

$$I = F(e) = e(t)^2 \qquad (1)$$

$$I = \int_0^T F(e)dt \qquad (2)$$

The quality index is a functional of the control signal if the optimum performance is obtained through the action of control signals, or it reduces to a function of the parameters c if parameter optimization is in question.

Dealing with this second case, the system is said to be optimum or to have adapted, as the case may be, if the parameters have been changed in such a way as to minimize the respective quality index. The main problem then is not to identify the present state but to contrive a way of changing it to a state nearer the optimum.

The usual procedure applied is based on the

"Superior numbers refer to similarly-numbered references at the end of this paper"

method of steepest descent. The method of steepest descent may be correctly applied under certain conditions depending on the rate of change of input characteristics and system parameters. To reach the optimum state, the parameters should be changed according to the relation

$$\dot{c}_i = -h \frac{\partial I}{\partial c_i} \qquad (3)$$

Of the methods employed in obtaining the gradient, the perturbation method and the sensitivity operator method are the most widely used. The perturbation method demands the injection of some kind of usually periodic test signals, to obtain the components of the gradient. The sensitivity operator method employs a sensitivity model of the system to obtain the same components. Hence the sensitivity operator model-reference adaptive systems may be treated as a subclass of the model-reference systems. Our work has been closely associated with the solution of problems connected with the application of the sensitivity operator method.

Whatever the form of the quality index I, the derivation of its components implies the need of obtaining the corresponding sensitivity functions $\frac{\partial e(s)}{\partial c_i}$

$$\frac{\partial I}{\partial c_i} = \frac{\partial F(e)}{\partial e} \cdot \frac{\partial e}{\partial c_i} \qquad (4)$$

Hence, a method enabling the simultaneous derivation of all, or most of the necessary sensitivity functions with a minimum of additional equipment would be most convenient.

This problem has been solved in a rather general manner at least for linear systems with slowly varying parameters.[5-7] The main result is that for a fairly general structure of system (plant plus controller) with feedback loops starting from the observable points of the plant, there exists a class of adjustable parameters of the controller, the sensitivity functions of which can be obtained simultaneously. The additional equipment consists of an exact replica of the system itself and filters starting from the sensitivity points. The filters are defined by the expression

$$B_i = \frac{\partial \ln G_i(s, p_i)}{\partial p_i}$$

or

$$D_i = \frac{\partial \ln H_i(s, p_i)}{\partial c_i}$$

When the output of the system $y(s)$ is brought to the input of the model (Fig. 1), the sensitivity functions are obtained at the output of

filters. If the adjustable parameters are gains of the blocks $G_i(s)$ or $H_i(s)$ the filters reduce to coefficients $1/c_i$ or $1/p_i$ - the reciprocal of the parameter values. In feedback loops the sensitivity functions may then be obtained directly by moving the sensitivity point in front of the parameter c. This procedure cannot be repeated for gains in the forward paths of the loops; the sensitivity functions cannot be obtained directly and the division by the corresponding value of the parameter must be performed. This is usually undesirable and sometimes impossible since the exact value of the parameter may not be known and is usually immaterial. However, for rather general structures of control systems, it is sufficient to have only one adjustable coefficient in the forward loop. The adjustable portion of the gain coefficient in the direct path can be situated in a minor feedback loop and the corresponding sensitivity function may be obtained directly.

The above results have culminated in the definition of sensitivity points (Fig. 2). If to a linear system with slowly varying parameters a sensitivity model identical to the system is added, the sensitivity functions $U_i(s)$ corresponding to the adjustable parameters of the controller will be obtained at the sensitivity points of the model. Our experience has shown that the signals at the sensitivity points may be, and should be, used as a means of adjusting the parameters of the controller even when one or more of the above stated conditions are not fulfilled.

Interestingly enough, the problem of choosing a compatible controller has received little attention in the field of model reference adaptive systems, especially when applying the sensitivity operator approach. The basic idea would be to choose such a controller that for the expected range of plant parameters variations, the controller parameters may always be adjusted in such a way that the quality index is below a prespecified value.

The individual controller components $H_i(s)$ may be determined by taking into account the most probable state of the plant, and an additional set of adjustable parameters c_i would be provided to cope with the actual situation at hand. The desired $H_i(s)$ may not be physically realizable and an approximation by a finite number of parallel filters $H_{ij}(s)$ may have to be used.[10] Every parallel filter would contain an adjustable gain factor intended to retain the optimum state not for the expected but the actual values of the process and input characteristics. The advantage of such a controller would of course be that all the sensitivity functions or signals corresponding to them could be obtained simultaneously on only one additional model of the system.

Such a method would be exact if the

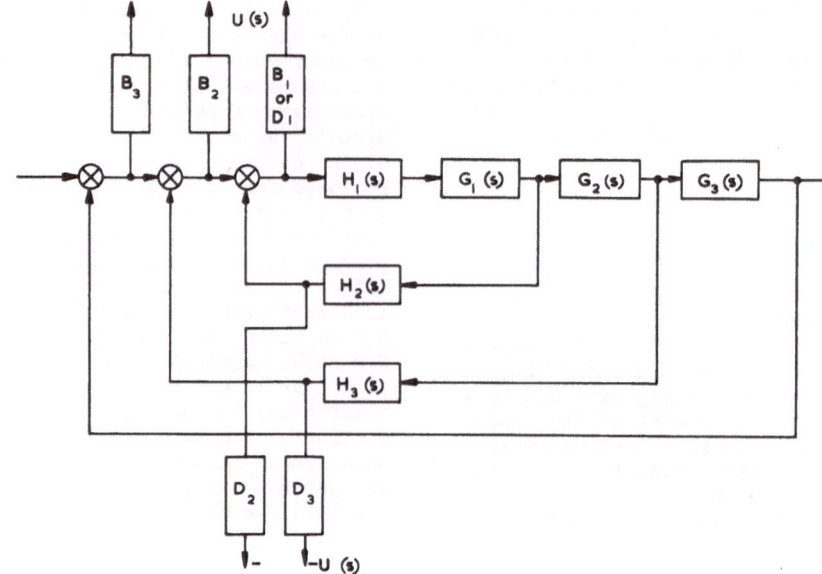

FIG. I

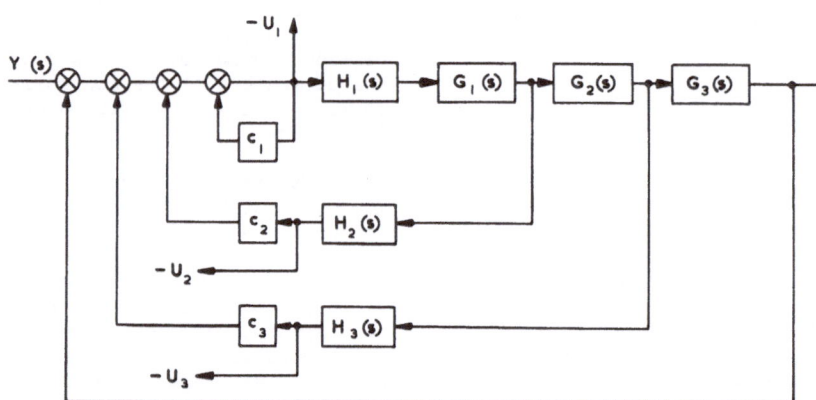

FIG. 2

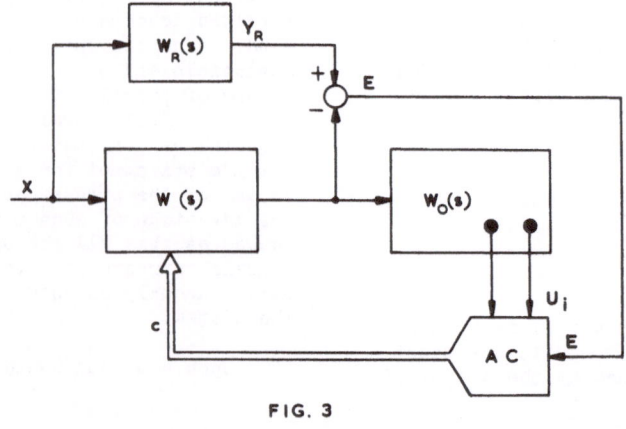

FIG. 3

sensitivity model could be made identical to the system. But ignorance about the state of the plant is precisely the reason why the adaptive approach was needed in the first place. Apparently, the sensitivity model can at best be an approximation of the system. Nevertheless, in the following sections we proceed to demonstrate some methods, and structures of self-optimizing and adaptive systems which take advantage of the sensitivity points even though the signals obtained at the points will not be the exact sensitivity functions.

2. The Process of Slow Adaptation

The procedure of investigating an adaptive system in any but the most trivial cases turns out to be the investigation of a non-linear and non-stationary process. The difficulties that an analytical approach to the problem would pose are overwhelming, and probably an analytical solution of such problems for any but the simplest of systems will be impossible to obtain. The path to be followed leads to numerical analysis and the application of computers. On the other hand, it is a disputable matter whether the exact solution of the optimization problem is a necessity, or whether the fact that the system is near the optimum is more important, especially since the state defined as optimum is the control engineer's and not the system's point of view.

What we are interested in, is the solution of the optimization or adaptation problem by the application of the sensitivity operator method, even though the method may not be exact. The restrictions on using the gradient method are quite strong. Nevertheless, we proceed to show how the signals at the sensitivity points of the system may be used in self-optimization and adaptation.

The exact sensitivity functions cannot be obtained in the first place since there exists some ignorance about the plant, even if the plant is linear and time-variant. For such a system the sensitivity functions needed for adaptation may be obtained in the following manner:

$$\frac{\partial E(s)}{\partial q_i} = \frac{\partial Y(s)}{\partial q_i} = \frac{\partial W(s)}{\partial q_i} X(s) =$$

$$\frac{1}{W(s)} \cdot \frac{\partial W(s)}{\partial q_i} W(s)X(s)$$

$$\therefore \quad \frac{\partial E(s)}{\partial q_i} = \frac{1}{W(s)} \cdot \frac{\partial W(s)}{\partial q_i} Y(s) \qquad (5)$$

and we define

$$S_i(s) = \frac{1}{W(s)} \cdot \frac{\partial W(s)}{\partial q_i} \qquad (6)$$

as the sensitivity operator, which must be applied to the output signal of the system in order to obtain the corresponding sensitivity functions. It is the equivalent of the transfer function from the input to the corresponding sensitivity point. The sensitivity points are in fact defined in this way.

Without identification of the plant we are not in position to know the exact value of its parameters and hence, of the adjustable sensitivity model which realizes simultaneously various sensitivity operators $S_i(s)$ ($i = 1,2,...$). At best we may pass the system output signal through a model - which we, for some reason, feel will be a satisfactory approximation of the exact sensitivity model. The obvious choice would be a fixed model of the system with the expected value of plant parameters and corresponding optimum values of the controller parameters. This would give the expected sensitivity model $W_0(s)$ (Fig. 3). The signal we are in position to obtain is then

$$U_i(s) = \frac{1}{W_0(s)} \cdot \frac{\partial W_0(s)}{\partial q_i} Y(s) \qquad (7)$$

On the other hand it may be shown that for the structure of the compatible controller proposed, possessing the set of adjustable parameters satisfying the simultaneity condition, with a variable plant and adjustable controller, the following relation holds:

$$\frac{1}{W_0(s)} \cdot \frac{\partial W_0(s)}{\partial q_i} = \frac{W_0(s)}{W(s)} \cdot \frac{1}{W(s)} \cdot \frac{\partial W(s)}{\partial q_i} \qquad (8)$$

Combining (7) and (8) the signal obtained at the sensitivity point may be written in the form

$$U_i(s) = \frac{W_0(s)}{W(s)} \cdot \frac{\partial E(s)}{\partial q_i} \qquad (9)$$

Evidently, the signals $U_i(s)$ will tend to the sensitivity functions if the adjustable parameters are changed in such a way that $W(s)$ tends to $W_0(s)$. So far we have only experimental evidence that the domain of parameter space within which the application of this "pseudogradient" brings the state of the system near the optimum is not smaller than the domain of the exact gradient.

If we follow this kind of reasoning, then the nominal or expected sensitivity model will have the characteristics similar to those of the reference model and $W_0(s)$ may be exchanged for a reference model. The objection to this is that the sensitivity points are lost since the reference model is much simpler and usually of the second or third order. At best, a separate and different filter would have to be used at

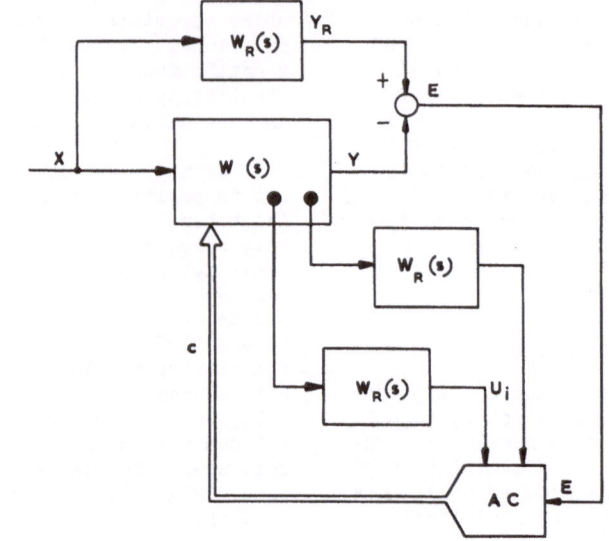

FIG. 4

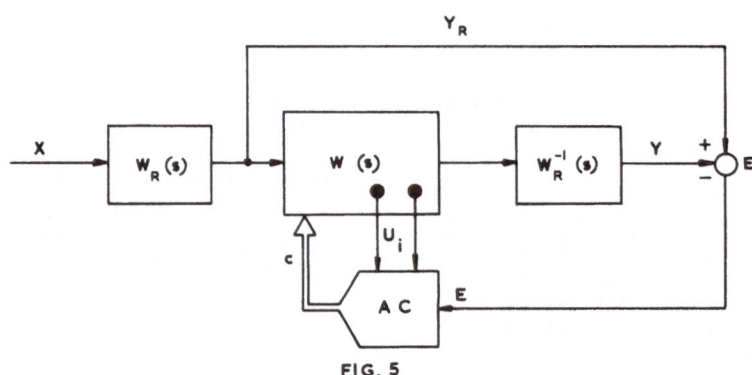

FIG. 5

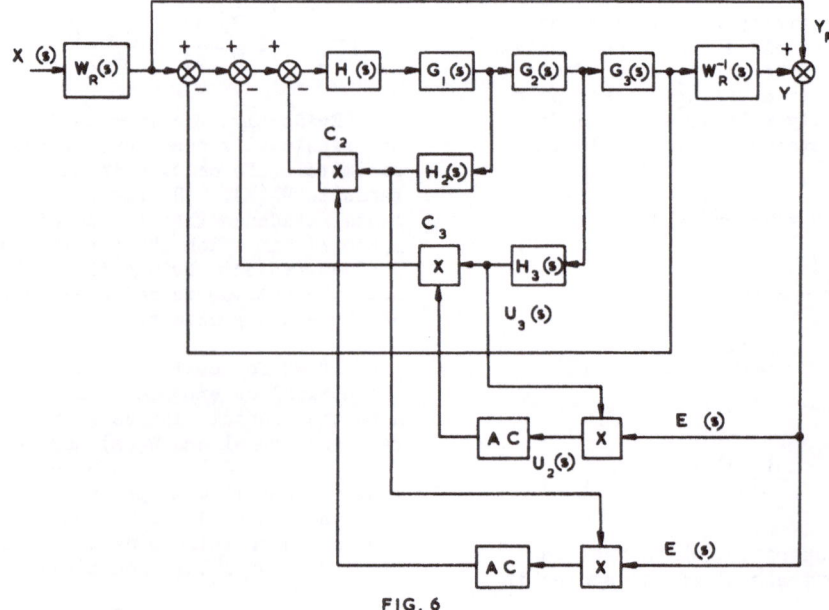

FIG. 6

208

the output to obtain each pseudo-sensitivity function. But the idea with this additional use of the reference model may be applied in a somewhat different manner. Expression (5) may be written in the following form:

$$\frac{\partial E(s)}{\partial q_i} = \frac{1}{W(s)} \cdot \frac{\partial W(s)}{\partial q_i} \, W(s) \, X(s)$$

$$= \frac{1}{W(s)} \cdot \frac{\partial W(s)}{\partial q_i} \, X(s) \cdot W(s) \qquad (10)$$

The signal

$$\frac{\partial W(s)}{\partial q_i} \cdot \frac{1}{W(s)} \, X(s) = S_i(s) \, X(s) = Z_i(s)$$

is the signal at the sensitivity point in the system itself. The signal has to be passed through the model of the system $W(s)$ to obtain the corresponding sensitivity function. As before, an objection is that $W(s)$ cannot be modelled, and besides that a separate model would be required for each sensitivity function. This objection, however, is not too severe if we now, as before, firstly place $W_0(s)$ instead of $W(s)$, and then exchange $W_0(s)$ for $W_R(s)$. For $W_0(s)$ we obtain the following signal

$$U_i(s) = \frac{1}{W(s)} \cdot \frac{\partial W(s)}{\partial q_i} \, X(s) \, W_0(s)$$

$$U_i(s) = -\frac{\partial E(s)}{\partial q_i} \cdot \frac{W_0(s)}{W(s)} \qquad (11)$$

the same as in (9), but without taking into account (8). If now $W_R(s)$ is placed instead of $W_0(s)$

$$U_i(s) = \frac{\partial E(s)}{\partial q_i} \cdot \frac{W_R(s)}{W(s)} \qquad (12)$$

and the structure first proposed by Osborn-Whitaker[2,3] is obtained (Fig. 4). For each separate sensitivity function a separate reference model is needed, but as the reference model is of a small order this solution may be tolerated.

If self-optimization is performed with analog computer simulation, the sensitivity model may be the exact replica of the system and the exact sensitivity functions and hence the real optimum may be found. But again, for analog simulation or real system self-optimization a very simple optimization structure may be obtained, evolving from the procedure already described for slow adaptation. In this case, the reference model may be in the direct path of the input signal giving additional freedom in solving the optimization problem. Instead of obtaining the sensitivity functions we are again satisfied with the

signals

$$U_i(s) = \frac{1}{W(s)} \cdot \frac{\partial W(s)}{\partial q_i} \, Y_R(s)$$

where $Y_R(s)$ is obtained by passing the input signal through $W_R(s)$. It is easily seen that

$$U_i(s) = \frac{\partial E(s)}{\partial q_i} \cdot \frac{W_R(s)}{W(s)} \qquad (13)$$

the same as in (12) but with a different structure (Fig. 5). The operator $W_R^{-1}(s)$ is needed to form the error $E(s)$, and is easily obtained, especially if the synthesis is performed by simulation in which case a few potentiometers and a summing point are sufficient. This would be the minimum of additional equipment still enabling a satisfactory optimization.

Up to now the sensitivity functions obtained were not exact due only to the ignorance of the state of the plant, all the other conditions necessary for the application of the gradient method being satisfied.

3. The Problems of Fast Adaptation

The real problem and one of the utmost theoretical as well as practical interest is the problem of fast adaptation of time varying and non-linear plants. Even if only linear non-stationary plants are considered, an analytical solution, except for very simple systems, is possible only for special cases, and usually may be obtained only by resorting to approximation.

On the other side, the procedure applied up to now will not be justified for a mass of reasons: the gradient method is not justified since the system is non-stationary and hence the quality index is a function of time; the sensitivity functions cannot be defined in the above sense since the parameters are adjusted instantly; the sensitivity functions cannot be obtained simultaneously since the system and its model are non-stationary.

Nevertheless, the idea of employing the information that may be obtained at the sensitivity points should not be deserted because of the mere fact that we are not able to express mathematically the phenomena taking place in such a system. The main question, anyhow, is to adapt the system and the secondary question is to find a mathematical justification for the procedure.

What we propose is to ignore the non-stationary characteristics of the plant and regard it as a stationary process, provide it with a compatible controller, in the sense that the sensitivity points may be obtained, and

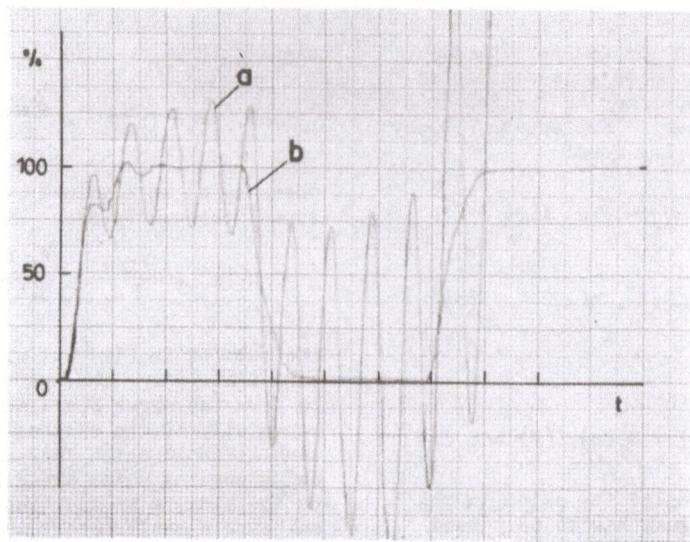

Fig. 7. Relationships with $K_0 = 8$, $m = 4$, $b = 0.125$, $n = 3$.

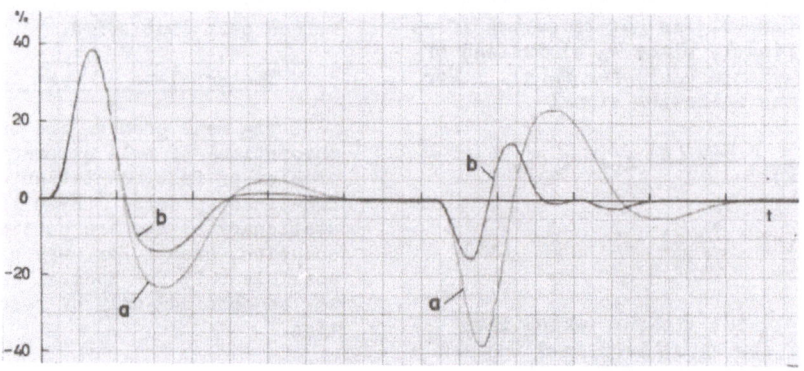

Fig. 8. Transient responses to ramp function input $X_0 = at$. ($K_0 = 8$, $m = 4$, $b = 0.125$, $n = 3$, $|A_M|/|A| = 2.56$)

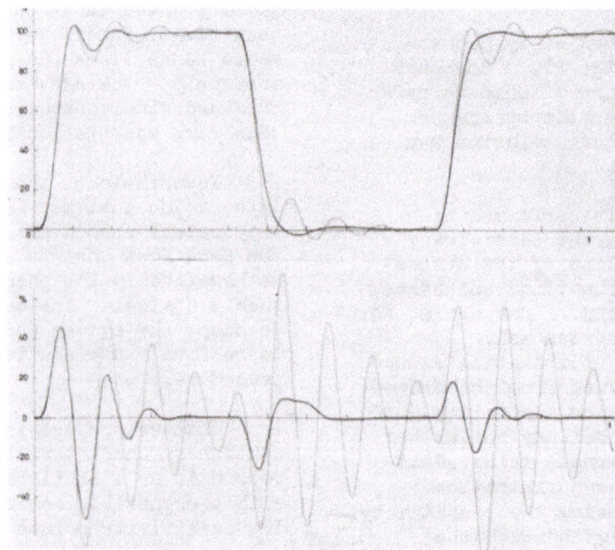

Fig. 9.

extend the methodology of slow adaptation to the process of fast adaptation. Of course, since the adaptation has to be executed instantly, the integral quality index does not come into consideration but an additional filter may be used to process the components of the would be gradient. In other words, the equivalents of the pseudo-sensitivity functions are multiplied by the error and passed through a shaping filter. The obtained signals are then used to adjust the parameters of the controller. Evidently this is far away from the gradient method serving as a starting point of the procedure.

The system obtained in this manner may be treated as a system in which a number of non-linear loops have been added with the intention of making its characteristics insensitive to the variation of the plant.

The selection of the filters in the adaptive loops is a question by itself and demands separate attention. In a preliminary study we have found it convenient to choose a controller filter of the form

$$A_i(s) = a_{1i} + \frac{a_{2i}}{a_{3i}s + 1} \qquad (14)$$

where the a_{ij} ($j = 1,2,3$) are chosen in such a way to give satisfactory adaptation. The reaction of such an adaptive system to a step response is given (Fig. 7) and as may be seen a satisfactory response is obtained even during the first step of the square wave response, the response to following steps being very good.

The possibilities and shortcomings of such non-linear and non-stationary systems have yet to be enlightened. The stability analysis will hardly be possible by the existing methods, and for now we have had to satisfy ourselves with the fact that the proposed system is much more insensitive to plant parameter variations than without adaptive loops.

The very statement above points out the criterion used in estimating the quality of the adaptive loops. The sensitivity functions with respect to the variable plant parameters have been obtained by the application of the structural method approach. The result was that the sensitivity functions (or a sensitivity index based on them) are much smaller with the adaptive loops than without them. With the system in the optimum state, the sensitivity to the plant parameter variations was lessened in a great degree in comparison to the sensitivity without the adaptive loops, by the sole presence of the loops (Fig. 8). The reduction in the sensitivity was even greater if the system was not in its optimum state, the adaptive loops being effective (Fig. 9). Incidentally, this sensitivity index was also used to select the optimum values of the filter in (14), the optimum values being those minimizing the

sensitivity to plant parameter variations.

Concluding Remarks

The sensitivity operator approach to the adaptation of variable plants gives perspective possibilities in the synthesis of model reference adaptive systems. The approach may well become useful as a procedure for setting-up of intentionally non-linear feedback loops with the purpose of diminishing the sensitivity of the system to plant parameter variations.

REFERENCES

1. Whitaker, H.P. "An Adaptive System for Control of the Dynamic Performance of Aircraft and Spacecraft. Institute of Aeronautical Sciences, Paper No. 59-100, June, 1959.

2. Osborn, P.V., Whitaker, H.P., Kezer, A. New Developments in the Design of Model Reference Adaptive Control System. IAS Paper No. 61-39, N.Y., Jan. 1961.

3. Whitaker, H.P. Design Capabilities of Model Reference Adaptive Systems. Proc. of NEC, 1962.

4. Li, Y.T., Whitaker, H.P. Performance Characterization for Adaptive Control Systems. IFAC Symposium on Self-Adjusting Systems, Rome, April 1962.

5. Kokotović, P. Simultaneous Obtaining of Sensitivity Functions. Pupin Control Lab. Report, July 1962, published in "Sensitivity Analysis of Dynamic Systems" by R. Tomović, Nolit, Belgrade 1963, and McGraw-Hill, New York, 1964.

6. Kokotović, P. The Structure of the Parameter Influence Analyser. Discussion, Second IFAC Congress, Basel, 1963.

7. Bingulac, S., Kokotović, P. Automatic Optimization of Linear Control System on an Analog Computer. Proceedings of the Inter. Assoc. for Analog Computation, Vol. VII, No. 1 January 1965 (also presented at 8th ETAN Conference, 1963).

8. Kokotović, P. Sensitivity Points Method in the Analysis and Optimization of Linear Control Systems. Avtomatika i Telemehanika, No. 12, Dec. 1964, Moscow.

9. Kokotović, P., Medanić, J., Bingulac, S., Vušković, M. Sensitivity Method in the Experimental Design of Adaptive Control Systems, submitted for the III IFAC Congress, London 1966.

10. Narendra, K.S., McBride, L.E. Multipara-
    meter Self-Optimizing Systems Using
    Correlation Techniques, IEEE Trans. AC-9,
    January 1964.

11. RISSANEN, J. On the Theory of Self-
    Adjusting Models, Automatica, Vol. 1,
    Pergamon Press, London 1963.

12. RISSANEN, J. Design of Self-Adjusting
    Control Systems by Use of a Functional
    Derivative Technique, IFAC/ETAN Symposium
    on Sensitivity Analysis, Dubrovnik, 1964.

13. Margolis, M., Leonides, C.T. A Parameter
    Tracking Servo for Adaptive Control
    Systems, Trans. IRE, AC-4, No. 2, 1959.

14. Margolis, M., Leonides, C.T. On the Theory
    of Adaptive Control Systems, the Learning
    Model Approach, Proc. First IFAC Congress,
    Moscow 1960.

15. Donaldson, D., Leonides, C.T. A Model
    Referenced Parameter Tracking Technique
    for Adaptive Control Systems, Trans. IEEE,
    Appl. and Ind., pt. 1 and 2, Sept. 1963.

16. Horrocks, T. Investigations into Model-
    Reference Adaptive Control Systems, Proc.
    IEE, Vol. 111, No. 11, November 1964.

17. Stear, E.B., Gregory, P.C. Capabilities
    and Limitations of Some Adaptive
    Techniques, IFAC Symposium on Optimizing
    and Adaptive Control, Rome, 1962.

18. Kazakov, I.E., Evlanov, L.G. On the Theory
    of Self-Adapting Systems with a Search of
    Gradient by the Method of Auxiliary
    Operator, Second IFAC Congress, Basle,
    1963.

ADAPTIVE DIRECT DIGITAL CONTROL WITH MULTI-PARAMETER ADJUSTMENT

Roger M. Bakke
Staff Engineer
International Business Machines Corporation
Systems Development Division
San Jose Laboratory
San Jose, California

ABSTRACT

An adaptive control scheme has been developed for the adjustment (i.e., tuning) of parameters in Direct Digital Control (DDC) algorithms.

This paper presents an extension of this adaptive control scheme to the multi-parameter adjustment problem. Multi-parameter adjustment is developed. The theoretical base of the identification scheme is presented. Practical limitations on the implementation of this adaptive control scheme into a DDC algorithm are discussed.

INTRODUCTION

The adaptive control scheme presented in this paper provides automatic correction of the coefficients in DDC algorithms regulating non-linear and/or time varying processes. These corrections are made in response to changes in the process dynamics resulting from time variance, or changes in the process operating conditions. Reference (1) is recommended to the reader interested in a discussion of the relation of this particular adaptive scheme to other adaptive control techniques that have been reported in the literature. This reference also provides results of the application of this scheme in the single parameter adjustment problem. References (2) and (3) are recommended for review of the adaptive control field in general and Reference (4) is recommended as an explanation of the relatively new field of DDC.

PERFORMANCE CRITERIA

In general, the performance criterion associated with this adaptive control system implies improved dynamic performance of the process under DDC regulation with a minimum number of process measurements. More specifically, the nonlinear time variant process (defined by equations 1, 2, 3 and 4):

$$\dot{y}(t) = f\big(y(t),\ v(t),\ t\big) \cdot y(t) + g\big(y(t),\ v(t),\ t\big) \cdot v(t) \tag{1}$$

$$\dot{z}(t) = f*\big(y(t),\ v(t),\ t\big) \cdot z(t) + g*\big(y(t),\ v(t),\ t\big) \cdot v(t) \tag{2}$$

$$z(t) = \begin{bmatrix} c(t) \\ o(t) \\ y(t) \end{bmatrix} \tag{3}$$

$$y(t) = \begin{bmatrix} m(t) \\ d\ (t) \end{bmatrix} \tag{4}$$

where:

$y(t)$ = the state vector of the process

$z(t)$ = the vector of instrumented variables

$v(t)$ = the vector of process inputs

$c(t)$ = the vector of controlled variables

$o(t)$ = the vector of observed variables

$m(t)$ = the vector of variables manipulated by the DDC

$d(t)$ = the process disturbance vector

t = the variable time

which is being controlled by the DDC algorithm (defined by equations 5, 6 and 7):

$$w(t + \Delta T) = \Phi_1 w(t) + \Phi_2\, \epsilon(t + \Delta T) \tag{5}$$

$$\epsilon(t) = r(t) - c(t) \tag{6}$$

$$m(t) = \Phi_3 w(t) + \Phi_4\, \epsilon(t) \tag{7}$$

where:

$w(t)$ = the state vector of the DDC controller

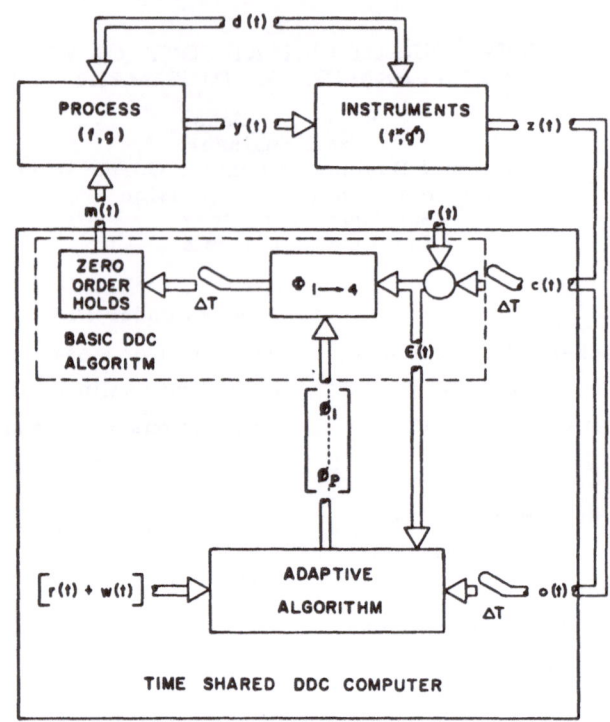

FIG. I. DDC-PROCESS SIGNAL FLOW

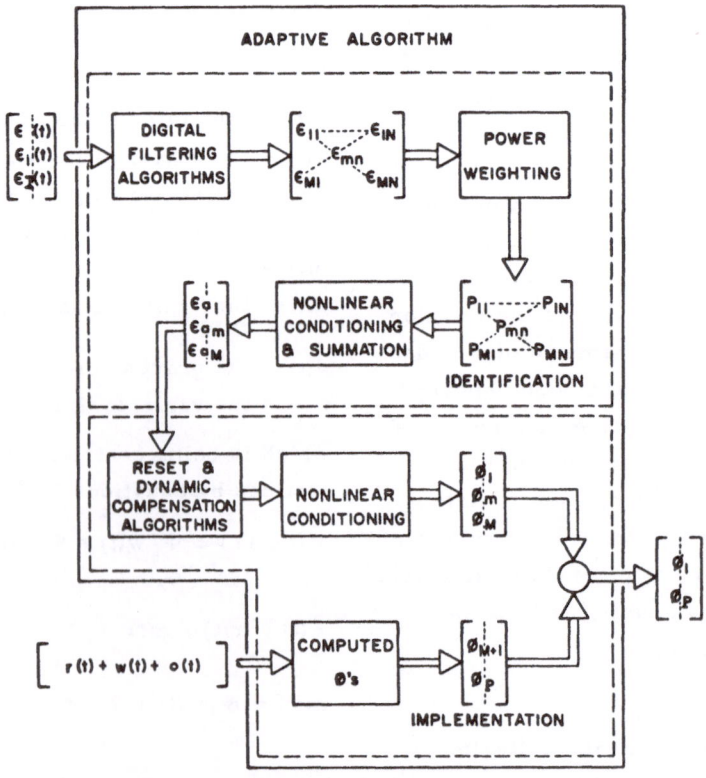

FIG. 2. ADAPTIVE ALGORITHM SIGNAL FLOW

214

Φ_{1-4} = the DDC controller matrices

$\epsilon(t)$ = the vector of controller errors

$r(t)$ = the reference (i.e., set-point) vector

ΔT = the basic sampling period of the DDC computation

is referred to as "the system". When linearized at each operating condition, the system is defined by equations 8 and 9.

$$x(t + \Delta T) = Ax(t) + Bu(t) \tag{8}$$

$$u(t) = \begin{bmatrix} r(t) \\ d(t) \end{bmatrix} \tag{9}$$

where:

$x(t)$ = the state vector of the system

$u(t)$ = the input vector of the system

A = the system transition matrix

B = the system input matrix

a = elements of A

b = elements of B

The adaptive algorithm, Figure (2), is used to cause the system to satisfy the following performance criteria:

- adjustment so that the deterministic response of c(t) to changes in r(t) about any system operating condition defined by the vectors x(t) and u(t), will be unique. This implies maintaining the elements a and b to constant values or maintaining equalities between these elements. The form of Φ and the values of a and b to be maintained are not implied by the adjustment technique.

- adjustment so that A and B are dependent on u(t) in a deterministic manner. Clearly this criterion would have to be applied separately of the first criterion.

- adjustment so that selected elements a and b are zero as required to eliminate interaction between the variables in c(t). This criterion is optional.

All of these criteria are to be applied in such a manner that the dimension of the vector o(t) is minimized. Reference 5 is recommended to readers not familiar with state vector notation.

IDENTIFICATION

Identification of parameter changes in A and B is derived in this scheme by recognizing that these changes affect the transmission of power in the disturbance vector to the error vector $\epsilon(t)$ in a non-uniform manner with respect to frequency. To apply this identification technique, digital filters are used to divide each variable $\epsilon_i(t)$ in the Ith order vector $[\epsilon(t)]$ into (MxN) signals $\epsilon_{mn}(t)$. Equations 10 and 11 are descriptive of the filtering algorithms that have been used to obtain these coefficients. Although these filtering algorithms are adequate, they are not implied to be optimum even from a computational standpoint.

$$\epsilon_{mn}(t) = c_1 \epsilon_i(t) + c_2 \epsilon_{mn}(t - \Delta T) + \_\_ + c_{(p+1)} \epsilon_{mn}(t - p \Delta T) \tag{10}$$

where:

p is an integer denoting the sharpness of filtering and c_1 through $c_{(p+1)}$ are constants that may be derived from equation 11.

$$\epsilon_{mn}(t) = \epsilon_i(t) \left[\left(\frac{\alpha_n}{1 - (1 - \alpha_n) Z^{-1}} \right)^p - \left(\frac{\alpha_{(n-1)}}{1 - (1 - \alpha_{(n-1)}) Z^{-1}} \right)^p \right] \tag{11}$$

where:

$\alpha_n = \omega_n \Delta T \leq 1.0$

$\omega_{(n-1)}$ to ω_n = the bandwidth of $\epsilon_{mn}(t)$, (rad./unit time)

ω_o = zero frequency

z^{-p} = an operator that denotes $\left(z^{-p} \epsilon_{mn}(t) \equiv \epsilon_{mn}(t - p\Delta T) \right)$

Having developed an MxN array of elements ϵ_{mn} each of these elements is converted to a power estimate (either by squaring or rectification) and weighted to develop an MxN array of elements p_{mn} as indicated in equation 12.

215

$$p_{mn} = (\underline{\pm}) \, w_{mn} |\epsilon_{mn}| \quad \text{or/}$$

$$p_{mn} = (\underline{\pm}) \, w_{mn} \, (\epsilon_{mn})^2 \tag{12}$$

where:

w_{mn} = linear weighting parameter in the adaptive algorithm.

The p_{mn} array is conditioned by the nonlinear weighting terms w^*_{mn} and a summation of these weighted elements is performed to develop an M dimensioned vector $[\epsilon a]$;

$$\epsilon a_m = \sum_{n=1}^{N} w^*_{mn} p_{mn} (\underline{\pm}) \, sd_m \tag{13}$$

where:

w^*_{mn} = weighting determined by a nonlinear algorithm

sd_m = stationary disturbance estimate

With the vector $[\epsilon a]$ defined, the identification is complete. Non-zero values for the element ϵa_m indicate an error in the parameter ϕ_m and the sign of this non-zero element will indicate if the magnitude of ϕ_m should be increased or decreased. Having formulated the structure of the identification scheme, the design for a particular application is reduced to establishing the dimensions of M, N and P; numerical values for α_n, w_{mn} and sd_m; and the algorithms desired to generate w^*_{mn}. Necessary conditions on the feasibility of the adaptive adjustment are expressed in the inequalities of equations 14 and 15.

It is necessary that a p_{mn} exists such that:

$$\frac{\partial P_{mn}}{\partial \phi_m} > \frac{\partial P_{m(n(\underline{\pm})c)}}{\partial \phi_m} \tag{14}$$

where:

c = is a nonzero integer
p^*_{mn} = the p_{mn} with the above property

(equation 14 implies that p^*_{mn} is the most sensitive p_{mn} with respect to variations in ϕm)

It is necessary that a p^*_{mn} exists such that:

$$\frac{\partial P^*_{mn}}{\partial \phi_m} > \frac{\partial P^*_{mn}}{\partial \phi(m(\underline{\pm})c)} \tag{15}$$

where:

c = is a nonzero integer
(equation 15 implies that p^*_{mn} is more sensitive to variations in the parameter it is being used to adjust than it is to variations in the other ϕ_m's)

To dimension M it is necessary to ascertain how many elements in the control matrices (Φ's) require adaptive identification for their adjustment. Furthermore, it is important to define a minimum M. It is computationally inefficient to identify adaptively elements which may be computed more directly using known relations between the adaptively computed variables and other variables available in the computer memory.

The dimensions required for N and P are influenced by the complexity of the performance criteria and the complexity of identifying the non-stationary d(t). The polarity of the w_{mn} associated with p^*_{mn} is of opposite sign to the polarity of all other w_{mn}'s, and sd_m. The scaling of α_n, w_{mn} and sd_m is performed so that $\epsilon a_m = 0$ at all operating conditions when ϕ_m is at the desired value.

The nonlinear terms w^*_{mn} provide potential nonlinear filtering, limiting functions and alarm checks to be used as needed. An example is a system where loss of control in the process can be detected by large numbers in the p_{mn} array. These large numbers can be used to stop the identification until the system is again in control.

The important question of the dimensions of N and P is resolved when the properties of d(t) are determined either analytically or experimentally. If d(t) is stationary both with respect to its power content and statistical description then N may be reduced to two and P to one. However, it appears that if a time shared filter routine is desired for general application on a variety of processes with a variety of non-sta-

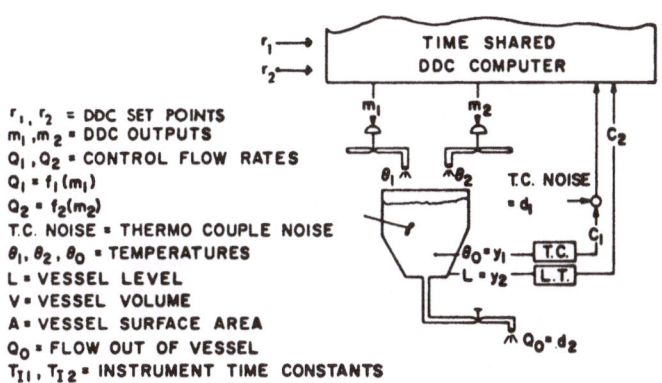

r_1, r_2 = DDC SET POINTS
m_1, m_2 = DDC OUTPUTS
Q_1, Q_2 = CONTROL FLOW RATES
$Q_1 = f_1(m_1)$
$Q_2 = f_2(m_2)$
T.C. NOISE = THERMO COUPLE NOISE
$\theta_1, \theta_2, \theta_0$ = TEMPERATURES
L = VESSEL LEVEL
V = VESSEL VOLUME
A = VESSEL SURFACE AREA
Q_0 = FLOW OUT OF VESSEL
T_{I1}, T_{I2} = INSTRUMENT TIME CONSTANTS

(PROCESS EQUATION) ASSUMING MIXING, NO EXTERNAL HEAT LOSS
& CONSTANT FLUID PROPERTIES

$$
\begin{bmatrix} \overset{o}{y_1} \\ y_2 \end{bmatrix} =
\underbrace{\begin{bmatrix} \left(-\dfrac{d_2}{V}\right) & 0 \\ 0 & 0 \end{bmatrix}}_{f(\ )}
\begin{bmatrix} y_1 \\ y_2 \end{bmatrix} +
\underbrace{\begin{bmatrix} \left(\dfrac{f_1(m_1)\theta_1}{V}\right) & \left(\dfrac{f_2(m_2)\theta_2}{V}\right) \\ \left(\dfrac{f_1(m_1)}{A}\right) & \left(\dfrac{f_2(m_2)}{A}\right) \end{bmatrix}}_{g(\ )}
\begin{bmatrix} m_1 \\ m_2 \end{bmatrix}
$$

(INSTRUMENT EQUATIONS)

$$
\begin{bmatrix} \overset{o}{c_1} \\ c_2 \end{bmatrix} =
\underbrace{\begin{bmatrix} \left(-\dfrac{1}{T_{I1}}\right) & 0 & \left(\dfrac{1}{T_{I1}}\right) & 0 \\ 0 & \left(-\dfrac{1}{T_{I2}}\right) & 0 & \left(\dfrac{1}{T_{I2}}\right) \end{bmatrix}}_{f^*(\ )}
\begin{bmatrix} c_1 \\ c_2 \\ y_1 \\ y_2 \end{bmatrix}
$$

FIG. 3. TEMPERATURE—LEVEL CONTROL EXAMPLE

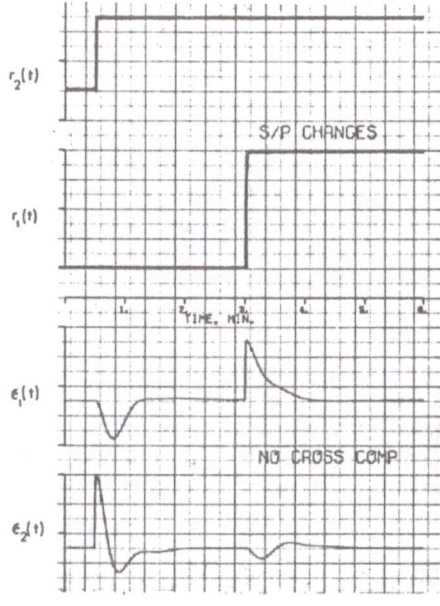

FIG. 4. DESIRED PROCESS RESPONSES WITHOUT
CONTROLLER CROSS COMPENSATION

217

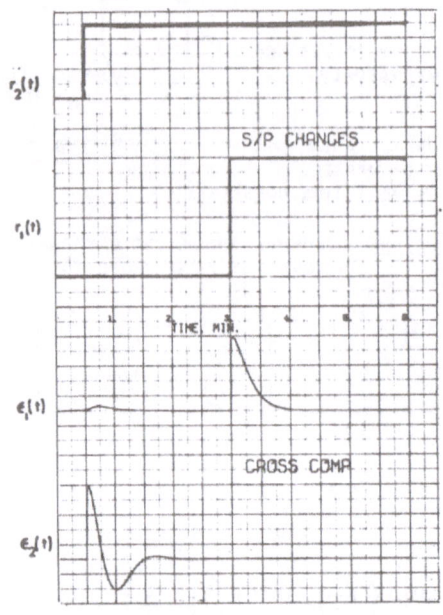

FIG. 5 DESIRED PROCESS RESPONSES WITH
CONTROLLER CROSS COMPENSATION

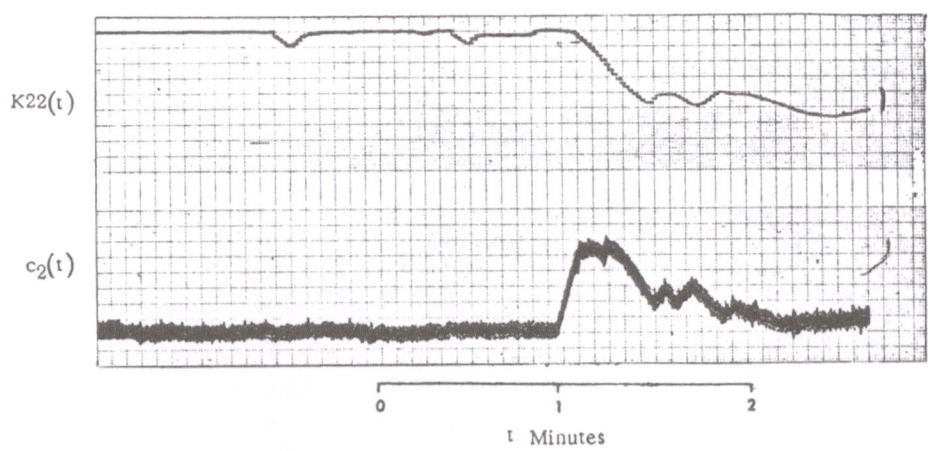

FIG. 6. ADAPTIVE ADJUSTMENT OF K22 FOR A CHANGE IN $\partial f_2(m_2)/\partial m_2$ RESULTING FROM
A STEP CHANGE IN $\partial_2(t)$ AT $(t) - 1.0$ minutes.

218

tionary d(t)'s the dimension of N and P will be greater than three.

IMPLEMENTATION

With the $[\epsilon\, a]$ vector established, M error signals are available for controlling M variables ϕ_m. DDC algorithms with reset and proportional action for lead compensation are used in conjunction with nonlinear output conditioning to close the adaptive control loops in a classical feedback fashion;

$$\phi_m(i\Delta T) = \left[\left(C_1\, \epsilon_m(i\Delta T) + \sum_i C_2\, \epsilon_m(i\Delta T)\right)^2\right]^{HB}_{LB}$$
(16)

where:

$i = 0, 1, 2 - - -$

c_1, c_2 are tuning constants ;

HB and LB, high and low bounds on the output ϕ_m.

The coefficients c_1, c_2 are adjusted heuristically until the speed of adjustment is adequate but slow enough to avoid a significant interaction with the basic system. Figure 2 provides a diagram of the total signal flow in this adaptive algorithm. This algorithm has been programmed for implementation on a binary computer with 16 bit word length and 4×10^{-6} sec. cycle time. The dimension of N was three and the dimension of P was three. This algorithm requires 300 plus 50M words in core. The run time per computation of the ϕ_m vector is approximately $3M*10^{-3}$ seconds.

EXAMPLE

The nonlinear, interacting process defined in Figure 3 will be used as an example to demonstrate this adaptive technique. The DDC algorithm controlling this process is defined by the following control matrices:

$$\Phi_1 = \left[(1)\ (1) \right], \quad \Phi_2 = \left[(1/T1)\ (1/T2) \right]$$

$$\Phi_3 = \begin{bmatrix} (K11) & (-K21) \\ (-K12) & (K22) \end{bmatrix}, \quad \Phi_4 = \Phi_3$$

The dimension of the system state vector x(t) in this example is six. The dimension of the input vector u(t) is four. Two performance criteria can be developed. The first criterion requires that the A and B matrices remain invariant so that the system will respond, at all operating conditions, as shown in Figure 4. This requires adaptive adjustment so that equations 17, 18 and 19 are satisfied.

$$L1G = K11 \cdot \left(\partial f_1(m_1)/\partial m_1\right) \cdot \left(\theta_1 - \theta_2\right) / \left(Q_o \cdot T1\right)$$
(17)

$$L2G = K22 \cdot \left(\partial f_2(m_2)/\partial m_2\right) \cdot \left((\theta_1 - \theta_2)/\theta_1\right) / \left(A \cdot T2\right)$$
(18)

$$T1 = V/Q_o$$
(19)

where:

L1G, L2G are constants

The second criterion requires that the A and B matrices remain invariant and are non-interacting with respect to the controlled variables so that the system will respond, at all operating conditions, as shown in Figure 5. This requires adaptive adjustment so that equations 17, 18, 19, 20 and 21 are satisfied.

$$K12 = \left((\partial f_1(m_1)/\partial m_1)/(\partial f_2(m_2)/\partial m_2)\right) \cdot K11$$
(20)

$$K21 = \left((\partial f_2(m_2)/\partial m_2)\theta_2/(\partial f_1(m_1)/\partial m_1) \cdot \theta_1\right) \cdot K22$$
(21)

REFERENCES

1. Bakke, R. M., "Adaptive Gain Tuning Applied to Process Control," I.S.A. Paper #3.2-1-64, October 1964.

2. Mishkin, Braun, and Ludwig, "Adaptive Control Systems," McGraw-Hill Book Co., New York, 1961.

3. Aseltine, J. A., Mancini, A. R., and Sarture, C. W., "A Survey of Adaptive Control Systems," I.R.E. Trans. on Automatic Control, December 1958.

4. Klock, H. F., and Schoeffler, J. D., "Direct Digital Control at the Threshold," Industrial Electronics, March 23, 1964.

5. Tou, J., "Modern Control Theory," McGraw-Hill Book Co., 1964.

OPTIMAL AND MULTILEVEL CONTROL SYSTEMS

SEMIDYNAMIC OPTIMAL CONTROL

by J.G. Balchen - ISA Member
Professor of Control Engineering
The Technical University of Norway
Trondheim, Norway

and A.B. Aune
Research Engineer
The Technical University of Norway
Trondheim, Norway

SUMMARY

The paper treats a method for developing a control strategy for a continuously operating nonlinear process. The strategy is shown to be optimal in the steady state and for small disturbances. It is near optimal for larger disturbances.

1. GENERAL PROBLEM

This paper has two purposes. First to show how a near optimal or "semidynamic" optimal control strategy can easily be formulated for a multivariable nonlinear process with a non-quadric performance index. Secondly it is shown how conventional control concepts can be extended to encompass the notion of a performance index and yield an optimal system. Thus the gap between the old and the new way of control system synthesis is being bridged.

A process described by the following equations is considered

$$\dot{\underline{x}} = \underline{f}(\underline{x}, \underline{u}, \underline{v}) \tag{1}$$

$$\underline{y} = \underline{g}(\underline{x}, \underline{u}, \underline{v}) \tag{2}$$

where
$\underline{x}$ = statevector
$\underline{y}$ = output vector
$\underline{u}$ = control vector
$\underline{v}$ = disturbance vector
$\underline{f}$ and $\underline{g}$ are vector functions.

The process is regarded as time invariant. Furthermore it is assumed appropriate to measure the performance of the process by means of a criterion of the form

$$I = \int_{t}^{t+\tau} q(\underline{y}, \underline{u}, \underline{v}) \cdot d\xi \tag{3}$$

The problem is to find the timefunction $\underline{u}(t)$ that maximizes (or minimizes) $I(t)$.

At present this general problem has not been given an exact solution which can be expressed in closed form.

2. LINEAR SYSTEM WITH QUADRATIC q

If however the process equations (1) and (2) are linear

$$\dot{\underline{x}} = A\underline{x} + B\underline{u} + C\underline{v} \tag{4}$$

$$\underline{y} = D\underline{x} + E\underline{u} + F\underline{v} \tag{5}$$

and $q(\underline{y}, \underline{u}, \underline{v})$ is quadratic and $\tau \to \infty$ the solution is simple and can be found by using variational principles [1]. One finds

$$\underline{u} = M\underline{x} + K\underline{v} \tag{6}$$

where

M is constant feedback controller matrix.

K is constant feedforward controller matrix.

In order to put this control law into practice both the state vector and the disturbance vector have to be measurable and the process characteristics have to be known.

The control law of (6) as shown in fig. 1 can be converted into an equivalent feedforward operation

$$\underline{u}(p) = H(p)\underline{v}(p) \qquad (p = \frac{d}{dt}) \tag{7}$$

where the dynamic matrix H(p) has the following properties

$H(p) \to Q_1$ = constant matrix when $p \to \infty$

$H(p) \to Q_2$ = constant matrix when $p \to 0$

The contents of (7) is shown in fig. 2.

Now, consider the process to be in the

Superior numbers refer to similarly-numbered references at the end of this paper

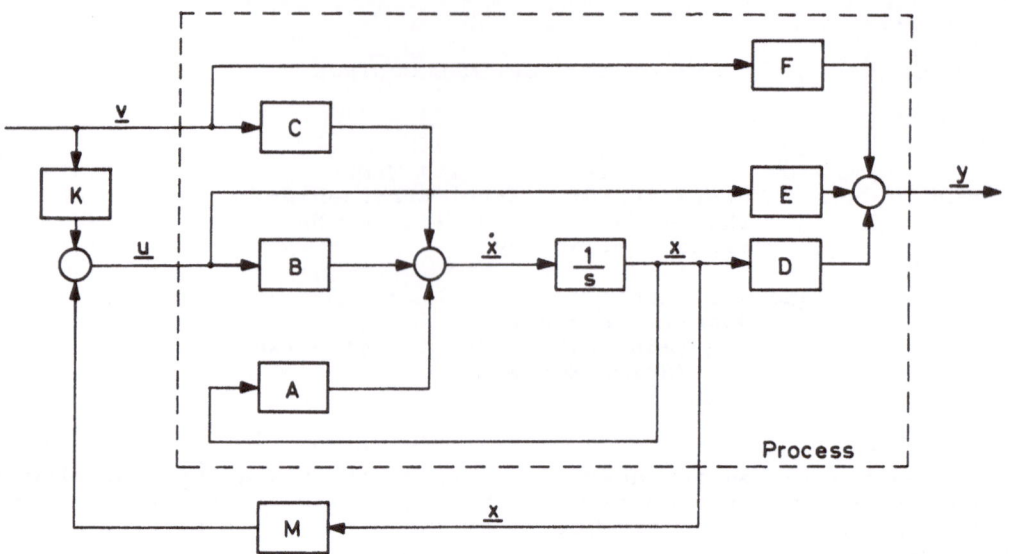

Fig. 1. Block diagram of optimal control
law for linear system.

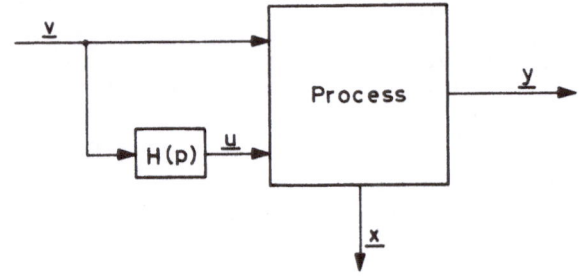

Fig. 2. The feedforward equivalent of
Fig. 1.

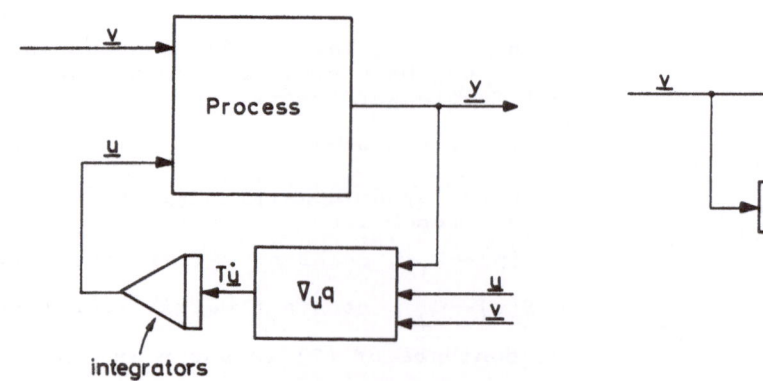

Fig. 3. Gradient feedback system
of eq. (10).

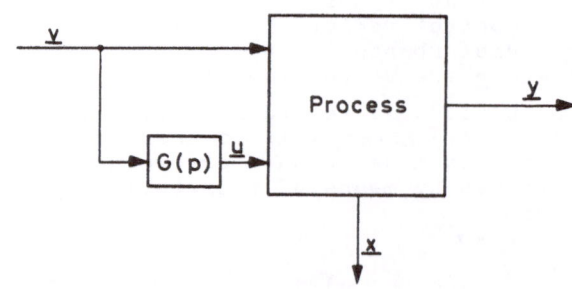

Fig. 4. The feedforward equivalent of
Fig. 3.

224

steady state. This means that $\underline{v}(t)$ is constant and thus $\underline{\dot{x}} = 0$. The optimal control vector under such circumstances can be found by simply requiring that the gradient of q with respect to $\underline{u}$ is zero

$$\nabla_u q = \left\{ \frac{dq}{du_j} \right\} = \underline{0} \qquad (8)$$

The elements of this gradient vector may be expressed in terms of the measurable output vector $\underline{y}$

$$\frac{dq}{du_j} = \sum_{i=1}^{m} \frac{\partial q}{\partial y_i} \cdot \frac{\partial y_i}{\partial u_j} + \frac{\partial q}{\partial u_j} \qquad (9)$$

where m is the number of outputvariables. The conditions of (8) can be satisfied by employing a steepest descent feedback control of the form

$$\nabla_u q = T\underline{\dot{u}} \qquad (T = \text{constant matrix}) \quad (10)$$

which in the steady state is identical to (8). Generally $\nabla_u q$ will be a nonlinear function of $\underline{y}$, $\underline{u}$ and $\underline{v}$. In the case of linear process and quadratic q the control law of (10) will become a set of linear equations.

Such a control law is shown schematically in fig. 3. If this system is being disturbed ($\underline{v}(t)$, its behaviour can be expressed in terms of an equivalent feedforward operation as shown in fig. 4.

$$\underline{u}(p) = G(p) \cdot \underline{v}(p) \qquad (11)$$

The dynamic matrix $G(p)$ now has the properties $G(p) \rightarrow P_1 = 0$ when $p \rightarrow \infty$

$$G(p) \rightarrow P_2 = \mathbf{Q}_2 \quad \text{when } p \rightarrow 0.$$

The last equality simply states that the systems of fig. 1 and fig. 3 give the same solution in the steady state.

Now the problem is how to modify the control system of fig. 3, so that it will yield optimal performance under dynamic conditions as well. Simple reasoning based on fig. 2 and fig. 4 leads to the result that the scheme of fig. 3 must be furnished with a feedforward operator of the form

$$R(p) = H(p) - G(p) \qquad (12)$$

as shown in fig. 5.

The new matrix operator $R(p)$ has the important properties that

$$R(p) \rightarrow H(p) \rightarrow \mathbf{Q}_1 \qquad \text{when } p \rightarrow \infty$$

$$R(p) \rightarrow 0 \qquad \text{when } p \rightarrow 0$$

This implies that corrective action generated from the disturbance vector has a transient character, i.e. $R(p)$ is a matrix of high pass filters.

Clearly $R(p)$ is dependent upon the integration rate of the integrators involved in fig. 5. If the integrator timeconstants $T = \{T_j\}$ as introduced in (10) are given large values the result is that the system of fig. 3 will be very stable and sluggish. Small values of $\{T_j\}$ on the other hand may lead to an oscillating system. Thus the rational functions of p constituting the matrix operator $R(p)$ will turn out to have complex zeroes when the T_j's are small.

Rather than compensating the effects of an oscillatory feedback loop by means of the "invers" operation in the feed-forward path the normal procedure is to introduce transient stabilizing feedback. The system of fig. 5 then can be converted into an equivalent system (fig. 6) which combines transient feedback and transient feedforward in addition to the "gradient"-loop. Clearly the two new matrix operators $R_1(p)$ and $R_2(p)$ can not be uniquely determined from $R(p)$. This is so because there exist an infinite number of combinations of $R_1(p)$ and $R_2(p)$ that can be converted back to an equivalent $R(p)$. Thus a sound method seems to be to choose a convenient form of the transient feedback operator $R_2(p)$ and put whatever is left neccesary (and possible) in $R_1(p)$.

Actually the operator $R_2(p)$ can be thought of as a set of ordinar proportional feedback controllers with "condensers" connected in series with the inputs. The operator $R_1(p)$ may be regarded as a speedup facility which helps the feedback loops in producing an appropriate fast action in the control vector $\underline{u}$ from a measurement of $\underline{v}$.

3. NON-LINEAR PROCESS WITH NON-QUADRA-
TIC CRITERION q

Now we may return to the original problem presented in eqs. (1), (2) and (3). In principle the steady state relations between the control- and disturbance vectors ($\underline{u}$ and $\underline{v}$) and the output vector ($\underline{y}$) may be found from (1) and (2) in the form

$$\underline{y} = h(\underline{u}, \underline{v}) \qquad (13)$$

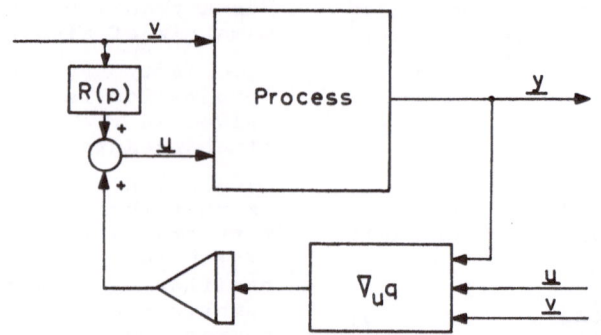

Fig. 5.   Optimal control law realized as
          combined gradient feedback and
          transient feedforward.

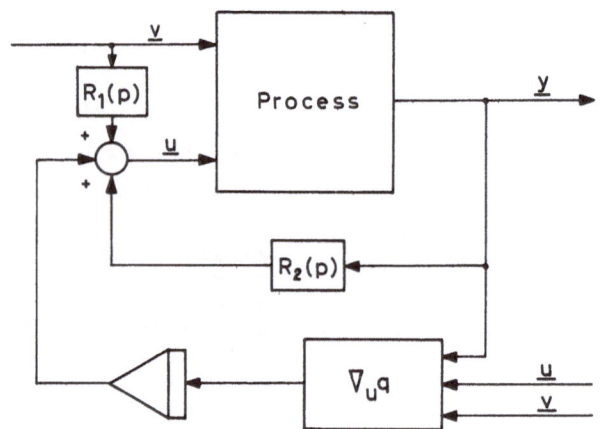

Fig. 6.   Same solution as Fig. 5 but
          transient action split in feed-
          back- and feedforward actions.

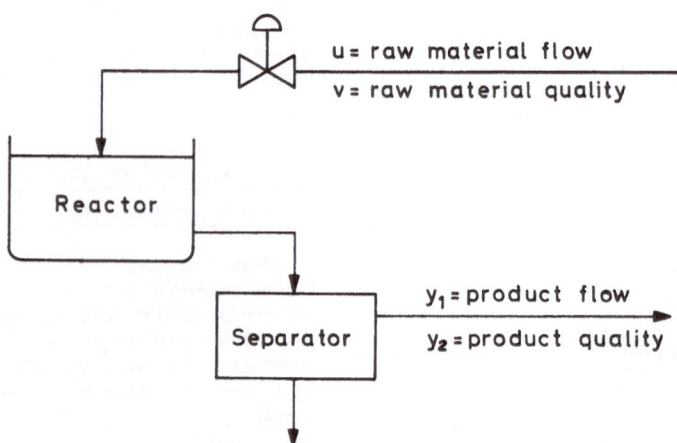

Fig. 7.   Example of reactor control.

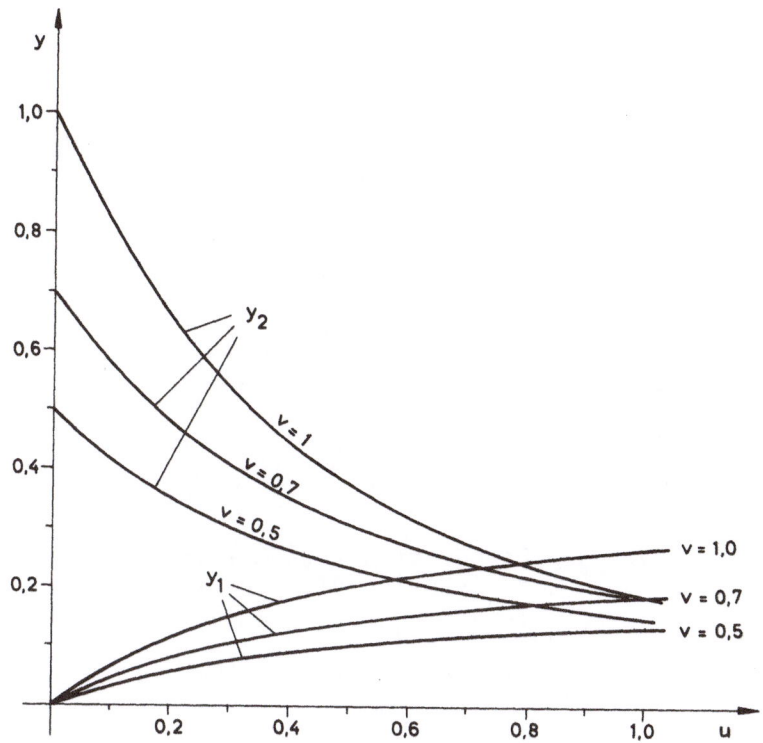

Fig. 8.   Steady state relations between
inputs and outputs.

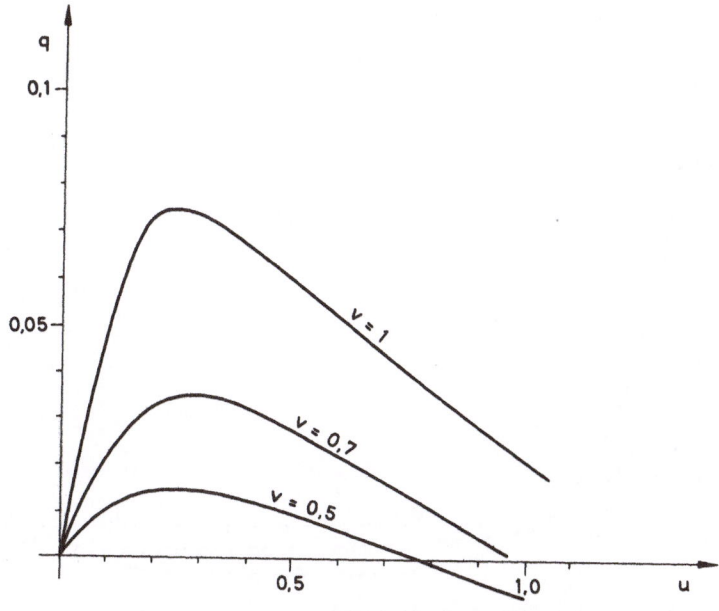

Fig. 9.   Steady state value of performance index $q = y_1 y_2 - 0.03 \cdot u$
as function of u and v.

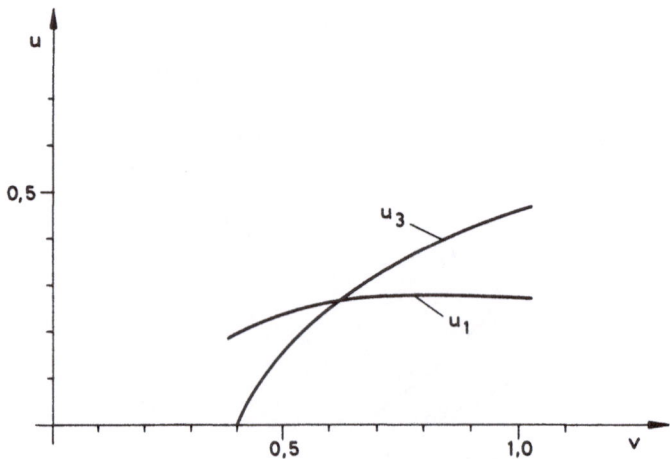

Fig. 10. Steady state optimal value of u
as function of v when:

$$(u_1) \; , \; q_1 = y_1 y_2 - 0{,}03 \cdot u$$

$$(u_3) \; , \; q_3 = y_1 - 40(y_2 - 0{,}4)^2$$

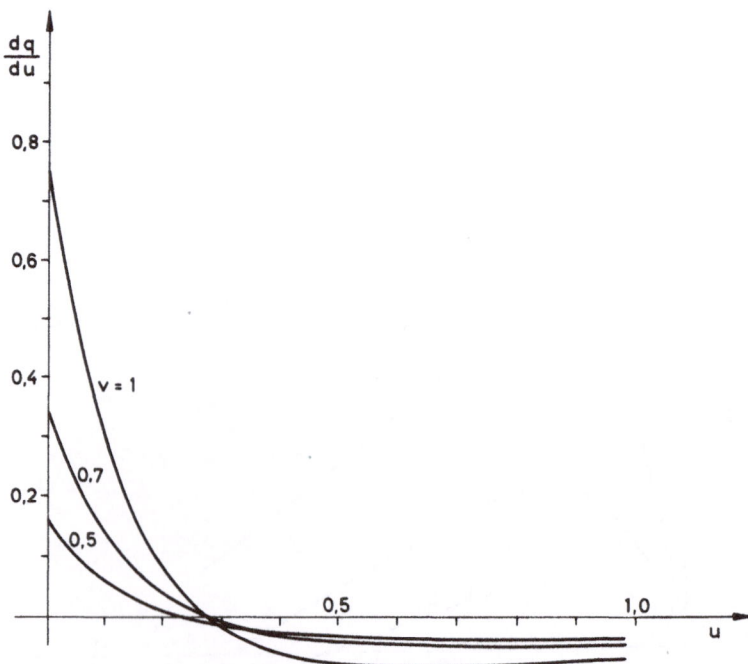

Fig. 11. The exact gradient as function
of u and v.

In practice it may be impossible to find $h(\underline{u}, \underline{v})$ as an exact analytic expression, but approximate expressions based partly on theoretical and partly on experimental data will in most cases be sufficient.

Clearly the gradient procedure of (10) applies equally well in case the process is nonlinear and the performance index q is nonquadratic provided (13) is formulated and $q(\underline{y}, \underline{u}, \underline{v})$ is given. Thus steady state optimal conditions may be found. Based on conventional linearization principles we now substitute a <u>linearized approximation</u> of (1) and (2) in the form of (4) and (5). Furthermore the non-quadratic performance index $q(\underline{y}, \underline{u}, \underline{v})$ is approximated by a <u>quadratic form</u> in the neighbourhood of normal operating conditions. Thus we may apply the method described in the preceeding section to find the dynamic feedforward matrix operator $R(p)$ or the equivalent operators $R_1(p)$ and $R_2(p)$.

This combination of a "gradient feedback" (based on the true steady state nonlinear process characteristics and the true performance index) and a transient matrix operator $R(p)$ (based on a linearized dynamic model) is called the "semi-dynamic optimal control law".

4. AN EXAMPLE

To illustrate this method consider the process sketched in fig. 7. This is a simplified chemical reactor assumed to operate at constant temperature. Based on a simplified description of the reaction mechanism the following matematical model may be formulated:

$$T_1 \dot{x}_1 = u(v - x_1) - k \cdot x_1 \qquad (14)$$

$$T_1 \dot{x}_2 = -ux_2 + kx_1 \qquad (15)$$

$$T_2 \dot{x}_3 = -x_3 + \eta u \cdot x_2 \qquad (16)$$

$$T_2 \dot{x}_4 = -x_4 + x_2(1 - \gamma x_1^2) \qquad (17)$$

$$y_1 = x_3 \qquad (18)$$

$$y_2 = x_4 \qquad (19)$$

where

$\{x_i\}$ = state variables

u = raw material input flow, controlable

v = raw material concentration, disturbance

y_1 = product flow

y_2 = product quality

T_1, T_2, k, η, γ are constants.

With the constant coefficients of (14), (15), (16) and (17), assuming the following numerical values:

$T_1 = 10$

$T_2 = 5$

$k = 0,5$

$\eta = 0,8$

$\gamma = 1$

the theoretical steady state relationship between inputs and outputs will be as shown in fig. 8. Real relationships as obtained from experimental tests may prove to be slightly different. If so corrective terms should be introduced in the process equations (14) - (19) (see ref [2]). A number of different performance indices may seem realistic for this kind of process:

1. $q_1 = y_1 \cdot y_2 - \alpha \cdot u \qquad \alpha = \text{constant} \quad (20)$
   yielding maximum profit assuming that changes in raw material quality does <u>not</u> change the price of raw material.

2. $q_2 = y_1 y_2 - \alpha \cdot u \cdot v \qquad (21)$
   when raw material price is proportional to quality.

3. $q_3 = y_1 - \beta \cdot (y_2 - y_{20})^2 \qquad (22)$
   $\beta = \text{constant}$
   yielding proper balance between maximum production and minimum deviation of quality from desired value (y_{20}).

The performance of the system is entirely dependent upon the type of performance index chosen.

In fig. 9 is shown how the system will perform (in the steady state) as judged by the index $q_1 = y_1 y_2 - \alpha u$ with $\alpha = 0.03$. In this case the controllable input flow u shall only be slightly adjusted in the range 0.23 to 0.28 when the input concentration v varies between 0.5 and 1.0. This is shown in fig. 10 by the curve labelled u_1. Nearly the same kind of behaviour would result if index $q_2 = y_1 y_2 - \alpha \cdot u \cdot v$ was used. This means that the process in this case is rather insensitive to the disturbance v. Or one may say that it has an inherent optimality.

If however the system is based on the index $q_3 = y_1 - \beta (y_2 - y_{20})^2$ with $\beta = 40$ and $y_{20} = 0.4$ another relationship between v and u as indicated in fig. 10

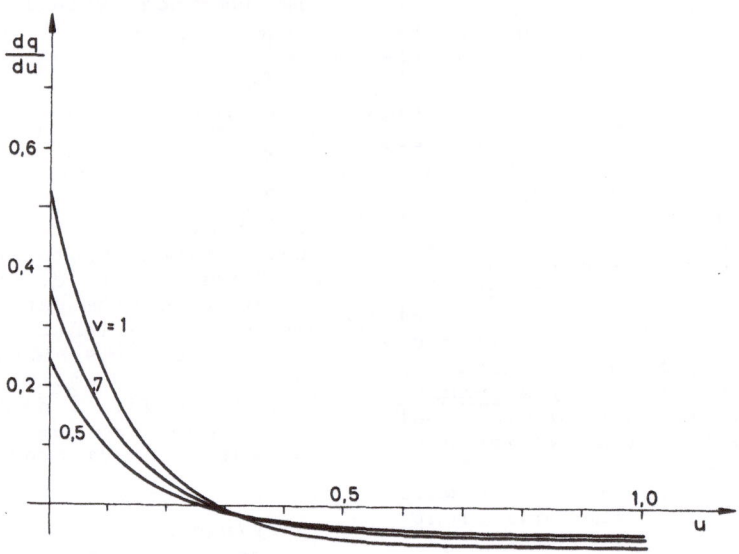

Fig. 12. The approximate gradient as
function of u and v when v is
estimated to be 0,7.

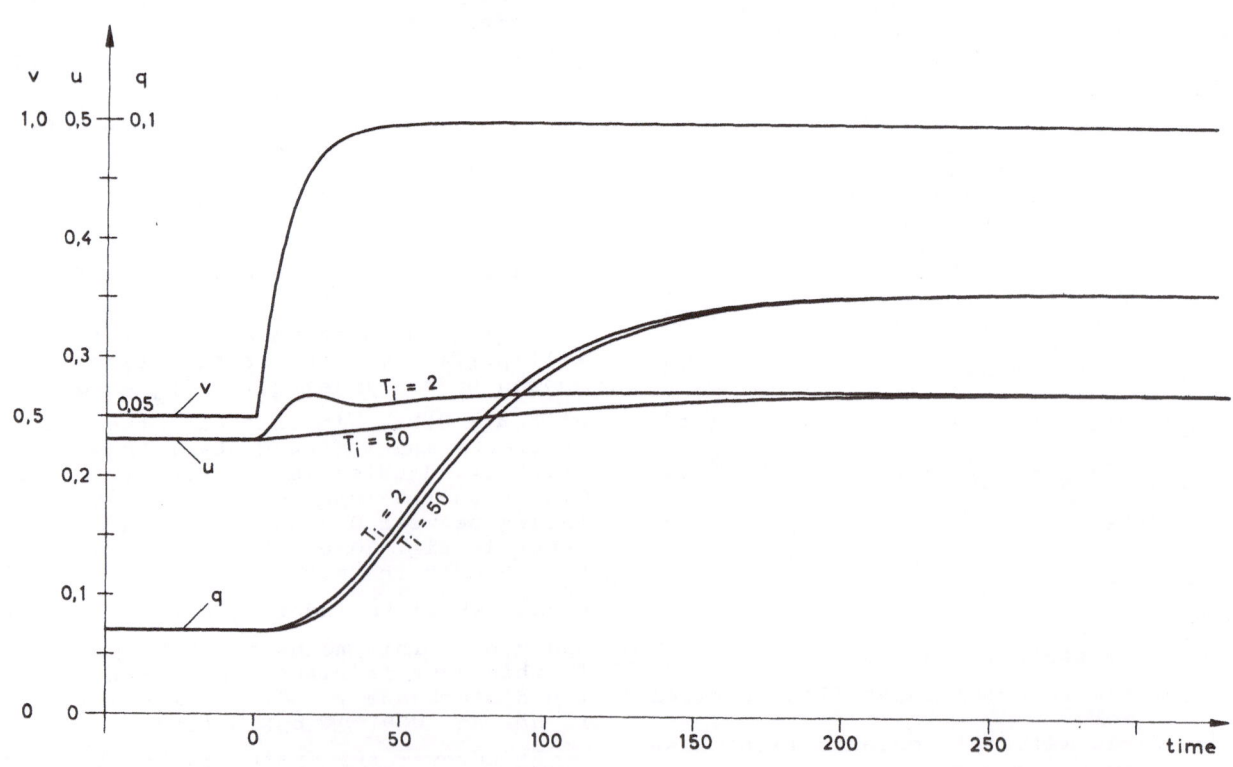

Fig. 13. Dynamic behaviour of semi-
dynamic optimal control law.

(curve u_3) is yielding steady state optimum.

Particular attention should be paid to the initial assumption that the disturbance v (or generally $\underline{v}$) is measurable. If v is not at all measurable we may still reach the steady state optimum. This will be the case provided v enters the process equations and the performance index in an additive manner. As the present method is based on a computational determination of the gradient $\nabla_u q$ the question arises how much error is introduced in the gradient due to imperfect detection of the disturbances.

Considering again the above example and using the index of (20) we obtain

$$\frac{dq}{du} = y_1 \cdot \frac{\partial y_2}{\partial u} + y_2 \cdot \frac{\partial y_1}{\partial u} - \alpha \qquad (23)$$

From the process equations we easily find

$$\frac{\partial y_1}{\partial u} = 0,2 \cdot v \cdot \frac{1}{(u + 0,5)^2} \qquad (24)$$

$$\frac{\partial y_2}{\partial u} = -0,5 \cdot v \cdot \frac{(u+0,5)^2 + u(1-u)v^2}{(u+0,5)^4} \qquad (25)$$

When no direct measurement of v is available we could use some estimated average value of v in (24) and (25). The result may still be a reasonably good estimate of the gradient (23) because the influence of v in y_1 and y_2 is measurable. The correct gradient $\frac{dq}{du}$ as a function of u is shown in fig.11.

In fig. 12 is shown the computed gradient according to (23) if v is not measurable but estimated to be v = 0,7. As is clearly seen this gradient computation leads to approximately the same result as the one shown in fig. 10 (curve u_1) when making $\frac{dq}{du} = 0$.

The dynamic behaviour of the above system should now be investigated. For this purpose the process equations (14)-(19) as well as the gradient feedback equations (10), (23)-(25) are programmed on a computer.

First of all the integrator timeconstant in the gradient loop should be established. Fig. 13 shows the responses in q(t) and u(t) when v(t) is a "soft step" from 0,5 to 1,0 (timeconstant = 10). Two sets of curves are given for integrator timeconstants T_i = 2 and T_i = 50. The value T_i = 2 apparently gives a slightly oscillatory gradient loop where-

as T_i = 50 yields a rather sluggish response. The differense between the two cases as measured in the q-response, however, is rather insignificant. This is quite typical and is certainly due to the above mentioned "inherent optimality". A reasonable choice thus seems to be T_i = 10.

Next the form of the feedforward operator R(p) should be chosen. Based on the fact that R(p) is a transient operator we assume a reasonable first approximation to be

$$R(p) = K_R \cdot \frac{T_R \cdot p}{1 + T_R \cdot p} \qquad (26)$$

Then we try different values of K_R and T_R and find that the system respons (q(t)) does not change noticably (within recorder line width). This again is so because the system has a strong self regulation. In other systems, however, the improving effect of R(p) may be more pronounced.

5. UPDATING OF SYSTEM PARAMETERS

In the computation of $\nabla_u q$ a number of system parameters are needed. These parameters refer to the steady state behaviour of the process. As stated in sec. 3 above an approximate analytic expression should be established for the relations between inputs and outputs. In this expression provisions should be made for experimental corrective updating. This corrective updating, if at all neccessary, may be performed in a number of ways.

One method is to adjust the corrective terms, observe the changes in q(t) and then hillclimb on the q-surface. Such a method would soon reveal which process parameters are critical to the performance of the system.

Another method is to update a programmed model of the process or parts of it against the actual behaviour of the process. The unknown system parameters may then be taken from the model.

At any rate parameter updating requires more computation. Therefore the possible improvements in system performance must be carefully valued against the added complexity.

6. CONCLUSIONS

The "semi-dynamic optimal control law" for a continuous process is believed to constitute a realizable compromise between strictly optimal control laws and

very primitive direct hillclimbing
methods. It is a comforting fact that
the method leads to a feedback system
which in special cases is identical to
a reasonable, conventional control scheme
for the same process. The dynamic properties of this control law are as near
the optimal as one would expect to come
in most practical cases.

ACKNOWLEDGEMENT

The research on which this paper is
based has been partly supported by the
Royal Norwegian Council for Scientific
and Industrial Research.

REFERENCES

[1] Aasma, F. and Balchen, J.G.:
    Optimal Control of Multivariable
    Processes.
    IFAC Symposium on the Theory of Ad-
    aptive and Optimal Systems,
    Rome, 1962
    (ISA-Publication)

[2] Aune, A.B.:
    On Near-Optimal Control of Continu-
    ous Processes.
    Thesis, Division of Automatic Con-
    trol, The Technical University of
    Norway,
    Trondheim, 1965

Discussion

Mr. M.J. Duckenfield (UK) asked for clarification
of equations 8 and 10 of the paper.

Prof. Balchen explained that his equation 8
expressed the condition of optimality in the
steady state, that the steady-state part of his
controller (equation 10) depended only on the
steady-state input-output relation of the process,
and that these relations, though complicated,
were much easier to find than the dynamically
optimal control law.

Mr. N.M. Dor (UK) asked what happened if the
overall gain of the steady-state relations was
unknown, as might happen in a practical chemical
reactor. Professor Balchen answered that an
estimate of the unknown gain would have to be
made, the cost of which usually obeyed a
square law in terms of accuracy. Perhaps if a
parameter was not known exactly it was best to
over-emphasise the transient part of the
controller.

Mr. A. Griffin (UK) asked about the stability
of the control law for large disturbances.
Prof. Balchen pointed out that the stability
depended on the gain T of the steady-state
feedback loop; we were free to choose T to
make the process stable; as G was a function
of T the transient part of the controller
would also change.

Written contributions to the discussion

Mr. N.M. Dor (U.K.)

Given a process P with control inputs $\underline{u}$ and
measurable outputs $\underline{y}$ together with the performance criterion (3) quoted in the paper, various
methods may be used to implement their scheme in
a practical application. Following the authors'
notation the semi-dynamic law (10) has validity
as an optimal control for the steady state and
requires that:

$$\frac{\partial q_i}{\partial u_i} = \sum_i \left(\frac{\partial q_i}{\partial y_i} \cdot \frac{\partial y_i}{\partial u_i} \right)$$

be found.

Now $\dfrac{\partial q_i}{\partial y_i}$ is usually known through the profit

structure. This in reality is an open problem of
supply and demand of materials, energy and
depreciation costs. It remains to find the

system gains $\dfrac{\partial y_i}{\partial u_i}$ and ensure an adequate response

to system changes. Three particular methods are
outlined below which have varying use according
to the time dependences of the variables concerned. These methods allow for automatic updating which for many of the gains is not
necessary.

232

[1] <u>Correlation analysis</u> according to the relationship:

$$\frac{\partial y_i}{\partial u_j} = \left\{ \frac{1}{T} \int_t^{T+t} y_i u_j dt \right\} \Big/ \left\{ \frac{1}{T} \int_t^{T+t} u_j^2 dt \right\} .$$

[2] <u>Model parameter updating</u> may be used where a speed-up of system response is required since parameters may often be used to reflect changes more quickly than the first method. The model is a partial system simulation since it need only calculate the particular gains required.

[3] <u>Feedback</u> can be used to alter the control variable (see figure)

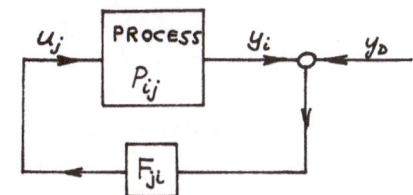

Then instead of

$$\frac{\partial y_i}{\partial u_j} = P_{ij}$$

we have $u = y_D$ and

$$\frac{\partial y_i}{\partial y_D} = \frac{P_{ij}F_{ji}}{1 + P_{ij}F_{ji}} \rightarrow 1$$

By choosing a high loop gain at low frequencies the gain may be maintained near unity despite system changes.

Where possible the last method shows most promise but can be misleading; deterioration of the system response requires changes in the integration times of the semi-dynamic optimal law (10). In these cases it is advisable to examine the possibilities in [2] or [1] which as open loop systems can have the fastest response times within the optimal closed loops.

Authors reply:

The authors highly appreciate the contribution by Mr. N. M. Dor and should like to make the following remarks.

The method [1] referred to would not seem to yield the proper value of the <u>steady state</u> gain $\partial y_i / \partial u_j$ except in cases of very slow changes in u_j. At any rate correlation methods will only yield a slowly responding measure of gain changes and do not take advantage of a priori knowledge about the process structure and parameters. Thus it seems more advantageous to insert a "constant" value for $\partial y_i / \partial u_j$ in the gradient computer and update this "constant" by some method of hillclimbing on the q-surface.

The authors feel that the method [2] is more meaningful and would give many advantages.

The method [3] seems to be misleading except when the control variables u_j do not enter into the q-function, i.e. when the control actions do not cost anything. Method [3] is a conventional control scheme where the setpoint y_D may be manipulated. This control scheme in fact would come out of the method given in the paper as a special case when u_j does not appear explicitly in the q-function. If as in method [3] the best thing is to bring y_i as close as possible to y_D then a reasonable q-function (for this part of the process) would be $q = \beta \cdot (y_i - y_D)^2$. The gradient feedback would then be given by

$$\frac{dq}{du_j} = 2\beta \cdot (y_i - y_D) \cdot \frac{\partial y_i}{\partial u_j} = T_j \cdot \dot{u}_j$$

Assuming $\frac{\partial y_i}{\partial u_j}$ to be some positive number (constant or varying) this relationship is identical to that of method [3] because

$$F_{ji}(p) = 2 \cdot \beta \cdot \frac{\partial y_i}{\partial u_j} \cdot \frac{1}{T_j \cdot p}$$

i.e. $F_{ji} \cdot (p)$ is an integral controller.

A METHOD FOR USING WEIGHTING FUNCTIONS AS SYSTEM DESCRIPTION IN OPTIMAL CONTROL

by J. Dawkins* and P. A. N. Briggs
National Physical Laboratory
Teddington, Middlesex, England

INTRODUCTION

A number of methods of linearised system dynamic identification, particularly those which do not require knowledge of the internal structure of the system, yield a system description in the form of weighting function (impulse response), or its Fourier transform frequency response.[1] However, for optimal control theory it is usual to assume a set of first order differential equations as system description, and there is very little information concerning the use of external descriptions like weighting functions in forming optimal control.

The method described in this paper follows that of Kramer,[2] which was developed originally for systems with time delays. With such systems a past history, that is, a set of values at previous times, rather than an instantaneous state vector value, is needed in order to predict the system behaviour. The same is true for a system specified in terms of a weighting function relating input and output values.

The essential feature of the method is to replace the problem of control of the current system output, by that of control of a predicted value of the output at some time, or times, in the future. The predicted value can be calculated from knowledge of the past inputs and the weighting function, and is used as input to a controller whose coefficients are derived by the familiar methods of Dynamic Programming.

A SIMPLE EXAMPLE

The method is best explained by means of a simple example, based on Kramer's paper. Consider the discrete time example shown in Figure 1, with equation

$$X_n = \sum_1^L W_s U_{n+s}$$

U_s is the input sequence; W_s is the set of weighting coefficients; and X_s is the output

sequence. As is usual in Dynamic Programming, the sequence points are numbered backwards from the terminal point (Figure 2). The criterion to be optimised is taken as

$$f_n^* = \min_{U_1 \cdots U_n} \left(c \sum_1^n U_s^2 + X_o^2 \right).$$

Define P_n as the value of X_o predicted from stage n assuming that no control is applied after that stage. Then

$$P_n = \sum_{n+1}^L W_s U_s$$

and

$$X_o = P_n + \sum_1^n W_s U_s.$$

This leads to a simple recurrence relation

$$P_{n-1} = P_n + W_n U_n.$$

It is now possible to perform optimisation stage by stage, starting at the terminal point, and using the properties of the predicted value P_n.

Stage 1:

$$f_1^* = \min_{U_1} \left(C U_1^2 + X_o^2 \right)$$

$$= \min_{U_1} \left(C U_1^2 + (P_1 + W_1 U_1)^2 \right).$$

This leads to

$$U_1^* = -\frac{W_1 P_1}{C+W_1^2}; \quad f_1^* = \frac{C P_1^2}{C+W_1^2}.$$

Superior numbers refer to similarly-numbered references at the end of this paper.

*Now with I.B.M. (United Kingdom) Limited

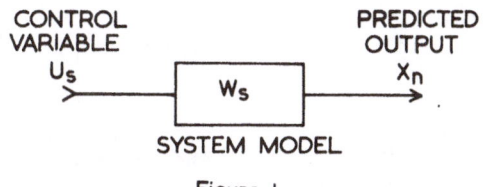

Figure 1.

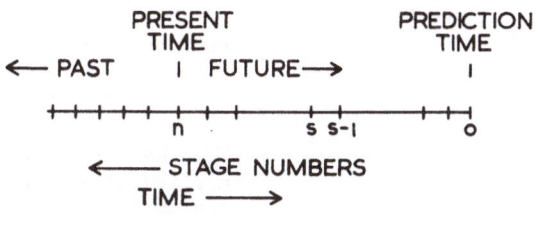

Figure 2.

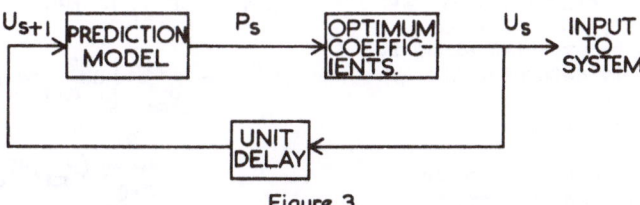

Figure 3.

Stage 2:

$$f_2^* = \min_{U_2} \left(CU_2^2 + f_1^* \right)$$

$$= \min_{U_2} \left(CU_2^2 + \frac{CP_1^2}{C+W_1^2} \right)$$

$$= \min_{U_2} \left(CU_2^2 + \frac{C(P_2+W_2U_2)^2}{C+W_1^2} \right).$$

This leads to

$$U_2^* = \frac{-W_2P_2}{C+W_1^2+W_2^2} \quad ; \quad f_2^* = \frac{CP_2^2}{C+W_1^2+W_2^2}.$$

By repeating the process, the general result may be written

$$U_n^* = \frac{-W_nP_n}{C+Q_n} \quad ; \quad f_n^* = \frac{CP_n^2}{C+Q_n}$$

where

$$Q_n = \sum_1^n W_s^2.$$

The control system based on this result, using P_n the predicted value of X, n stages ahead without control, is shown diagrammatically in Figure 3.

CONTROL IN THE PRESENCE OF DISTURBANCES

The controller shown in Figure 3 has, with respect to the actual system, an open loop configuration. This is to be expected since it was developed for an entirely deterministic system is not deterministic, the predictions of system behaviour will be inaccurate.

An identification procedure using actual system performance will allocate the error to uncertainty in the knowledge of parameters; including the system weighting coefficients and the parameters of random disturbances. An updating routine for these parameters can be devised[3] and it is by this means that feedback is built into the controller.

Thus, the optimal control system in the presence of random noise must include parameter identification and updating, and this is shown in block diagram form in Figure 4. The effective feedback gain will depend on both the actual disturbance in the system and on the expected disturbance.

MULTIPLE PREDICTION POINTS

Closer control of the system might be obtained by using a criterion function involving predicted values of X at two or more times in the future. For example, in the system of Figure 1, consider the use of the criterion function:

$$f_n^* = \min_{U_1 \dots U_n} \left(C \sum_1^n U_s^2 + X_m^2 + X_o^2 \right) \qquad n > m$$

This problem may be evaluated as before, if the value of X_m predicted at stage $n > m$, assuming no control is applied after that stage, is called R_n. Then

$$R_n = \sum_{n-m}^L W_s U_{s+m}$$

and there is a recurrence relation

$$R_{n-1} = R_n + W_{n-m}U_n.$$

Dynamic Programming methods can now be applied as before. For $s \leq m$ the results are the same

$$U_s^* = \frac{-W_sP_s}{C+Q_s} \quad ; \quad f_s^* = \frac{CP_s^2}{C+Q_s}$$

but at the $m+1$ stage X_m must be included in the criterion.

$$f_{m+1}^* = \min_{U_{m+1}} \left\{ CU_{m+1}^2 + X_m^2 + f_m^* \right\}$$

$$= \min_{U_{m+1}} \left\{ CU_{m+1}^2 + (R_{m+1}+W_1U_{m+1})^2 \right.$$

$$\left. + \frac{C}{C+Q_m}(P_{m+1}+W_{m+1}U_{m+1})^2 \right\}.$$

This leads to the optimal values

$$U_{m+1}^* = -\frac{CW_{m+1}P_{m+1} + (C+Q_m)R_{m+1}}{C^2 + C(Q_1+Q_{m+1}) + Q_1Q_m}$$

$$f_{m+1}^* = C\frac{\left\{ \begin{array}{c} (C+Q_1)P_{m+1}^2 - W_1W_{m+1}R_{m+1}P_{m+1} \\ + (C+Q_{m+1})R_{m+1}^2 \end{array} \right\}}{C^2 + C(Q_1 + Q_{m+1}) + Q_1Q_m}.$$

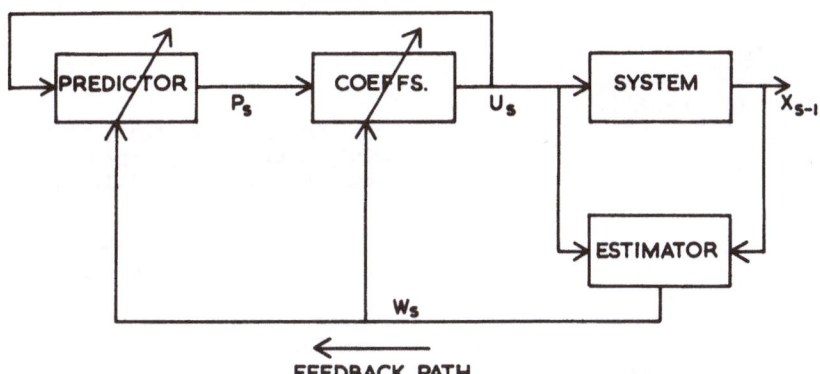

Figure 4.

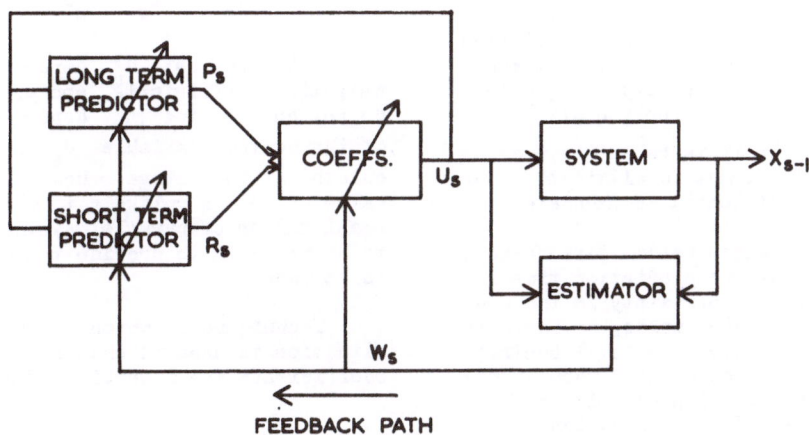

Figure 5.

237

The general form can be obtained as

$$U^*_{m+s} = -\frac{\left\{\begin{array}{l}\left[(C+Q_s)W_{m+s}-S_sW_s\right]P_{m+s} \\ + \left[(C+Q_{m+s})W_s-S_sW_{m+s}\right]R_{m+s}\end{array}\right\}}{\left[(C+Q_s)(C+Q_{m+s})-S_s^2\right]}$$

$$f^*_{m+s} = C\frac{\left\{\begin{array}{l}(C+Q_s)P^2_{m+s}-2S_sR_{m+s}P_{m+s} \\ + (C+Q_{m+s})R^2_{m+s}\end{array}\right\}}{\left[(C+Q_s)(C+Q_{m+s})-S_s^2\right]}$$

where

$$S_s = \sum_1^s W_s W_{m+s} \ .$$

Although these expressions are more complicated than in the single prediction case, it will be seen that they are of similar form.

A controller based on multiple prediction is shown in Figure 5. Again, feedback will be introduced through the system parameter identification procedure.

CHOICE OF PREDICTION TIME

So far the choice of a suitable prediction time has not been considered. If it is taken too far in the future, the corresponding value of W_n will be so small that only negligible control is exerted. If it is taken too near, then the system will attempt to eliminate errors so quickly that control demand is excessive.

In the practical case, random disturbances are present so that, as the prediction time is increased there is more uncertainty in the prediction. Without the random noise, an increase in prediction time decreases the total control power required since it can be used more effectively. Thus, for a criterion function which reflects both the cost of control action and the uncertainty of the prediction, the optimum values of the criterion will show a minimum with respect to prediction time. This would appear to give the most suitable value for a single prediction system

Additive noise may be accounted for in the following way, taken from Kramer (reference 2). Since now the model is

$$X_k = \sum_1^L W_s(U_{k+s} + V_{k+s})$$

where the V_s are random samples from a source of mean zero and variance σ^2, the value at terminal point may be taken in the form

$$X_o = P_n + \sum_1^n W_s(U_s + V_s)$$

where P_n is the expected value of X_o predicted at stage n assuming no further control. The recurrence relation for P_n becomes

$$P_{n-1} = P_n + W_n(U_n + V_n) \ .$$

Kramer shows that, if stage by stage minimisation using expected values is carried out, then the control law is unaffected

$$U^*_s = -\frac{W_s P_s}{C + Q_s}$$

whilst the cost function includes an additional term due to noise

$$f^*_s = \frac{CP^2_s}{C + Q_s} + \sigma^2 \sum_1^s \frac{CW^2_r}{C + Q_r}$$

which may be written as

$$f^*_s = A_s + B_s \sigma^2 \ .$$

While the set B_s depend only on the weighting coefficients, and are constant if these do not change, the A_s depend on the past values of the control variable U_s and must be recalculated at each stage. However, as the best value of n is unlikely to vary greatly, only a small number of parallel synthesisers are likely to be required to provide the data for the choice to be made.

Techniques in which the uncertainty of the situation is used to change the controllers coefficients are deserving of more attention.

CONCLUSIONS

Where linearised system descriptions in the form of weighting functions are available, optimal control policies may be obtained through the methods of dynamic programming, if the criterion is based on a predicted future value of system output. In order to provide feedback control, the system description must be updated by an identification process. Both optimal control and identification will require the use of computers in any practical system.

Computer investigation of the method is required in order to evaluate its various aspects. It is hoped to present some results of such studies in a future paper.

ACKNOWLEDGEMENT

The work described in this paper forms part of the research programme of the National Physical Laboratory and is published by permission of the Director of the Laboratory.

REFERENCES

(1) Briggs, P. A. N., Hammond, P. H., Hughes, M. T. G. and Plumb, G. O., "Correlation Analysis of Process Dynamics Using Pseudo-Random Binary Test Perturbations". Proc. Convention on Advances in Automatic Control, April 1965 Proc. Inst. Mech. Engrs.Vol. 179, part 3H, 1965

(2) Kramer, Jr., J. D. R., "On Control of Linear Systems with Time Lags". Information and Control 3 299-326 (1960).

(3) Dawkins, J., "Control Measurement in the Presence of Noise". Pts I and II, Journal of Electronics and Control. Vol. 15, no. 3, September 1963.

Discussion

Dr. D. Graupe (UK) commented that he had recently described a procedure (ref. 1) similar to that of the authors for finding the response to a step or impulse function. He had had difficulty in stabilising the procedure. What computational procedure did the authors recommend; could they give details of the range of information on which his predictions were based, and did they introduce a filter to eliminate noise?

Mr. Briggs replied that he had not yet made any computer studies. In a practical application he expected filtering to be important. One of the reasons for using the weighting function directly was that the weighting functions measured by the authors on a distillation column were noisy and not suitable for Fourier-transform analysis.

Reference

1. Graupe, D. and Cassir, G. "Adaptive Control System utilising sampling techniques and Aitkens Approximation", I.E.E. Electronics Letters, Vol. 1, No. 5, p.126, 1965

AN OPTIMAL CONTROL SYSTEM HAVING A NON-UNIQUE SOLUTION OF PONTRYAGIN'S EQUATIONS

by M. Masubuchi and H. Kanoh
Yokohama National University
Yokohama, Japan

ABSTRACT

L. S. Pontryagin's Maximum Principle is a very straightforward method for obtaining optimal control, but gives only solutions which satisfy a necessary condition of optimality. Therefore, solutions which are not truly optimal may be obtained. This paper presents a method of obtaining optimal trajectories, by means of a simple analytical example.

INTRODUCTION

Generally speaking, L. S. Pontryagin's Maximum Principle only gives solutions which satisfy a necessary condition of optimality, in optimal control problems. Therefore solutions which are not truly optimal may be included and they have to be eliminated. This problem often arises not merely in nonlinear systems but also in linear systems, depending on the selection of performance functionals and the boundary conditions[1].

This paper presents a method of obtaining a unique optimal solution in an analytical example. In the example the performance functional $\int_0^T |x|dt$ is minimized in the system

$$\ddot{x} = u , \quad |u| \leq 1 .$$

This system was previously analyzed by Fuller[2] and has been discussed by several authors[3-8] by means of the maximum principle or dynamic programming. But the system will be generalized by considering a performance functional of the form

$$\int_0^T (1 - w + w|x|)dt ,$$

where w is an arbitrary weighting, $0 \leq w \leq 1$. This functional will indicate the essential character of optimal solutions.

The results of this analysis show that when the solution by maximum principle is not unique there may be besides switching lines a separatrix which separates the control policy into two strategies: determination of this separatrix gives unique optimal trajectory.

ANALYSIS OF THE PROBLEM

Consider a simple linear system characterized by

$$\ddot{x} = u \tag{1}$$

where u is the control signal and is constrained by $|u| \leq 1$.

Determine the control signal u which minimizes the following performance functional, from the initial state $x(0)$ to the origin $x(T)$ of the state plane:

$$J = \int_0^T (1 - w + w|x|)dt , \quad 0 \leq w \leq 1 . \tag{2}$$

Let $x_1 = x$, then the differential equations for the system are

$$\dot{x}_1 = x_2 , \quad \dot{x}_2 = u \tag{3}$$

and $x_1(T) = x_2(T) = 0$.

The Hamiltonian for this system is[9]

$$H = p_1 x_2 + p_2 u - 1 + w - w|x_1| = 0 \tag{4}$$

and

$$\dot{p}_1 = w \, \text{sgn} \, x_1 , \quad \dot{p}_2 = -p_1 . \tag{5}$$

To apply the maximum principle, the Hamiltonian of the system is maximized with respect to u; that is

$$u = \text{sign} \, p_2 . \tag{6}$$

From eqn (4), at time T, namely $H_{(t=T)} = 0$, we have

$$p_2(T) = 1 - w, \quad \text{if} \quad u(T) = 1; \quad p_2(T) = -1 + w$$
$$\text{if} \quad u(T) = -1 . \tag{7}$$

For convenience, consider the case for $u(T) = 1$, and analyze the system from the origin in reverse time. In Fig. 1, start from 0, 0', which are respectively on the x- and p- planes, and assume that t_1 is the first switching time at points A, A'.

$$x_{1A} = \tfrac{1}{2}t_1^2 , \quad x_{2A} = t_1 \tag{8}$$

$$p_{10'} = \frac{1-w}{t_1} - \tfrac{1}{2}wt_1 , \quad p_{1A'} = \frac{1-w}{t_1} + \tfrac{1}{2}wt_1 , \quad p_{2A'} = 0 . \tag{9}$$

Next, let the time from the point A to the point B where the trajectory crosses the x_2 axis be t_2, then

$$x_{1B} = 0 , \quad x_{2B} = -\sqrt{2}t_1 \tag{10}$$

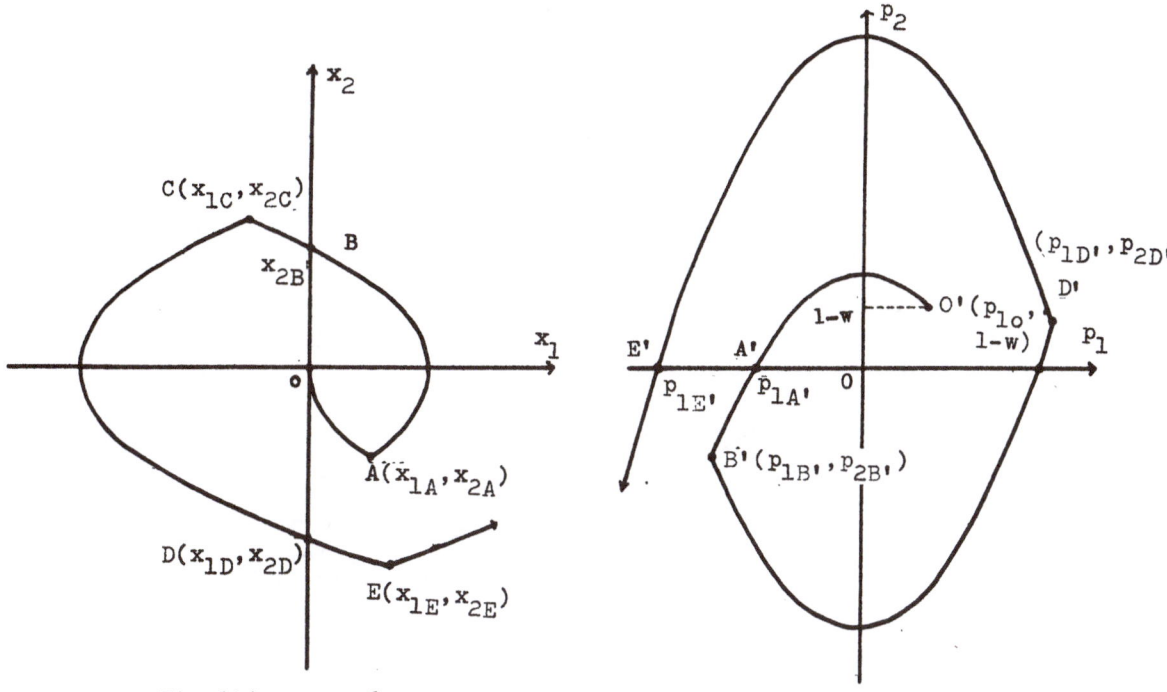

Fig. 1(a). $x_1 - x_2$ plane.

Fig. 1(b). $p_1 - p_2$ plane.

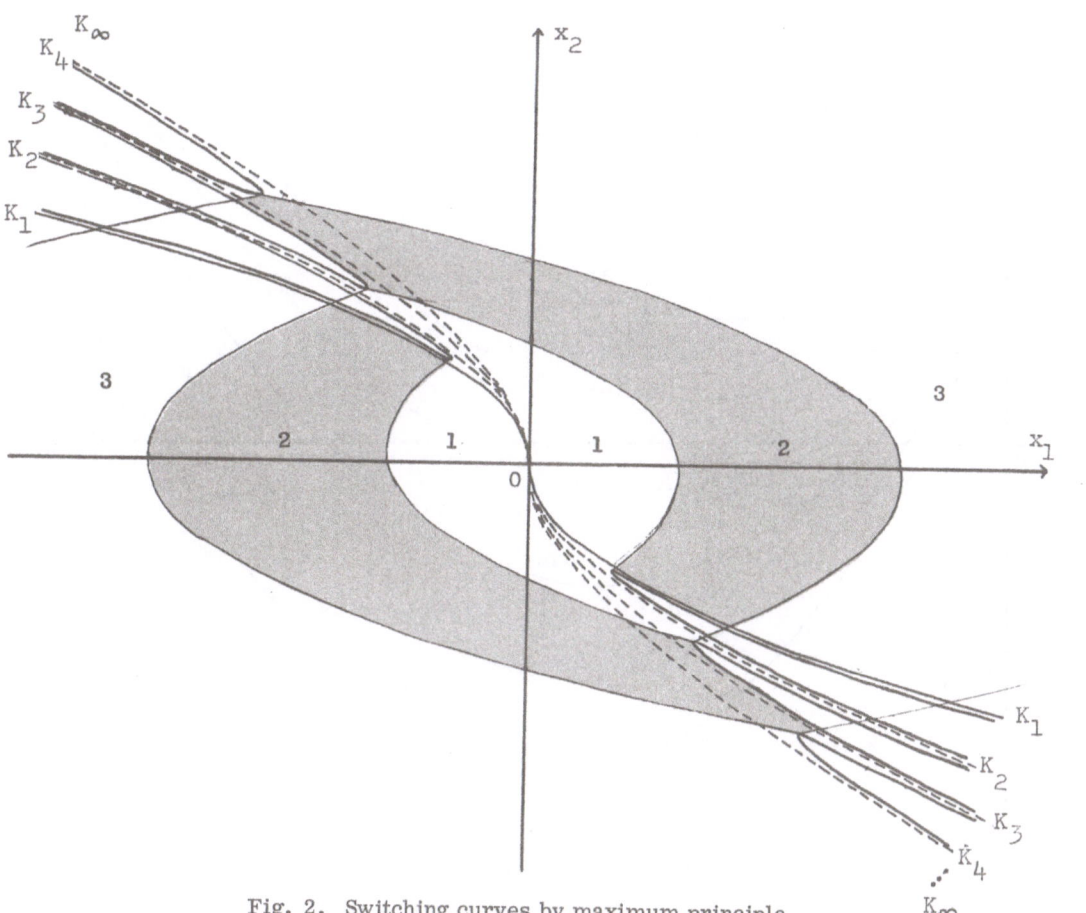

Fig. 2. Switching curves by maximum principle.

241

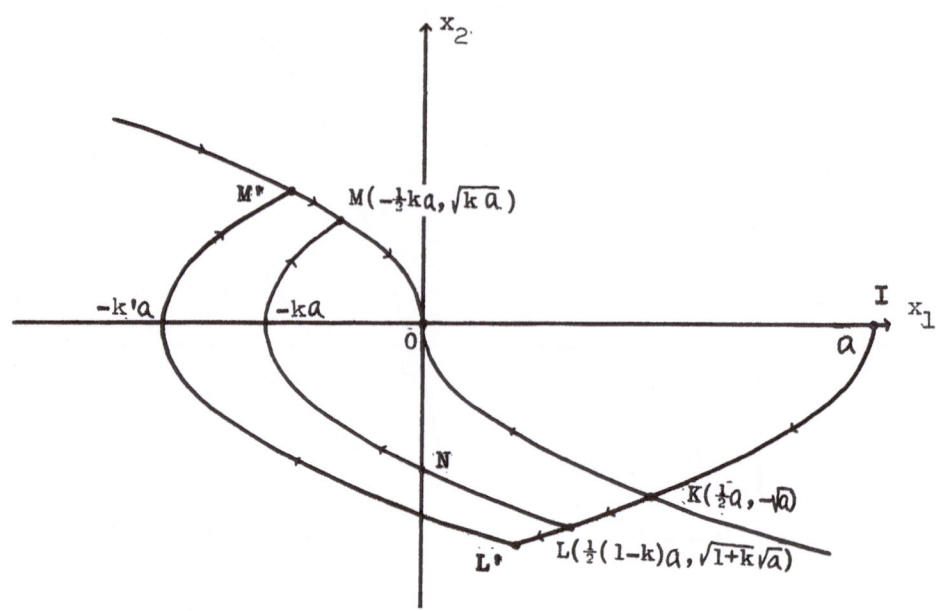

Fig. 3. Comparison of performance functional.

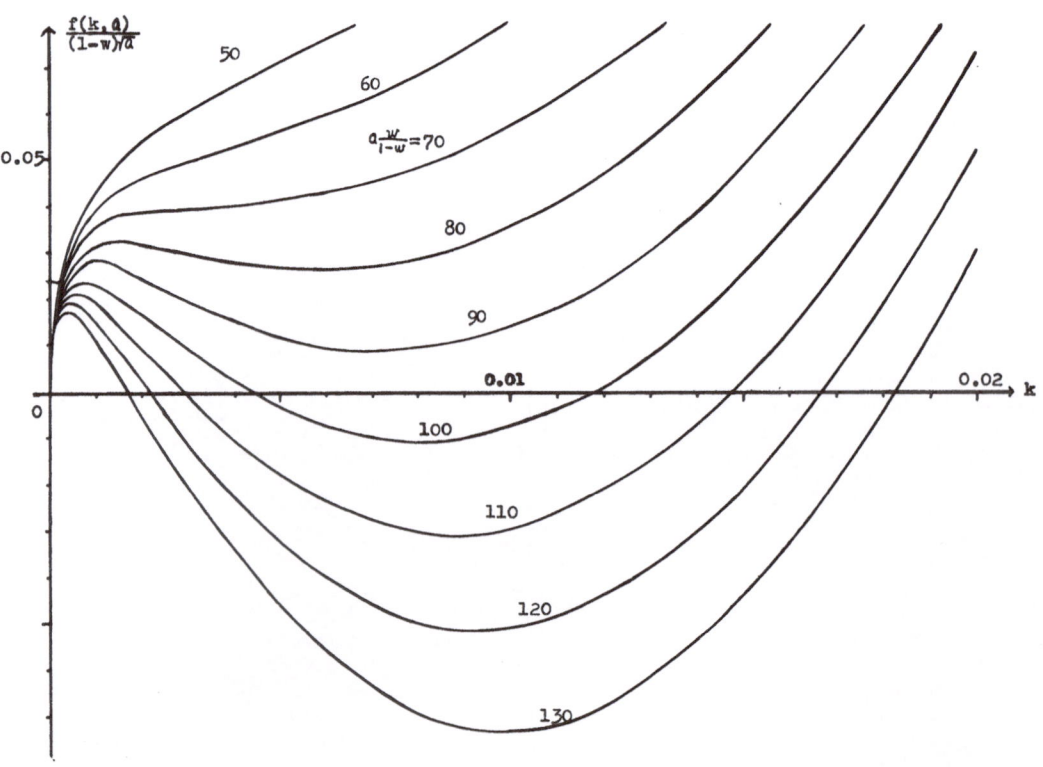

Fig. 4. Maximum and minimum values of f(k, a).

$$t_2 = (1 + \sqrt{2})t_1 \tag{11}$$

$$P_{1B'} = \frac{1-w}{t_1} + \frac{w}{2}(3 + 2\sqrt{2})t_1$$

$$P_{2B'} = -(1 + \sqrt{2})(1-w) - \frac{w}{2}(4 + 3\sqrt{2})t_1^2 \tag{12}$$

Also, let the time from the point B' to the point C' where the trajectory on the p_1–p_2 plane crosses the p_1 axis be t_3. Then from $P_{2o'} = 0$

we get

$$t_3 = \frac{1-w}{w}\frac{1}{t_1} + \frac{1}{2}(3 + 2\sqrt{2})t_1 + \frac{1}{t_1}\sqrt{\frac{1}{4}(33 + 24\sqrt{2})t_1^4 + \frac{1-w}{w}(5 + 4\sqrt{2})t_1^2 + \left(\frac{1-w}{w}\right)^2} \tag{13}$$

$$P_{1c'} = -\frac{w}{t_1}\sqrt{\frac{1}{4}(33 + 24\sqrt{2})t_1^4 + \frac{1-w}{w}(5 + 4\sqrt{2})t_1^2 + \left(\frac{1-w}{w}\right)^2} \tag{14}$$

$$x_{1c} = -(1+\sqrt{2})\frac{1-w}{w} - \frac{1}{2}(4+3\sqrt{2})t_1^2 - \left(\frac{1-w}{w}\frac{1}{t_1} + \frac{1}{2}(3+4\sqrt{2})t_1\right)t_3$$

$$x_{2c} = -\frac{1-w}{w}\frac{1}{t_1} - \frac{1}{2}(3+4\sqrt{2})t_1 - \frac{1}{t_1}\sqrt{\frac{1}{4}(33+24\sqrt{2})t_1^4 + \frac{1-w}{w}(5+4\sqrt{2})t_1^2 + \left(\frac{1-w}{w}\right)^2}. \tag{15}$$

From these equations, we get the locus of the point (x_{1c}, x_{2c}) with t_1 as parameter. This locus is the second switching line reached on tracing back from the origin, and from eqn (15), this switching line approaches

$$x_{1c} = -\frac{1}{2}x_{2c}^2 \qquad \text{for } t_1 \to 0$$
$$x_{1c} = -0.485893\, x_{2c}^2 \qquad \text{for } t_1 \to -\infty \tag{16*}$$

When this analysis goes on to the points D, E, etc. we get switching lines that approach third, fourth, etc., parabolas (see Fig. 2), and K_∞ on the figure is a known value[2].

$$K_1 = 0.5, \quad K_2 = 0.485892--, \quad K_\infty = \sqrt{\sqrt{5}-2}$$

Analysis with $u = -1$ gives similar results as shown on Fig. 2.

As can be seen from Fig. 2, there may be several trajectories which satisfies the maximum principle from an arbitrary initial point to the origin. Among these, the absolutely optimal one must be obtained. Figures in each region indicated on Fig. 2 show the number of these trajectories leading to the origin.

Thus, consider the region where there are two trajectories. This region does not include the origin, because near the origin there is a unique solution of Pontryagin's equations.

In Fig. 3, let us assume the phase point to go from an arbitrary point I (with coordinate $x_1 = a$) on the x_1 axis with control functions $u = -1$ as far as a point L, then with $u = 1$ to a point M, and with $u = -1$ to the origin. Let the coordinate of the intersection of trajectory LM and the x_1 axis be $x_1 = -ka$, where k is a parameter. By changing the value of k from 0 to ∞, we can get the value of k for which the performance functional is minimum.

Consider the two performance functionals J_1 for trajectory $K \to 0$ and J_2 for trajectory $K \to L \to M \to 0$,

$$J_1 = \sqrt{a}\left(1 - w + \frac{1}{6}wa\right) \tag{17}$$

$$J_2 = \sqrt{a}\left\{(1-w)\left[2\sqrt{k} + 2\sqrt{k+1} - 1\right] + \frac{1}{6}wa\left[(6+8\sqrt{2})k\sqrt{k}+6(1-k)\sqrt{1+k}-5\right]\right\}. \tag{18}$$

By taking the difference as

$$f(k,a) = J_2 - J_1$$

we get

$$f(k,a) = 2\sqrt{a}(1-w)(\sqrt{k} + \sqrt{k+1} - 1) + \frac{1}{6}wa\sqrt{a}\{(6+8\sqrt{2})k\sqrt{k} + 6(1-k)\sqrt{1+k} - 6\} \tag{19}$$

and by minimizing the difference for $0 \le k < \infty$, we shall get an optimal trajectory. Fig. 4 shows the value of $f(k,a)$, with a as a parameter.

From $df(k,a)/dk = 0$

we get

$$\frac{1-w}{w}\left\{\frac{1}{\sqrt{k}} + \frac{1}{\sqrt{1+k}}\right\} + \frac{1}{2}a\left\{(3 + 4\sqrt{2})\sqrt{k} - \frac{1+3k}{\sqrt{1+k}}\right\} = 0 \tag{20}$$

*See Appendix

243

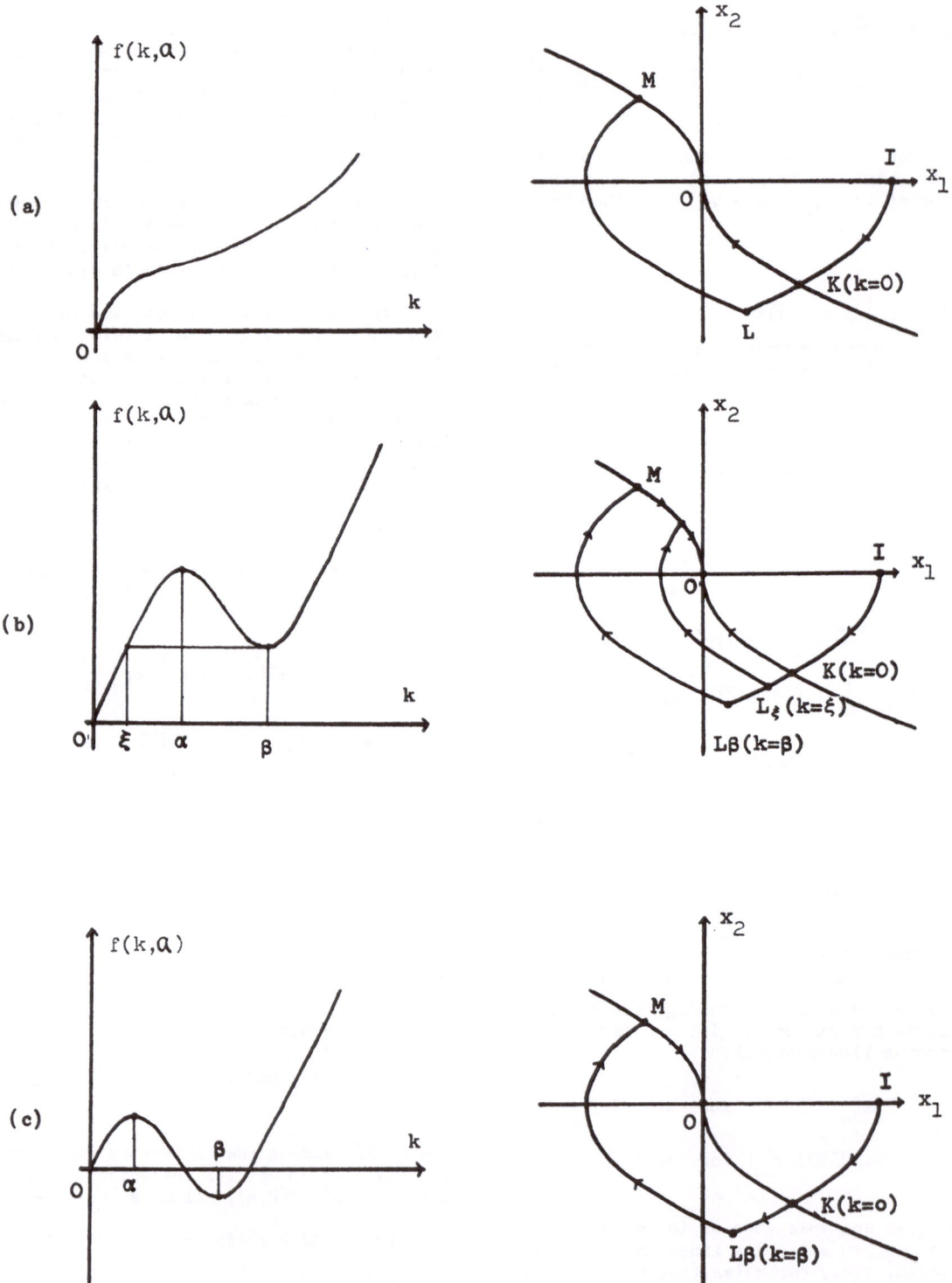

Fig. 5. f(k,a) patterns and optimal switching condition.

244

$f(k,a)$ is the difference between the performance functional of the trajectory from a point on IK, switched at K and going directly to the origin, and the performance functional of the trajectory switched at L and further switched at M, then going to the origin.

But, in Fig. 3, if the initial point is on L, we must further consider L' shown in the figure. There may be two trajectories: L→M→0 and L→L'→M'→0.

Taking the performance functional along the trajectory KL as J(KL), the one along LMO as J(LMO) and so on, and taking the similar form of difference as $f(k,a)$, we get

$$f(k,a) = J(KL) + J(LMO) - J(KO)$$

$$f(k',a) = J(KL) + J(LL'M'O) - J(KO)$$

$$f(k',a) - f(k,a) = J(LL'M'O) - J(LMO) \qquad (21)$$

If we can make clear the character of $f(k,a)$, the character of $f(k',a) - f(k,a)$ will also be known.

There are three types of patterns (a), (b) and (c) as shown on Fig. 5 according to the nature of $f(k,a)$. The optimal trajectory can be determined as follows:

Case (a): $f(k,a)$ is monotonically increasing and for all the trajectories starting from the initial points on KI, point K is the best switching point. If the initial point is on L, trajectory LMO is optimal.

Case (b): If the initial point is on KI, the best switching point is K, but if the initial point is on $L_\xi L_\beta$, the best switching point is L_β and the trajectory $L_\beta MO$ is optimal.

Case (c): If the initial point is on IL_β, the best switching point is L_β, and trajectory $L_\beta MO$ is optimal.

Therefore, point L_β corresponds to the minimum value of $f(k,a)$ and is the optimal switching point. Point L_ξ is the separating point which separates the two control policies. Point L_α corresponds to the maximum value of $f(k,a)$ and must not be used here, so L_α is not shown on the right-side figures.

Now, by taking the various values of a, points L_α, L_β, L_ξ form a locus on the state plane. Also, a locus of points L_α, L_β can be obtained by eqn (20). Let us investigate what kind of relations can exist between this locus and the second switching line obtainable from the maximum principle. We name the times for a

state point to move along the trajectories OM, MN, NL each as t_1, t_2, t_3, and find

$$t_1 = \pm\sqrt{k}\,\sqrt{a}$$

$$t_2 = -(1+\sqrt{2})\sqrt{k}\sqrt{a} = (1+\sqrt{2})t_1 \qquad (22)$$

$$t_3 = -\sqrt{a}(\sqrt{k+1} - \sqrt{2}\sqrt{k}) .$$

By substituting eqn (22) into eqn (13), we get eqn (20). Therefore, the previous locus of L_α, L_β can be found to be the second switching line given by the maximum principle. That is, the switching line obtained from the maximum principle corresponds to an extremum.

From the above procedure, we get the optimal solution as on Fig. 6. Line CAD is a switching line obtained from the maximum principle, and line CA corresponds to a minimum value and line AD to a maximum value. The latter should not be used as a switching line. Line AB on the figure is a separatrix. If the initial point is on the line AB, there are two optimal trajectories leading to the origin, these two trajectories have the same value of performance functional. Although the curve OBE is the first switching line, sub-curve BE should not be used here. Thus, we get the region of optimal trajectories shown on Fig. 7.

Next, we must consider the region where the third switching is required. In Fig. 8, considering a trajectory ILNQO, we modify the difference of the performance functionals to

$$J_2(ILNQO) - J_1(IKO)$$

$$= J(ILMO) - J(IKO) + J(I'NQO) - J(I'MO)$$

$$J_2 - J_1 = f(k_1,a) + f(k_2, ak_1) . \qquad (23)$$

Therefore, we can calculate optimal conditions using the results of the previous $f(k,a)$.

Generally, when the switching is required n times, we get the relation

$$J_2 - J_1 = \sum_{i=1}^{n-1} f(R_i, ak_1 . k_2 \ldots k_{i-1}) \qquad (24)$$

where, J_2 is the performance functional with n switchings, J_1 with only one.

In this case, if we take the value k_{n-1} which minimizes the $J_2 - J_1$ with fixed $k_1, \ldots, k_{n-2}$, we get $k_{n-1} = \phi(a, k_1, \ldots, k_{n-2})$ by the principles of optimality. Next, we determine the values k_{n-2} which minimizes $J_2 - J_1$, and so on. All $k_1, \ldots, k_{n-1}$ are found as functions of a. For example, in eqn (23), we must first determine $k_2 = \phi(ak_1)$ which minimizes $f(k_2, ak_1)$, then determine $k_1 = \phi(a)$ which

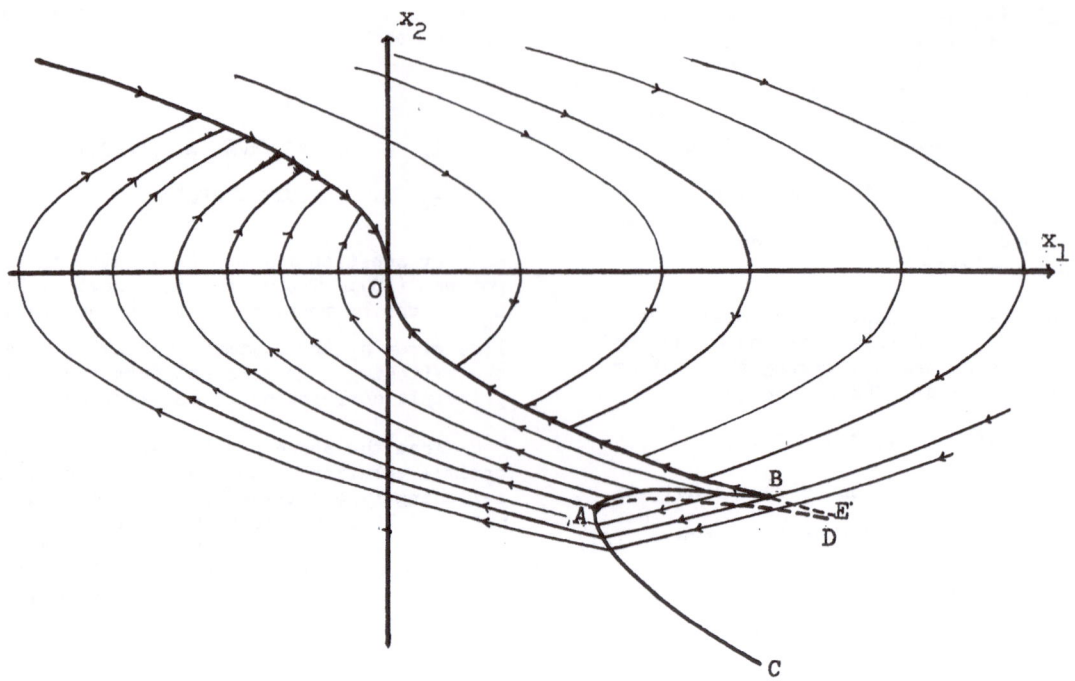

Fig. 6. Separatrix AB.

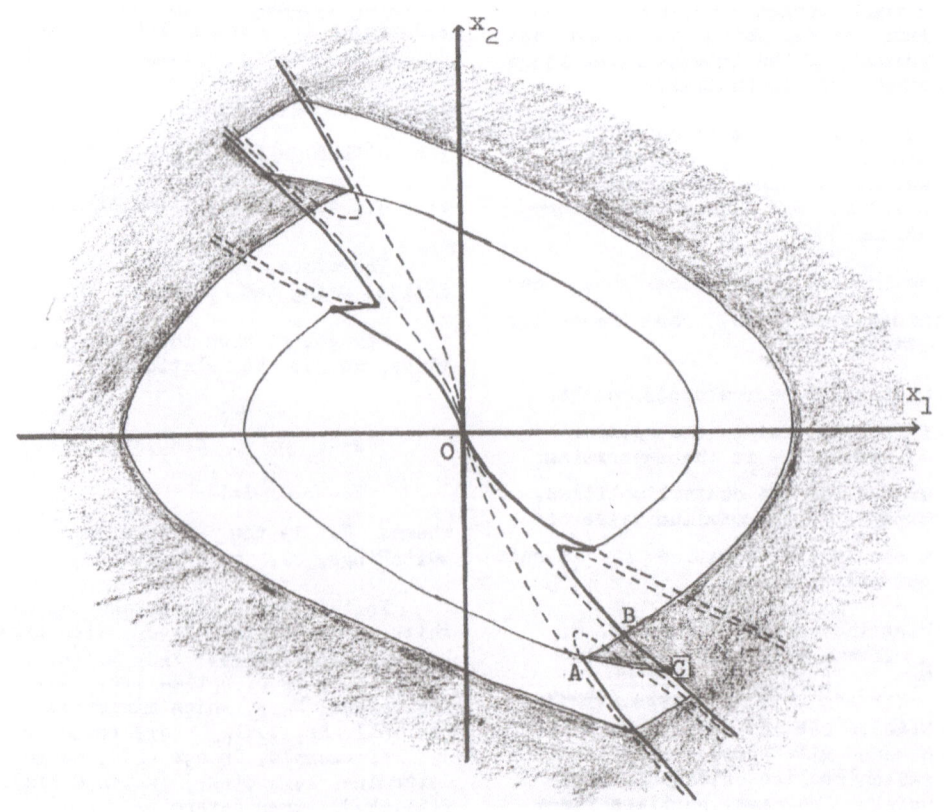

Fig. 7. Region of optimal control.

246

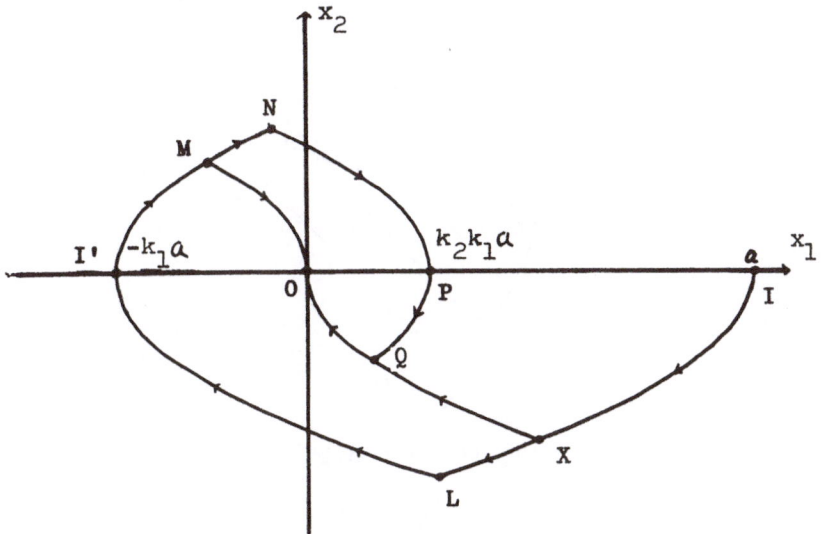

Fig. 8. Trajectory with third switching.

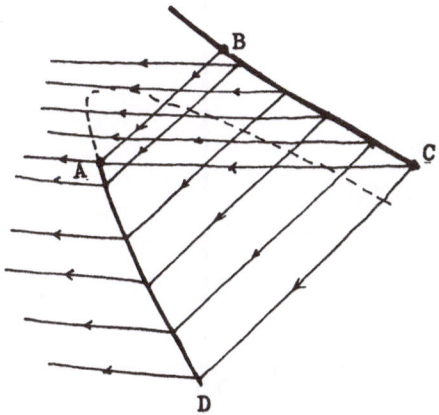

Fig. 9. Crossing of trajectories.

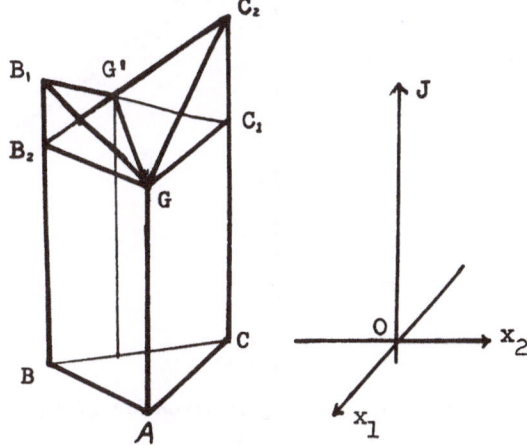

Fig. 10. Crossing line of two surfaces.

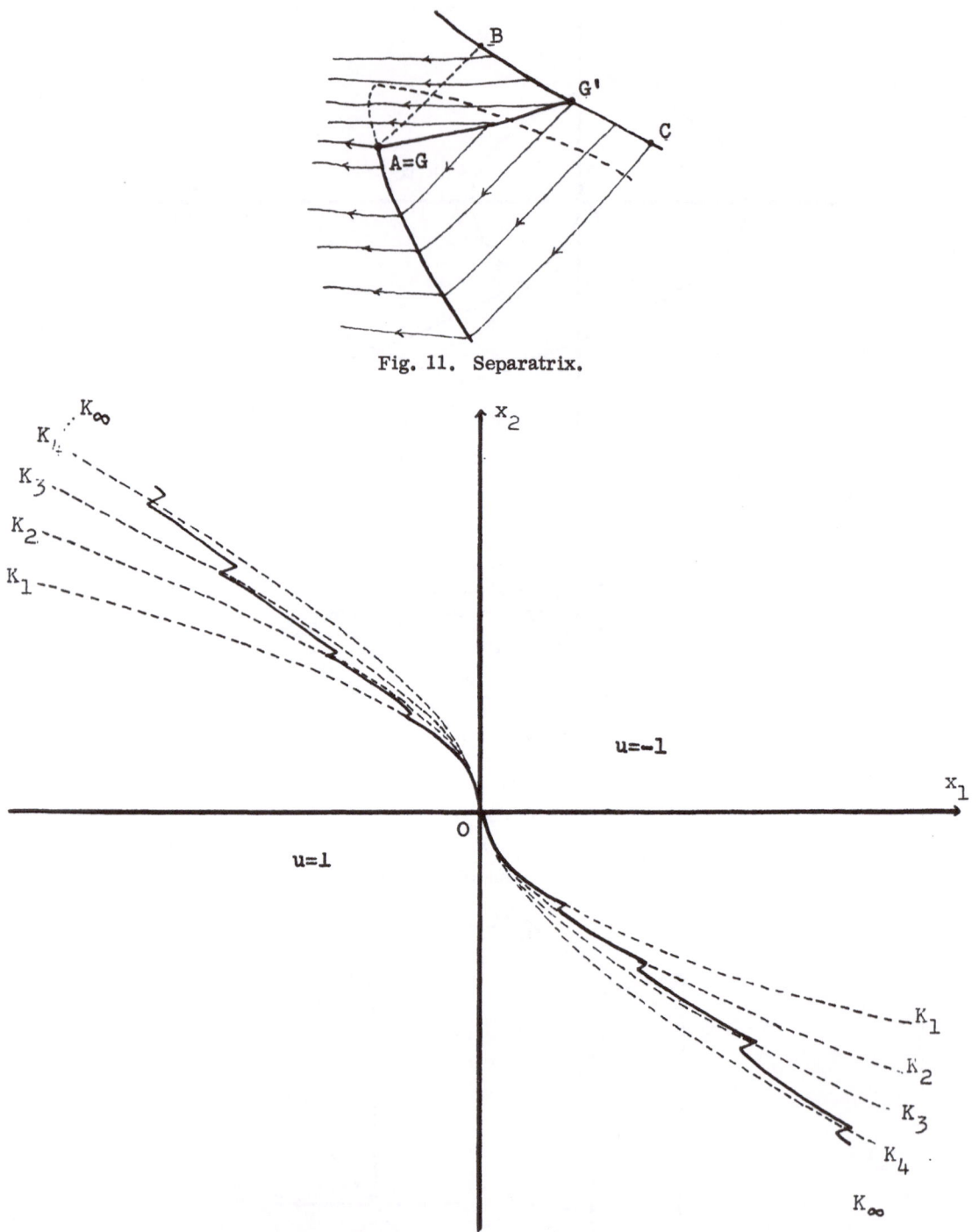

Fig. 11. Separatrix.

Fig. 12. Switching lines and separatrixes for optimal control.

minimizes $f(k_1,a) + f(\phi(ak_1), ak_1)$. Here $k_2 = \phi(ak_1)$ is the relation found on Fig. 6, and k_1 must be found on Fig. 7.

The part ABC on Fig. 7 must then be considered, and it is shown on Fig. 9 enlarged. There must be separatrix within this region ABC and it is a curve connecting point A with one point on BC, because all the trajectories starting from an initial point on AB will reach the origin with one switching with smaller value of performance functional than the trajectories with switching at point A, and because the trajectories with initial points on AC have smaller values of performance functional when switched at points on AD.

In Fig. 10, the value of performance functional is taken on the vertical axis and the separatrix is shown to be a intersection line of GG' of the two surfaces having the values of performance functional of the two control policies. Point G on this figure coincides with the point A on Fig. 9 and G' is on BC. In this way, we get separatrix GG' as shown on Fig. 11.

As described thus, optimal trajectories can be determined from the neighbourhood of the origin, and we get the switching lines shown on Fig. 12. The switching lines are discontinuous and between them there are separatrixes.

In a number of researches since Fuller, the analysis was carried out for the case when $w = 1$. When w tends to 1, the switching line between two parabolas will asymptotically coincide with these two parabolas. Therefore, the parabols $x_1 = K_1 x_2^2$ will become a switching line which gives both minimum and maximum values. In Fig. 12 discontinuous parts of the switching line and separatrixes will degenerate into the origin as w tends to 1, and only most outward parabola $x_1 = K_\infty x_2^2$ remains as a n optimal switching line. This result coincides with that of many authors.

CONCLUSIONS

Through the above analytical example, it has been shown that by modifying the performance functional from

$$J = \int_0^T |X| dt \quad \text{to} \quad J = \int_0^T (1-w+w|X|) dt$$

thus with time added, the real nature of uniqueness of solutions become clear. By calculating the minimum condition of the difference of performance functionals absolutely optimal switching lines can be obtained and the separatrix becomes necessary for unique trajectories.

APPENDIX

Part 1: Asymptotic line of the locus of point C.

This asymptotic line can be obtained by considering $t_1 \to 0$ and $t_1 \to -\infty$.

For $t_1 \to 0$, we use

$$
\begin{aligned}
x_{1C} &= x_{2B} t_3 - \tfrac{1}{2} t_3^2 \\
x_{2C} &= x_{2B} - t_3
\end{aligned}
\tag{25}
$$

Therefore,

$$x_{1C} = -\tfrac{1}{2} x_{2C}^2 + \tfrac{1}{2} x_{2B}^2 . \tag{26}$$

From eqn (10), for $t_1 \to 0$ we get $x_{2B} \to 0$ and so

$$x_{1C} = -\tfrac{1}{2} x_{2C}^2 . \tag{27}$$

For $t_1 \to -\infty$, we get from eqn (13)

$$t_3 \to \tfrac{1}{2}\{(3+2\sqrt{2}) + \sqrt{33+24\sqrt{2}}\} t_1 . \tag{28}$$

Substituting this into eqn (15), we get

$$x_{1C} \to -\tfrac{1}{2}\{4+3\sqrt{3} + \tfrac{1}{2}(3+4\sqrt{2})(3+2\sqrt{2} + \sqrt{33+24\sqrt{2}})\} t_1^2$$
$$x_{2C} \to -\tfrac{1}{2}\{3+4\sqrt{2} + \sqrt{33+24\sqrt{2}}\} t_1$$
$$\tag{29}$$

Hence

$$
x_{1C} = - \frac{\sqrt{33+24\sqrt{2}}}{3+4\sqrt{2} + \sqrt{33+24\sqrt{2}}} x_{2C}^2
$$
$$= -0.485893 \, x_{2C}^2 . \tag{16}$$

Thus, we have eqn (16).

Part 2: Asymptotic line of the locus of point E.

At point D where the trajectory crosses the x_2 axis,

$$
\begin{aligned}
x_{1D} &= x_{1C} + x_{2C} t_4 + \tfrac{1}{2} t_4^2 = 0 \\
x_{2D} &= t_1 - t_2 - t_3 + t_4
\end{aligned}
\tag{30}
$$

From the first of these, we find

$$t_4^2 + 2(t_1 - t_2 - t_3) t_4 - (2\sqrt{2} t_1 t_3 + t_3^2) = 0 . \tag{31}$$

Also

$$
\begin{aligned}
P_{1D'} &= P_{1C'} - w t_4 \\
P_{2D'} &= -P_{1C'} t_4 + \tfrac{1}{2} w t_4^2
\end{aligned}
\tag{32}
$$

249

At point E' where the trajectory crosses the p_1 axis,

$$P_{1E'} = P_{2D'} + wt_5$$

$$P_{2E'} = P_{2D'} - P_{1D'} t_5 - \tfrac{1}{2}wt_5^2 = 0 \qquad (33)$$

Also

$$x_{1E} = (t_1 - t_2 - t_3 + t_4)t_5 + \tfrac{1}{2}t_5^2$$

$$x_{2E} = t_1 - t_2 - t_3 + t_4 + t_5 \qquad (34)$$

The locus of point E becomes the switching line shown on Fig. 3. Let us obtain the asymptotic line of the locus of point E for $t_1 \to 0$.

For $t_1 \to 0$ we obtain the following results:

From eqn (9)

$$P_{1A} \to \frac{1-w}{t_1} \qquad (35)$$

From eqn (10)

$$x_{2B} \to 0 \qquad (36)$$

From eqn (12)

$$P_{1B'} \to \frac{1-w}{t_1}$$

$$P_{2B'} \to (1+\sqrt{2})(1-w) \qquad (37)$$

From eqn (13)

$$t_3 \to \frac{2}{t_1} \frac{1-w}{w} \qquad (38)$$

From eqn (14)

$$P_{1C'} \to -\frac{1-w}{t_1} \qquad (39)$$

From eqn (15) and (38)

$$x_{1C} \to -\tfrac{1}{2}t_3^2 = -\frac{2}{t_1^2}\left(\frac{1-w}{w}\right)^2$$

$$x_{2C} \to -t_3 = -\frac{2}{t_1}\left(\frac{1-w}{w}\right) \qquad (40)$$

From eqn (31) and (11)

$$t_4 \to (1+\sqrt{2})t_3 = \frac{2+2\sqrt{2}}{t_1}\frac{1-w}{w} \qquad (41)$$

From eqn (32), (39) and (41)

$$P_{1D'} \to -\frac{3+2\sqrt{2}}{t_1}(1-w)$$

$$P_{2D'} \to \frac{8+6\sqrt{2}}{t_1^2}\frac{(1-w)^2}{w} \qquad (42)$$

From eqn (33) and (42)

$$t_5 \to (3+2\sqrt{2} + \sqrt{33+24\sqrt{2}}) \frac{1}{t_1} \frac{1-w}{w} \qquad (43)$$

From eqn (34), (38), (41) and (43)

$$x_{1E} \to \tfrac{1}{2}(3+2\sqrt{2} + \sqrt{33+24\sqrt{2}})(3+6\sqrt{2} + \sqrt{33+24\sqrt{2}})$$
$$\times \frac{1}{t_1^2}\left(\frac{1-w}{w}\right)^2 \qquad (44)$$

$$x_{2E} \to (3+4\sqrt{2} + \sqrt{33+24\sqrt{2}}) \frac{1}{t_1} \frac{1-w}{w}$$

From eqn (44)

$$x_{1E} = \frac{\sqrt{33 + 24\sqrt{2}}}{3 + 4\sqrt{2} + \sqrt{33 + 24\sqrt{2}}} x_{2E}^2 \qquad (45)$$

Compare (45) with the first equation after (29).

Thus, it has been shown that the parabola to which the locus of point C approaches asymptotically for $t_1 \to -\infty$ is the same as the parabola to which the locus of point E approaches asymptotically for $t_1 \to 0$.

BIBLIOGRAPHY

(1) M. Masubuchi, T. Sekiguchi and H. Kanoh: _Time-Optimal Control of Second Order Systems with Both Forcing Term and Parameter Variations_, Jour. of the Soc. of Instr. and Control Engineers, Japan, Vol. 3, No. 10, 1964, pp.750-758.

(2) A. T. Fuller: _Relay Control Systems Optimized for Various Performance Criteria_, Proc. of the 1st IFAC Congress, Moscow, 1960.

(3) J. D. Pearson: _Note on a Solution due to A. T. Fuller_, Jour. of Elect. and Contr., Vol. X, No. 4, 1961, pp.323-332.

(4) P. J. Brennan, A. P. Roberts: _Use of an Analogue Computer in the Application of Pontryagin's Maximum Principle to the Design of Control Systems with Optimum Transient Response_, Jour. of Elect. and Contr., Vol. XII, No. 4, 1962, pp.345-352.

(5) W. M. Wonham: _Note on a Problem in Optimal Non-linear Control_, Jour. of Elect. and Contr., Vol. XV, No. 4, 1963, pp.59-62.

(6) A. T. Fuller: _Study of an Optimum Non-linear Control System_, Jour. of Elect. and Contr., Vol. XV, No. 4, 1963, pp.63-71.

(7) D. M. Eggleston: Trans ASME, Sept. 1963, pp.478-480.

(8) Y. Sawaragi, Y. Sunahara and T. Ono: A
     Study on the Synthesis of Optimum Control
     Systems with Magnitude Constraint, Control
     Engineering, Japan, Vol. 8, No. 6, 1964,
     pp.281-290.

(9) L. S. Pontryagin and others: The
     Mathematical Theory of Optimal Processes,
     Interscience Pub., 1962.

MULTI-LEVEL CONTROL SYSTEMS

by J. D. Pearson
Systems Research Center
Case Institute of Technology
Cleveland 6, Ohio, U.S.A.

ABSTRACT

A decomposition technique applied to the optimum control problem for interacting dynamic systems is used to generate an equivalent two level hierarchy of sub-problem. Application to the synthesis of controllers is discussed and an important special case for "slack" systems is indicated with an example relating conventional integral action control with coordination.

INTRODUCTION

Large organizations for example of economic or biological nature, inevitably appear to have a hierarchial chain of command. Characteristic of such organizations is that the hierarchy is a pyramid like structure of decision problems and goals which vary in complexity. Problems at the base of the structure are usually fairly simple though numerous. Each of these problems is solved relative to a few intervention parameters which are themselves manipulated by higher, more complex problems. This structure of parametrized sub-problems repeats itself up the hierarchy until at the apex there are one or more sophisticated problems upon whose outcome the whole system depends.

In this paper some questions relative to hierarchies of problems associated with control systems are investigated. In connection with the control of large scale systems a natural desire is to somehow "organize" the control of small parts of the system so that in an overall sense this decentralized control is as good as its centralized conventional equivalent.

These decentralized systems are often referred to as "multi-level" systems[1] because the hierarchy divides naturally into levels upon each of which roughly the same activity proceeds. Each level is coordinated by the level above and itself coordinates the level below. Thus level of control is a natural concept, corresponding to military rank for example. The kind of problem activity associated with the level is not necessarily unique, but decisions reached by that level are always binding on a lower level upon which the same activity may proceed. These terms are defined elsewhere.[2]

Given a system consisting of many interacting sub-systems, for which a hierarchial organization is required, there are essentially two approaches which could be pursued, relevant to its study:

(i) Analysis (decomposition)

Here a problem could be posed relevant to the whole integrated system and analysis is attempted which indicates how to decompose the problem into independent parts relevant to the sub-systems, together with a means for coordinating them. This automatically generates an equivalent two level structure whose additional properties can then be examined.

(ii) Synthesis (coordination)

Alternatively a hierarchy could be directly synthesized by means which guarantee its resulting properties in terms of some definition of simplicity or cost, but which might not yield a system of equivalent performance.[3]

There exist a large body of results applied to both specific techniques described above. This paper will be concerned primarily with the first approach.

Decomposition techniques

A uniform discussion of decomposition techniques can be made by presentation of a saddle value argument stated for the appropriate control problem in variational form.[4] Note that in the static case the Kuhn-Tucker theorem provides the relevant result.[5]

Suppose we are concerned with a linear system which is a collection of N sub-systems.

$$\frac{dy_i}{dt} = A_i y_i + B_i x_i + C_i m_i \quad i=1,2,\ldots N \quad (1)$$

$$N_i d_i + \sum_j N_{ij} y_j = x_i \; ; \quad N_{ii} = 0 \quad j,i=1,2,\ldots N \quad (2)$$

Here y_i, m_i are respectively the state and control column vectors of dimension n_i, r_i. x_i, d_i are interaction and disturbance column vectors of dimension s_i, t_i respectively. These equations simply define N interacting sub-systems with control and disturbance inputs.

Suppose m_i, $i = 1,2,\ldots N$ must be selected on $[0, t_1]$ so as to minimize a performance functional -

$$v(d) = \sum_i [g_i(y_i(t_1)) + \int_o^{t_1} f_i(y_i, x_i, m_i,)dt] \quad (3)$$

$(d' \underline{\Delta}(d_1, d_2, \ldots d_N)'$, d' denotes transpose d, a row vector).

- subject to inequality constrains:

$$R_i(y_i, x_i, m_i) \leq 0 \qquad (4)$$

Certain conditions must be satisfied, namely that f_i, g_i, R_i be convex C'' functions of their arguments with f_i strict. This is to avoid discussion of singular cases and provides continuity of the decision variable, m_i.

Associated with this control problem, a Lagrangian functional $J(y,x,m;p,\mu,\pi,e)$ can be defined by: $j,i = 1,2,\ldots N$:

$$J \triangleq \sum_i [(g_i + \int_o^{t_1} (f_i + p_i'(A_i y_i + B_i x_i + C_i m_i - dy_i/dt)$$

$$+ \pi_i'(\sum_j N_{ij} y_j + N_i d_i - x_i) + \mu_i' R_i) dt + e_i'(y_i - c_i))] \qquad (5)$$

(p_i, π_i, μ_i, e_i are multipliers defined by the first order necessary conditions). It is quite easy to show [2,6] that the optimum value of J, J^o, defined by the J above with y, x, m, p, μ, π, e taking on their optimum values y^o, x^o, m^o, p^o, μ^o, π^o, e^o can be realized by a saddle value operation.

$$J^o = \underset{p,\mu,\pi,e;y,x,m}{\text{Max}} [\text{ Min } J(y,x,m;p,\mu,\pi,e)] \qquad (6)$$

($p' \triangleq (p_1, p_2 \cdots p_N)'$, $y' \triangleq (y_1, y_2, \ldots y_N)'$ etc.)

Actually J^o is of little value but the saddle value can be utilized as follows. The multipliers p,π,e are associated with the dynamic and inequality constraints. They can be eliminated from J simply by enforcing these constraints, i.e.

$$J^o = \underset{\pi}{\text{Max}}[\underset{y,x,m\epsilon A}{\text{ Min }} [\sum_i (g_i + \int_o^{t_1} (f_i + \pi_i'(\sum_{j=1} N_{ij} y_j +$$

$$+ N_i d_i - x_i)) dt]] \qquad (7)$$

The set A of admissible functions are those satisfying (1), and (4) under the influence of control functions C'' on $[0,t_1]$.

A rearrangement of the integrand of (7) indicates that the minimization required in (7) can be carried out for each sub-system independently. In fact, performance functionals $v_i(\pi)$ can be defined such that given the parameters π defined on $[0,t_1]$ and N independent subsystems on the first level, defined by:

$$\frac{dy_i}{dt} = A_i y_i + B_i x_i + C_i m_i \ ; \ y_i(0) = c_i$$

Control variables m_i, x_i must be found which minimize -

$$v_i(\pi) = g_i + \int_o^{t_1} (f_i + y_{ij}' \Sigma N_{ij}' \pi_j - x_i' \pi_i) dt \qquad (8)$$

$$i,j = 1, \ldots N$$

- subject to inequality constraints:

$$R_i(y_i, x_i, m_i) \leq 0$$

Addition of all N sub-problems is equivalent to the summand of (7). Define $y_i^o(\pi)$, $x_i^o(\pi)$, $m_i^o(\pi)$ as the optimum solutions to the first level subproblems (8). Since no special concern was taken over the interaction constraint (2), it follows that in general $y_i^o(\pi)$ etc. will not satisfy (2) for a given π. However, in view of the existence of an optimal solution to (6) a value π^o exists such that corresponding values $y_i^o(\pi^o)$ do satisfy (6) the interaction constraint (2). From (7), this value π^o can be found by completing the saddle value operation as the second level problem:

$$J^o = \underset{\pi}{\text{Max}}[\sum_{i=1}^N (v_i^o(\pi) + \int_o^{t_1} (\pi_i' N_i d_i) dt)] \qquad (9)$$

Here $v_i^o(\pi)$ is the optimum value of $v_i(\pi)$, i.e. the value of $v_i(\pi)$ subject to the first order necessary conditions (10)-(12) for the minimization of (8).

$$\frac{dp_i}{dt} + A_i' p_i + \frac{\partial}{\partial y_i} (h_i) = 0 \ ; \ p_i(t_1) - \frac{\partial g_i}{\partial y_i} = 0 \qquad (10)$$

where: $h_i \triangleq f_i + \mu_i' R_i$

together with: $\frac{\partial}{\partial x_i} (h_i) + B_i' p_i = 0 \qquad (11)$

$$\frac{\partial}{\partial m_i} (h_i) + C_i' p_i = 0 \qquad (12)$$

In principle a solution of (9) completes the coordination of the N sub-problem so that the interconnection constraint is finally satisfied.

A simple technique for solving (9) can be obtained by a gradient method. It turns out that the argument of (9) is in fact the dual[6] of the previous variational problem, denoted $\omega(\pi,d)$.

$$\omega(\pi,d) = \sum_i (v_i^o(\pi) + \int_o^{t_1} (\pi_i' N_i d_i) dt) \qquad (13)$$

253

If we denote $\pi(t,0)$ as an initial guess and compute $d\pi(t,\sigma)/d\sigma$ as σ, a real scalar parameter, increases, we find that:

$$\frac{d\omega}{d\sigma}(\pi(\sigma),d) = \sum_i \int_o^{t_1} (\sum_j (N_{ij}y_j^o(\pi) + N_i d_i - x_i^o(\pi)), \frac{d\pi}{d\sigma})dt \qquad (14)$$

Thus the simple coordination rule defined by:

$$\frac{d\pi}{d\sigma}(t,\sigma) \underline{\Delta} \; \epsilon \cdot (\sum_j N_{ij}y_j^o(\pi) + N_i d_i - x_i^o(\pi)) \quad (15)$$

$$1 >> \epsilon > 0$$

is such that $d\omega(\pi(\sigma), d)/d\sigma > 0$ everywhere, whenever the interaction constraint is not satisfied. As a result ω increases monotonically with σ until the constraint (2) is satisfied, $d\pi/d\sigma = 0$. However ω is bounded above by J^o when (2) is satisfied,[6] thus ω achieves its maximum value J^o which solves (9), as required.

DISCUSSION

The essential ideas behind the decomposition method as developed here, are related to that of Dantzig and Wolfe's "Decomposition Principle".[7] However it must be remarked that it is not possible to produce a finite procedure analogous to the linear programming case.

Basically what is proposed here is a two-level solution of complex dynamic optimization problems.

Level 1. Consists of solving N parametric sub-problems containing initally unknown functions $\pi(t)$.

Level 2. Consists of coordinating the N independent sub-problems by choosing the parameters $\pi(t)$.

There is one sub-problem for each sub-system, and one coordinating problem for each group of interacting sub-systems.

The principal difficulty is one of solving a sub-problem containing arbitrary time functions $\pi(t)$. This is circumvented in practice by using an iterative technique to adjust an initial guess $\pi_o(t,0)$ until both, the first level sub-problems and the second level coordinating problem are solved. Experience with first order gradient methods defined by (15) indicate that they are slow. Second order methods are more efficient and experiments indicate that the first level problems need not be completely solved initially for the accuracy is cumulative in the sense that corrections for changes in $\pi(t,\sigma)$ can improve the optimality of the sub-problems as measured by the degree to which (12) is satisfied.

Slack Systems[8]

A simplification is possible for a certain class of systems whose interconnections are true in an average sense only. Equation (2) is then replaced by:

$$\int_o^{t_1} (N_i d_i + \sum_j N_{ij}y_j - x_i)dt = 0 \qquad (16)$$

$$N_{ii} \underline{\Delta} \; 0 . \quad i,j = 1,2,\dots N .$$

This kind of interconnection is not uncommon. Most inventory systems have stock, chemical processing systems have storage tanks between processes, economic systems have credit; all of which can be interpreted as slack between sub-systems.

The effect of a constraint (16) is to replace the unknown price function $\pi(t)$ by an unknown constant π . Thus the sub-problems are generally simpler to solve and it is often possible to synthesize the control policy as a function of π .

At the second level the constant π can be found by a simple approximation device of interpreting the iteration time scale σ as real time t.

The first order gradient method (15) defines $\pi(\sigma)$ as:

$$\pi(\sigma) = \pi(0) + \epsilon \int_o^\sigma [\sum_j N_{ij}y_j^o(t,\pi) + N_i d_i(t) - x_i^o(t,\pi)]dt$$

ϵ determines convergence rate, $\pi(0)$ is the initial estimate. The effect of interpreting σ as real time t , is as follows.

(i) The sub-problem solutions are correct only for constant π and it will now change

(ii) The larger ϵ , the more rapidly $\pi(\sigma) \equiv \pi(t)$ changes, gradually introducing instabilities

(iii) The second coordinating algorithm can be interpreted as integral control action with ϵ chosen by the conventional stability analysis.

Thus, without pursuing a general stability analysis it is evident that if ϵ is small, the sub-problem solution error will be small and chances of instability will be reduced.

Example

In Figure 1, consider N rollers of mass M_i , radius R_i rolling strip at velocity V_i into each roller and tension T_i out of each roller. Suppose each roller pair is driven by a motor exerting torque $R_i m_i$.

Each pair of rollers has inertia $M_i R_i^2$ and operates at angular velocity V_i/R_i from which:

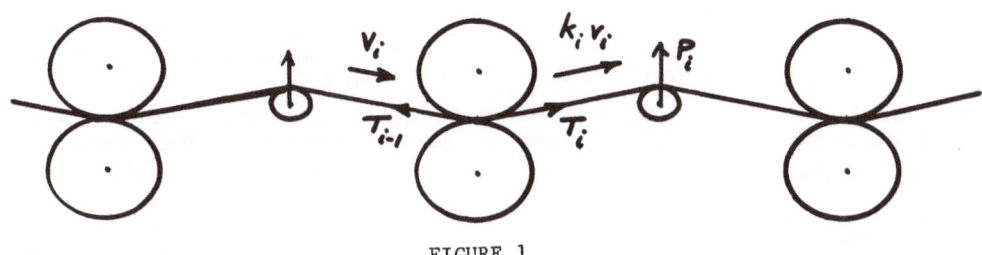

FIGURE 1

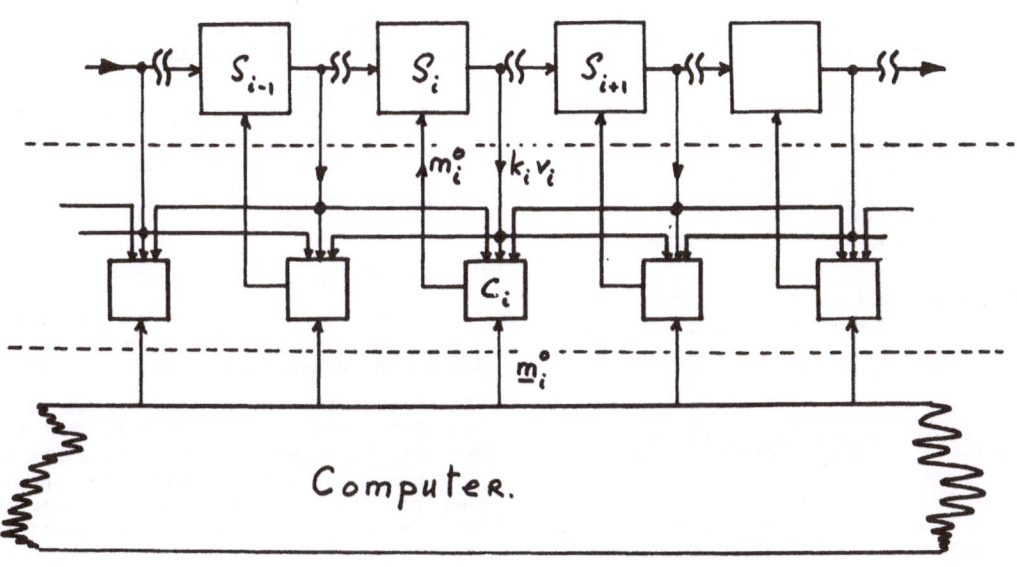

Computer.

FIGURE 2

255

$$M_i \cdot \frac{dV_i}{dt} = T_i - T_{i-1} + m_i$$

$$V_i(0) = C_i . \quad i = 1, 2, \ldots N$$

T_o is defined.

Assume that the strip mass, and the looper mass and inertia are small compared with the roller mass. The tensions along the strip between rollers are constant and equal.

$$2T_i = (2P_i / \sin\phi_i) . \sin\phi_i = 2P_i$$

ϕ_i is the inclination of each strip section with the horizontal and the upward tensioning force is $(2P_i / \sin\phi_i)$ designed to give constant strip tension for any degree of slack.

Suppose that the problem is to control an initial transient error situation due to random strip temperature fluctuations (for example), expressed by minimizing:

$$\sum_{i=1}^{N} \frac{1}{2} \int_o^\infty [(k_i V_i - V_i^*)^2 + (m_i / \lambda_i)^2] dt$$

with:

$$\int_o^\infty (k_i V_i - V_{i+1}) dt = 0 \quad i = 1, 2, \ldots N-1 .$$

Here V_i^* are desired output velocities and $k_i V_i$ is the actual output velocity, $k \geq 1$ allows for a velocity increment due to strip extension due to rolling for a given roller setting (and rigid rollers).

Applying the decomposition algorithm reveals N simpler individual roller control problems given π, minimize.

$$\frac{1}{2} \int_o^\infty [(k_i V_i - V_i^*)^2 + 2\pi_i V_i - 2\pi_{i-1} V_i + (m_i / \lambda_i)^2] dt$$

In view of the linearity of the system models the solution can be expressed in terms of constant π values directly.

$$m_i^o = -\lambda_i [k_i V_i - V_i^* + \pi_i - \pi_{i-1}] + T_{i-1} - T_i$$

The approximate coordination algorithm to find the constant multiplier π_i recovers them as time functions

$$\pi_i(t) = \pi_i(0) + \varepsilon \int_o^t (k_i V_i - V_{i+1}) dt$$

ε must be chosen to give most rapid convergence.

Figure 2 indicates the final two level structure of coordinated proportional controllers with feed forward terms π_i provided by integral action coupling between roller units. The example, though highly idealized, retains some semblance of the "slack control" problem and indicates that the obvious practical solution is an approximation to the ideal two-level control structure.

BIBLIOGRAPHY

C.I.T.   Case Institute of Technology

1. M.D. Mesarovic, "A general systems approach to organizational theory." SRC 2-A-61-2 . C.I.T.

2. M.D. Mesarovic, I. Lefkowitz, J.D. Pearson, "Advances in Multilevel Control." IFAC Symposium, Tokyo, 1965.

3. J.D. Pearson, Decomposition, Coordination and Multilevel Systems. 1965 Systems Science Conference (IEEE) C.I.T., Cleveland, USA.

4. Robert Wilson, Private communication. Graduate School of Business, Stanford University.

5. L.S. Lasdon, J.D. Schoeffler, Decentralized Plant Control, I.S.A. Preprint 3.3-3-64

6. J.D. Pearson, "Duality and a Decomposition Technique," to appear Journal of SIAM Series A Control, 1966.

7. G. Dantzig, P. Wolfe, "Decomposition principle for linear programs," J. Operations Research, 8, No. 1., pp. 101-111, 1960.

8. S. Coopermen, MS Thesis, C.I.T.

Discussion

An unidentified speaker questioned the statement of continuity ascribed to the effort function $\phi(n_i)$. He noted that it was a function of dimension and that continuity followed only in the distributed case. Further, this function was also dependent on the control law and so is more complicated than shown. The questioner suggested that this complication necessitated transfer of the functional optimization to the control parameter space. Dr. Pearson did not agree and said that further elaboration would be forthcoming in a co-worker's paper. He added that the example was intended to convey the meaning of the terms used and to help illustrate the general conclusions drawn.

Dr. P. Alper (USA) inquired about the significance of the order of the max-min operation noting that in general, the operation is irreversible. The speaker replied that the reverse order may hold in this particular case but that he had not established this point.

ON A METHOD OF COMPENSATION AGAINST PARAMETRIC DISTURBANCES

by S.V. Yemel'yanov, M.A. Bermant, N.E. Kostylova, and V.I. Utkin
USSR

This paper deals with the design of an automatic system having low sensitivity to wide range variations of the controlled plant parameters. The design uses the properties of sliding modes (see Editors note paper 4.2) which can appear in any dynamic system, described by a set of differential equations with discontinuous right-hand parts. The problem is solved by using a class of variable structure systems, i.e. systems in which structure and parameters of the controller change stepwise during the transition process according to a logical relationship depending on the system's state.

1. Selection of the Control Function.

Let us assume that the controlled plant dynamics can be described by a linear differential equation of the following type

$$Q(D)x = P(D)U \qquad (1.1)$$

where x is the coordinate of error

U is the coordinate of control

and $Q(D)$, $P(D)$ are operational polynominals, $\left(D = \dfrac{d}{dt}\right)$ given by:-

$$Q(D) = \sum_{i=0}^{n} a_{i+1} D^i , \quad P(D) = \sum_{i=0}^{m-1} b_{i+1} D^i , \quad n \geqslant m$$

a_i, b_i are the parameters of the plant, varying within the ranges given by the following inequalities:-

$$a_{i\,min} \leqslant a_i \leqslant a_{i\,max} \quad i = 1,\ldots,n$$

$$b_{i\,min} \leqslant b_i \leqslant b_{i\,max} \quad i = 1,\ldots,m$$

$$a_{n+1} = 1 \qquad (1.2)$$

$$b_m > 0$$

In further discussion Equation (1.1) will be represented for convenience in the vectorial form

$$\frac{d\bar{x}}{dt} = \bar{f}(\bar{x},U)$$

where

$$\bar{x} = (x_1,\ldots,x_n) , \quad \bar{f} = (f_1,\ldots,f_n)$$

$$f_i = x_{i+1} \qquad i = 1,\ldots,n-1 \qquad (1.3)$$

$$f_n = -\sum_{i=1}^{n} a_i x_i - \left(\sum_{i=0}^{m-1} b_{i+1} D^i\right) U .$$

Consider a space X of the system coordinates $(x_1 \ldots,x_n)$ and assume in this space some hyperplane S defined by the equation

$$S = 0 , \quad S = \sum_{i=1}^{n} C_i x_i , \quad C_i = const , \quad C_n = 1 \qquad (1.4)$$

let us suppose that, with the selected control function U, the quantity f (1.3) undergoes discontinuities of the first kind. Then when the representative point reaches S, in the dynamic system (1.3) sliding mode dynamics may occur. The characteristics of this motion is that the phase trajectories belong to the hyperplane S (ref. 1). This motion is described by the $(n-1)$th order linear homogeneous differential equation (ref. 2) of the following type:

$$x_1^{(n-1)} + C_{n-1} x_1^{(n-2)} + \ldots + C_1 x_1 = 0 \qquad (1.5)$$

Obviously, the solution of (1.5) does not depend on the controlled plant characteristics.

The property of the sliding mode dynamics mentioned above will be used as a basis for the design of a system having low sensitivity to variation of the parameters a_i, b_i within the ranges specified in (1.2). The essence of the design is that at some time the control system shall exhibit continuous dynamics in the sliding mode. In terms of phase representations of the motion this means, that, firstly, the hyperplane is a hyperplane of sliding, i.e. in any point of S there is a sliding mode, and, secondly, the representative point from any initial position will always reach S. Note, that a sliding mode appears in those points of the hyperplane S, in whose neighbourhood the following inequalities

"Superior numbers refer to similarly-numbered references at the end of this paper"

apply:-

$$\bar{C} \frac{d\bar{x}}{dt} \geq 0 \quad \text{when} \quad S < 0$$

$$\bar{C} \frac{d\bar{x}}{dt} \leq 0 \quad \text{when} \quad S > 0 \tag{1.6}$$

where the vector $\bar{C} = (C_1,\ldots,C_n)$ is the normal to the hyperplane S.

As was previously stated the method of synthesis suggested implies that the function f_n has discontinuities of the first kind. The above constraint is valid if the control satisfies the differential equation

$$\frac{d\bar{z}}{dt} = \bar{h}(\bar{z}) \tag{1.7}$$

where

$$\bar{Z} = (Z_1,\ldots,Z_{m-1}), \quad \bar{h} = (h_1,\ldots,h_{m-1})$$

$$h_i = \frac{1}{T_i}(Z_{i-1} - Z_i), \quad i = 1,\ldots,m-1$$

T_i are constant factors

$$Z_{m-1} = U, \quad Z_0 = V$$

V is an arbitrary function with discontinuities of the first type.

From the point of view of practical implementation the function U for $T_i > 0$ can be easily obtained with a passive filter, consisting of $m-1$ series connected first orderlags.

2. Conditions for existence of hyperplane of sliding

According to the above assumptions the dynamics of the system considered is characterised by the following equation

$$\frac{d\bar{R}}{dt} = \bar{G}(\bar{R}) \tag{2.1}$$

which describes the state of the controlled process in $(n-m+1)$-dimensional phase space of the vector $\bar{R}$, where

$$\bar{R} = (x_1,\ldots,x_n;\ z_1,\ldots,z_{m-1})$$

$$\bar{G} = (f_1,\ldots,f_n;\ h_1,\ldots,h_{m-1}).$$

By (1.7), the component f_n of the vector $\bar{G}$ may be written in the following form:

$$f_n = -\sum_{i=1}^{n} a_i x_i - \sum_{j=1}^{m-2}\left(\sum_{i=j}^{m-1} b_{i+1} A_{j,i}\right) Z_{m-1-j}$$

$$- \sum_{i=1}^{m-1}(b_{i+1}A_{0,i} + b_i)\ Z_{m-1} - b_m A_{m-1,m-1} V$$

Here all $A_{i,j}$ are found from the recurrent relations

$$A_{j,i+1} = A_{j-1,i}\frac{1}{T_{m-j}} - A_{j,1}\cdot\frac{1}{T_{m-j-1}},$$

$$A_{-1,i} = 0,\ A_{i+1,i} = 0,\ A_{0,1} = -\frac{1}{T_{m-1}},\ A_{1,1} = \frac{1}{T_{m-1}}.$$

Owing to the discontinuities of the first kind in the function V the variable f_n also has discontinuities of the first kind, therefore in the dynamic system (2.1) a sliding motion may occur. To make these motions independent of the plant parameters the function V should have discontinuities on the hyperplane W, defined in the space of the vector $\bar{R}$ by the equation

$$S = 0,\quad S = \sum_{i=1}^{n} C_i x_i. \tag{2.2}$$

By the proper selection of the coefficient C_i one may obtain the desired dynamic properties of the sliding mode, described by the differential equation (1.5). In accordance with the method of design suggested, the auxiliary function should be selected from the given range (1.2) for any selected value C_i and arbitrary a_i and b_i in such a way that the hyperplane Q is the hyperplane of sliding. The problem can be solved if the function V is formed as a linear combination of the system coordinates with stepwise changing coefficients, in the following way

$$V = \sum_{i=1}^{n-1}\psi_i x_i + \sum_{i=1}^{m-1}\phi_i Z_i \tag{2.3}$$

$$\psi_i = \begin{cases} \omega_i^x & \text{when } x_i S > 0 \\ \lambda_i^x & \text{when } x_i S < 0 \end{cases} \quad i=1,\ldots,n-1$$

$$\phi_i = \begin{cases} \omega_i^z & \text{when } Z_i S > 0 \\ \lambda_i^z & \text{when } Z_i S < 0. \end{cases} \quad i=1,\ldots,m-1$$

The necessary and sufficient conditions for the existence of the hyperplane of sliding are:-

$$A_{m-1,m-1}\,\omega_1^x \;\geqslant\; \max_{a_i a_n b_m b_m} \; \frac{1}{-} \;(C_{i-1}-a_i-C_i C_{n-1}+C_i a_n)$$
$$i=1,\ldots,n-1$$

$$A_{m-1,m-1}\,\lambda_1^z \;\leqslant\; \min_{a_i a_n b_m b_m} \; \frac{1}{-} \;(C_{i-1}-a_i-C_i C_{n-1}+C_i a_n)$$

$$A_{m-1,m-1}\,\omega_1^z \;\geqslant\; \max_{b_2,\ldots,b_m}\left(-\frac{1}{b_m}\sum_{i=j}^{m-1} b_{i+1}A_{j,i}\right)$$
$$j=1,\ldots,m-2$$
$$(2.4)$$

$$A_{m-1,m-1}\,\lambda_1^z \;\leqslant\; \min_{b_2,\ldots,b_m}\left(-\frac{1}{b_m}\sum_{i=j}^{m-1} b_{i+1}A_{j,i}\right)$$

$$A_{m-1,m-1}\,\omega_{m-1}^z \;\geqslant\; \max_{b_1,\ldots,b_m}\frac{1}{b_m}\left(-\sum_{i=1}^{m-1} b_{i+1}A_{0,i}-b_i\right)$$

$$A_{m-1,m-1}\,\lambda_{m-1}^z \;\leqslant\; \min_{b_1,\ldots,b_m}\frac{1}{b_m}\left(-\sum_{i=1}^{m-1} b_{i+1}\,A_{0,i}-b_i\right).$$

Thus, if inequalities (2.4) apply, the solution of the system (2.1) depends on the parameters and b_i until such time as $\bar{R}$ reaches the hyperplanes Q after which, by (1.5), the solution is determined only by the factors C_i.

If all the C_i satisfy the Hurwitz criterion then the coordinate of error in the sliding regime will asymptotically tend to zero, which is the purpose of control.

The design of a control algorithm for the system, which has low sensitivity to variations of the plant's parameters over a wide range, can be considered complete only if the representative point of the system (2.1) can reach the hyperplane Q from any initial position. By the theorem formulated below for control laws of the type (2.3) there is always a control law for which any motion enters the sliding mode after a certain time.

Theorem

Assume, that in (2.3)

$$\lambda_1^x \;=\; -\omega_1^x\,, \qquad \omega_1^x > 0$$

and the conditions under which the sliding mode starts over the whole hyperplane Q are met.

If all $C_i > 0$ then, there is a value Ω such that at $\omega_1^x > \Omega$ the representative point in the space of the system (2.1) coordinates starting from any initial position reaches the hyperplane Q within a finite interval of time.

Thus, if there is a sliding mode at any point of hyperplane Q then when the condition for the representative point to reach this hyperplane is met, the system (2.1) will be of low sensitivity to the variations of the controlled plant parameters over a wide range.

It is important to note that the control algorithm suggested is implemented by the use of terminal gains and does not require the introduction of adaptive elements in the controller.

References

1. Philippov, A.F., Differential equations with discontinuous right(hand) sides. Math. Sym. (Math. Sbor), Vol. 61, No. 1, 1960.

2. Emel'yanov, S.V., High quality control using certain nonlinear devices with variable parameters. Proc. Acad. Sci. USSR, No. 4, 1962 (Energetuka i Automatika).

CONTROL USING PERTURBATION TECHNIQUES

A MULTI-DIMENSIONAL SELF-OPTIMIZING CONTROL SYSTEM INVOLVING

DYNAMICS AND DISTURBANCES, EMPLOYING RELAY EXTREMUM CONTROL

by G. Broekstra, C.J.D.M. Verhagen, J.A. van Arkel
Instrumentation Laboratory, Applied Physics Department,
Technological University, Delft, Netherlands

ABSTRACT

Investigated is the applicability of the relay- self-oscillating method, in which the optimiser generates its own testsignals, composed of ramps. Various kinds of instrumentation schemes for one- and three-dimensional systems are applied. Experimental results are discussed for different types of non-linear performance surfaces, input and output dynamics (critically , over- and underdamped), noise, disturbances and constraints. A very wide class of systems and disturbances produced a proper working as to convergence speed and hunting loss. An application to a real one-dimensional system is given.

INTRODUCTION

The less that is known about a system and about the disturbances the more experimental procedures are essential to find a control law which is expected to optimize a static performance index. As to applications in connection with simple and cheap processes, we state that only a simple control law and an inexpensive data processing unit are admissible.

In this paper a control system is described, which generates its own test signals, employing hill climbing techniques, using a so called relay extremum method. The test signals are composed from positive and negative rampfunctions. By varying the instants at which the ramp changes sign, one can influence the mean value of the output signal, around which hunting occurs. The ramps are obtained by integrating a constant, e.g. a constant positive or negative voltage, produced by a relay. (This explains the name "relay control"). This method has the advantage that roughly speaking one only needs to know whether the extremum (supposedly unique) of the static performance surface possesses a maximum or a minimum. Moreover the instrumentation is simple indeed.

Other test signals (e.g. binary noise) may have advantages from a theoretical point of view, but when it comes to "systems engineering", low cost, easy installation and maintainability must also be considered.

Though the relay extremum method is not new at all, up to now it has not been sufficiently investigated for what class of systems and disturbances this method will operate properly. Several aspects of this problem are treated below.

The performance P is supposed to be a nonlinear function $P = f(\underline{x})$ of a number of parameters indicated by the vector $\underline{x}$ (Fig. 1). The function $f(\underline{x})$ is assumed to have a unique extremum. The magnitude of performance P has to be measured via more or less unknown linear dynamics $H_2(s)$, giving the signal y. The multidimensional input $\underline{u}$ influences the parameters $\underline{x}$ via more or less unknown dynamics $H_1(s)$. The disturbances may in general enter the system anywhere. The information from y must be analysed in order to generate a $\underline{u}$ according to some control law, so as to find and approximately maintain P at its extremum. The control law is implemented by the optimizing controller OC. Each of the components u_k of $\underline{u}$ is only allowed to vary with a constant rate: $\dot{u}_k = c_k A_k$, $c_k = \pm 1$, where the constant value of A_k is still to be selected.

The controller determines the sign of c_k according to the control law. Some results for a one- and a three-dimensional system will be presented, but the fundamentals will be explained for a one-dimensional system.

A ONE-DIMENSIONAL SYSTEM

In a one-dimensional system $\underline{u}$ and $\underline{x}$ have one component u and x. In the simplest system without dynamics, dead time and disturbances, $H_1(s) = 1$, $H_2(s) = 1$, u = x, y = P. In this case the control law for reaching the extremum of P = f(x) is trivial. If the extremum is a minimum the control law is found as follows: if $\dot{y} > 0$, the performance criterion increases and hence the direction of speed of the input u must be reversed. Thus the sign of c is changed only if $\dot{y}$ becomes positive.

"Superior numbers refer to similarly-numbered references at the end of this paper"

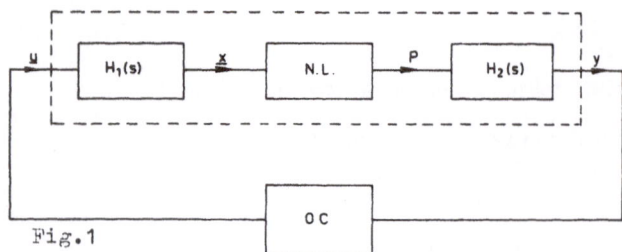

Fig.1

General set-up of the self-optimizing
control system.

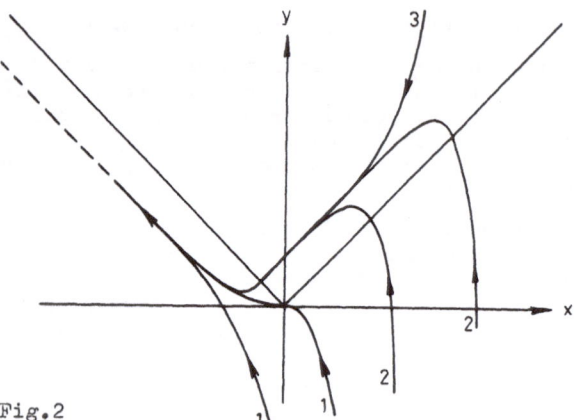

Fig.2

Some y vs x trajectories for a one-
dimensional system; absolute value device;
2nd-order critically damped dynamics
(timeconst. 0.4-0.6s); c=-1; A=10.

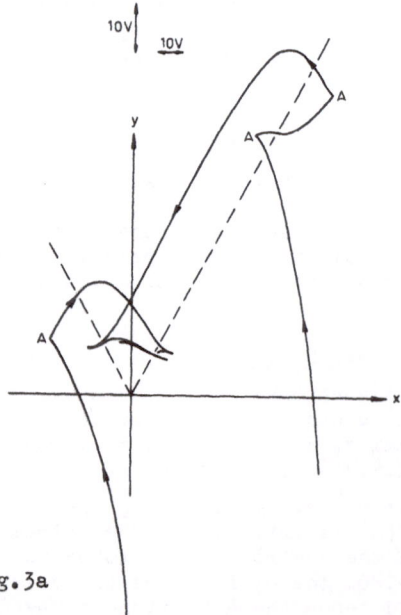

Fig.3a

Some examples of y vs x for a one-dimensional
system; absolute value device;3rd-order
critically damped dynamics
(timeconst. 0.3-0.4-0.6s); A=10; T=1 s.

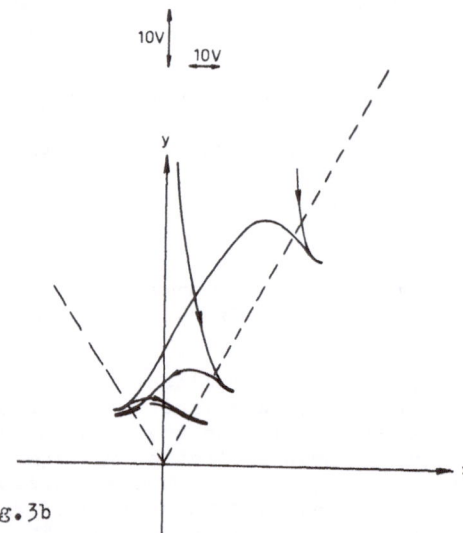

Fig.3b

Some examples of y vs x for a one-dimensional
system;same conditions as in Fig. 3a.

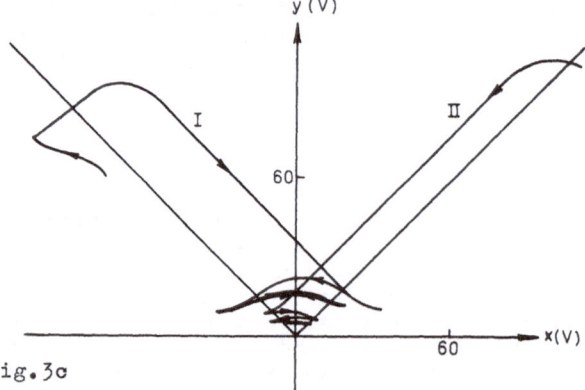

Fig.3c

Some examples of y vs x for a one-dimensional
system; absolute value device;
I 10th-order critically damped dynamics;A=5;
II 3rd-order critically damped dynamics;A=10.

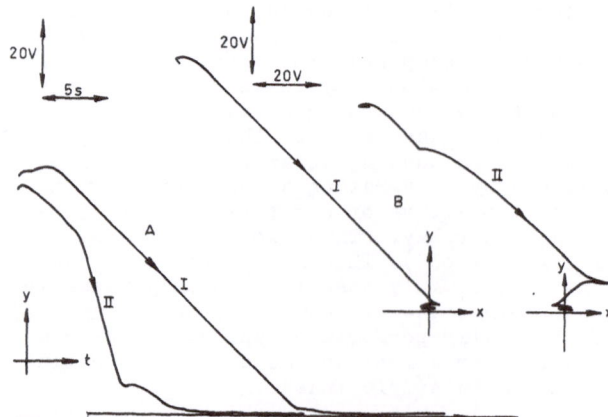

Fig.4

y vs t$^{(A)}$and y vs x(B) for a one-dimensional
system; absolute value device;2nd-order
critically damped dynamics (timeconst.0.5-
0.6s); I integration-timeconst. 2 s.
     II integration-timeconst. 0.5-2 s

When $H_2(s)$ differs from 1, but is known, and when $P = f(x)$ is also known, one can compute the trajectories $y = y(x)$ for $\dot{x} = c A$, $c = \pm 1$, given the initial conditions. For several simple situations results have been published[1,2,3]. Fig. 2 gives one more example. Now as x changes with a constant rate to the left ($c = -1$), starting in point 1, the corresponding y does not go through an extremum. Starting in point 3 it assumes one and in point 2 two extrema. So three different modes of behaviour exist. For $c = +1$ a similar figure exists. As has been mentioned before[1] a rather large class of functions $P = f(x)$ with one extremum (including functions with breaks and inflection points) exhibits these three modes of behaviour. Also the exact knowledge of the dynamics $H_2(s)$ is of little importance; with a pure time delay and critically damped or overdamped systems of all order, one obtains similar trajectories. The precise shape of the curves depends upon $P = f(x)$, $H_2(s)$ and A.

It is obvious that for systems with trajectories analogous to those of Fig. 2, the sign of $\dot{y}$ is not sufficient to determine which action to take to approach the minimum, e.g. a positive value of $\dot{y}$ in the lower part of the curve from point 2 is useful; the same situation in the curve from point 1 leads to divergence.

In order to obtain convergence it is possible to add higher derivatives of y, considering a more general criterion, e.g. the decision function $d = a\dot{y} + b\ddot{y} + \ldots + r$ (where $a, b, \ldots, r$ are constants), in combination with so-called "forced switching"[1,2,3]. One may also employ quite another criterion, e.g. the discontinuity in an appropriate derivative[4,5]. For well known reasons, to be considered below, the use of higher-order derivatives has disadvantages in systems with noise. So in this investigation only the first derivative is used in the decision function, which usually has the form $d = a\dot{y}$. An additive constant r in $d = a\dot{y} + r$ will not be considered here, because it does not give - unlike[2] $d = b\ddot{y} + r$ - any improvement.

For the type of trajectories shown in Fig. 2 it is plausible (and for simple well-defined cases it can be proven) that the following control law does not lead to divergent situations, but produces hunting near the extremum. The law is formulated for a minimum.
a. If $\dot{y}$ changes from negative to positive, c has to change sign.
b. If $\dot{y}$ stays positive longer than a certain time T, the sign of c has to change at

$t = nT$ ($n = 1,2,\ldots$). This arrangement avoids divergence via trajectories like the one originating from point 1 in Fig. 2 and is known as "forced switching".
c. If $\dot{y}$ changes from positive to negative, c does not change sign.
d. If $\dot{y}$ stays negative, the sign of c does not change either.

Some examples of the application of this control law are given in Figs. 3a, 3b and 3c. At points A in Figs. 3a and 3c forced switching occurs. Starting from any initial condition, y moves towards the static characteristic in the x-y plane, approaches the minimum and starts hunting near it.

To apply the control law, first the values of T and A must be chosen.

Concerning T the following remarks can be made. Forced switching is not desired when hunting around the minimum, since this will result in an irregular hunting pattern with a larger mean hunting loss. To prevent this, T must be larger than roughly a quarter of the hunting period. As to the highest value of T no stringent conditions exist. Too large excursions in the wrong direction before forced switching occurs, should be prevented, so T must not be too large. To be on the safe side (also in view of possible variations of the system), T can be chosen about half the hunting period.

Concerning the slope of the ramp the following should be considered. A fast approach to the minimum after a big disturbance requires the slope A to be large. This, however, may also produce a big excursion in the wrong direction before forced switching occurs. Furthermore, as is well known, a large slope causes a big hunting loss. To reduce this loss a slow rate of change of x is desirable. A compromise can be made. With a little extra instrumentation one can switch to a larger value of A, when during a certain time interval $\dot{y}$ stays negative. As soon as $\dot{y}$ becomes positive, the original smaller value of A is switched on again.

In Fig. 4 curves I show y as a function of x and of t following a big disturbance when a small A (or a large integration time constant) is used. The hunting loss is small but the approach to the minimum is slow. In curves II after a certain time interval the slope is made four times as large, resulting in a fast approach to the minimum. Since this would produce a big hunting loss, the small A is used again as soon as $\dot{y}$ changes sign.

Other methods can be applied to avoid or to reduce the hunting loss. One of these is to compute the mean of the extreme values of x during a hunting period and to use this as a constant input signal (so without hunting) given certain conditions [5].

This general survey makes it plausible - and it has been broadly verified - that the given control law can lead to convergence near the optimum under quite mild conditions on T and the slope of the ramp for many (if not all) one-dimensional static performance curves with one extremum provided that the static part of the system is only followed by critically damped or overdamped linear dynamics and that there are no disturbances or constraints. It remains now to study the performance when these restrictions do not hold.

a. Underdamped Output Dynamics

For underdamped output dynamics the trajectories in the x-y plane will show an oscillatory behaviour. For small damping the trajectories have additional relative minima. With the control law employed this means that after a disturbance reversal of the input may take place at the wrong time. This affects the speed of convergence unfavourably, but in the cases which were tried, the optimum could still be reached with the same control law. Fig. 5 gives an example of a one-dimensional system with underdamped dynamics (relative damping $\xi_1 = 0.25$; $\xi_2 = 0.085$). The behaviour for $\xi = 0.085$ shows several relative minima. The convergence is slowed down. The mean hunting loss in this example is quite small, however.

b. Noise, Disturbances and Constraints

In most processes noise is present. Noise may enter into the system at several points. The least convenient is to have noise, especially high frequency noise, added to the output y, before differentiation. It is advisable to decrease the high frequency component by filtering before or (and) after differentiation. We have to determine the kind and amount of noise which can be added while proper operation is maintained. The meaning of proper operation has to be defined. Both the wanted speed of convergence and the admissible hunting loss should be accounted for. In one of the experiments noise, with a spectrum flat to about 1.5 c.p.s. and decreasing with 6 db/oct. thereafter, was added at point y, so that the RMS value of the noise at the point $\dot{y}$ was approximately equal to the amplitude of $\dot{y}$ during hunting without noise. The hunting became more erratic and the average hunting loss was about four times higher than in

the noise-free system (Fig. 6). Convergence to the minimum still occurred but the convergence was about twice as slow. If in the control law higher derivatives of y were used, in the presence of noise less satisfactory results have been achieved than by only using $\dot{y}$. When noise is absent, however, it was found that when $d = b\ddot{y}$ is used, the speed of convergence sometimes is higher for critically damped systems. This does not hold anymore when the output dynamics exhibit an underdamped behaviour. For these reasons it seems that the use of $d = a\dot{y}$ as a criterion is more favourable, when the knowledge about the system is very restricted and noise is present.

A changing performance curve, and so a moving extremum, is to be expected in time-varying systems. This has been simulated by adding a sinusoidal component to the performance curve $(P = f(x) + a \sin\omega t)$; up to a frequency of about half the hunting frequency, the system keeps hunting around the shifting extremum, independent of the amplitude a.

Often the range of permissible values of x is restricted. Several simple restrictions have been tried, e.g. x must always be smaller than a certain value. The extremum may lie within or outside this range of x. In the former case hunting around the extremum resulted; in the latter the system keeps hunting on the boundary of the range of x which is closest to the extremum.

c. Linear Input Dynamics

To study the influence of input dynamics $H_1(s)$ preceding the non-linear function, we consider the case when part of the output dynamics is moved over to the input[7,8]. Two effects result. The triangular output of the integrator is somewhat smoothed, so that a smaller hunting loss might be expected. At the same time, however, a frequency change may take place. When the resulting frequency is lower, this produces a larger integrator output signal. This was e.g. the case with an absolute value device as non-linear function, followed by a third-order system with time constants of 0.5-0.5-0.5 s, when one of these time constants was moved over to the input. For this special case, the net result was a reduction of the mean hunting loss with a factor 0.9. In all cases tried with input dynamics a proper working of the control law has been observed.

MULTI-DIMENSIONAL SYSTEMS

The present method may be extended to

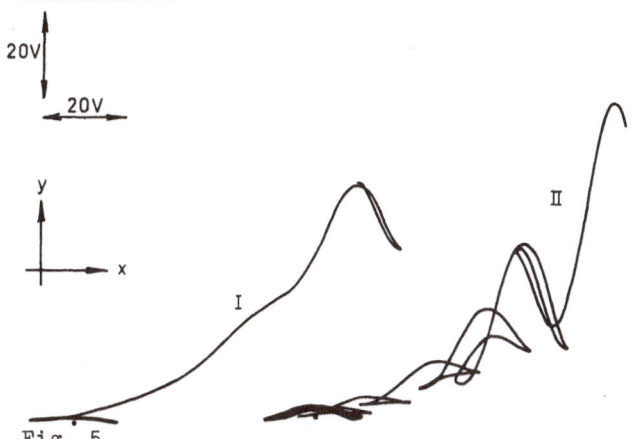

Fig. 5
y vs x for a one-dimensional system;square
value device; 3rd-order dynamics
I timeconst. 0.9 s + 2nd-order (ω_n=2;ξ=0.25);
II timeconst. 0.9 s + 2nd-order (ω_n=2;ξ=0.085)

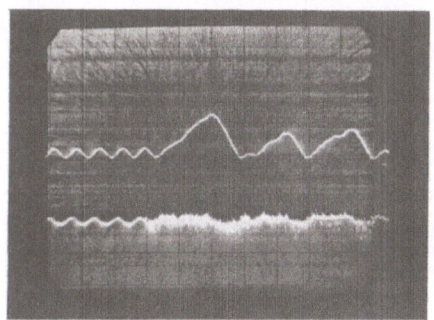

Fig.6
y vs t (above) and ẏ vs t (below) during
hunting for a one-dimensional system, with
noise (right) and without noise(left).

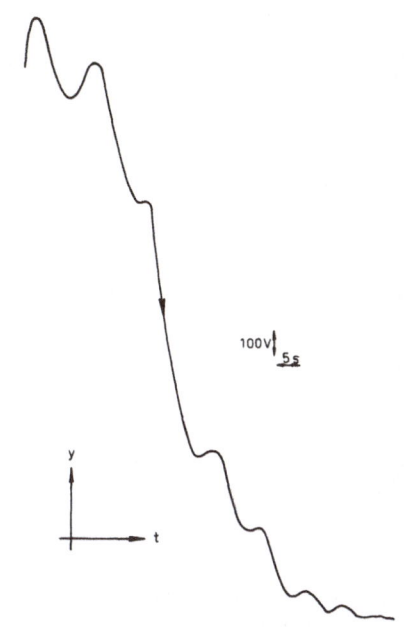

Fig. 7b. y vs t for a three-dimensional
system with simultaneous search; same
conditions as in Fig. 7a.

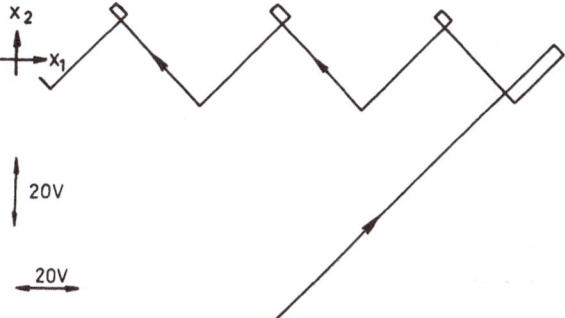

Fig. 8a. x_1 vs x_2 for a three-dimensional
system with simultaneous search;
P=|x_1|+|x_2|+|x_3|;3rd-order critically damped
dynamics
(timeconst. 0.3-0.4-0.6 s); A=5-20;T= 1 s

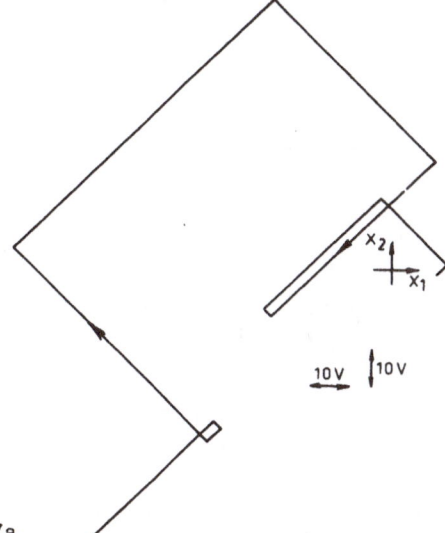

Fig. 7a
x_1 vs x_2 for a three-dimensional system;with
simultaneous search P=x_1^2+x_2^2+x_3^2; 3rd-order
critically damped dynamics (timeconst.0.3-
0.4-0.6s) A=5-20 ;T=2 s.

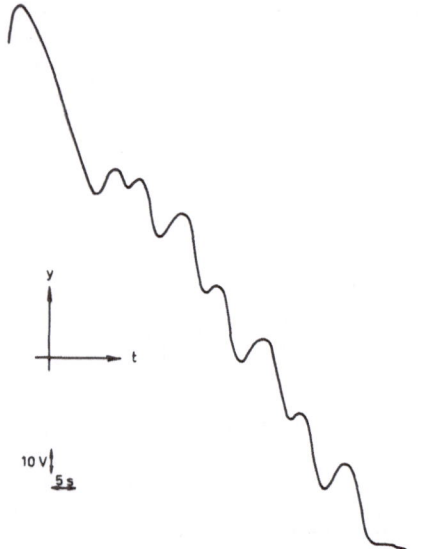

Fig. 8b. y vs t for a three-dimensional
system with simultaneous search; same
conditions as in Fig. 8a.

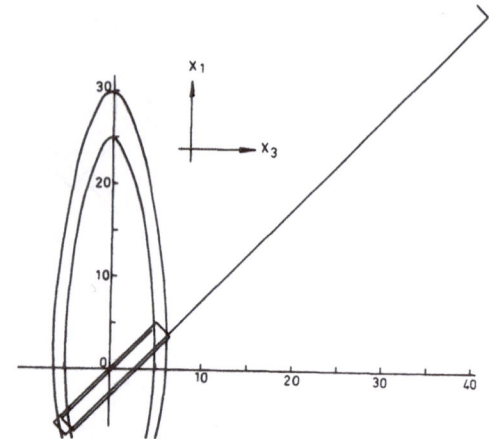

Fig . 9a x_1 vs x_3 for a three-dimensional
system with simultaneous search;
$$P = \frac{x_1^2}{100} + \frac{x_2^2}{4} + \frac{x_3^2}{4} \; ; \; 4^{th}\text{-order dynamics}$$
(timeconst. 0.3-0.5 s + 2nd-order
(ω_n=2.24; ξ=0.45))
A= 5-20; T= 2 s

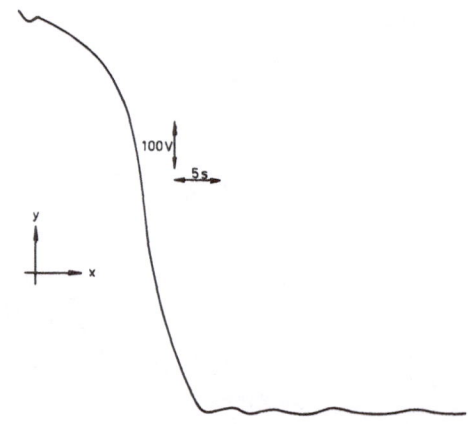

Fig. 9b. y vs t for three-dimensional
system with simultaneous search; same
conditions as in Fig. 9a.

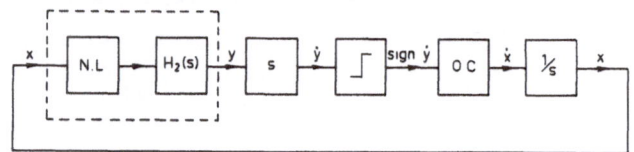

Fig. 10 A realization of a one-dimensional
self-optimizing control system.

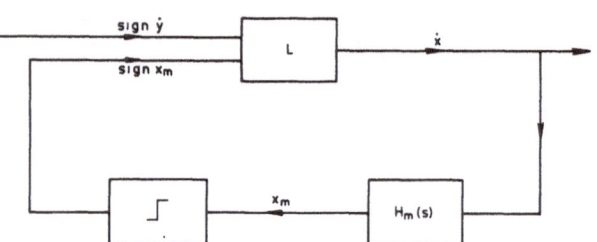

Fig. 11 A realization of the optimizing
controller.

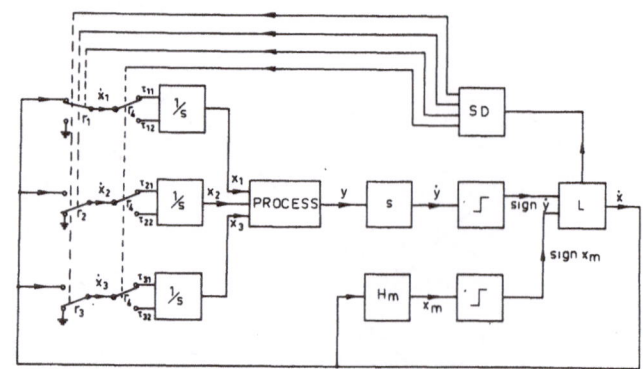

Fig. 12 Blockdiagram of a three-dimensional
self-optimizing control system with
sequential search.

multi-dimensional processes, in particular to those having a static performance surface with one extremum and without saddlepoints, ridges or other pathological features. The components of the input may be adjusted either simultaneously or sequentially. Again sign $\dot{y}$ is used as a switching criterion. When the search is simultaneous all u_k vary continuously according to $\dot{u}_k = c_k A_k$ with $c_k = \pm 1$.

If - in the case of a minimum - sign $\dot{y}$ changes from negative to positive or stays positive longer than a certain time T, at $t = nT$ (n = 1,2,.) the speed of one or more of the inputs is reversed. The order of succession in which the inputs are changed may be cyclical (e.g. successively x_1, x_2, x_2, x_3, x_1,...). When the search is sequential only one input is varied, while the other ones are held constant. The behaviour of sign $\dot{y}$ is used in a similar way to activate the inputs.

INSTRUMENTATION AND EXPERIMENTAL RESULTS

The optimizing system has been implemented in a number of ways.

a. To study the behaviour under ideal conditions, a digital simulation on the Delft University TR4 computer proved very useful. The dynamics of the process could easily be programmed using a special procedure (DISAR, Digitale Simulatie van Analoge Rekenmachine) by which the behaviour of an analogue computer is simulated. Noise can be added at different points by a random number generator. A TR4 simulation of one- and three-dimensional systems using various performance surfaces and dynamics was tried with good results. Figs. 7-9 give some examples of simultaneous search with several non-linear characteristics and dynamical behaviours.

b. A straightforward physical realization of a one-dimensional system using analogue computer components is sketched in Fig. 10. Except for the optimizing controller OC all elements are ordinary analogue computer components. If sign $\dot{y}$ changes from negative to positive, in OC a bistable digital multivibrator is triggered, which starts running at a frequency $\frac{1}{2T}$ producing an alternating output voltage $\dot{x}$ with levels +A and -A volts. As soon as sign $\dot{y}$ changes from positive to negative the multivibrator stops, holding $\dot{x}$ at the voltage +A or -A.

Fig. 5 , treated before, has been obtained in this way.

c. Slightly less straightforward than the OC of b, but intimately connected with other systems reported[9] is the OC of Fig. 11. It constists of some logic L, again producing an $\dot{x}$ which is either +A or -A, a dynamic system H_m with output x_m and a sign-relay giving sign x_m. If sign $\dot{y}$ is positive L produces an $\dot{x}$ with the opposite value of sign x_m. The loop with L, H_m and the sign-device produces an oscillation of which the frequency $\frac{1}{2T}$ and the amplitude can be calculated (e.g. by the describing function method). H_m has to be at least a third-order system; it may be composed from R-C networks. If sign $\dot{y}$ changes from positive to negative, L maintains $\dot{x}$ at the existing voltage +A or -A, and prevents a change of $\dot{x}$ by sign x_m as long as sign $\dot{y}$ is negative as required by the control law. So instead of a digital multivibrator an analogue one is used to produce forced switching when sign $\dot{y}$ is positive.

The instrumentation scheme of a three-dimensional system with sequential search is given in Fig. 12[10]. The switching device SD activates the relays r_1, r_2, r_3 in a cyclic way each time sign $\dot{y}$ changes from negative to positive. The SD also activates the relay r_4 in order to obtain a reduction of the hunting loss and a high speed of convergence by using two different integration-constants (τ_{ij}; i, j = 1,2,3) which is equivalent to the use of different values of A.

Fig. 13 gives an example of the operation of the system after a step disturbance in x_1 and x_2.

A similar instrumentation scheme can be obtained for the case of simultaneous search[10]. Fig. 14 shows the behaviour of such a system after a step disturbance in x_1 and x_2.

With implimentation according to a, b and c comparable results have been obtained.

d. The instrumentation used in c and indicated in Figs. 9 and 10 can be redrawn as in Fig. 15 for which quite another interpretation can be given. Here $\dot{x}$ does not only act upon the process but also on a model H_m. The sign of the output x_m of the model is used in L together with sign $\dot{y}$. The idea of the model H_m is primary to compensate for the dynamics of the process. The use of such a model has been advocated from the very

269

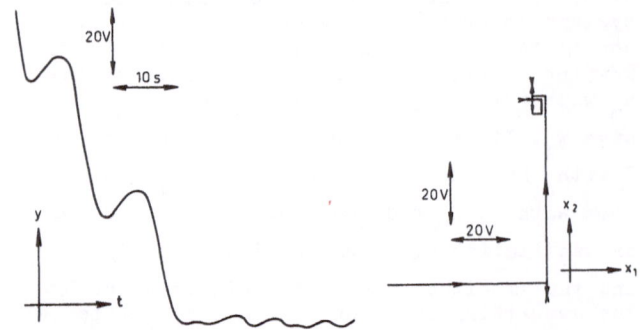

Fig. 13   y vs t and x_1 vs x_2 for a three-dimensional system with sequential search; $P=|x_1|+|x_2|+|x_3|$; critically damped dynamics (timeconst. 0.6-1-2 s); A=5-20; H_m(timeconst. 2-3 s)

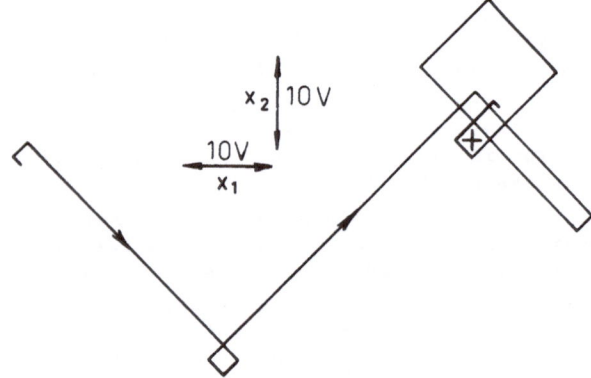

Fig. 14 x_1 vs x_2 for a three-dimensional system with simultaneous search; $P = x_1^2 + x_2^2 + x_3^2$; critically damped dynamics (timeconst. 0.3-0.4-0.6 s)

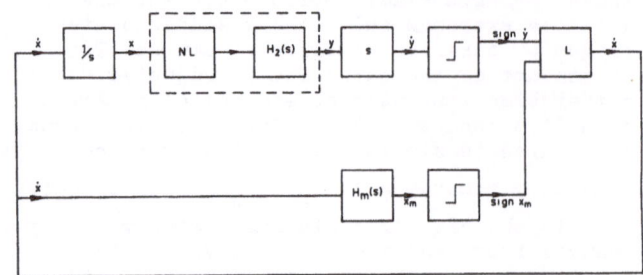

Fig. 15 Blockdiagram of a one-dimensional self-optimizing control system.

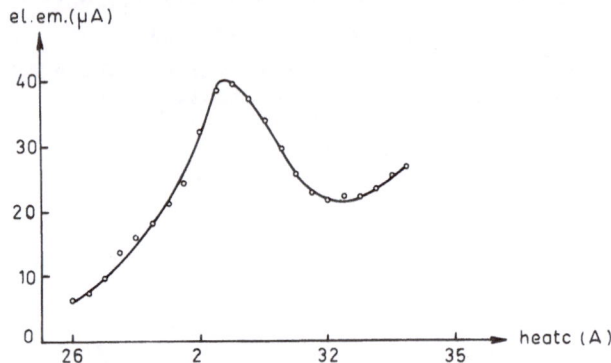

Fig. 16   Electronemission vs heating current for the evaporation of a NiCr-wire.

start of the study of this type of optimizing systems. In that concept a partly different control law is applied allowing a change of $\dot{x}$ by a change of sign x_m also when sign $\dot{y}$ is negative.

Among others Voskamp[9] has proved the proper operation of such a system. By further analysis of the properties of those systems [11], however, it has been found that a better speed of convergence was obtained when the model was only active when sign $\dot{y}$ was positive, which resulted in the control law treated. Thus this type of law has been found independently and along quite different lines from other research workers[1,2].

e. As a simple real process the evaporation of a NiCr-wire to produce thin-film resistors has been used. As a function of the temperature of the wire, determined by the heating current, there exists a maximum in the electron emission (Fig. 16). The position of the maximum varies heavily with time in an erratic way and is strongly dependent on the situation at the surface of the wire. For reasons connected with the evaporation process it seemed appropriate to keep the wire at maximum electron emission during its effective life time. The levels +A and –A of $\dot{x}$ activated a motor that varied a resistor in series with the wire in order to regulate the heating current. Schotanus[12] demonstrated that the maximum was attained sufficiently fast and could easily be followed. For this simple process a simple instrumentation of the control law proved to be adequate and very easy to install.

f. Many of the schemes treated in a, b and c have been tried with noise, time-varying performance surfaces, input dynamics, constraints etc. Convergence to the extremum has always been found. No significant difference seemed to exist between simultaneous and sequential search in the case of three-dimensional systems with the performance surfaces and dynamics used. A point of further research is the question what performance surfaces, system dynamics and disturbances are admissible for the given control law and for given requirements of speed of convergence and hunting loss.

CONCLUSIONS

It has been found that the control law treated produces convergence to the extremum and hunting around it with many different static performance surfaces with one minimum, with many different processes and disturbances and with various kinds of instrumentation. It is still impossible to state exactly what are the admissible systems and disturbances for proper operation. It is clear, however, that the class of systems and disturbances is quite wide. So, when little is known about a system, only that one single extremum of the static performance index is to be expected, it may be worth while to start with a relay-self-oscillating procedure; the instrumentation is simple and cheap and little risk is taken from an economic point of view. Afterwards, when more is known about the system, and it seems economically justified, more sophisticated control laws, possibly involving identification procedures and computer control might be applied.

REFERENCES

(1) Bobrov Yu.I., Kornilov R.V., Putsillo V.P., "Determination of the optimizer control law by taking into account the relaxation of the system". Automat. Telemech., XXIV, no. 2 (1963), 185-192.

(2) Perret R., Rouxel R., "Principle and application of an extremal computer". Proc. 2nd IFAC Congress, Basle, 1963, 527-540. London; Butterworths.

(3) Perret R., Rouxel R., "Study of an algorithm for dynamic optimization. 1964. Computing Methods in Optimization Problems; ed. by A.V. Balakrishnan, et al, Academic Press

(4) Hamza M.H., Discussions ad (2) 538.

(5) Hamza M.H., "Extremum control in the presence of variable pure delay" Soc. Instr. Techn. IFAC (Teddington, September 1965) Symp. on the Theory of Selfadaptive Control Systems.

(6) Carmassi M., Helein J., Rouxel R., "Optimization of a sulphur recovery plant", Preprints IFAC (Tokyo, August 1965) Symposium on Systems Engineering For Control System Design

(7) Morosanov I.S., "Methods of extremum control", Automat. Telemech., XVIII, no. 11 (1957) 1077-1092.

(8) Tsien H.S., Engeneering Cybernetics McGraw-Hill Publishing Company LTD, New York, 1954.

(9) Voskamp J.H., "An optimalizing control system " Trans. Soc. Instr. Techn., XIV, no. 3 (1962) 192-200.

(10) Broekstra G., "Optimum zoekend systeem voor processen met meerdere ingangen", report Instrumentation Laboratory, Technological University, Delft, April 1965.

(11) Kooymans W., "Een optimum zoekend systeem voor processen met twee in-gangen", report Instrumentation Labo-ratory, Technological University, Delft April 1964.

(12) Schotanus D.J., "Het verdampen van een nichroomdraad", report Instrumentation Laboratory, Technological University, Delft, October 1964.

DISCUSSION

Dr. M.H. Hamza (Switzerland) pointed out that in an extreme case the optimum could be reached through a succession of steady state measurements although fairly obviously this would not be a good controller because of the time consumed. An objective of dynamic self-optimising controllers is to provide a means of reaching the optimum as accurately and rapidly as possible. The author was asked to comment on the performance of this system from the point of view of the time taken to reach the extremum, especially since diagrams had been shown of two dimensional systems with the controlled trajectory spiralling in to the origin. The suggestion was that such a spiral was a waste of time and that a better perform-ance could easily be obtained by making simult-aneous adjustment of the variables.

The authors agreed that the more one knew about a system the better one could design the control procedure. One flexible feature of the method described is the slope of the ramp perturbation so that by increasing the slope the system could be made to approach the optimum more rapidly. However such action also results in an increased hunting loss in the region of the optimum. Therefore a compromise was suggested where-in a large value of slope would be used initially and then reduced in the region of the optimum. In this way a rapid approach to the optimum would be combined with reasonable hunting losses once the optimum was reached.

Written contribution to the discussion
Dr. J.D. Roberts (U.K.)

I would like to ask Professor Verhagen two questions about his very interesting physical application. These questions arise out of certain theoretical results about optimum extremum regulation[1],[2] of randomly drifting processes. The results are asymptotically valid for slow variations, and are relevent when the effect of the initial transient search is finished.

If a process can be represented as a quad-ratic non-linearity disturbed by Brownian motion drift followed by a linear operator $G(s)$ and measurement noise, what of all possible ways of closing the loop produces the minimum mean square error e ?

The optimal extremum regulator uses a sinusoidal perturbation. The output of the plant filtered by a linear operator $H(s)$ (cut-ting off low frequencies) is multiplied by the perturbation (possibly phase shifted) and integrated before adding the perturbation.

The optimum frequency of the perturbation is the resonant frequency of $G(s) H(s)$. This does not mean that the only system which pro-duces a performance nearly as good as this system is made according to this structure, but the per-formance of a system is limited by the waveform of the fluctuation of the input signal about the best estimate of the peak input at any time. If a different kind of perturbation is used, the performance depends on the frequency spectrum of the perturbation and the flatness of $S(\omega) = G(j\omega) H(j\omega)^2$. If there are no dynamics, $S(\omega)$ is flat. If for example $G(s)$ represents a simple lag $1/(1 + T s)$ and the measurement noise is white, and if T is made very large, then $S(\omega)$ approaches a certain shape asymptotically. The mean square errors for various types of perturbation exceed the mean square error for a sine wave by factors.

1.01 for a triangular (sawtooth) wave

1.07 for a square wave

1.7 for a random telegraph signal

<1.7 for a chain code

(when the extremum regulator is designed optimally in other respects).

If there are two input signals to be opti-mised, the best searching strategy use a pair of sine waves. This cannot however be extended to three or more inputs because a further sine wave of the same frequency could not be linearly independent. The practical solution would be a compromise between properties of the individual frequency spectra and of the correlation spectra of the signals.

Thus a ramp searching strategy is good because of its closeness to a sine wave perturba-tion. For this reason however we would expect interesting problems when three or more inputs had to be optimised simultaneously by a ramp searching strategy.

Could I please therefore ask Professor Verhagen:

(i) It is possible to say at all what the random disturbances to the optimum are like?

(ii) What does the timing of the ramp searching strategy look like when three variables are optimised simultaneously?

References

1. Roberts, J.D. Extremum or 'hill-climbing' regulation: a statistical theory involving lags, disturbances and noise, Proc. Instn. Elect. Engrs. 1965 (Jan.) 112, 137.

2. Roberts, J.D. On the design of optimal extremum regulators, 'Advances in Automatic Control', U.K.A.C. Convention (Nottingham, England, 1965), (Instn. Mech. Engrs. London).

Author's reply:

I thank Dr. Roberts for his contribution to one of the problems encountered in self-optimizing control systems. I should like to state, however, that his approach differs essentially from our method. We do not use external perturbations, nor linear operators for filtering, multipliers, phase correctors etc. As a consequence no such difficulties as the use of non-interrelated external perturbation signals are encountered in multi-dimensional systems.

As has been described, self-oscillation produces in our system positive and negative ramps, each with a length that is automatically adjusted by the control law used, resulting in an approach to the optimum and a hunting around it.

The sequence of switching in more-dimensional systems is not critical at all, but can be based e.g. on practical considerations, depending on the type of disturbances and the shape of the performance surface. An example of the switching for a 3-dimensional process ($P = |x_1| + |x_2| + |x_3|$, 2-nd order dynamics, time constants $0.3-0.5s$ $A=2-10$, with simultaneous search is given in the following map:

sign $\hat{y}$	sign x_1	sign x_2	sign x_3
0	0	0	0
1	1	0	0
0	1	0	0
1	1	1	0
0	1	1	0
1	0	1	0
0	0	1	0
1	0	1	1
0	0	1	1
1	1	1	1
0	1	1	1
1	1	0	1
0	1	0	1
1	0	0	1
0	0	0	1
1	0	0	0

Other sequences, e.g. pure cyclically, do werk succesfully too.

As to the first question, many kinds of noise and a sinusoidal shift of the non-linear characteristic have been applied. Some specific examples are given in the paper. As has been mentioned it is certainly a point of futher consideration, what are the admissible disturbances in relation to the process parameters to obtain a certain convergence speed and hunting loss.

THE USE OF PSEUDO-RANDOM SIGNALS
IN ADAPTIVE CONTROL

by J. L. Douce and K. C. Ng
School of Engineering Science
University of Warwick,
Coventry, England

1. INTRODUCTION

Previous work on adaptive control has treated the problem as being distinct from the allied subject of process identification. Thus entirely separate techniques have been developed for the solution of these two related topics. Periodic test signals have received wide acceptance as the probe for hill-climbing systems and sinusoidal perturbation is claimed to be the optimum in particular situations[1]. The most recently proposed input for identification purposes is the pseudo-random binary sequence (PRBS), which can be chosen to give whatever detail is required concerning the impulse response of the system.

This paper shows how the same binary sequence can be used for simultaneous hill-climbing, with a response time comparable with the previously proposed optimum system using high-frequency perturbation. The technique is easily realised, using either a digital or a hybrid computer.

Extension of the technique to simultaneous optimisation of several parameters of a multi-parameter system requires a corresponding number of uncorrelated PRBS's or 'chain codes'. These signals also find important applications in the determination of the impulse responses of multiple input (and output) systems. The existing PRBS's do not seem to possess this desirable zero cross-correlation function. A search for uncorrelated codes has been undertaken and several approaches for synthesising new codes are considered, but none have proved successful to date.

2. DEVELOPMENT

The use of a random perturbation for hill-climbing or gradient estimation has been described by van der Grinten[2]. This reference shows that the slope of the performance index P as a function of parameter K is given by

$$\frac{\int_{-\infty}^{\infty} \phi_{xy}(s)\,ds}{\int_{-\infty}^{\infty} \phi_{xx}(s)\,ds} = \int_{-\infty}^{\infty} h(\tau)\,d\tau \qquad (1)$$

where $\phi_{xx}(s)$ is the auto-correlation function of the random perturbation signal and $\phi_{xy}(s)$ is the cross-correlation function between the perturbation x and the performance index signal y. $h(\tau)$ is the impulse response of the control system from parameter adjustment to performance index.

If the random signal x is stationary then $\int_{-\infty}^{\infty} \phi_{xx}(s)\,ds$ is constant so that only the cross-correlation function must be measured. Furthermore the integral may be approximated by

$$I = \int_{0}^{T_2} \phi_{xy}(s)\,ds = \int_{0}^{T_2} \int_{-\infty}^{\infty} y(t)\,x(t-s)\,dt\,ds$$

where T_2 is finite such that $\int_{T_2}^{\infty} \phi_{xy}(s)\,ds$ is small. Finally, interchanging the order of integration gives

$$I = \int_{-\infty}^{\infty} y(t) \left\{ \int_{0}^{T_2} x(t-s)\,ds \right\} dt . \qquad (2)$$

Thus the slope is found by taking the running average of the random input x, multiplying this by the performance measure y and integrating with respect to time.

The major difficulty with this technique is the statistical fluctuations in the measurements of the random signals. These fluctuations fundamentally limit the maximum rate at which the parameter can be adjusted towards the optimum value. Another difficulty is the generation of accurate delayed versions of the random signal x. It is also necessary to ensure that the random signal has constant statistical properties.

The use of a pseudo-random binary signal as the probing perturbation removes each of these practical difficulties. Since this signal is deterministic, by measuring the slope over one repetition period, all statistical errors are eliminated. The signal itself is easily generated using shift-registers and delayed versions of this signal for correlation measurements are readily obtained using simple logic elements. The running average operation is performed using a reversible binary counter and the digital output of this element forms one input to a hybrid multiplier, the analogue output feeding an analogue integrator. A block diagram of the system is shown in Figure 1.

It is very important in this technique that the mean value of x, y and of the running average of x be eliminated by passing these signals through high-pass filters with low cut-off frequencies. Otherwise, if the signal y has mean value $\bar{y}$ and the running average has mean

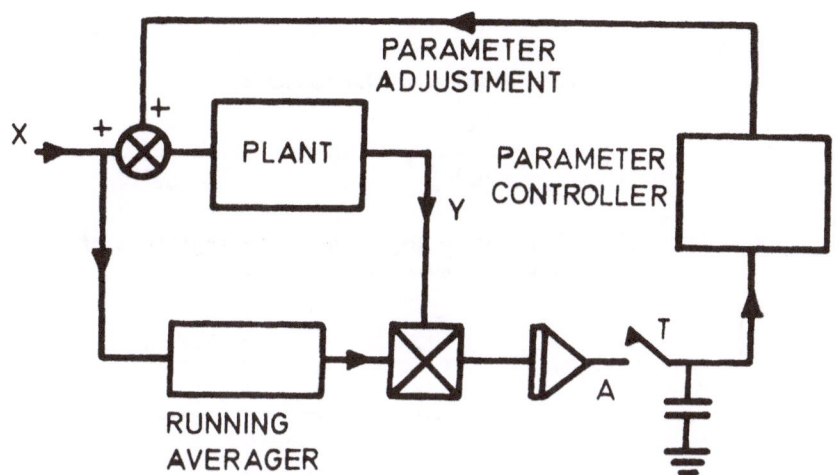

Fig. 1.   Block Diagram of the Proposed System

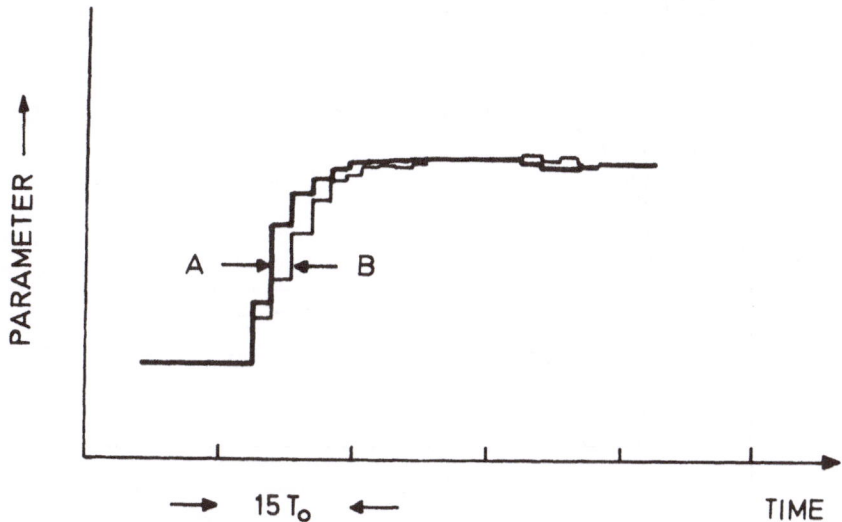

Fig. 2.   Typical Responses

(a) Proposed System

(b) Conventional System with phase compensation.

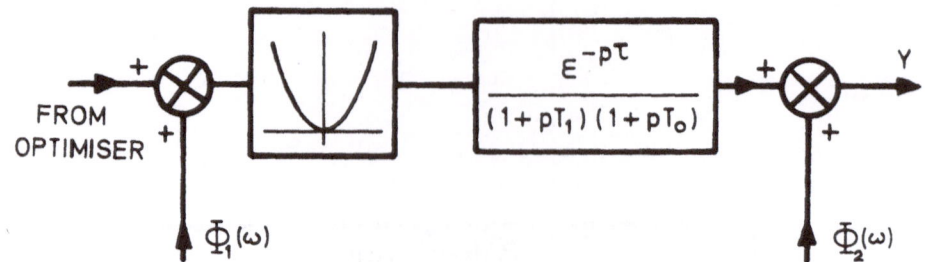

Fig. 3.   2nd-Order Plant with Disturbances

$$\Phi_1(\omega) = \frac{1}{1+2\omega^2 T_0^2} , \qquad \Phi_2(\omega) = \frac{1}{1+36\omega^2 T_0^2}$$

$\sigma_1^2 = H^2$, $\sigma_2^2 = \frac{1}{2}H^2$ where   H = amplitude of perturbation signal $\tau = T_1 = 0.2T_0$.

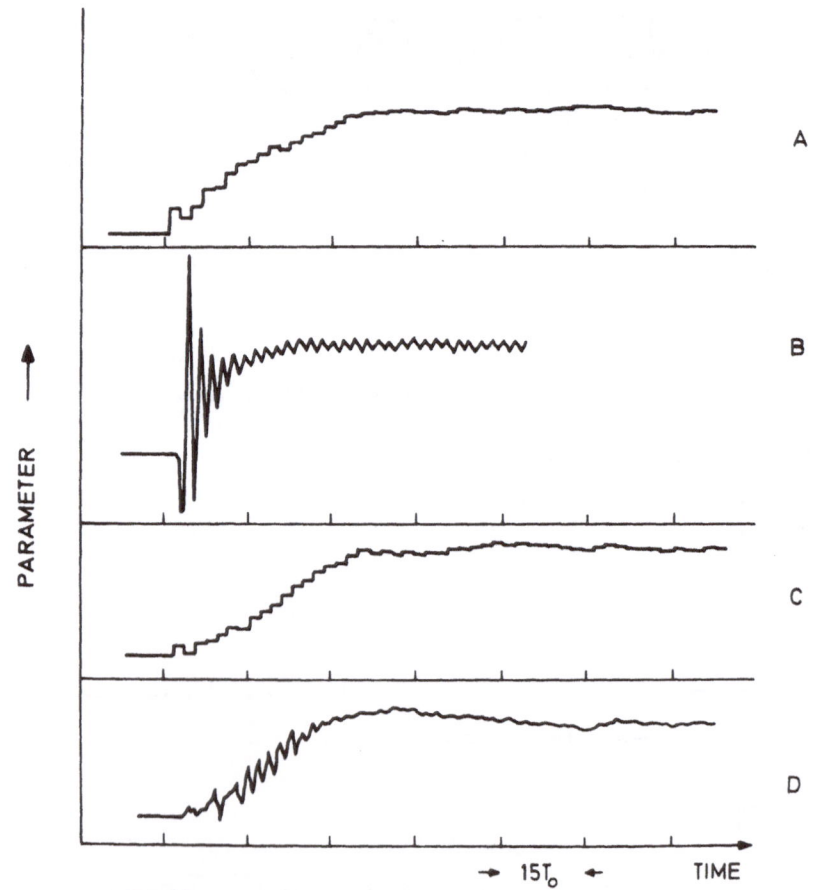

Fig. 4.   System Responses with Plant in Fig. 3.

(i) Conventional System, phase compensated:- (a) with sample-and-hold, (b) without sample-and-hold.

(ii) Proposed System:- (c) with sample-and-hold, (d) without sample-and-hold.

276

value $\overline{X_{RA}}$, then the signal $\int_0^{T_2} \phi_{xy}(s)\,ds$ will contain error terms

$$\int_{-\infty}^{\infty} \overline{y} \left\{ \int_0^{T_2} x(t-s)\,ds \right\} dt + \int_{-\infty}^{\infty} \overline{y}\ \overline{X_{RA}}\,dt + \int_{-\infty}^{\infty} \overline{X_{RA}} \cdot y(t)\,dt$$

These unwanted signals can cause the parameter to drift away from the optimum value. The presence of slow drift terms (of the form k.t) in y introduces errors proportional to the first moment integral of the perturbation signal, i.e.

$$k \int_{-\infty}^{\infty} t \cdot \left\{ \int_0^{T_2} x(t-s)\,ds \right\} dt \ .$$

This has been found to have significant effect on the accuracy of measurements of the cross-correlation functions.

3. COMPARISON WITH OTHER SYSTEMS

The system discussed has been compared, analytically and experimentally, with a conventional system using square-wave perturbation. The features considered important are speed of approach to the optimum and amplitude of steady-stat fluctuations about the optimum, including stability investigations. Analysis is possible only for the simplest plants, for example a parameter-performance index relationship consisting of a square-law characteristic followed by a simple lag.

The two systems are compared theoretically by determining the signal at the point A in figure 1 after a time T and at a similar point in the conventional system, over the same time interval with the same perturbation amplitude. For the simple plant of figure 2, the ratio of the signals at point A of the two systems is given by

$$\frac{G_{PRBS}}{G_{SW}} = \frac{4\pi^2}{127} \sqrt{1 + \frac{4\pi^2 T_0^2}{T^2}} \cdot \frac{(1 - \varepsilon^{-T/2T_0})}{\cos(\phi_n + \theta_n)}$$

where ϕ_n is the phase shift in the conventional system, equal to $\tan^{-1}\left(-\frac{2\pi T_0}{T}\right)$ and θ_n is the compensating phase angle. The system developed is slightly inferior, in that less amplitude is given at this point. This analysis assumes, however, that the phase shift of the plant is known when periodic perturbation is applied and the resulting advantage may not be realised in practice.

The steady state fluctuations are smaller in the system proposed here, due, it is considered, to the inserted sampler. A complete analysis is not possible as yet but the experimental results discussed below fully demonstrate the form of the responses.

4. EXPERIMENTAL RESULTS

In the experiments, the pseudo-random binary signal used is a maximal length sequence of 127 bits. The period T is chosen equal to twice the plant time constant T_0 and the running average is taken over 63 bit intervals.

Figure 2 compares the behaviour of the proposed system with that of a conventional hill climber using square-wave perturbation, of period T, followed by a sample-and-hold unit. In each case the gain is adjusted for optimum results. The plant is a simple lag of time constant T_0 preceeded by a square-law device. No disturbance signals are present. The response of the two schemes are very similar and in each case the steady state is reached after a time equal to about $12T_0$.

The plant shown in figure 3 is a simple approximation to the transfer function of a part of a distillation column, with appropriate disturbance signals present. The performance of the two adaptive systems are shown in figure 4. This figure again demonstrates the similarity of behaviour of the two techniques. One advantage of the new system is that no knowledge is required of the magnitude of the phase shift introduced by the plant to the perturbation signal. This figure also illustrates the advantage of inserting the sample-and-hold unit shown in figure 1, even in the conventional system[4], although this is not predicted by one derivation of the optimal adaptive system[1].

5. EXTENSION TO MULTI-PARAMETER SYSTEMS

The technique can be extended to simultaneous optimisation of multi-parameter systems by employing uncorrelated PRBS's as the perturbation signals. Uncorrelated signals are desirable to reduce cross-coupling between the several optimising loops, thus increasing the accuracy of measurements and reducing the probability of instability. These signals also find important applications in the determination of the impulse responses of multiple input (and output) systems. These signals must

(i) be repetitive

(ii) possess an auto-correlation function which approximates to an impulse for small τ and which is small and constant for large τ.

Codes having the above properties are already known and mostly belong to two major classes known as "maximal length null sequences" and "quadratic residues codes".

Unfortunately, none of the existing codes have so far been shown to be uncorrelated to any other. A search has therefore been undertaken to find codes with this desired property. The initial step is to generate new codes and several approaches have been proposed. This section outlines some methods already considered which have proved unsuccessful so far but may serve to encourage further work in this subject.

5.1 The first method is a straight-forward search involving the use of a digital computer. If the two levels of the signal are represented by +1 and 0, then it can be shown by considering the desired auto-correlation function that the following relationship must be satisfied for a code to exist.

$$\frac{m(m-1)}{N-1} = c \quad \text{(an integer)} \qquad (3)$$

where N is the length of the code, m is the number of "ones" and c is the constant value of the auto-correlation function for large τ. Two other relationships[3] useful in this work are defined as follows. If n_1 is the number of groups of consecutive ones, n_{11} (n_{10}) the number of groups of single ones (zeros), then

$$n_1 = \frac{m(N-m)}{N-1} \qquad (4)$$

$$n_1 = n_{11} + n_{10} . \qquad (5)$$

Given the data on N and m the computer tries all the possible combinations of m ones in N places, subject to the constraints on n_1, n_{11} and n_{10}, which give an auto-correlation function of the desired form. Two codes of length 13 have been found by this method, one of which has not been tabulated before, although it is isomorphous to a known sequence[5].

The method becomes unwieldy as N gets large and careful elimination of certain combinations by inspection must be exercised to cut down computing time.

5.2 The power spectrum of the code is a line spectra with components at frequencies $\alpha\omega_c$, where $T_c = \frac{2\pi}{\omega_c}$ is the basic even time, and of magnitude proportional to $\left(\frac{\sin \alpha\omega/N}{\alpha\,\omega/N}\right)^2$. If therefore a signal consisting of the sum of sine waves of equal amplitude and random phase and of frequencies $\alpha\omega_c$ up to $\frac{N-1}{2}\omega_c$ is passed through

a sample-and-hold element, the resultant signal will have the required power spectrum. In general, the signal will be a multi-level signal. Only by adjusting the phases of the original sine waves can the levels be made equal to ± 1. This procedure is readily implemented on a digital computer, but no results are available to date.

6. CONCLUSIONS

An optimising system employing a pseudo-random binary signal as perturbation waveform has been described. The method has many advantages over the systems employing either stochastic or periodic signals as regards accuracy and speed of optimisations and also has practical advantages. In particular, one test signal can be used for simultaneous hill-climbing and plant identification. The significant disadvantage of high-frequency perturbation systems, requiring a knowledge of phase-shift introduced by the plant dynamics, is overcome in an elegant manner. The extension of the technique to multi-parameter systems is briefly considered.

7. REFERENCES

(1) Roberts, J. D. "Extremum or hill-climbing regulation: a statistical theory involving lags, disturbances and noise". Proc. I.E.E. Volume 112, No. 1 January, 1965.

(2) Van der Grinten, P. M. E. M. "The Application of Random Test Signals in Process Identification". I.F.A.C. 1963, Paper 263. Basle Congress

(3) Private correspondence with Mr. D. Everett of the General Electric Company Ltd., Wembley.

(4) Douce, J. L. and Bond, A. D. "The Development and Performance of a Self-optimising System". Proc. I.E.E. 110(3), 1963.

(5) Marshall Hall, Jr., "A Survey of Difference Sets". Proc. Amer. Math. Soc., Vol. 7, No. 6, December 1956, pp.975-987.

8. ACKNOWLEDGEMENT

The authors gratefully acknowledge the assistance of the National Physical Laboratory Autonomics Division both for fruitful discussions and for financial assistance.

DISCUSSION

Dr. O. L. R. Jacobs (U.K.) asked about the choice of amplitudes of the periodic square wave and pseudo random binary sequence used in the comparative study. Dr. Ng stated that they were made equal to one another. Dr. K. R. Godfrey (U.K.), commenting on section 5 of the paper on the extension of the method to multi-parameter systems mentioned that work had been done at the National Physical Laboratory synthesising approximately uncorrelated multi-level sequences having the idealised impulse-like autocorrelation function (ref. 1). Dr. Godfrey also asked about the frequency of updating and about the effect of noise.

Dr. Ng replied that random disturbances certainly had been injected into the system described. Unfortunately he did not have any quantitative analysis of the effects of such disturbances on the system performance. Nevertheless it was suggested that the disturbances used were quite large and that therefore the results were pessimistic.

On the question of frequency of updating the author stated that although he had so far only considered systems with updating every period of the test sequence, he felt optimal updating would probably be every p subintervals of the sequence, where p is an integer less than the number of subintervals in each period of the sequence. However a method for choosing p could not be given.

Dr. H. A. Barker (U.K.) raised several points as to the theoretical advantage of using pseudo random binary sequences, how the sequence period and number of subintervals should be chosen, and whether in fact it was even necessary that the sequence have an impulse like autocorrelation function. This last point was based on the observation that each control step was a function of the integral of the cross-correlation function.

Discussing the points Dr. Ng suggested that the theoretical advantage of his method using pseudo random binary sequences as a perturbation lay in its ability to account for plant dynamics directly without the need for extra information such as the phase compensation required by other techniques. The choice of the sequence parameters depends largely on the type of system that one is trying to optimise. The period of the sequence should be chosen two or three times the dominant plant time constant, while the number of subintervals must be chosen so that the test signal spectrum is wideband relative to that of the plant.

Regarding the form of the test sequence autocorrelation function the author agreed that it need not necessarily approximate to an impulse provided that it had suitable frequency characteristics.

Mr. C. S. Berger (U.K.) commenting on systems with random disturbances, suggested that given a certain signal to noise ratio there is an optimum sinusoidal perturbation for determining the hill gradient. Thus for a given signal power it would appear a sinusoid has certain advantages over PRBS which distributes the power uniformly over a wide frequency band. Dr. Ng in reply said he had no quantitative argument to put forward on this matter.

Mr. A. Griffin (U.K.) suggested the use of information theory as a possible basis for studying the system described. The system is a loop into which is injected a probing signal. There is a measuring device which is essentially a filter which extracts information about the hill via the probing signal. The filter will have a resonant frequency so that if the probe is arranged at this frequency maximum information will be forthcoming. It was queried whether this is a useful approach. Dr. Ng thought his system could indeed be regarded in this way.

Written contributions to the discussion

Dr. V. W. Eveleigh (USA)

I would like to make several comments on this paper. First it is only fair to say that the use of pseudo random signals for perturbing the parameters in adaptive loops is an interesting approach worth consideration. I feel, however, that the authors have made an unfair comparison with more conventional methods using square wave or sinusoidal perturbations and have drawn erroneous conclusions about the practical prospects of their proposed method.

In the initial derivation, a function denoted $h(\tau)$ is introduced and it is pointed out that $h(\tau)$ is "the impulse response of the control system from parameter adjustment to performance index". Although this function proves useful in the subsequent development, the adaptive system is typically highly nonlinear between the two points considered and the impulse response certainly depends upon the magnitude of the test signal as well as the parameter setting when the test signal is applied. This is not, however, my major criticism.

The authors point out that their system is "slightly inferior (to that using square wave perturbations) in that less amplitude (error corrective signal) is developed". Yet the two responses shown by the authors in Figs. 4B and 4D of the paper illustrate approximately a 10:1 advantage in response time for the square wave system. The authors also claim that changes in plant phase shift cause significant degradation in performance of the square wave system. In fact the square wave system is relatively insensitive to phase changes, with adaptive loop gain changing by only about 30 per cent for a 45° change

279

in steady state phase shift. Independence of the proposed system to phase changes is over-emphasized.

Another advantage of their system claimed by the authors is that its response is much less erratic near the optimum and during the transient period than that of the square wave system. This is certainly true for the relatively unsophisticated square wave system chosen by the authors, but not true in general. The simple addition of a synchronous sampler and first order hold circuit to the square wave system (note that the authors deem its use advisable in their system) will eliminate this discrepancy. Alternately, a carefully chosen continuous filter following the detector will accomplish the same purpose.

It seems even less practical, to consider the use of pseudo-random codes as perturbation signals for multi-parameter adaptive controllers. The authors point out that it is only necessary to find n codes which are uncorrelated to control n parameters without loop inter-coupling. This is not completely true. What is actually desired are n codes each of which will be uncorrelated with the measured performance criterion components due to all other codes as they appear at the detection point. Since the system is highly nonlinear between the parameter input and detection points, the equivalent requirements on the codes themselves are not readily obtained. One advantage of the sinusoidal perturbation system in the n parameter case is the relative ease in isolating the various adaptive loops by judicious choice of perturbation frequencies. Although the performance criterion measurement process introduces extraneous signal components, they are subsequently eliminated by narrow band filters and only the desired information appears in each parameter control loop. This is likewise true for square wave perturbations, although additional harmonics are present and filtering is more important. No such frequency domain isolation is possible in the authors' proposed system, decoupling depending entirely upon lack of correlation in the time domain.

Perturbation of the parameters about the optimum setting results in performance degradation or hunting loss. It seems intuitively obvious from an information theory point of view that all such perturbations should be chosen to be as efficient in making gradient measurements as possible. The authors' proposed method places much of the energy in frequencies which are unable to pass through the measurement filter and this explains their relatively smaller corrective signal. Particularly in the presence of noise, such inefficient use of perturbation energy and the corresponding wide bandwidth requirement results in significant adaptive loop performance degradation. The authors claim as an advantage the potential use of the same pseudo-random signal for identification and parameter perturbation. Although isolated cases may exist where this is practical, it seems much more reasonable to choose the most efficient signal for

each job rather than compromise both sets of objectives.

The authors replied as follows:

With regard to the specific points raised, by Dr. Eveleigh we make the following comments:

1. The use of the linearised impulse response from parameter adjustment to performance index is perfectly valid if the test signal is small; the effect of non-linear terms can be evaluated, but for parameter changes of less than 10% of the current value, experimental results, not given in this paper, show excellent agreement with prediction based on a linear model.

This is our justification for stating that uncorrelated codes will enable simultaneous adjustment of several parameters to be achieved.

2. As was explained at the Symposium, and as shown in the final version of the paper, the speed of adjustment with the method developed is very nearly equal to the rate of adjustment using sinusoidal (or square wave) perturbation. A synchronous chopper, or sample and hold, was included in the comparison system.

3. The model considered in one example was chosen simply because this is the most favourable for the sinusoidal perturbation system, since perfect phase compensation can be introduced for all parameter values. It should be expected that the proposed system will show a marked superiority in examples of practical importance where exact phase compensation cannot be built into the optimising units.

References

1. Briggs, P.A.N. and Godfrey, K.R. Pseudo Random Signals for the dynamic analysis of multivariable systems. Proc. I.E.E. July 1966.

LIMIT CYCLE CONDITIONS IN OPTIMALIZING CONTROLLERS

Virgil W. Eveleigh
Associate Professor of Electrical Engineering
Syracuse University
Syracuse, New York

ABSTRACT

A method is developed for determining search
limit cycle frequency and magnitude in single
dimensional optimalizing (peak holding) systems.
The performance criterion measurement process
is modeled by a nonlinear function (often
adequately approximated by a parabola) in cascade with a linear filter, denoted $G(j\omega)$,
representing measurement delay. The analysis is
based upon a single describing function (DF) of
the logic circuitry, the parameter adjustment
integrator, and the performance criterion non-
linearity. The DF is frequency dependent, adding some complexity to the standard, although
seldom applied, graphical solution method, but
the procedure is otherwise straightforward.
Limit cycle amplitude and frequency are determined directly from DF and G locus intersection
values in the usual way. Prediction accuracy
averages approximately 5 percent for the
examples considered, and is often even better.
As should be expected, greatest predictive
accuracy is obtained when considerable smoothing (measurement filtering) is present in the
feedback loop. When measurement filtering is
negligible, limit cycle conditions are readily
obtained by more direct methods.[1]

INTRODUCTION

Control systems which automatically adjust
one or more parameters to seek out the maximum
or minimum value of some index of performance
(hereafter abbreviated IP for convenience) are
rather common in the adaptive control area. A
variety of parameter adjustment strategies have
been suggested and the study of such strategies
to evaluate their relative merits in specific
situations constitutes a major adaptive control
research area. In the special case of a single
adaptive parameter one common approach is to
use a constant rate parameter drive, noting the
location of the minimum (although a minimum
value is assumed, similar statements apply for
a maximum) IP value as it is passed and revers-
ing the drive direction when performance degradation, as measured by an increase in the IP from
its minimum, exceeds a threshold level. This
approach was initially proposed in a paper by
Draper and Li[2] and has since been considered in
many texts and papers on adaptive control.
Draper and Li suggested that such systems be
referred to as "optimalizing systems" and they
are so designated throughout this paper.

For the most part, prior discussions of
optimalizing systems have been restricted to
consideration of the special case arising when
the IP can be measured with no appreciable delay
or phase lag. In such cases the limit cycle
amplitude and frequency and the corresponding
hunting loss, or average performance degradation
due to that limit cycle, are readily determined
if the IP is known as a general function of the
controlled parameter. Such idealized conditions
are seldom encountered in practice. A problem of
practical interest is the development of a general technique for determining the hunting limit
cycle conditions for the more general case arising
when significant phase lag is introduced by the
IP measurement process. A technique for solving
this problem is presented herein assuming an
arbitrary IP function nonlinearity providing a
unique relative minimum as the desired optimum
operating point in the parameter search region
followed by a linear filter representing measurement phase lag.

One solution of the optimalizing system limit
cycle problem is provided by Morosonov,[3] who
assumes the IP function is parabolic in form
throughout the operating region, or is adequately approximated as such. Measurement delay is
represented by two single time constant lag
filters, one preceding and the second following
the IP nonlinearity. The approach developed by
Morosonov is adequate for this particular situation, is rigorously and clearly developed, and
could be extended to include more complex models,
but with considerable difficulty. The basic
method involves determining a describing function
(DF) for the drive reversal logic circuitry, a
separate DF for the IP nonlinearity, and solving
the algebraic equations implied by the requirements that loop transmittance magnitude equal
unity and total phase shift equal 360° to
sustain steady state oscillation of the adaptive
loop. For the particular IP measurement model
assumed the algebraic difficulty is not pro-
hibitive, but it soon becomes so for more complex
situations.

The approach proposed herein differs from
that used by Morosonov in several respects.
Perhaps the most significant distinction is
the use of a single DF. The nonlinear logic
circuitry, the nonlinear IP function, and the
frequency dependent transfer functions between
them are treated as a single frequency dependent nonlinearity. The DF combination is the

---

Superior numbers refer to similarly-numbered references at the end of this paper.

only nonlinear characteristic which must be considered. Although the resulting single DF is more complex than that of either nonlinear element treated separately, the total effort required to obtain it is not appreciably more than that involved in obtaining the other two, and the result is more generally useful in describing system performance. Reduction to a single DF allows application of the usual graphical techniques for predicting limit cycle amplitudes and frequencies as a function of loop parameters, thereby eliminating the necessity for further algebraic analysis and extending the methods applicability to systems with higher order smoothing filters and/or time delay in the IP measurement process. Since the IP nonlinearity and the associated smoothing are constrained only to provide proper adaptive loop operation, the approach is extremely flexible.

The paper is broken down as follows. First, a brief review of the fundamental principles of optimalizing systems is provided. This is followed by the development of a system model, emphasizing in particular the IP measurement process and the drive reversal threshold logic. Using the mathematical model thereby formulated, the details of the analysis method are developed, including a summary of the graphical techniques required in working with frequency sensitive DF's. Limit cycle amplitude and frequency results as predicted by the proposed method, and as obtained by simulation of the corresponding adaptive loop on the analog computer, are compared for several specific problems. It is shown that frequency can be predicted with 2 to 4 per cent accuracy and amplitude can be predicted with 5 to 10 per cent accuracy. It is generally found that amplitudes predicted from DF analyses are more in error than the corresponding frequency predictions, since the first frequency correction term in the higher order representation is zero, and these results illustrate this phenomenon. It is assumed throughout the development that the reader is generally familiar with adaptive control technology as it has been presented in the literature and also with basic analysis techniques such as the DF method and the Nyquist stability criterion to which DF analysis is so closely related. Only a minimum of background material is provided.

THE OPTIMALIZING ADAPTIVE SYSTEM

A typical optimalizing system is shown in the block diagram of Fig. 1. Note that the adaptive control loop is appended to a controller which may or may not be of the closed loop type. When the adaptive loop is in operation, the controller is time varying, its input to output response characteristics depending upon the instantaneous value of parameter x, and perhaps upon other factors as well. Since x is swept at constant rate, $g(t,\tau,x)$ is continuously changing, where $g(t,\tau,x)$ is the controller impulse response function relating system output to control input signals. (Characterizing the system by its time dependent impulse response function implies that the system is linear. Linearity is not necessary to the general development, and no such assumption should be construed from this representation, which is used only for symbolic convenience.) In most applications, however, the parameter adjustments made by the adaptive loop due to steady state hunting about the optimum point are kept small, and should have little average effect upon controller transfer characteristics. Typical parameters available for adaptive control are gain and time constant values. Interest here is centered on general characteristics rather than a specific application.

When external effects tend to modify controller performance, a new value of x minimizes the IP and the adaptive loop is designed to automatically drive x toward the new optimum point. Thus near optimal performance of the controller, as measured by the specific IP used, is maintained even in the presence of changing operating conditions by modifying x accordingly. If the adaptive loop is broken at the x input to the controller, adaptive action is eliminated and the controller operates with x as a fixed parameter, although with its performance degraded accordingly.

Optimalizing techniques apply to systems using an IP with a maximum or minimum value at the desired operating point. Assume as before that a minimum value applies. Mean squared error is a common example of such an IP. Figure 2 shows a typical IP function and further illustrates the time variation of x about the optimum point as caused by the adaptive controller. To simplify the initial explanation, assume the IP can be measured with zero phase lag. Operation of the corresponding adaptive loop is particularly simple. Parameter x is swept linearly (the output of the adaptive controller is held constant at either +1 or -1) until the IP, functionally denoted by ϕ for convenience, passes through its minimum value, ϕ_{min}. ϕ_{min} is retained in some form of memory. As ϕ increases its instantaneous value is compared to ϕ_{min} and when $\phi - \phi_{min}$ exceeds a threshold value, denoted Δ, the adaptive controller output changes sign and the process repeats. Thus the adaptive loop continuously limit cycles x about that value providing $\phi = \phi_{min}$.

Obviously there is no difficulty in determining limit cycle amplitude and frequency in those cases where variations in ϕ caused by changes in x are measurable with zero time or phase lag and with no corrupting noise. Practical situations in which instantaneous measurement is possible are seldom encountered, however, so a general technique is desired for predicting limit cycle conditions including the effect of measurement smoothing and/or delay. The problem is complicated by the presence of two nonlinear operations in the loop, but can be solved if an acceptable mathematical model of the IP measurement process can be developed.

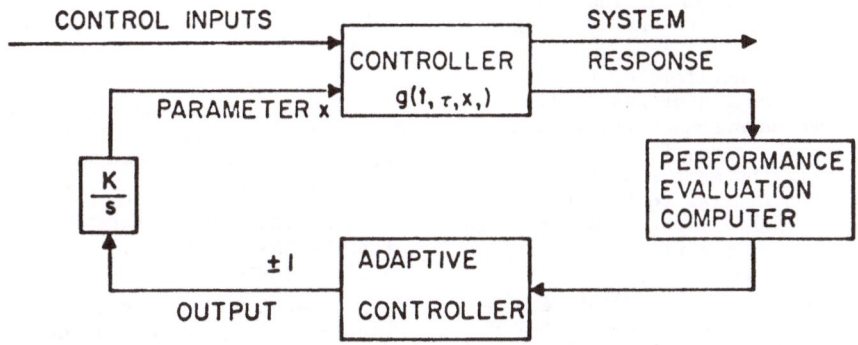

Figure I. The Adaptive Loop for Controlling Parameter x.

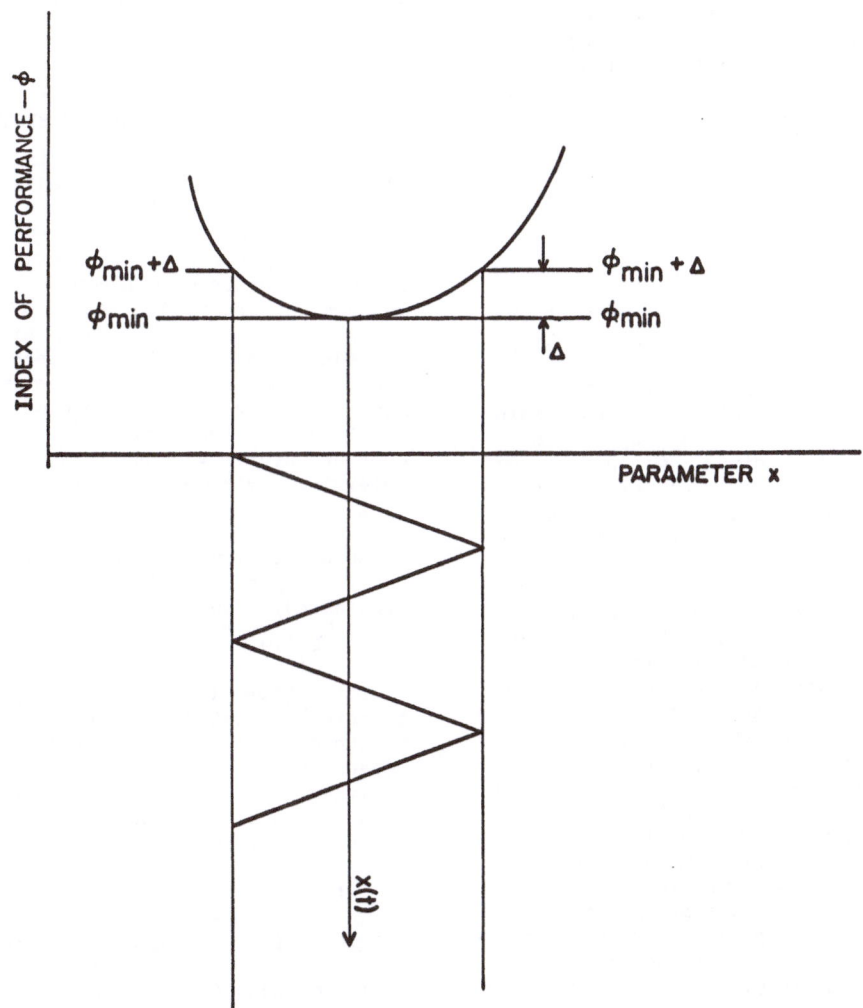

Figure 2. A Typical Optimalizing System IP Function And The Corresponding x(t).

THE IP MEASUREMENT PROCESS

Development of a complete mathematical model of the optimizing system requires that the transfer relationship between the parameter x input and measured IP output be approximated. When measurement delay is negligible, this relationship is a general nonlinear function, as shown in Fig. 2. When measurement lag is appreciable, as more often the case, it should be adequate to approximate the measurement process by a nonlinear function similar to that shown in Fig. 2 followed by a linear filter. This approximation is used throughout the present development, since some assumption must be made, although the approach proposed for predicting limit cycle conditions is not limited to any such restricted model. The author has often used a model of this form.[4] As previously mentioned, Morosonov[3] uses a single section lag filter, followed by the nonlinear function, assumed to be quadratic in form, followed in turn by a second single section lag filter. The proposed model is more general in that the nonlinear function is not constrained to be quadratic, although simplifications result when that is the case, and the linear filter may take any form. It is less general in not including a filter prior to the nonlinearity. The basic approach proposed herein applies to a model including a pre-filter, although this adds considerable computational complexity to the procedure.

It should be pointed out in passing that the IP measurement process is often noisy. Take, for example, the measurement of mean square error (E_{ms}). The nonlinear element in the system model would normally be developed by determining analytically the dependence of E_{ms} upon x for the anticipated input spectrum. Measurement of E_{ms}, however, might be made by actually squaring the error signal and smoothing over a period sufficiently long to assure a reasonable sample. Any finite smoothing period results in significant fluctuations at the measured output, however, constituting a noise input to the adaptive control system. Increasing smoothing time (and thereby phase lag in the adaptive loop) reduces noise effects, but also increases adaptive loop response time, so some compromise must be established. The proposed analytical approach will prove useful in resolving this dilemma. For simplicity, the effects of noise are not considered further. It is apparent that random modulation of the switching times is the primary effect of noise, the relative amount of modulation depending upon noise power level. Adaptive loop operating characteristics are not appreciably changed from those predicted using a deterministic model if the noise variance is small relative to threshold level Δ.

THE OPTIMALIZING SYSTEM MODEL

The adaptive control loop model is shown in Fig. 3. The NL element representing the IP function is constrained only in that it exhibits a single relative minimum in the region of interest. G(s) is assumed linear and time invariant. The logic function is also nonlinear and performs according to the following restrictions:

1. The output of the logic unit is either +1 or -1 at all times.

2. The logic unit contains a monitor which follows all decreasing signals at the input, stores the minimum when it is reached, and switches the output when the input level exceeds the minimum level by threshold value Δ.

3. A successive reversal is allowed only after ϕ_m has decreased, except when the system is first turned on.

(Requirement 3 is imposed to prevent two or more drive reversals caused by the same positive swing of ϕ, thereby leading to instability.) Thus, the logic unit consists of positive and negative peak detectors, a hold circuit, and a threshold element, all of which are readily instrumented or simulated using standard components.

ANALYSIS OF THE OPTIMALIZING SYSTEM

The basic method of analysis is as follows. There are two nonlinear elements in the adaptive loop, the control logic and the IP function, separated by an integrator with gain K(or more generally by a linear transfer function if a pre-filter is required to adequately model the IP measurement process). The DF of this combination can be obtained and although this DF is frequency dependent, complicating the mechanics somewhat, the usual DF techniques may be applied to determine frequency and amplitude of the resulting limit cycle. The predicted results thereby obtained are very close to actual conditions determined by simulation of the corresponding system.

Assume a sinusoidal excitation input to the control logic block in Fig. 3 (the standard DF procedure). Thus

$$\phi_m(t) = A \sin \omega t \qquad (1)$$

Let $N_e(A,\omega)$ denote the steady state sinusoidal transfer characteristic relating $\phi_m(x)$, the input to the control logic, to $\phi(x)$, the IP output. Because of the integrator between the two nonlinearities N_e is a function of both input amplitude A and frequency ω. The loop transfer function can be expressed as

$$A(j\omega) = G(j\omega) \, N_e(A,\omega) \qquad (2)$$

The instability conditions are defined by

$$1 - G(j\omega) \, N_e(A,j\omega) = 0$$

or

$$G(j\omega) = \frac{1}{N_e(A,j\omega)} \qquad (3)$$

Thus $G(j\omega)$ and $1/N_e(A,j\omega)$ may be plotted on the complex plane and intersections of the two loci define all possible conditions of instability.

284

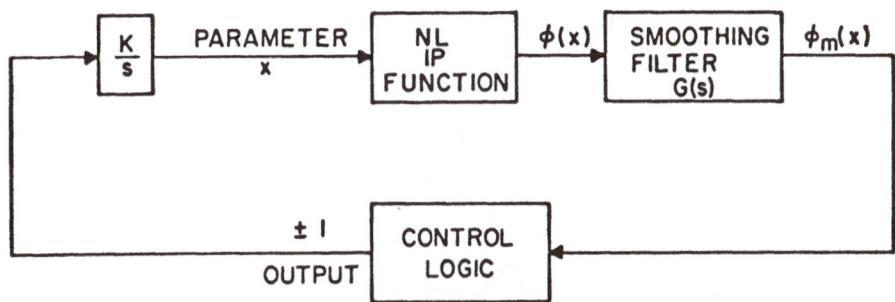

Figure 3. The Optimalizing Control Loop.

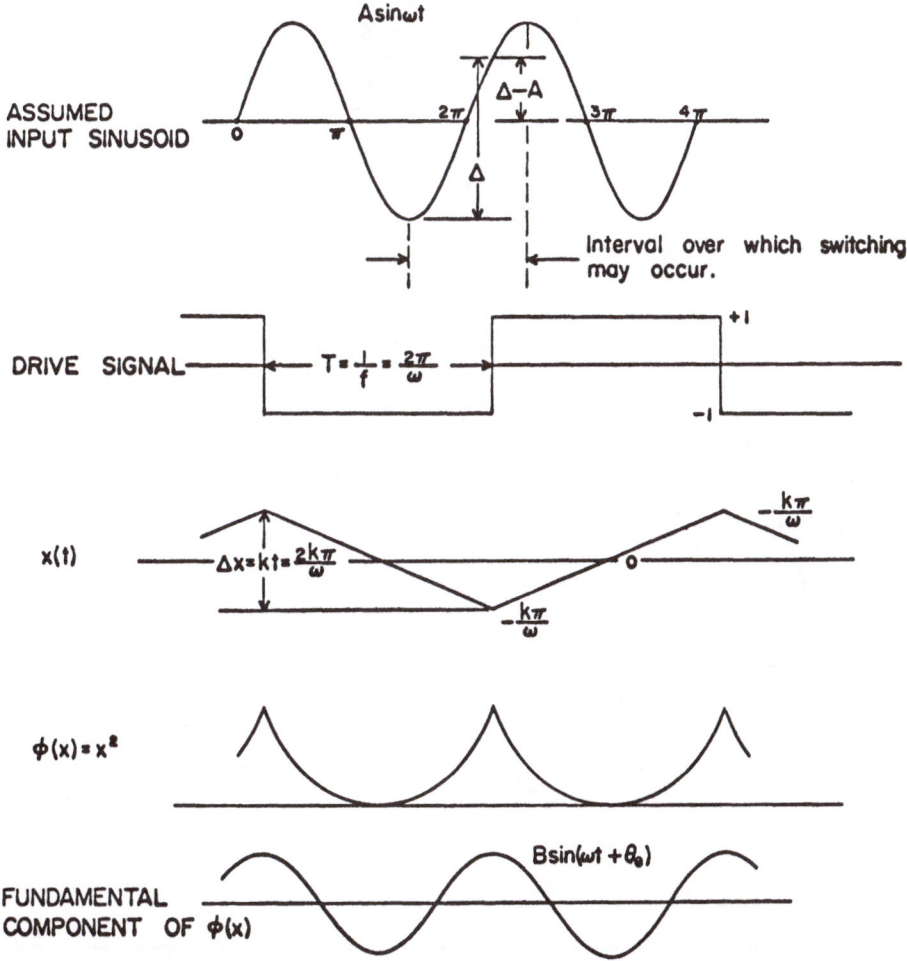

Figure 4.   Critical Waveforms From The System Shown In Fig. 3

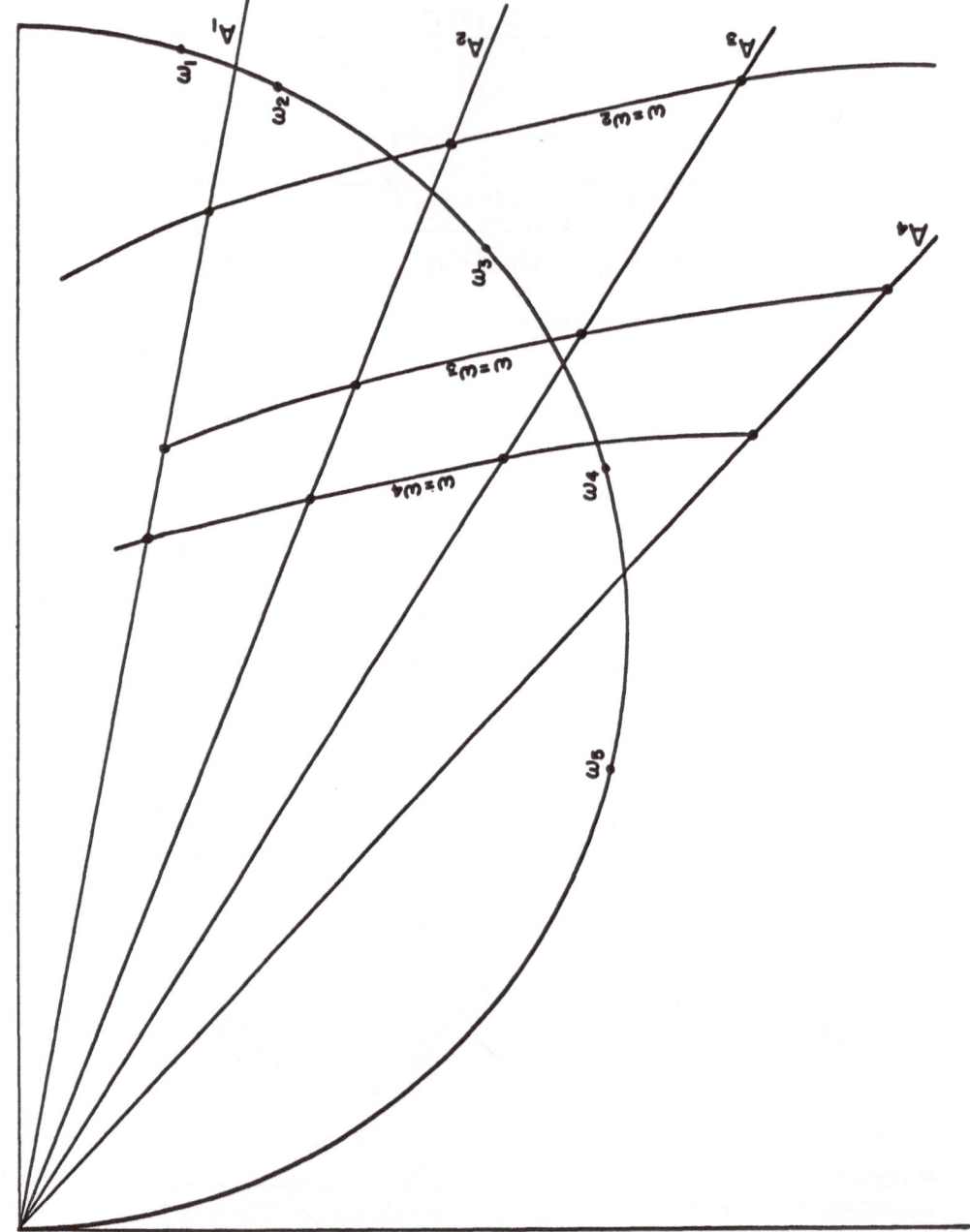

Figure 5. Suggested Solution Procedure

286

In this particular case, since N_e depends upon both A and ω, families of plots are required to determine intersections where ω is identical on both the G and N_e curves. This is the "standard" frequency sensitive DF approach, but it is seldom put to practical use.

Figure 4 shows signals taken from critical points in the adaptive loop of Fig. 3. Assume for convenience that

$$\phi(x) = x^2 \qquad (4)$$

In the general case $\phi(x)$ could be a more complex function, perhaps defined only graphically, but the basic procedure would be the same. Directly from Fig. 4, it is apparent that the fundamental component of $\phi(x)$ is related in phase to the input sinusoid by

$$\Theta_e = -\frac{3\pi}{2} - \sin^{-1}\frac{\Delta - A}{A} \qquad (5)$$

where Θ_e denotes the DF phase. For the $\phi(x)$ defined in (4) and illustrated in Fig. 4 the magnitude of the fundamental component in $\phi(x)$ is readily found to be

$$B = 4K^2/\omega^2 \qquad (6)$$

Thus the DF gain is

$$|N_e| = \frac{B}{A} = \frac{4K^2}{A\omega^2} \qquad (7)$$

and

$$N_e(A,\omega) = \frac{4K^2}{A\omega^2} \Big/ -\frac{3\pi}{2} - \sin^{-1}\frac{\Delta - A}{A} \qquad (8)$$

For the special case assumed $|N_e|$ depends upon both amplitude and frequency, whereas the phase angle of N_e depends upon amplitude alone. When $\phi(x) \neq x^2$, a similar analysis applies, but the solution is seldom as simple in form, which is the main motive for the parabolic assumption. It should generally be possible to adequately approximate IP's of the type under consideration by equivalent parabolas in the operating region. When such is not the case, a somewhat less convenient analysis must be carried out. In all cases where $\phi(x_o + \delta x) = \phi(x_o - \delta x)$ for arbitrary δx, x_o denoting the optimum x, DF phase depends only upon A.

The following procedure is recommended for determining limit cycle amplitude and frequency. Since the angle of N_e depends only upon amplitude values covering the range anticipated and plot $1/N_e$ vs. ω for each amplitude. On the same scale, plot $G(j\omega)$. Sketch loci of constant ω between the various members of the $1/N_e$ family, and note intersections of these loci with the $G(j\omega)$ curve where the ω coordinates correspond. These points (there should generally be only one) define operating amplitude and frequency for the conditions chosen.

Figure 5 illustrates the suggested procedure for

$$G(j\omega) = \frac{1}{1 + j\tau\omega} \qquad (9)$$

In this case a semicircular locus is obtained for $G(j\omega)$, and the loci of $1/N_e(A,j\omega)$ are radial lines for fixed values of A, with ω as a parameter on each locus. Constant ω loci are sketched from data points on the constant amplitude loci and interpolation from the curves in Fig. 5 shows that

$$\omega \cong \omega_3 + 0.7(\omega_4 - \omega_3) \qquad (10)$$

and

$$A \cong A_3 + 0.4(A_4 - A_3) \qquad (11)$$

Improved accuracy is readily obtained by sketching additional loci.

EXAMPLES AND SUPPORTING DATA

As in all approximate analyses of this type, prediction accuracy is the most important criterion of true value. To establish this, three sets of data were obtained using a general purpose analog computer to simulate the optimalizing system. In each case it is assumed that $\phi(x) = x^2$, so the previous analysis applies directly, and K = 2. Table 1 shows predicted and actual amplitude and frequency results for each of three systems considered. The predicted results were obtained graphically as outlined in the previous section.

It should be noted that the predicted frequency results are consistently within 2 - 4 percent of the actual values and amplitude predictions, although somewhat poorer as is usually true for DF analyses, are consistently within 10 percent and often much better. These accuracies are well within the usual limits anticipated for approximate analyses of this form. Since the frequency value completely defines system operation, more exact predictions of amplitude could, if desired, be made from the predicted frequency values. This is hardly worthwhile considering the minor improvement potentially available.

The value $\Delta = 5.2$ deserves a few words of explanation. The simulation was carried out in real time using a threshold circuit to close and open a relay at level Δ. Initially the bias was set from an external regulated source at 5.0 volts, and only after two runs were complete at that level was it found that the threshold circuit had a small residual bias (0.2 volts) which caused the relay to switch at $\Delta = 5.2$ volts. The value $\Delta = 10$ was obtained by setting the actual switching point at 10.0 volts.

CONCLUSIONS

A general approach for determining limit cycle amplitude and frequency in optimalizing systems is developed herein, and experimental data provided which indicates that prediction accuracy is consistently within 2 - 4 percent in frequency

Table 1. Comparison of Predicted and Actual Values

Δ	G(s)	T	Predicted Values		Actual Values	
			ω	A	ω	A
5.2	$\frac{1}{1+Ts}$	0.1	2.45	2.64	2.50	2.80
5.2	$\frac{1}{1+Ts}$	0.2	2.30	2.73	2.30	2.95
5.2	$\frac{1}{1+Ts}$	0.5	1.95	3.02	1.93	3.25
5.2	$\frac{1}{1+Ts}$	1.0	1.60	3.38	1.56	3.60
5.2	$\frac{1}{1+Ts}$	2.0	1.25	3.80	1.23	3.95
5.2	$\frac{1}{1+Ts}$	5.0	0.90	4.32	0.89	4.40
10	$\frac{1}{1+Ts}$	0.1	1.77	5.02	1.84	5.25
10	$\frac{1}{1+Ts}$	0.2	1.72	5.15	1.75	5.50
10	$\frac{1}{1+Ts}$	0.5	1.52	5.60	1.51	6.00
10	$\frac{1}{1+Ts}$	1.0	1.27	6.20	1.26	6.50
10	$\frac{1}{1+Ts}$	2.0	1.00	7.00	0.996	7.25
10	$\frac{1}{1+Ts}$	5.0	0.730	7.90	0.720	8.25
5.2	$\frac{1}{(1+Ts)^2}$	0.1	2.40	2.76	2.42	3.00
5.2	$\frac{1}{(1+Ts)^2}$	0.2	2.15	3.07	2.10	3.40
5.2	$\frac{1}{(1+Ts)^2}$	0.5	1.57	4.32	1.53	4.70
5.2	$\frac{1}{(1+Ts)^2}$	1.0	1.12	5.98	1.10	6.50
5.2	$\frac{1}{(1+Ts)^2}$	2.0	0.780	9.36	.744	9.75

and 5 - 10 percent in amplitude. Such accuracy makes the method readily acceptable as a design and analysis tool. Limits are often imposed upon adaptive system hunting loss and the proposed analysis technique allows direct system synthesis within such a constraint.

Careful observation of the data presented in Table 1 makes it clear that predicted amplitudes are consistently low relative to those obtained from the simulation. This is due in part to a small delay in loop elements which were assumed to operate instantaneously, such as the switching relay and the multiplier. Inclusion of these effects in the analysis is possible, but adds considerable complexity. For this reason they were not included in the analysis to allow illustration of the basic principles without a distractive level of analytical and graphical complexity. The resulting accuracy is only slightly penalized.

The approach presented is not limited to the particular IP measurement model used, although considerable simplification accrues from its use. A general nonlinearity in series with a linear filter has often been used in the past to represent the IP measurement process, establishing ample precedent for such a model. The nonlinearity has also often been assumed parabolic near the optimum, as used in the examples presented.

The most important concept utilized in this development is the representation of two nonlinear elements separated by a frequency dependent transfer function (an ideal integration in the system model assumed) by a single frequency and amplitude dependent describing function. The remainder of the analysis is based upon standard principles. The data presented provide adequate proof of the utility and accuracy of the approach.

ACKNOWLEDGEMENTS

The author is indebted to Syracuse University and the Syracuse University Research Corporation for support of research efforts leading to the results presented. Thanks are also due to the Heavy Military Department of General Electric Company, which provided the analog computer facility for generating supporting data. The assistance of H. Hildebrand, T. Kohler, L. Spafford, and L. Bauer in providing access to the computer and in obtaining data for the examples is deeply appreciated.

REFERENCES

1. H. S. Tsien, "Engineering Cybernetics," McGraw-Hill, Inc., New York, 1954.

2. C. S. Draper and Y. T. Li, "Principles of Optimalizing Control Systems and an Application to the Internal Combustion Engine," ASME, New York, 1951.

3. I. S. Morosonov, "Methods of Extremum Control," Automation and Remote Control, Vol. 18, No. 11, November, 1957

4. V. W. Eveleigh, "General Stability Analysis of Sinusoidal Perturbation Adaptive Systems," IFAC Conference, Basel, Switzerland, August, 1963.

Discussion

Mr. T. Horrocks (UK) asked for confirmation that in general the graphical procedures were more complicated than that shown for a single lag in Figure 5 of the paper.

Dr. Eveleigh confirmed that this was so and went on to state that the dynamics might easily be as complex as a 2nd or 3rd order Butterworth smoothing filter. However provided the dynamics followed the non-linearity the graphical procedure was quite manageable. Unfortunately if it is essential to have pre-filtering, dynamics before the non-linearity, then that does add a great deal of complexity.

Mr. L. Haller (UK) observed that all the predicted values in Table I of the paper are less than the actual values and asked the significance of this.

Dr. Eveleigh replied that in the interest of simplicity of presentation he had neglected small delays associated with switching logic. The effects of these delays would be to increase the limit cycle amplitude and degrade the accuracy of preduction.

Dr. C.J.D.M. Verhagen (Netherlands) asked about the errors of approximation occurring when pre-filtering is present. In reply the author said that he had no information to give but that it would certainly be worthwhile to obtain some.

EXTREMUM CONTROL IN THE PRESENCE OF VARIABLE PURE DELAY

by Dr. M.H. Hamza
The University of Zurich
Zurich, Switzerland

ABSTRACT

An extremum control system is considered which consists of a parabolic function in cascade with a first-order system having a pure delay. All parameters of the system are allowed to vary including the pure delay. A method is given for determining the extremum in the presence of disturbances and digital computer results are provided.

INTRODUCTION

To control an industrial process when complete information about its characteristics is available, the statistical control methods are to be recommended [1]. In many cases such " a priori " information is absent, or if it is available initially it loses authenticity as time progresses due to changes in the system parameters. Under such conditions it is advisable to employ self-adjusting controllers [2,3,4].

In this paper a study is made of a continuous extremum control system having variable pure delay. Such systems occur often in practice and especially so in the chemical industry. System identification in the method proposed is obtained by measuring the sign and magnitude of a discontinuity in an appropriate derivative of the system output. This information is used in the control law proposed. Convergence and stability are studied and computer results verifying the theory are given.

The method suggested is an extension of previous work by the author [5].

THE SYSTEM

The system considered consists of the following in cascade: an integrator, a function of the parabolic type and a first-order system having a variable pure delay. A symbolic representation of such a system is given in Fig.1. K_2, α, τ and δ of Fig.1 are permitted to vary with time. The pure delay may be a distributed one.

The system of Fig.1 without the pure delay, is the one most widely covered in the literature on extremum control [2,3,4] [6].

Assuming no pure delay to be present, the equations of the extremum control system studied are the following:

$$r(t) = \epsilon K_1 \qquad (1)$$

$$x(t) = \int_0^t r(\xi)\, d\xi + x(0) \qquad (2)$$

$$y = \alpha x^2, \quad \alpha > 0 \qquad (3)$$

$$\tau \frac{dc}{dt} + c = K_2 y \qquad (4)$$

The above equations are valid during sufficiently short intervals of time to allow K_1, K_2, τ and α to be considered constant. ϵ is a discrete variable which belongs to the set of permissible values ϵ_r, $r=1,2,\ldots,n$, and is governed by the extremum controller.

Starting from permissible initial

Superior numbers refer to similarly-numbered references at the end of this paper.

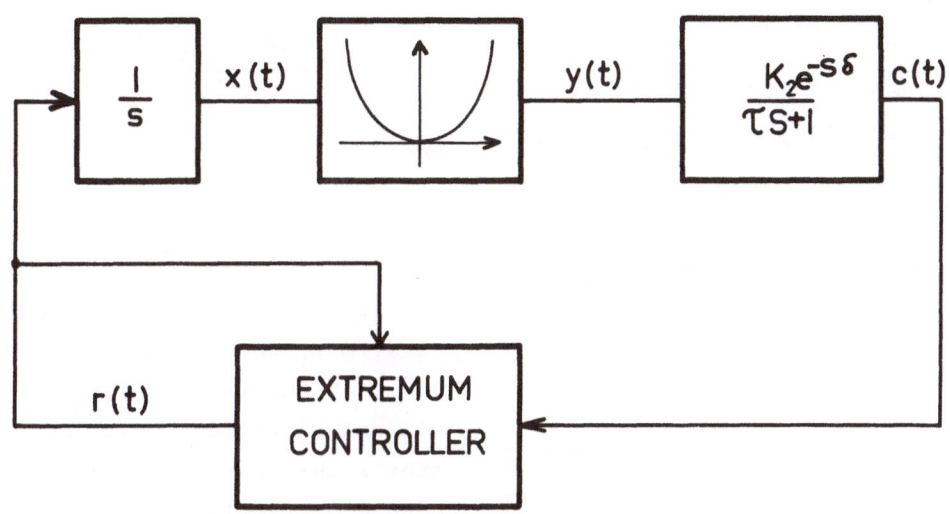

Fig.(1)   The extremum control system

conditions of $c(t)$ and $x(t)$ and in the presence of disturbances, the aim is to determine ϵ such that $f(t)$ is a minimum, where

$$f(t) = \int_o^t c(\xi)\,d\xi \qquad (5)$$

subject to the constraint that ϵ belongs to the set ϵ_r.

THE CONTROL LAW

Solving equations (1) to (4) we obtain

$$c(x) = 2\alpha K_2 \left\{ 0.5x^2 - \epsilon K_1 \tau x + (\epsilon K_1 \tau)^2 \left[1 + \lambda \exp\left(\frac{-x}{\epsilon K_1 \tau}\right) \right] \right\}$$
$$\dots\dots\dots (6)$$

where λ is an integration constant. Equation (6) can be expressed as

$$c(x) = 2\alpha K_2 \left[0.5x^2 - \epsilon K_1 \tau x + \frac{(\epsilon K_1 \tau)^2}{2\alpha K_2}\frac{d^2 c}{dx^2} \right] \qquad (7)$$

Using equations (1) and (2) it can be shown that

$$\frac{d^2 c}{dx^2} = \frac{1}{(\epsilon K_1)^2}\frac{d^2 c}{dt^2} \qquad (8)$$

Substituting equation (8) into equation (7) gives

$$\frac{d^2 c}{dt^2} = \frac{2\alpha K_2}{\tau^2}\left\{\frac{c[x(t)]}{2\alpha K_2} - \frac{[x(t)]^2}{2}\right.$$
$$\left. + \epsilon K_1 \tau x(t) \right\} \qquad (9)$$

If ϵ is changed suddenly at time t_p

from ϵ_i to ϵ_f, a discontinuity in $\frac{d^2 c}{dt^2}$ results. This discontinuity will be denoted by J_t. From equation (9) and noting that $c[x(t)]$ is a continuous function of time, the following relation can be derived

$$J_t = \frac{2\alpha K_1 K_2}{\tau}(\epsilon_f - \epsilon_i)\,x(t_p) \qquad (10)$$

or that

$$J_t = A.\eta.\,x(t_p) \qquad (11)$$

where

$$A = \frac{2\alpha K_1 K_2}{\tau} \qquad (12)$$

and

$$\eta = \epsilon_f - \epsilon_i \qquad (13)$$

It should be clear from the above equations, that even if A varies with time, at the instant $t=t_p$ it can be considered to be constant. Equation (11) states that the discontinuity in the second derivative of the system output with respect to time is proportional to the independent variable x at time t_p with respect to its required value which in this case is zero. Extremum controllers whose operation is based on equation (11) have been successfully designed[7,8].

If the pure delay δ is constant and nonzero, equation (11) becomes

$$J_t = A.\eta.\,x(t_p - \delta) \qquad (14)$$

This implies that a perturbation in the input at time $t=t_p$ results in a discontinuity in the second derivative of the system output δ units of time later.

Because a system with pure delay does not respond instantaneously to an input applied to it, it is difficult to control[9]. Often, if the pure delay is known, its effect can be compensated[10]. It might be worth noting that in some control systems pure delay is intentionally introduced[10,11].

To determine the extremum when the system has a pure delay[3,6], the following method which is based on equation (14) is recommended.

It is suggested that the input consist of pulses each having a duration h_1, where $h_1 > \delta_{max}$. Applying such a pulse to the system, two discontinuities in $\frac{d^2c}{dt^2}$ result. These discontinuities will be denoted by J_{t1} and J_{t2} and their time of occurrence by t_1 and t_2 respectively, Fig.2. If $|J_{t2}| < \delta$, where δ is an appropriately chosen constant, $r(t)$ is kept equal to zero until the end of an interval h_2, Fig.2, whereupon a pulse similar to the previous one is applied to the system and the procedure is repeated. If $|J_{t2}| \geqslant \delta$, the latter pulse is applied at time t_2 with its sign given by $\epsilon_{req.}$, where

$$\epsilon_{req.} = (\text{sign } J_{t2}) . \epsilon_i \qquad (15)$$

ϵ is permitted to assume the values ± 1 and zero.

If the pure delay is a variable one, the same procedure can be still applied. The pure delay δ must satisfy the condition

$$h_1 > \delta \geqslant 0 \qquad (16)$$

CONVERGENCE AND STABILITY

Starting from an initial condition 'a' which is not close to the extremum, typical behaviour of the system is as sketched in Fig.3,I,II. If the system moves initially in the false direction, Fig.3,II, it will correct its path as time progresses, as may be readily verified by applying the control law (15).

Close to the extremum oscillations around the latter take place. These oscillations will not be usually symmetrical with respect to the c(x) axis, Fig.3. Their peak to peak variation in the x-direction will be equal to $K_1 h_1$. Thus they can be reduced by decreasing K_1. Unfortunately, this has the disadvantage of reducing the magnitude of the discontinuities and thus making it difficult to measure them. Consequently, K_1 should not be reduced beyond a value which depends upon the system characteristics and its working condition.

The performance of the system will be in general satisfactory provided $\delta < \tau$. If $\delta \gg \tau$, h_1 will be large and K_1 small and it will be difficult to measure the discontinuities in the second derivative of the system output. A modification of the method suggested having K_1 large and h_1 small can be used, but the system response will then follow close to its static characteristics.

COMPUTER RESULTS

For convenience nondimensional variables were used in the system simulated on the digital computer. Let

$$R = \frac{r}{k_1} \qquad (17)$$

$$T = \frac{t}{\tau} \qquad (18)$$

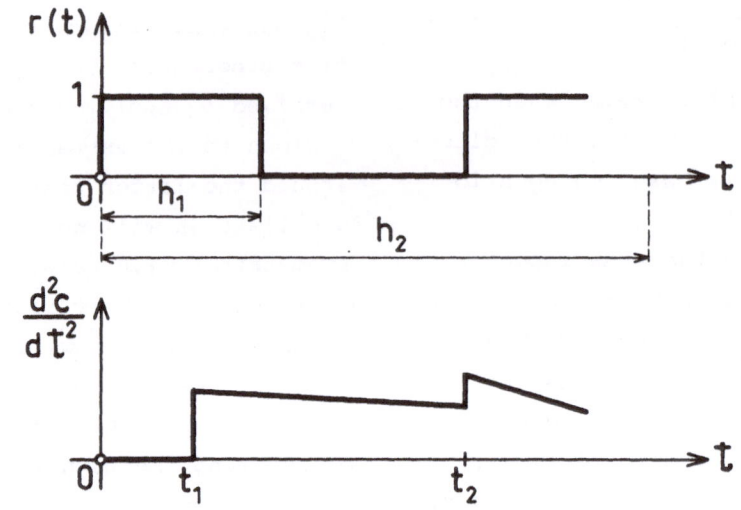

Fig.(2) $r(t)$ and $\dfrac{d^2c}{dt^2}$ when the system has a pure delay

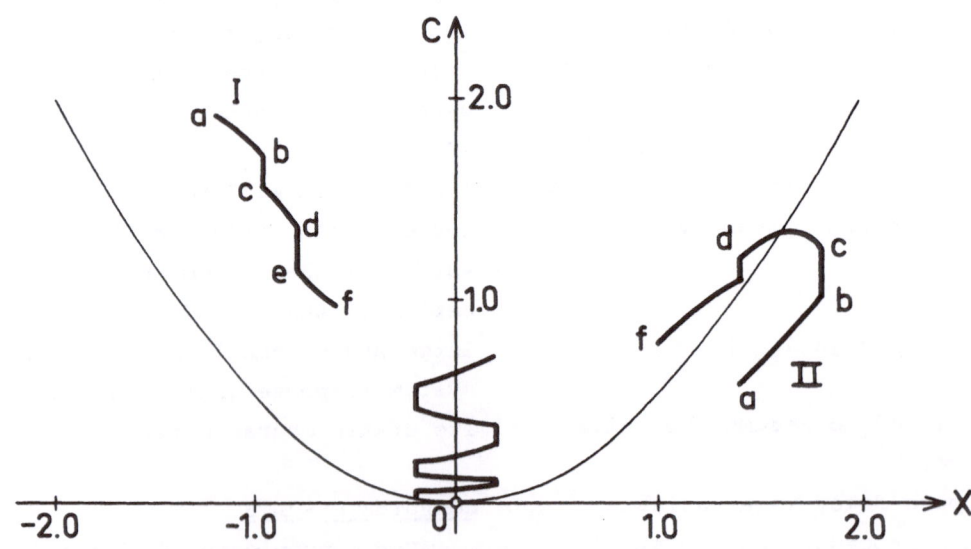

Fig.(3) Typical behaviour of the system

294

$$X = \frac{x}{K_1 \tau} \qquad (19)$$

$$Y = \frac{y}{2\alpha (K_1 \tau)^2} \qquad (20)$$

$$C = \frac{c}{2\alpha K_2 (K_1 \tau)^2} \qquad (21)$$

$$\Delta = \frac{\delta}{\tau} \qquad (22)$$

For $\Delta = 0$, the equations describing the system are

$$R(T) = \epsilon \qquad (23)$$

$$X(T) = \int_O^T R(\zeta) \, d\zeta + X(O) \qquad (24)$$

$$Y(t) = \frac{x^2}{2} \qquad (25)$$

$$\frac{dC}{dT} + C = Y \qquad (26)$$

and as may be readily verified, the control law, equation (14), when $\Delta \neq 0$, becomes

$$J_T = \eta \cdot X(T_p - \Delta) \qquad (27)$$

Equation (15) remains unchanged, except that now J_{T2} is a nondimensional variable.

The normalized system was simulated on a digital computer and a sample of the results obtained is given in Fig.4. In (a) the pure delay is constant and in (b) and (c) it varies sinusoidally. In all cases the extremum was identified.

CONCLUSION

An extremum control system having a variable pure delay has been described. The system presented converges to the extremum independent of the initial conditions. In general, the system performance will be satisfactory, unless there is too much noise or the pure delay is large compared to the system time constant.

REFERENCES

(1) Pugachev,V.S.,"Statistical methods of Automatic Control". Proc. of the IFAC Congress, Basel,1963. Butterworths, London.

(2) Feldbaum,A.A.,"Computers in Automatic Control". Oldenbourg,Munich,1962.

(3) Kazakevich,V.V.,"Extreme Control Systems and their Improvement". Automatic control and computer engineering. Pergamon Press,1961.

(4) Westcott,J.H.,"An Exposition of Adaptive Control". Pergamon Press, 1962.

(5) Hamza,M.H.,"Synthesis of Extremum-seeking Regulators". Automation and Remote Control, vol.XXV, part 8, 1964.

(6) Perret,R. and Rouxel,R.,"Principle and Application of an Extremal Computer". Proc. of the IFAC Congress, Basel, 1963. Butterworths,London.

(7) Discussions of reference (6).

(8) Hamza,M.H."Extremum Control of Continuous Systems". To be published.

(9) Krasovsky,N.N.,"Optimum Processes in Systems with Time Delay". Proc. of the IFAC Congress,Basel,1963. Butterworths,London.

(10) Smith,O.J.M.,"Feedback Control Systems". McGraw-Hill Book Company, Inc.,New York, 1958.

(11) Hamza,M.H.,"Adaptive System Controllers using Delay Elements". Proc. of the IFAC Symposium on Optimizing and Adaptive Control,Rome, 1962. Instrument Society of America.

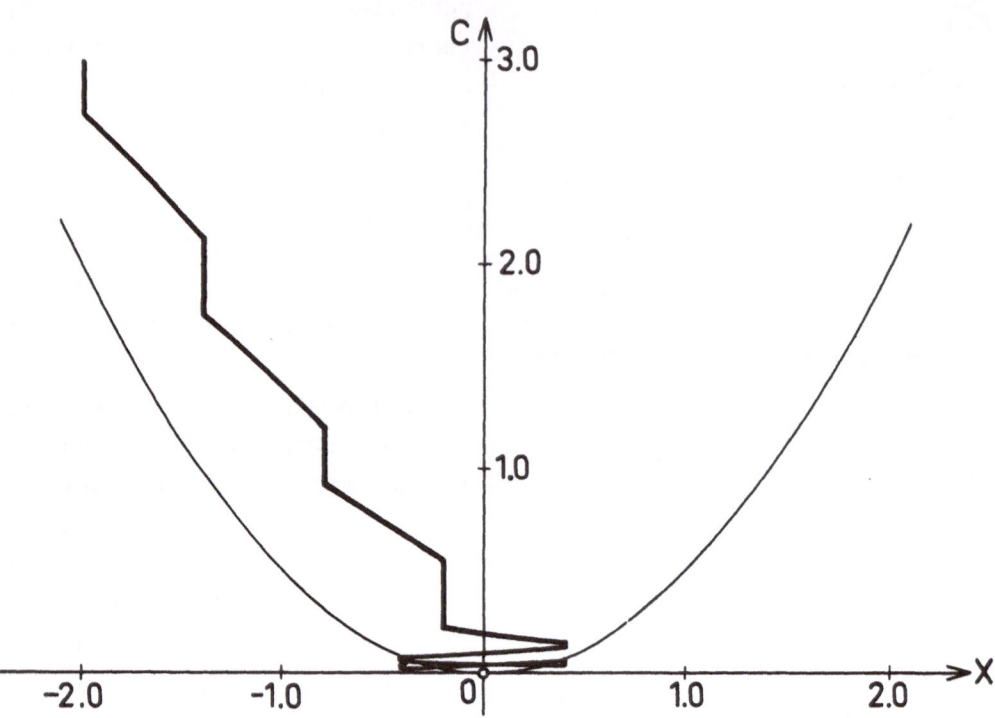

Fig.(4) Digital computer results

(a) $\epsilon_0 = 1$ $C_0 = 3$ $h_2 = 1.5$ $\gamma = 0.2$

     $X_0 = -2$ $h_1 = 0.6$ $\Delta = 0.4$

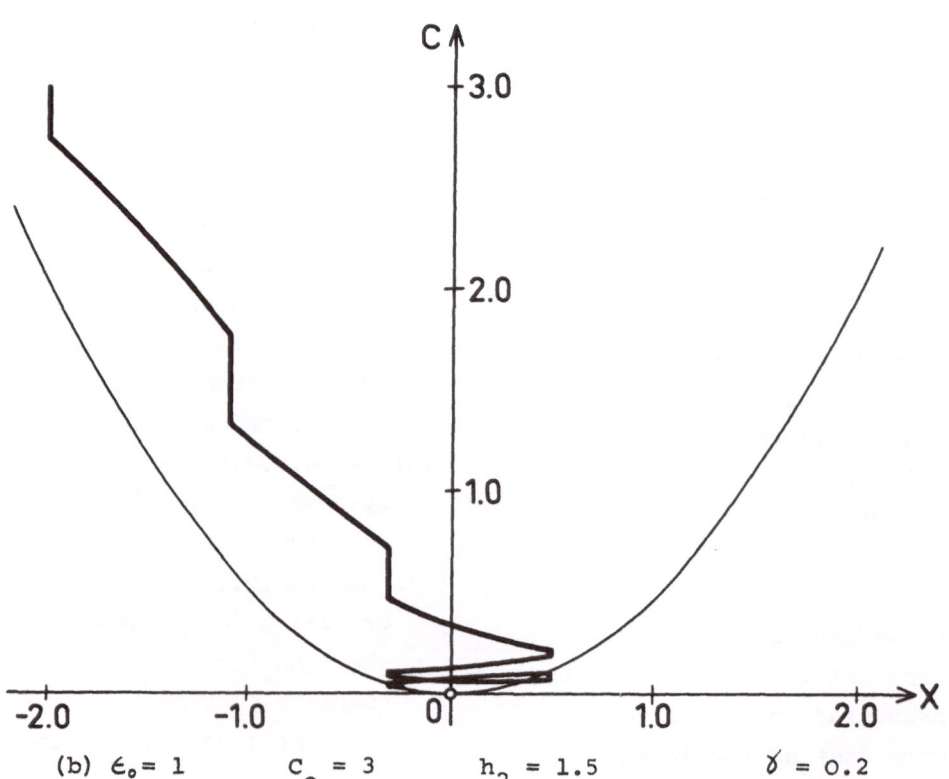

(b) $\epsilon_0 = 1$ $C_0 = 3$ $h_2 = 1.5$ $\gamma = 0.2$

     $X_0 = -2$ $h_1 = 0.8$ $\Delta = 0.2 \sin 0.1T + 0.4$

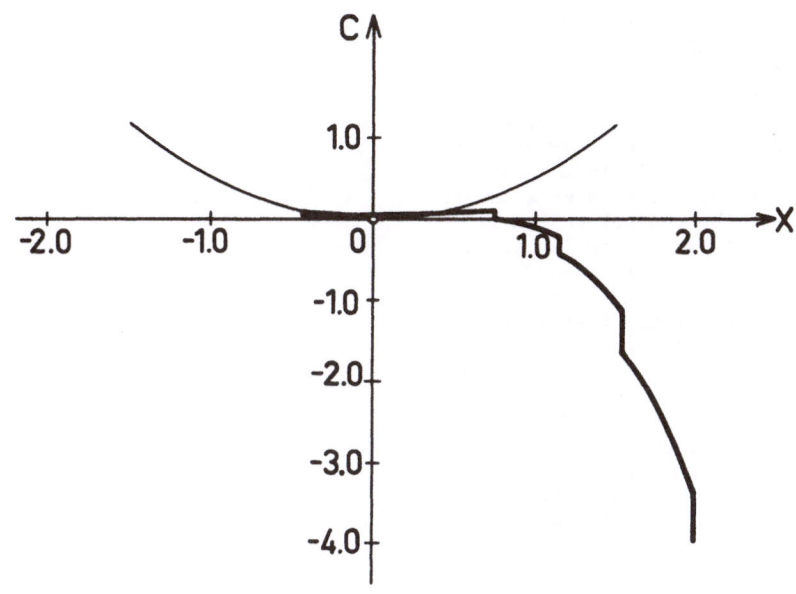

(c) $\epsilon_o = -1$ $C_o = -4$ $h_2 = 0.75$ $\gamma = 0.36$

      $X_o = 2$ $h_1 = 0.4$ $\Delta = 0.1 \sin 0.2T + 0.15$

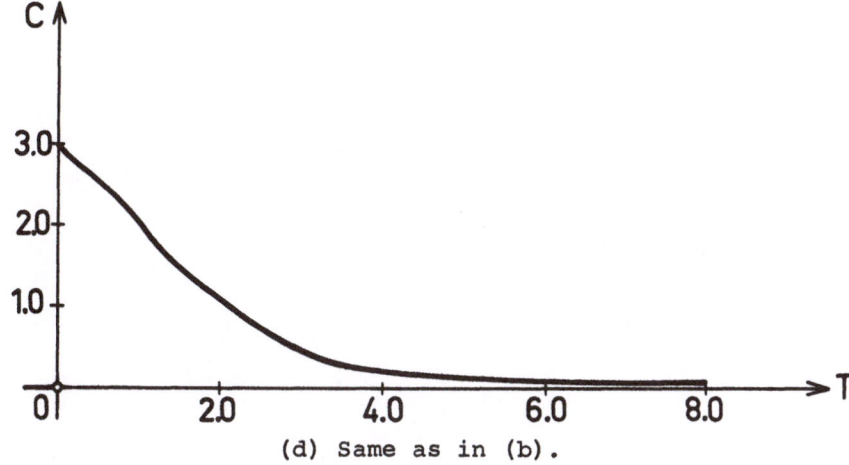

(d) Same as in (b).

DISCUSSION

Prof. C. J. D. M. Verhagen (Netherlands) questioned the author about the effects of noise on his system. Dr. Hamza stressed that he had been more concerned with mathematical details in the absence of any noise. However he did suggest that by employing suitable averages it might be possible to use the method in noisy situations.

Mr. P. H. Hammond (U.K.) asked the author how his method compared with that of R. Perret and R. Rouxel as described in reference 6 of the paper. The author showed how his control equation was based on the use of a derivative, one order higher than that used by Perret and Rouxel. In their method use is made of loci in the x-y plane on which $\dot{y}$ is zero. As soon as a trajectory crosses such a locus the control action is switched and the system moves along the locus to the origin. The method described in the present paper however has nothing to do with loci of constant or zero derivatives. Instead a perturbation is made at invervals and the sign of the change in the 2nd derivative of the response is noted. The sign of this change is used to decide in which direction movement is to continue. In this way the origin is reached. Since the method described by Perret and Rouxel uses a derivative one order lower than that used in the present paper it would be preferable in a noisy system.

DESIGN OF THE ADAPTIVE FEEDBACK LOOP IN PARAMETER-PERTURBATION ADAPTIVE CONTROLS

Dr. R. E. Kronauer
Professor of Mechanical Engineering
Harvard University
Cambridge, Massachusetts

Dr. P. G. Drew
Staff Engineer
Arthur D. Little, Inc.
Cambridge, Massachusetts

ABSTRACT

Prior analyses of parameter-perturbation adaptive control systems have demonstrated that the idea will work, but have not investigated the design of the adaptive loop. They require that the perturbations be small and slow and that the adaptive feedback be slow, but do not address quantitatively the questions of how small is small enough and how slow is slow enough. These are the questions dealt with in this paper.

INTRODUCTION

Parameter-perturbation adaptive control systems were first discussed by Draper and Li (Ref. 1) who applied the idea to the control of a throttle on a gasoline engine. Subsequently, McGrath and Rideout (Ref. 4) in the U. S. and Krasovski (Ref. 3) in Russia made more detailed analyses. The general idea is to learn the effect of a parameter of the plant on a performance criterion by perturbing the parameter, usually by oscillating it sinusoidally. This information can then be used to adjust automatically the parameter to the value which minimizes the performance criterion.

A block diagram of the general system is shown in Figure 1. The input x excites both a model G_d, which produces the desired output y_d, and the plant G, which produces the output y. (Upper case letters denote operators, which may be time-varying, while lower case letters denote signals.) The difference between y_d and y is the error e.

The perturbations are imposed on the adjustable parameter, a, thus modulating G and hence y and e. For purposes of analyzing the adaptive system, e is the only output of interest. The effect of the perturbations on the performance criterion is detected by multiplying the criterion by the perturbations to give z, which is then averaged by the correlating filter. It should be noted that e^2 is not the performance criterion until it has been averaged, but, since the correlating filter is an averaging device, there is nothing to be gained by averaging prior to multiplication by the perturbations. If the perturbations are small enough and slow enough, and the correlating filter averages for long enough, the output of the filter is proportional to the gradient of the performance criterion versus the parameter a. (Ref. 3 and 4) The signal proportional to the gradient can be integrated and fed back negatively to correct a.

The use of parameter-perturbation adaptive controls is possible for non-linear plants and non-quadratic performance criteria. However, since our goal is to analyze the adaptive loop in order to select its design parameters, and such an analysis is difficult enough even for linear plants and quadratic performance criteria, these restrictions will be imposed.

In order to proceed with the analysis, it is necessary to allow parameter changes of two types: One is a small but possibly rapid variation, typified by the perturbations, and the other is a slow but possibly large variation, typified by the drift due to environmental changes and the adaptive feedback. The error e is expanded in a double series in two small expansion parameters: α characterizing the small signals and β characterizing slow signals. The latter distinction, that between slow and fast time scales, is a crucial one for adaptive controls. It is on that basis that a varying quantity can legitimately be called a parameter or varying statistics of an input signal be defined.

THE α β EXPANSION

The linear system of Figure 1 is characterized by the differential equation

$$Qe(\widetilde{t},\hat{t}) = Px(\widetilde{t},\hat{t}) \tag{1}$$

where P is a differential operator of order m and Q is a differential operator of order n, $\widetilde{t}$ denotes the "slow" time variable. The symbol p denotes the total derivative with respect to time,

$$p = \frac{d}{dt} = \frac{\partial}{\partial\widetilde{t}} + \beta\frac{\partial}{\partial\hat{t}} = \widetilde{p} + \beta\hat{p} \tag{2}$$

and $\widetilde{p}$ and $\hat{p}$ the corresponding partial derivatives with respect to the fast and slow independent variables. (Parentheses are used to designate arguments and brackets, where necessary, to designate factors in multiplications.)

In general, the coefficients of the operators P and Q will be slowly varying due to environmental drift. In addition, they will depend on the single parameter, a, available for adaptation.

$$Q(a,\hat{t},p) = \sum_{k=o}^{n} a_k(a,\hat{t})p^k \tag{3a}$$

$$P(a,\hat{t},p) = \sum_{\ell=o}^{m} b_\ell(a,\hat{t})p^\ell \tag{3b}$$

Superior numbers refer to similarly-numbered references at the end of this paper.

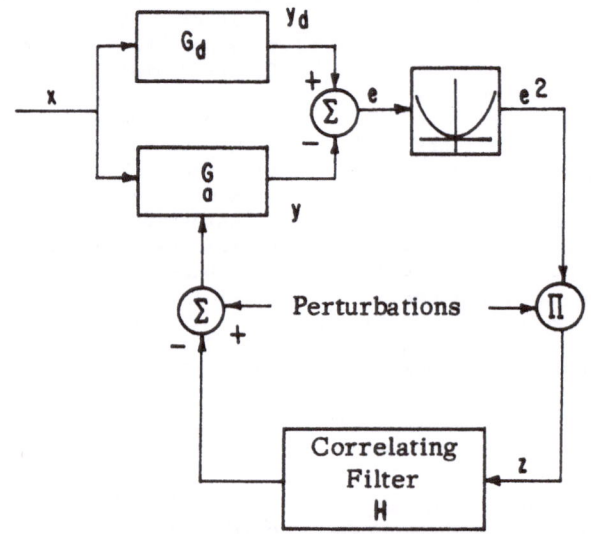

FIGURE 1 BLOCK DIAGRAM OF ADAPTIVE SYSTEM

where the explicit dependence on $\hat{t}$ represents environmental effects. The adaptation parameter is assumed to consist of a small, but not necessarily slow, part; and a slow, but not necessarily small, part:

$$a = \alpha \, \rho(\tilde{t}) + \mu(\hat{t}) \qquad (4)$$

Substituting (2) and (4) into (3), Q and P are seen to be formally functions of α and β, and as such may be expanded in a double Taylor Series about $\alpha = 0$, $\beta = 0$:

$$Q(a,\hat{t},p) = \sum_{i,j} [\alpha \, \rho]^i \, Q_{ij} \, [\beta \, \hat{p}]^j \qquad (5a)$$

$$P(a,\hat{t},p) = \sum_{i,j} [\alpha \, \rho]^i \, P_{ij} \, [\beta \, \hat{p}]^j \qquad (5b)$$

where

$$Q_{ij}(\mu,\hat{t},\tilde{p}) = \frac{1}{i!\,j!} \left. \frac{\partial^{i+j}}{\partial a^i \partial p^j} Q \right|_{\substack{a = \mu \\ p = \tilde{p}}} \qquad (6)$$

and similarly for P_{ij}. Note that the operators Q_{ij} and P_{ij} differentiate only in the variable $\tilde{t}$ while their coefficients are functions of $\hat{t}$.

The dependence on $\hat{t}$ arises explicitly through drift as previously noted and also implicitly through μ. Since we wish to maintain the possibility that ρ vary rapidly, it must remain to the left of the Q_{ij} and P_{ij} operators in (5), just as the operator in turn must stand to the left of $\hat{p}$. The expansions (5) clearly terminate in β at the powers n and m, respectively. They need not terminate in α, although in many practical cases they will. For example, if the coefficients a_k or b_ℓ depend linearly on a, the expansions terminate at $i = 1$. The operators Q_{ij} are interrelated as follows:

$$Q_{ij}' = \frac{1}{i} \frac{\partial}{\partial \mu} Q_{(i-1)j} = \frac{i}{j} \frac{\partial}{\partial \tilde{p}} Q_{i(j-1)} \qquad (7a)$$

$$Q_{ij} = \frac{1}{i!\,j!} \frac{\partial^{i+1}}{\partial \mu^i \partial \tilde{p}^j} Q_{00} \qquad (7b)$$

and similarly for P_{ij}. It should be noted that the Q_{0j} are the time domain equivalents of the time-varying system transfer function when coefficients are changing slowly, found by Zadeh (Ref. 5).

Next, the solution is expanded in the same way

$$e(\tilde{t},\hat{t}) = \sum_{i,j} \alpha^i \beta^j e_{ij} \qquad (8)$$

and together with (5) is substituted in (1) to yield a sequence of differential equations:

$$\alpha^0\beta^0: \quad Q_{00}e_{00} = P_{00}x \qquad (9a)$$

$$\alpha^1\beta^0: \quad Q_{00}e_{10} = -\rho Q_{10}e_{00} + \rho P_{10}x \qquad (9b)$$

$$\alpha^2\beta^0: \quad Q_{00}e_{20} = -\rho^2 Q_{20}e_{00} - \rho Q_{10}e_{10} + \rho^2 P_{20}x \qquad (9c)$$

. . . . .

$$\alpha^0\beta^1: \quad Q_{00}e_{01} = -Q_{01}\hat{p}e_{00} + P_{01}\hat{p}x \qquad (10a)$$

$$\alpha^1\beta^1: \quad Q_{00}e_{11} = -\rho Q_{11}\hat{p}e_{00} - Q_{01}\hat{p}e_{10} - \rho Q_{10}e_{01} + \rho P_{11}\hat{p}x \qquad (10b)$$

$$\alpha^2\beta^1: \quad Q_{00}e_{21} = -\rho^2 Q_{21}\hat{p}e_{00} - \rho Q_{11}\hat{p}e_{10} - Q_{01}\hat{p}e_{20} - \rho^2 Q_{20}e_{01} - \rho Q_{10}e_{11} + \rho^2 P_{21}\hat{p}x \qquad (10c)$$

. . . . .

The equation set (9) may be solved in the sequence shown by integration in the variable $\tilde{t}$ (with $\hat{t}$ held constant). These solutions may then be used in the right hand sides of (10) and this set may be solved similarly. The operator Q_{00} on the left of each of equations (9) and (10) is a constant coefficient operator for the purpose of integration, since $\hat{t}$ is held constant.

Proceeding with the solutions formally, we write

$$e_{00} = \frac{P_{00}}{Q_{00}} x \qquad (11a)$$

where by $1/Q_{00}$ we imply the usual convolution for the constant coefficient operator. Continuing,

$$e_{10} = -\frac{1}{Q_{00}} \left[\rho \frac{Q_{10}}{Q_{00}} P_{00}x - \rho P_{10}x \right] \qquad (11b)$$

$$e_{01} = -\frac{1}{Q_{00}} \left[Q_{01}\hat{p} \frac{P_{00}}{Q_{00}}x - P_{01}\hat{p}x \right] \qquad (11c)$$

. . . . .

The differentiation with respect to $\hat{t}$ implied by $\hat{p}$ in equations (11) can be carried out explicitly. For example,

$$\hat{p} \left(\frac{P_{00}}{Q_{00}} x \right) = \frac{1}{Q_{00}} \left[\left. \frac{\partial P_{00}}{\partial \hat{t}} \right|_\mu + \frac{\partial P_{00}}{\partial \mu} \frac{\partial \mu}{\partial \hat{t}} \right] x \qquad (12)$$

$$- \frac{P_{00}}{Q_{00}^2} \left[\left. \frac{\partial Q_{00}}{\partial \hat{t}} \right|_\mu + \frac{\partial Q_{00}}{\partial \mu} \frac{\partial \mu}{\partial \hat{t}} \right] x + \frac{P_{00}}{Q_{00}} \frac{\partial x}{\partial \hat{t}}$$

where the partial derivatives of P_{00} and Q_{00} in

$\hat{t}$ with μ held constant relate only to the explicit dependence on $\hat{t}$ (i.e., the environmental drifts, which, it should also be noted, are allowed in x). Some simplification of (12) results from using (7).

Note that in equation set (10) $\hat{p}$ operates on solutions from the set (9) only, and, in particular, as equation (12) illustrates, it introduces derivatives of $\mu(\hat{t})$ with respect to $\hat{t}$. It also introduces derivatives of $x(\tilde{t},\hat{t})$ with respect to $\hat{t}$. Although we shall not elaborate the point here, we have in mind that the random process $x(\tilde{t},\hat{t})$ has been characterized as an ensemble for which the dependence on $\tilde{t}$ disappears; that is, the hierarchy of distribution functions for x show x to be ergodic with regard to the "fast" time scale. At the same time, these distribution functions do depend on $\hat{t}$, but the dependence is such that the variables in $\hat{t}$ can be treated as parameters. This is an unfamiliar way of looking at ergodic processes because it is not strictly possible to retain any time dependence in an ergodic process. We suggest that such processes might be called "practically ergodic" because they are not strictly ergodic although they can be treated as if they were for the practical purpose of designing a control system.

The terms in the expanded solution which are linear in β are also linear in $\hat{p}$ and therefore linear in $\partial\mu/\partial\hat{t}$. For this reason the first correction, accounted for by β-terms, appear linearly in the equation of adaptation, and reflect a lag in the response of the system to changes in μ. The terms quadratic in β involve two differentiations by $\hat{p}$, and lead to terms in e_{12}, some multiplied by $\partial^2\mu/\partial\hat{t}^2$ and some multiplied by $(\partial\mu/\partial\hat{t})^2$. Similarly, terms cubic in β involve three differentiations with respect to $\hat{p}$, and lead to terms in e_{13}, some multiplied by $\partial^3\mu/\partial\hat{t}^3$, some by $(\partial\mu/\partial\hat{t})(\partial^2\mu/\partial\hat{t}^2)$, and some by $(\partial\mu/\partial\hat{t})^3$. The pattern is clear, but because of the loss of linearity, the utility of terms quadratic in β and higher is diminished.

GENERAL ADAPTIVE LOOP DESIGN CONSIDERATIONS

Development

When the adaptive loop is closed, the parameter, a, should approach its optimum through the changes of μ. Environmental drifts provide driving terms which require continual adjustment of μ. For the present, we will analyze only the transient, unforced behavior of μ. This is, after all, the representative behavior and contains all the information about stability and residual error generation in the adaptive process. The addition of the drifts can be accomplished later, but at this point it only obscures more vital features. With drifts suppressed,

$$e_{00} = \frac{P_{00}}{Q_{00}} x \tag{13a}$$

$$e_{10} = -\frac{1}{Q_{00}}[\rho Q_{10}e_{00}-\rho P_{10}x] = \frac{1}{Q_{00}}\rho Q_{00}\frac{\partial e_{00}}{\partial\mu} \tag{13b}$$

$$e_{01} = -\frac{1}{Q_{00}}[Q_{01}\hat{p}e_{00}-P_{01}\hat{p}x] = -\frac{\partial\mu}{\partial\hat{t}}\frac{Q_{01}}{Q_{00}}\frac{\partial e_{00}}{\partial\mu} \tag{13c}$$

. . . .

Proceeding through the adaptive loop, the error is squared and then multiplied by the detection signal to yield the signal z. The detection signal will be taken to be $\rho(\tilde{t})$. Some improvement might be realized by filtering ρ in a manner similar to that in which the plant filters the perturbations, but this refinement is too encumbering here.

$$e^2(\tilde{t},\hat{t}) = e_{00}^2 + \alpha[2e_{00}e_{10}] + \beta[2e_{00}e_{01}] + \dots \tag{14}$$
$$+ \alpha\beta[2e_{00}e_{11}+2e_{10}e_{01}] + \dots$$

$$z(\tilde{t},\hat{t}) = \rho(\tilde{t})e^2(\tilde{t},\hat{t}) \tag{15}$$
$$= z_{00} + \alpha z_{10} + \beta z_{01} + \alpha\beta z_{11} + \dots$$

Anticipating a result to follow, we take $\rho(\tilde{t}) = \cos\lambda\tilde{t}$, where λ is the perturbation frequency. We note that in the error signal, ρ always appears inside a convolution $1/Q_{00}$, as in (13b), for example. If ρ does not vary too rapidly, then it could be taken outside the convolution with only a small error, the size of which will be examined below. This will be termed the transposition of ρ.

Then, for example,

$$z_{10} = 2\rho e_{00}\frac{1}{Q_{00}}\rho Q_{00}\frac{\partial e_{00}}{\partial\mu} \tag{16}$$

$$\simeq \rho^2\frac{\partial e_{00}^2}{\partial\mu} = [\frac{1}{2}+\frac{1}{2}\cos 2\lambda\tilde{t}]\frac{\partial e_{00}^2}{\partial\mu} \tag{17}$$

We now wish to consider a statistical average over both ρ and x as contained in e. Since $\rho(\tilde{t}) = \cos\lambda\tilde{t}$, which is deterministic, and taking its statistical average merely yields $\cos\lambda\tilde{t}$ again. However, we intend to use the statistical average as the description of the output of an averaging filter which will, of course, average ρ as well as x -- the averaging will not be perfect since it is a finite time average, and we will also have to consider the variance of the output to determine how good an average it is. With this fact in mind, we arbitrarily introduce a random phase $\emptyset$, uniformly distributed over the interval 0 to 2π, $\rho(\tilde{t}) = \cos(\lambda\tilde{t} + \emptyset)$. This makes it possible to treat ρ and x consistently in the statistical averaging, even though the random phase need not actually be introduced in the real system.

We do not average over functions of μ because we expect to design our averaging filter so that, although it affects functions of μ, it does not form a good average of such functions.

$$\overline{e^2} = \overline{e_{00}^2} + \beta[\overline{2e_{00}e_{01}}] +\ldots \tag{18a}$$

$$+ \frac{1}{2}\alpha^2[\overline{2e_{00}e_{20} + e_{10}^2}] +\ldots$$

$$\overline{z} = \alpha\overline{z_{10}} + \alpha\beta\overline{z_{11}} +\ldots \tag{19a}$$

Transposing ρ as in (17) permits the approximations which follow:

$$\overline{e^2} \cong \overline{e_{00}^2} + \beta[\overline{2e_{00}e_{01}}] +\ldots \tag{18b}$$

$$+ \frac{1}{2}\alpha^2\overline{\rho^2}\,\frac{\partial\overline{e_{00}^2}}{\partial\mu^2} + \frac{1}{2}\alpha^2\beta\overline{\rho^2}\,\frac{\partial[\overline{2e_{00}e_{01}}]}{\partial\mu^2} +\ldots$$

$$\overline{z} \cong \alpha\,\overline{\rho^2}\,\frac{\partial\overline{e_{00}^2}}{\partial\mu} + \alpha\beta\,\overline{\rho^2}\,\frac{\partial[\overline{2e_{00}e_{01}}]}{\partial\mu} +\ldots \tag{19b}$$

In $\overline{z}$ we have neglected terms $O(\alpha\beta^2)$ and $O(\alpha^3)$ and higher.

In an adaptive system we have, of course, only a single sample of each of the random processes x and ρ. If we assume x to be ergodic, then the statistical averages in (18) and (19) are equal to the infinite time averages with μ held constant. However, here we demand a finite time average because we do not wish to hold μ fixed. The correlating filter H provides the finite time average. We require that the filter output be a good measurement of $\overline{z}$ because the leading term in $\overline{z}$ is approximately the one desired for proper adaptive action.

The principal filter output is $\overline{z}$ operated on by H plus measurement noise, resulting from using a finite time average. Some of the noise is contributed by every term in z, but the major contributor is z_{00}. It has a spectrum consisting of two lines at $\pm\lambda$ arising from the modulation of $\overline{e_{00}^2}$ by the detection signal plus a more or

less uniform part extending over twice the range of the error spectrum. Evidently H must be a low pass filter if it is to pass $\overline{z}$, a slowly changing function of μ, but reject most of this broadband noise.

The Equation of Adaptation

If the filter H passes the desired signal $\overline{z_{10}}$, it will also pass other components of $\overline{z}$ which are similarly functions of μ and slowly varying. The next important term, $\overline{z_{11}}$, is proportional to $\partial\mu/\partial\hat{t}$ and arises from a lag in response of e to a change in μ. Neglecting for the present the noise input to the filter, closing the adaptive loop consists of identifying μ with the output of the filter

$$\mu = -\alpha\,H(p)\,[\overline{z_{10}} + \beta\,\overline{z_{11}} +\ldots] \tag{20}$$

Since $\overline{z}$ is slowly varying, p can be regarded as $\beta\hat{p}$.

Consider (20) linearized about μ_0 where μ_0 is the optimum value of μ.

$$\mu = -\alpha\,H(\beta\hat{p})\,\{K_0[\mu - \mu_0] + K_1\beta\hat{p}\,\mu\} \tag{21}$$

where

$$K_0 = \frac{\partial\overline{z_{10}}}{\partial\mu}\bigg|_{\mu=\mu_0} \cong \frac{1}{2}\frac{\partial^2\overline{e_{00}^2}}{\partial\mu^2}\bigg|_{\mu=\mu_0}$$

$K_1 = \overline{z_{11}}$ with the factor $\partial\mu/\partial\hat{t}$ removed, evaluated at μ_0.

Suppose that we use the simplest possible averaging filter, an integrator, $H(\beta\hat{p}) = A/\beta\hat{p}$. Then (21) becomes

$$\mu = \frac{\alpha A K_0}{[1 + \alpha AK_1]\beta\hat{p} + \alpha AK_0}\,\mu_0 = \frac{\sigma}{\beta\hat{p} + \sigma}\,\mu_0 \tag{22}$$

Thus μ approaches μ_0 exponentially with a time constant $1/\sigma = [1 + \alpha AK_1]/\alpha AK_0$ measured in "fast" time or β/σ measured in "slow" time. We observe that "slow" time can be converted to "fast" time wherever we wish by using $\tilde{p} = \beta\hat{p}$ and $\tilde{t} = \hat{t}/\beta$.

The Noise Feedback

The noise feedback provides the considerations which fix A, the feedback gain. The noise input to H is denoted by ζ, which is taken to be composed of two parts ζ' and ζ''. The first, ζ', consists of the frequencies in ζ which are below a limit frequency ν and hence may be considered μ-type*. The high frequency noise ζ'' must be attenuated by H sufficiently to be a ρ-type signal. In fact, any H which reduces the ζ' component to tolerable limits will ordinarily make ζ'' negligible. Possible exceptions are the line spectra at $\pm \lambda$ in z_{00}. If necessary, these can easily be eliminated by including a bandstop filter in H -- a chief reason for using sinusoidal perturbations.

The term in curly brackets on the right side of (21) is the linearized output of the plant with noise disregarded. The noise ζ' is simply an addition to the plant output which is affected by the adaptive loop just as μ is. Thus, ζ' adds to μ a component μ', which, it should be noted, is partly compensated by the adaptive action.

$$\mu' = - H(p) \{\alpha K_0 \mu' + \alpha K_1 p \mu' + \zeta'\} \quad (23)$$

$$= - \frac{1}{\alpha K_0} \frac{\sigma}{\tilde{p} + \sigma} \zeta'$$

where we have converted to "fast" time. The variance of μ' is the mean-squared value of the uncompensated noise and constitutes an unavoidable "hunting" of μ.

$$\overline{\mu'^2} = \frac{1}{2\pi \alpha^2 K_0^2} \int_{-\nu}^{\nu} d\omega \frac{\sigma^2}{\omega^2 + \sigma^2} S_{\zeta'}(\omega) \quad (24)$$

$$\cong \frac{\sigma}{2\alpha^2 K_0^2} S$$

where $S_{\zeta'}$ is the spectral density of ζ', S is $S_{\zeta'}(0)$, and the approximate solution is correct if $\sigma \ll \nu$ and $S_{\zeta'}$ is flat out to ν. The exact expression can be used if necessary. We intend to select A so that $\overline{\mu'^2}$ will be $0(\alpha^2)$. Therefore its contribution to $\overline{e^2}$ can be found in the same manner as the contribution due to the perturbations, and can be shown to be $\overline{\mu'^2} K_0$.

Now both the perturbations $\alpha \rho$ and the noise feedback μ' contribute to excursions of the parameter, a, from its optimum. The increase in $\overline{e^2}$ due to the perturbations has been shown to be

---

*The signal μ must either be band limited or possess an exponential spectrum for the β series to converge. In the case of band limitation the frequency ν is a property of the plant alone (not the adaptive elements) and can be recognized by considering e^2. No such restriction applies to ρ-type signals in the α expansion.

nearly $\alpha^2 K_0/2$; that due to the noise feedback has been found to be approximately $\sigma S/2\alpha^2 K_0$. For simplicity we assume $|\alpha AK_1| \ll 1$ (the usual case) so that $\sigma \cong \alpha AK_0$. Thus the increase in $\overline{e^2}$ is

$$\overline{e^2} - \overline{e_{00}^2} = \frac{K_0}{2} \left[\alpha^2 + \frac{AS}{\alpha K_0}\right] \quad (25)$$

We elect to make the power in each contribution equal. This is a purely arbitrary decision although a plausible one. Any other choice is perfectly allowable and does not affect the analysis. Then the increase in $\overline{e^2}$ is

$$\overline{e^2} - \overline{e_{00}^2} = \alpha^2 K_0 = \frac{AS}{\alpha} \quad (26)$$

The integrator gain is then

$$A = \frac{\alpha^3 K_0}{S} \quad (27)$$

and

$$\sigma = \frac{\alpha^4 K_0^2}{S} \quad (28)$$

For the special case of a Gaussian input x, e_{00} is Gaussian, and it is well known that one-third of the power in e_{00}^2 lies in $\overline{e_{00}^2}^2$ while two-thirds lies in the broadband spectrum, which, roughly, we can take to extend uniformly between plus and minus $2\omega_0$, where ω_0 is the bandwidth of the error. Neglecting the shifts produced by the detection signal, we can state that $\omega_0 S \cong \pi \overline{e_{00}^2}^2$. Using this approximation, (26) and (28) can be combined to show that

$$\frac{\overline{e^2} - \overline{e_{00}^2}}{\overline{e_{00}^2}} \cong \left[\frac{\pi}{\omega_0} \sigma\right]^{1/2} \quad (29)$$

Considerations in Selecting the Frequency of the Perturbation

The line spectra of z_{00} at $\pm \lambda$ must be rejected by H(p) which suggests that λ should be well above the passband of H. On the other hand, the equivalence of $\overline{z_{10}}$ and the gradient of $\overline{e_{00}^2}$ in μ in (19) rests on λ not being too high. This last qualification must be made more precise. Examining (16) with ρ inside the convolution

304

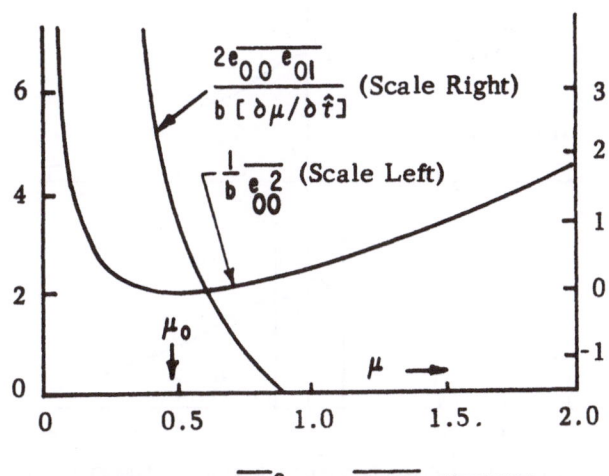

FIGURE 2 $\overline{e_{oo}}^2$ AND $\overline{2e_{oo}e_{o1}}$ VERSUS μ

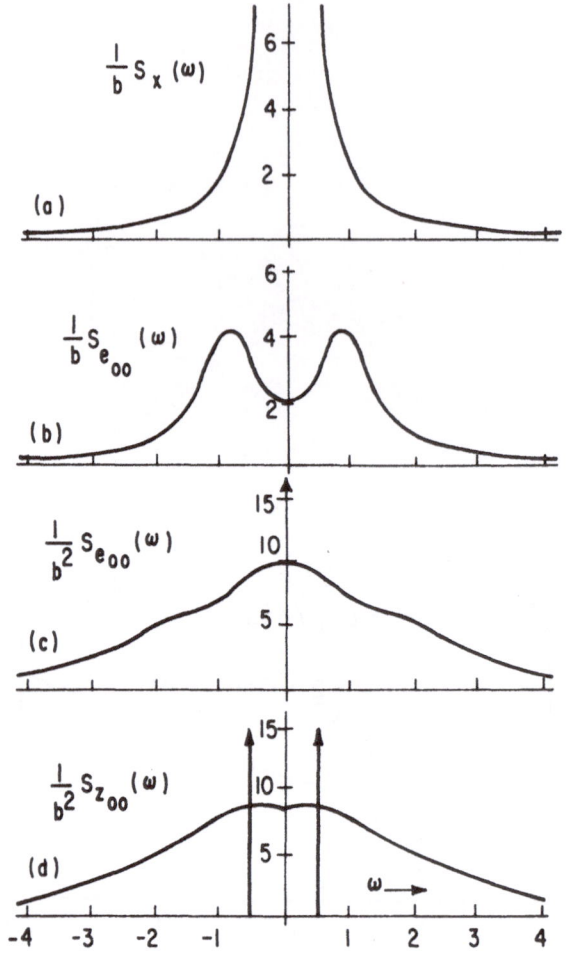

FIGURE 3   COMPOSITION OF SPECTRUM OF z_{00}

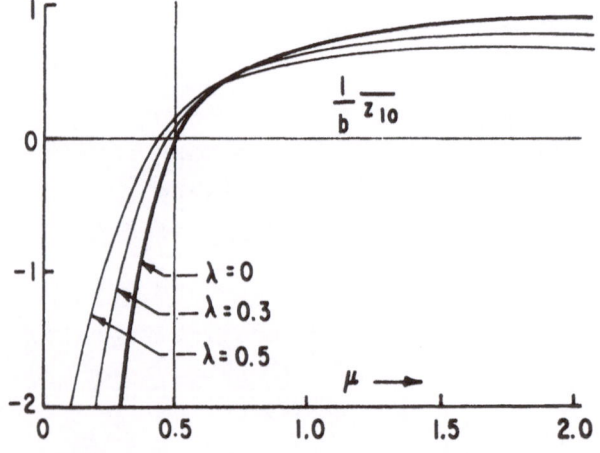

FIGURE 4 $\overline{z_{10}}$ VERSUS μ FOR VARIOUS λ

the inner multiplication by $\rho = \cos \lambda \tilde{t}$ splits the spectrum of the function on which it operates into two halves, one of which is shifted upwards by λ and the other downwards by λ. The outer multiplication by ρ has a similar effect, returning two-fourths of the spectrum to its original frequency and leaving two parts, each one-fourth of the spectrum, at displacement of $\pm 2\lambda$. The latter, when multiplied by e_{00} as in (16) and averaged are, in general, not correlated with e_{00}. Of the two-fourths returned to the original frequency, one will have been operated on by $1/Q_{00}$ while displaced $+\lambda$, and the other while displaced $-\lambda$. Thus, the components contributing to $\overline{z_{10}}$ are those which are correlated with e_{00}. In the frequency domain they have been multiplied by

$$\left[\frac{1}{Q_{00}(j\omega + j\lambda)} + \frac{1}{Q_{00}(j\omega - j\lambda)}\right] \frac{Q_{00}(j\omega)}{4}$$

The difference between this expression and its value when $\lambda = 0$ produces a discrepancy between $\overline{z_{10}}$ and the true gradient. It is evident by expanding this expression that the discrepancy is $0(\lambda^2)$; that is, it depends principally on the curvature of $1/Q_{00}$ in ω. In cases where the poles are reasonably well damped, the discrepancies are remarkably small even for quite rapid perturbations. A similar discussion applied to other terms in e^2 and z where ρ was taken outside the operator $1/Q_{00}$.

SPECIAL RESULTS FOR A SECOND ORDER POSITIONING SERVO

We consider a second order positioning servo with an adjustable damping coefficient, a, and a natural frequency of unity. The desired output is taken to be the input x, in which case the error is, from (1)

$$Q(a,\hat{t},p) = p^2 + 2ap + 1 \tag{30a}$$

$$P(a,\hat{t},p) = p^2 + 2ap \tag{30b}$$

For this plant, equations (6), for example, are as follows:

$$\left.\begin{array}{ll} Q_{00} = \tilde{p}^2 + 2\mu\tilde{p} + 1 & P_{00} = \tilde{p}^2 + 2\mu\tilde{p} \\[2mm] Q_{10} = P_{10} = 2\tilde{p} & Q_{01} = P_{01} = 2\tilde{p} + 2\mu \\[2mm] Q_{02} = P_{02} = 2 \end{array}\right\} \tag{31}$$

All further P_{ij} and Q_{ij} are zero.

Suppose we consider a Gaussian random input x with a spectral density $2b/[\omega^2 + b^2]$, $b \ll 1$. We find $\overline{e_{00}^2} = b[4\mu^2 + 1]/2\mu$, as shown in Figure 2. The gradient $\partial \overline{e_{00}^2}/\partial\mu$ is $b[4\mu^2 - 1]/2\mu^2$, which vanishes at $\mu_0 = 1/2$, for which $\overline{e_{00}^2} = 2b$. The curvature $\partial^2 \overline{e_{00}^2}/\partial\mu^2$ is b/μ^3, and at μ_0 this is $8b$; hence $K_0 = 4b$.

The term $0(\beta)$ in $\overline{e^2}$ is $b[\partial\mu/\partial\tilde{t}][1-8\mu^4]/4\mu^3$, which is also shown in Figure 2. At μ_0 this is $b[\partial\mu/\partial\hat{t}]$. The frequency ν, which is the maximum frequency for which the feedback noise can be considered a μ-type signal, depends on the ratio of $\overline{e_{00}^2}$ to this term. The ratio is 2, and hence $\nu \ll 2$ is a suitable bound. By differentiating the term $0(\beta)$ in $\overline{e^2}$ and evaluating at $\mu = \mu_0$, we find K_1 appearing in (21) is $-7b$. Note that for this system K_1 is negative, and, if $|\alpha A K_1|$ is not less than unity, the equation of adaptation is unstable.

The composition of the spectrum of $z_{00} \cong \zeta$ is shown in Figure 3. Its spectral density S at zero frequency is $8.6b^2$. Suppose we are willing to accept at a 5% increase in $\overline{e^2}$ due to the perturbations and noise feedback. Then, from (26), the perturbation amplitude α is 0.16. From (27), the integrator gain A is $0.0018/b$, and from (28), the adaptation break frequency σ is 0.0012. Since $|\alpha A K_1|$ is 0.002, neglecting it in (25) and subsequent equations was justified. We also observe that the requirement $\sigma \ll \nu \ll 2$ used in (24) and subsequent equations was justified as well. Finally, we consider the frequency of perturbation λ. Curves of $\overline{z_{10}}$ versus μ for various values of λ are shown in Figure 4. The discrepancy between actual values of $\overline{z_{10}}$ and the true gradient (occuring for $\lambda = 0$) at μ_0 is not large even for λ as high as 0.5. Although discrepancies are large for small values of μ, adaptive action is not impaired because $\overline{z_{10}}$ still has the proper sign. We note that the sinusoid in the noise feedback arising from $\overline{e_{00}^2} \cos \lambda \tilde{t}$ is reduced by the factor A/λ. Hence, if we use $\lambda = 0.5$, its amplitude is 0.004 or 2.4% of α. This would probably not be troublesome.

CONCLUSIONS

The design of the adaptive loop is controlled primarily by the conflicting goals of feeding back as large an identification signal as possible for rapid adaptive action, yet no more noise than is tolerable from the standpoint of increased error when the optimum condition is achieved. While this has been recognized by previous authors, this is the first presentation of quantitative design procedures. The techniques of filter optimization could be brought to bear to improve design of H(p), but the various analytic approximations make this of secondary importance. The effect of lags in the adaptive response of the plant and the effect of the frequency of the perturbations can both be calculated by the methods given here.

While we have not specifically considered environmental drifts (including statistical drifts of the input signal), the mechanism for doing so is worked out. The extension to multi-parameter adjustment is likewise straightforward, requiring an application of the $\alpha\beta$ series to each parameter and a sufficient separation of perturbation frequencies to permit rejection of the unwanted identification signals in each separate correlation filter. One point is clear from all this analysis: To design an efficient adaptive system requires a rather complete and exact knowledge of the plant, its input and its environment. This is a long way from some of the basic motivations behind adaptive control.

We have not investigated convergence of the $\alpha\beta$-series, although clearly it yields the correct solution in the limit when α and β approach zero, and yields the exact solution when the complete (non-terminating) series is considered. Also we have not considered possible correlations between the input and the adaptive feedback signal. Doing so displays the essential nonlinearity of the adaptive process, and is a very difficult problem. The fact that there are many phase shifts resulting from squaring the error, multiplying it by the detecting signal, and integrating it, gives hope that any such correlations are destroyed.

The second order system was simulated on an analog computer and gave results substantially in agreement with those presented here (Ref. 2).

ACKNOWLEDGMENT

This work was supported in part by the Joint Services Electronics Program (U.S. Army, U.S. Navy, and U.S. Air Force) under Contract Nonr-1866(16), and by the Division of Engineering and Applied Physics, Harvard University.

REFERENCES

1. C. S. Draper and Y. T. Li, "Principles of Optimalizing Control Systems and an Application to the Internal Combustion Engine," ASME publication, September, 1951.

2. P. G. Drew, "A Rapid-Acting Parameter-Perturbation Adaptive Control," Ph.D. Thesis, Harvard University, January, 1964.

3. A. A. Krasovski, "The Dynamics of Continuous Automatic Control Systems with Extremal Self-Adjustment of the Correcting Devices," Proc. of the First International Congress of the IFAC, Moscow, 1960, pp. 679-685, published by Butterworth's, London, 1961.

4. R. J. McGrath and V. C. Rideout, "A Simulator Study of a Two Parameter Adaptive Control System," IRE Trans. of PGAC, Vol. AC-6, pp. 35-42, Feb., 1961.

5. L. A. Zadeh, "Correlation Functions and Power Spectra in Variable Networks," Proc. IRE, Vol. 38, pp. 1342-1345, Nov., 1950.

SOME QUESTIONS OF LEARNING OPTIMIZATION OF LARGE-SCALE PROCESSES.

by M. Orbán
Institute for Automation of the Hungarian Academy of Sciences
Budapest, Hungary

ABSTRACT

So far the problem of optimization of large-scale processes is not solved on a general and effective basis, although such a basis exists in connection with the adaptivity of the living organisms . The subject of this paper is such a method of learning optimization which simulates the living adaptivity to a great extent.

1./ THE DESCRIPTION OF THE PROBLEM.

The object of major scientific trends of the control engineering in the recent years was to solve the optimization of processes characterized as follows:

1./ The course of the process is determined by a relatively large number of input variables representing the disturbing effects of external origin and the control activity.

2./ The causal relations between the changes of the input and output variables of the process can be described satisfactorily, but not by very precise and deterministic mathematical expressions, because of the number of nonlinearities and stochastic phenomena.

3./ The variables of the process are continuous ones, the disturbances of stochastic nature.

4./ The processes are established to accomplish a well determined aim, for example such as profit.

5./ The properties of the process which can be described more or less by the functional relations between the input and output variables are time - variant and therefore the mathematical model of the plant, based on the existing laws of the process at a given time are valid only for a limited time interval. The input and output variables of a process can be classified according to the block - scheme shown in Figure 1, in the following way:

1./ The input vector X / the components of which are $x_1/t/$, $x_2/t/...x_i/t/$ / represents the disturbances, or in other words the influence of the environment on the plant.

2./ The input vector Y - with components $y_1/t/$, $y_2/t/,...y_j/t/$ - is the set of the control variables. These are set according to the policy of optimum control.

3./ The single variable C is the "object function" / yield function /. Generally this is a function of both input and output vectors established arbitrarily from the control-engineering viewpoint. Assuming, that

$$Z = Z \ /X,Y,t/ \qquad 1./$$

where t denotes time, the object function is

$$C = C \ /X,Y,t/ \qquad 2./$$

The above mentioned task concerning the optimization was to set the control vector /Y/ to achieve the optimal value /maximum or minimum/ of the object - variable, in other words the aim of the optimum-control-policy is to accomplish the necessary condition for the object function

$$\frac{\partial C}{\partial y_1} = \frac{\partial C}{\partial y_2} = \dots = \frac{\partial C}{\partial y_j} = 0 \qquad 3./$$

with regard to the possible constraints:

$$\begin{aligned}
\alpha_i &\leq y_i \leq \beta_i, \\
g_1 \ &/ X,Y \ / \leq 0, \\
g_2 \ &/ X,Y \ / \geq 0
\end{aligned} \qquad 4./$$

2./ THE PRESENT STATE OF PROCESS - OPTIMIZATION

So far there are only a few practical solutions of the optimal control of processes. The most famous of all is the computer control of the Mon Santo ammonia plant. Following this very significant initiation the optimizing methods based on the approximating mathematical models of the plants have spread on several fields of process control. The optimal control using mathematical model has the advantage of feedforward operation, but

Superior numbers refer to similarly-numbered references at the end of this paper

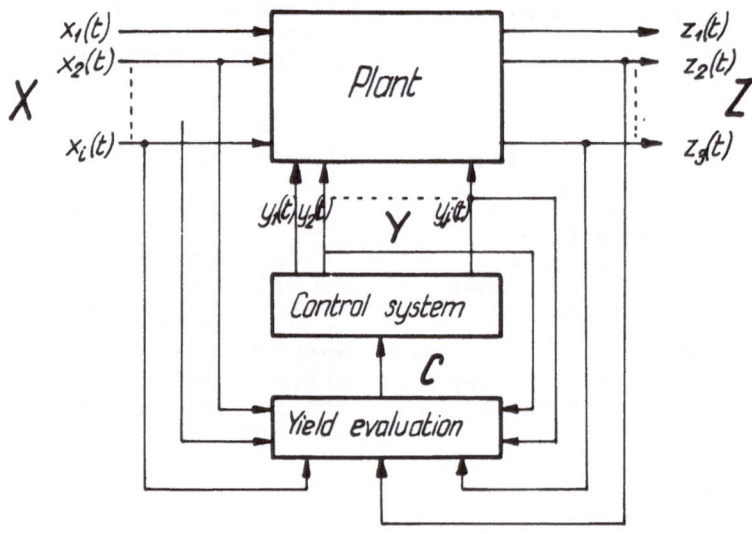

Fig. 1

310

usually does not accomplish the theoretically available steady state and dynamic optimum of the system, because of the fact, that the laws of the processes are to a great extent unknown and therefore can not be described precisely. A further disadvantage of these methods is, that they do not usually adapt themselves to the changing properties of the processes. As a consequence of this, the performance of the optimization gradually worsens in time.

There are optimizing procedures which fit for rather small processes only and are based on statistical principles[1,2,3,4,5]. Further methods carry out the optimization by search[6,7]. Some detailed summaries of the possible optimizing procedures can be found in the technical litterature often in connection with function minimization or maximization[8,9].

Reviewing the available information on the optimal process-control the following conclusion arises: the optimization of large - scale processes can be successfully carried out by the use of mathematical models, however this method is usually complicated and operates regardless of the time variant properties of the process. The economic improvement with regard to the costs of the hard- and software of the optimizat zation is usually modest mainly because of the fact that this method can be used only for steady state and not for dynamic optimization. The other methods can not be applied but to processes with a few input variables.

3./ THE BASIC PRINCIPLES OF THE LEARNING ADAPTIVE CONTROL.

In contradiction to the poor technical and scientific results on the field of optimal control, the living organisms show a perfect solution of optimization in the form of general adaptivity. First of all the living organisms can be characterized by the following features:

1./ the organisms are under the influence of their environment through a very large number of variables and they react to those by a similar multitude of variables.

2./ From the viewpoint of the organism the environment can not be described by the means of the mathematical analysis: its laws can be regarded as "a priori" unknown ones.

3./ The organism and its environment together form a time variant system.

4./ The organism possesses a basic aim, the projections of which in the subspaces of the activity - space of the organism give secondary aims. These are followed by the activities of the parts of the organism / organs /.

Now we can enlighten some of the basic concepts occuring in connection with the theme of adaptivity in the following way: the self-adaptation is the behaviour of the organism directed towards the optimal realization of the aim of the organism in the situation determined by the environment. The self - adaptation of orgainsms of high level is a compound process containing as a part of its the learning. Through the learning the organism collects and fixes the data describing the impressions of the environment on itself evaluating these from that point of view, to what extent are the external effects and the reactions of the organism itself in accordance with the basic aim.

As far as the structure of the nervous system is known we can obtain a fairly good picture of its control mechanism. The effects of the environment are transmitted by the sensory organs through the nerves to the brain and there these events are recorded, together with the reaction of the organism and a quantity characterizing their value from the viewpoint of the aim of the adaptive control. All the reactions of the organism are set according to the earlier experiences trying to achieve better yield value. It is a very important fea - ture of the living organism that the content of its memory gets continuously refreshed; new data are stored in the place of earlier experiences. Another feature worth mentioning is the hierarchic structure of the living control system, which means that there are adaptive sub - systems behaving independently of each other and the whole organism consists of these sub - systems.

Returning to our original problem of process-optimization we can ascertain that the relation between a process of "a priori" unknown characteristics and its control system, is similar to a great extent to that between the environment and living organism. This is why the control procedures of the latter systems can be applied in the case of the former ones.

It is to be emphasized, that further important features of the living self adaptation are: the complete absence of any mathematical or logical abstraction, and it is optimal in steady state and dynamic sense too.

4./ THE STATIC OR STEADY STATE OPTIMIZATION.

The learning adaptive method to be described is equally suitable for static and dynamic optimization but the details of its operation are clearly demonstrated in the simple static case.

The process of the learning optimization consists of two parts: the data collection and control. The task of the data-collection is to build up a function table in the memory of the system through singular points of observation. These points represent the vector X of external - disturing - variables which describe the state of the environment from the viwwpoint of the process and the control vector Y as the set of data corresponding to the reaction of the control system in the given environmental situation. These data are supplemented by the value of a single variable, the yield function, as a consequence of the input variables X and Y. The set of connected data of X, Y, C can be regarded as a point of the surface of the function C = C /X, Y/. Provided that the number of components of X is "i" and that of Y is "j", the space containing the yieldsurface is an i+j+1 dimensional one. Because of the fact, that the points of this surface represent situations or events in the course of the process, this space - otherwise numerical model - can be named as the "space of events". The detailed structure of a point looks like the following series of data:

$$x_1/n/, \ldots x_i/n/, \; y_1/n/, \ldots y_j/n/, \; C/n/$$

A limited number of such points are stored in the memory of the system. Since - according to the algorithm ot the learning optimization developed in the Institute for Automation - the collection of the observed data is a continuous process after filling all the available capacity of the memory, the earliest data are to be replaced by the newest ones. Meanwhile the data-collection is carried out without any control activity, the distribution of points in the space of events is almost completely random with flat frequency surface and this does not vary with the refreshing of the points. On the other hand, when the optimal control activity is in operation a tendency can be observed in connection with the rearrangement of the point-population due to the refreshing process. As a consequence of this, the distribution of points becomes less random: the density of their population increases near a space

curve, the trajectory of optimal control. This is one of the most significant features of the learning optimizing method and gives the basis of its adaptability, namely the trajectory of optimal control always reflects the even then existing characteristics of the process.

The control activity can be described as follows: At a given time, the vector X/t/ characterizes the existing situation of the environment. From the closest vicinity of this vector a given number of points of similar X vectors are looked up. The control vector parts /Y/ of these points are used for the evaluation of a new step i.e. a new vector Y/t/. In this evaluation the direction and the length of the new step are determined. It is worth mentioning, that only a few points, representing the previous experience of the system, are used for the evaluation of the new step despite of the fact that the control vector is usually a multi-dimensional one. If a continuous surface were available for the optimization, it would be very easy to get satisfactory information about the shape of the surface near by a given point. However in the present case points relatively far from each other have to supply information about the yield function at a given point. The degree of uncertainty caused by the use of such points can be estimated in the following way;

Assuming normalized variables and denoting the number of points in the space of events by "N", the number of components of the X and Y vectors by "i" and "j" the uncertainty is given by the following expression:

$$\triangle = \frac{1}{{}^{i+j}\sqrt{N}} \qquad\qquad 5./$$

meaning, that the component of the average distance measured between two neighbouring points, provided a completely random distribution is $\triangle$. The quality of the step-evaluation is characterized by this quantity. In the Table 1. for the sake of illustration some numerical examples are shown for the value of uncertainty.

Table 1.

N i+j	10^2	10^4	10^8
40	0.89	0.79	0.63
20	0.79	0.63	0.40
10	0.63	0.40	0.16

It can be observed that the increase of the number of points results in a very small decrease of the uncertainty. Because of this uncertainty the learning optimization is an iterative process.

The next question to be answered is: whether is the learning optimization a convergent process with regard to the uncertainty. The convergence of such an iterative optimization presents itself in the tendency, that the steps done by the optimizing procedure are always getting closer to the trajectory of optimal control. There are two important criteria of the convergence: the correctness of the direction and that of the length of the steps. In the case, when these steps are carried out in a random way without any optimal policy, it can be expected that half of the components of the new direction are correct. This means, that the randomness of the point - distribution does not change. Now applying the optimal policy which evaluates the new direction making use of the data of points found in the neighbourhood of a base - point we find so that the ratio of the number of correctly determined components to that of all components - provided that k is the number of the control variables /otherwise: all the components/ - is as follows:

$$\eta = \frac{\nu + k - 1}{2k} \qquad 6./$$

It can be quite easily understood, that the majority of components of a new direction are correct in that case too, when beside a base-point only one further neighbouring point is chosen /ν=2/ for the evaluation. The length of steps can be set in an adaptive manner observing through a time T the value of a function similar to the yield function in the case of dynamic optimization:

$$E = \frac{1}{T} \int_{t}^{t+T} C dt \qquad 7./$$

E is to be maximized by varying the length of steps.

Concerning the details of the procedure for the determination of the steps it is to be mentioned, that the vector of the new step is derived from vectors pointing from the base point towards the neighbouring points weighted by the values of their ascents.

In the practice, the learning optimization starts after an interval during which only the data collection is carried out and a small portion of the memory gets filled. Meanwhile the plant is controlled by the operators. If they try

to carry out an optimal control, the points will approach the optimal trajectory in the event - space and therefore, when the automatic control begins to operate, it will show a much faster convergence than after non-optimal manual control.

The system, remembers the properties of the process for a time proportional to the capacity of its memory and the time interval between two consecutive observations of event-points. In the case of a process of rapidly changing nature it is advisable to keep low the value of remembering time by decreasing the memory-capacity or increasing the speed of data collection. In this way the time-basis of the adaptivity will be reduced.

5./ DYNAMIC OPTIMIZATION WITH THE LEARNING METHOD.

The dynamics of large-scale processes can be described first by time-delays and time-lags. In most cases the time-delays are very significant. The steady state optimization loses much in effectivity when the time-delays are greater or in the order of time-constants of the disturbances.

The task of dynamic optimization can be given in the following general form:

$$\max E = \frac{1}{T_c} \int_{t-T_c}^{t} C[X/D_x, t/, Y/D_y, t/, t] dt \qquad 8./$$

where C is the yield function of the input vectors X,Y and D_x, D_y are symbols representing in short form notation the dynamic parameters of the concerning time-functions. The $X/D_x, t/$ vector- function is the set of independent variables from the viewpoint of the control system and is continuously observed by it. The task of the optimal control is to set the absolute values and dynamic parameters as functions of time to maximize the time integral of the yield function over a period of T_c. We regard this time-interval as such in which all the effects of the input changes wholly develop. The value of T_c can be arbitrarily established, but this means that the causal relations are arbitrarily limited.

The learning optimization possesses the advantageous property of being capable to operate with regard to the time-delays. These time-delays are due to the advance of the material through the plant. The problem of the delays arises in both phases of the learning optimization: first in the data-collection, second in the

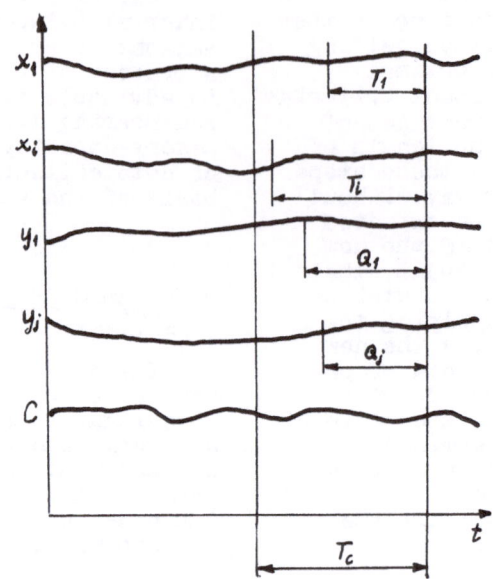

Fig. 2

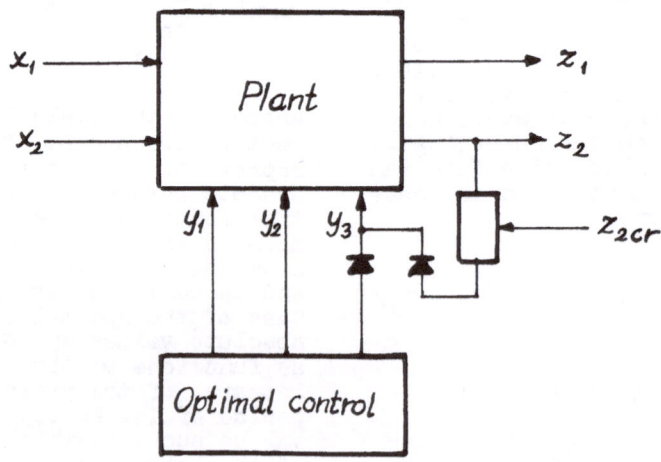

Fig. 3

314

control.

In the case of data-collection the task is to coordinate the related values of input and output variables as causes and consequences in the time interval T_c. According to the fact, that the input variables are stochastic quantities, the most suitable description of the causal relations could be carried out by the correlation - analysis. On the basis of the analogy of living adaptivity the correlation between the input and output quantities can be determined by independent optimization technique. In the practice the correlation can be determined with radioactive trace - technique.

In the second case - as mentioned above - the problem is the maximization of the time-integral of the yield function through the dynamic control vector $Y/D_y,t/$. A general solution can be applied for the static and dynamic optimization making use of the extended space of events. This has the point structure as follows:

$$x_1/t/, \ T_1/t/, \ldots x_i/t/, \ T_i/t/,$$

$$y_1/t/, \ Q_1/t/, \ldots y_j/t/, \ Q_j/t/$$

In this simplified set of data the time-delays $T_1,\ldots T_i, \ Q_1,\ldots Q_j$ are to be understood as time-intervals which elapse from a change in the value of the related input variable till the moment when its effect developes in the value of yield function at time t /Fig. 2./. Now the process of optimization gets modified in the following way: the dynamic parameters of the input variables $/D_x,D_y/$ are attached to the control variables of the static optimization /Y/, and are dealt with as those. Their values are varied to achieve:

$$\frac{\partial E}{\partial y_1} = \ldots \frac{\partial E}{\partial y_j} = \frac{\partial E}{\partial T_1} = \ldots \frac{\partial E}{\partial Q_j} = 0 \qquad 9./$$

This general form of optimization can be carried out on certain restricting conditions.

6./ LEARNING OPTIMIZATION IN THE PRESENCE OF CONSTRAINTS.[10]

Although the method to be shown has been developed in connection with the learning optimization, its basic ideas are general and can be used together with other optimizing procedures.

For the sake of simplicity we regard the problem in the case of static optimization. The task is the optimiza-

tion of the function $C/X,Y/$ with regard to one or more constraints of this type /the other types - because of their simplicity - may be neglected here /:

$$g \ / \ X,Y \ / \ \leq 0 \qquad 10./$$

As long as the inequality is valid, the optimization is carried out regardless of the constraint. When the equality turns to be valid, the structure of the control system changes in the following way: the variable representing the constraint becomes a simple controlled variable in an independent feedback loop. Its value is set by one of the former control - variables withdrawn from the sphere of activity of the optimal control. This variable is selected for the control loop as that having the strongest influence upon the new controlled variable.

This procedure is shown with regard to a real technological problem in Fig. 3. The block diagram represents a process - section of three parallel units. The output variable z_2 is constrained from above:

$$z_2 = z_2/x_1,x_2,y_1,y_2,y_3/ \ \leq \ z_{2cr}$$

Until the $z_2 < z_{2cr}$ holds, the optimization is carried out in the usual way but as soon as z_2 reaches the critical value z_{2cr}, a structural change takes place and the control variable y_3 will form an independent feedback loop together with z_2. The set-value of this control loop is z_{2cr}. After this only the remaining two control-vectors y_1, y_2, will carry out the optimization . The rectifiers in the diagram indicate that the variable y_3 is set either through the optimization or through the control loop. As soon as a step of the optimization will be directed inside the permitted area of the constrained space, the structure of the system will be set back in its original form.

In this way the optimization with constraints is very easily solved. The dimensionality of the optimization decreases with the increase of number of constraints and because of this the convergence gets accelerated.

7./ THE HIERARCHIC STRUCTURE OF LEARNING OPTIMIZATION.

Referring to the values of uncertainty presented in Table 1. it is interesting to point out that the uncertainty in the case of the human nervous system is very high with regard to the $10^8 - 10^9$ sensory nerve-cells and $10^{13} - 10^{15}$ memory cells. At the same time it is evident that these organisms display the most perfect self

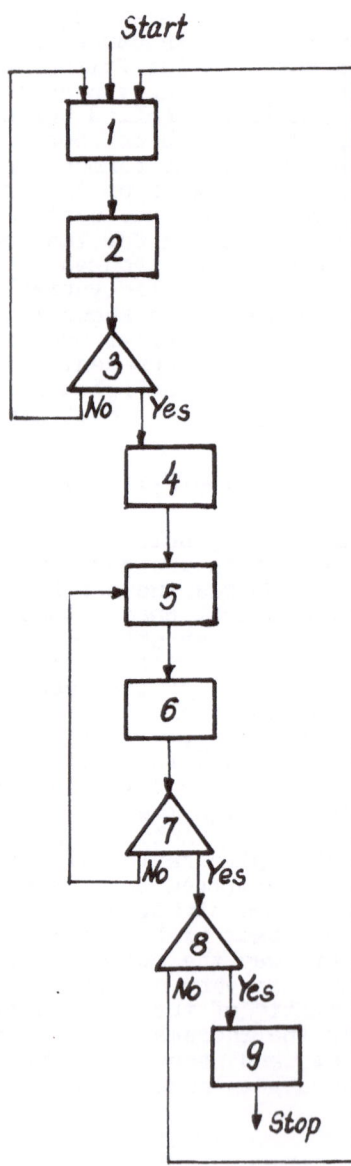

Start

1. Establishing of a random point

2. Function evaluation

3. Does the number of points reach the preset number of neighbouring points?

4. Determination of a new direction

5. Determination of the step-length

6. Function evaluation

7. Does the difference of the last and former dependent variables decrease below a given limit?

8. Does the number of consecutive differences less than the given minimum reach a preset value?

9. The optimum is found

Fig. 4

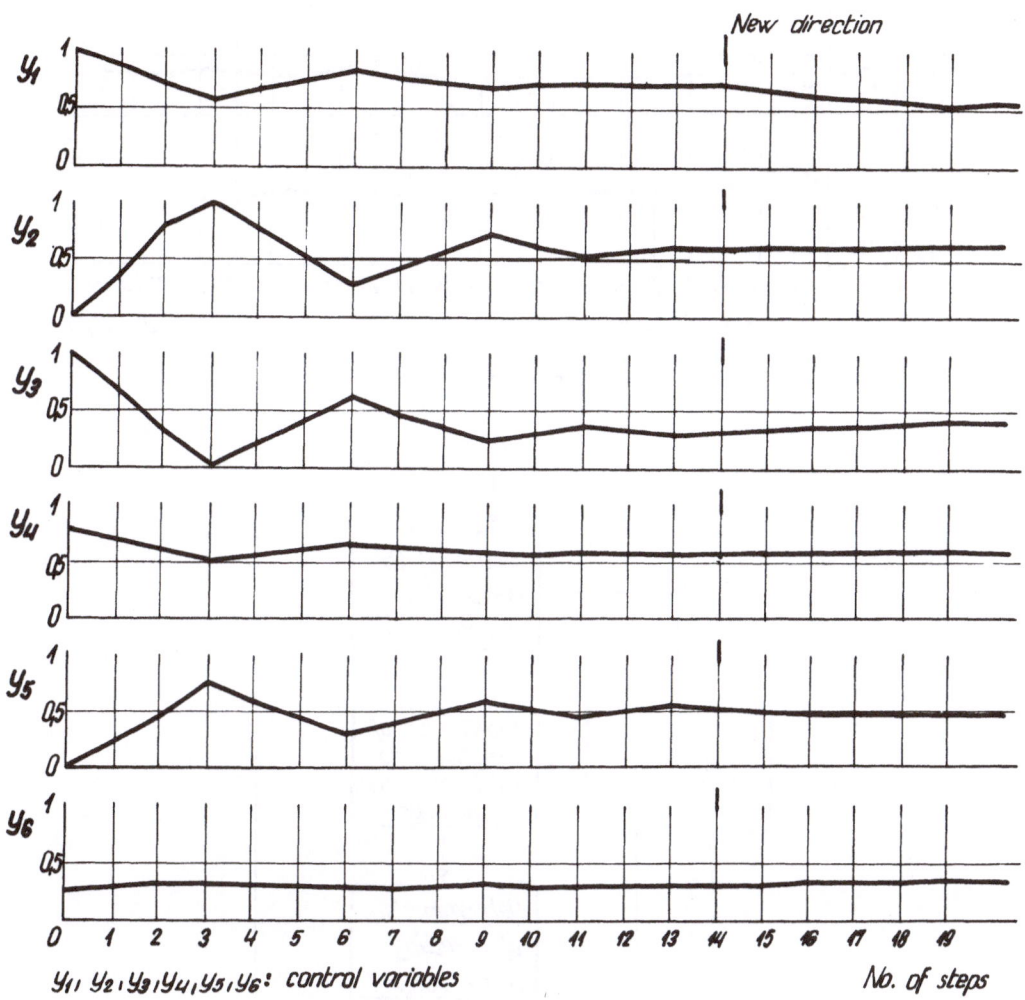

Fig. 5

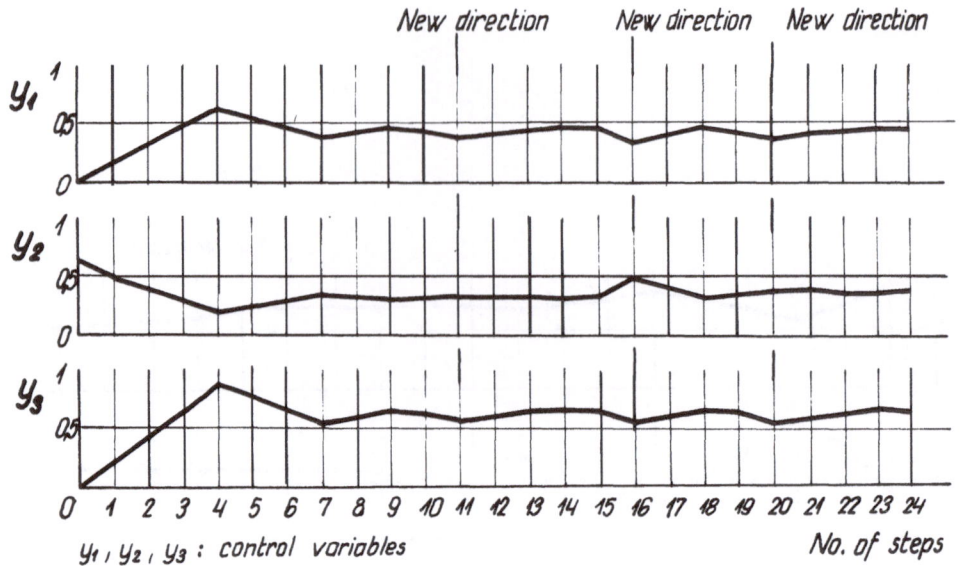

y_1, y_2, y_3 : control variables   No. of steps

Fig. 6

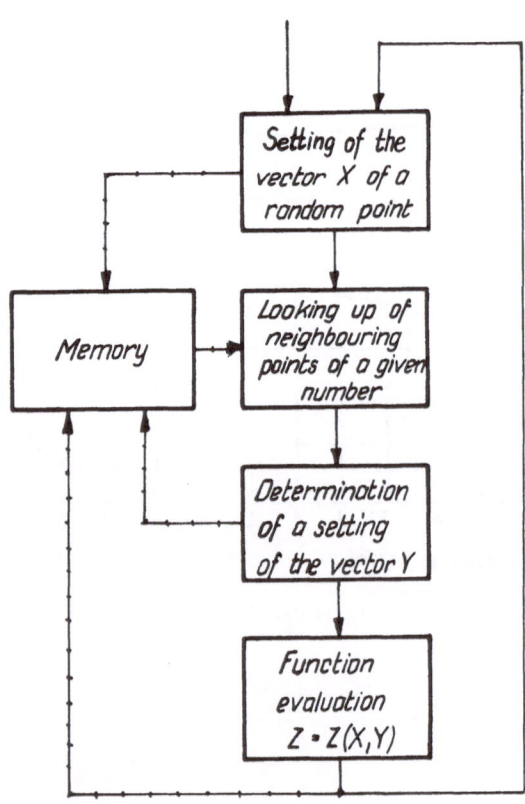

Fig. 7

318

adaptation. This could not be possible with a value of uncertainty close to 1 / as in the case of the human nervous system /. The point is, that the structure of living adaptive systems give a multilevel hierarchy. In other words the main object of the adaptive control is divided to "subobjects" and so the adaptive activity takes place in more or less independent subsystems. In these subsystems the local uncertainty - parameters are far better than in the onelevel system of the same outer characteristics.

The learning optimization can be applied to a much broader field of problems by splitting the process into small parts. This can be done in almost all practical cases. The object function can be given in the following detailed form owing to the fact that in functions of many variables subfunctions can generally be established:

$$C = C/X,Y/ = C\left[f_1/X,Y,Y/,f_2/X,Y,Y/..\right]$$

where the following relations can be given for the sets of the variables:

$$\underline{Y}_\nu \; \underline{Y}_\mu \; \underline{Y}; \qquad \underline{Y}_x = \underline{Y}-\underline{Y}_\nu-\underline{Y}_\mu$$

but   11./

$$\underline{Y} \neq \underline{Y}$$

The components of the cectors Y, Y are: $y_\nu ..y_{\nu e}, y_{\mu 1}..y_{\mu g}$. The first criterion of the optimum

$$\frac{\partial C}{\partial y_{\nu_i}} = \frac{\partial C}{\partial f_1} \cdot \frac{\partial f_1}{\partial y_{\mu_i}} = 0 \qquad 12./$$

The necessary condition for the existence of optimum is satisfied if

$$\frac{\partial f_1}{\partial y_{\nu_i}} = 0 \qquad 13./$$

According to the relations introduced above, the basic idea of the hierarchic operation can be interpreted as follows; In the first stage the independent optimization of the subfunctions f_1, f_2.. takes place Through the related control vectors Y_ν, Y_μ... simultaneously. In the second stage the optimization of the whole system is carried out through the rest of the control variables $/Y_x/$.

In most of the practical cases not only two but more stages of the hierarchy can be established and therefore the optimization is carried out in more than two tacts. The following data are worth mentioning; in the case of an ammonia plant it is necessary to store 10^5 points in the memory without hierarchic structure to achieve a prescribed value of the uncertainty, but with the same uncertainty in hierarchic structure the number of stored points can be diminished to about 200.

8./ SOME EXPERIMENTAL INVESTIGATIONS.

The learning optimization operates in two different ways. In the case, when the results of the input-variations develop sooner than a new variation arises, the optimization can be accomplished in a manner similar to the hill-climbing-techniques. In the second case, the consequences of the input variations slowly develop and therefore the hill-climbing method can not be applied. In this case the control system makes steps evaluating the earlier experiences and the results of these steps will be observed later, after a more or less long series of steps.

The experiments so far accomplished have been to simulate both types of learning optimization mentioned above. / An Elliott 803 digital and a Solartron Sc 30 analog computer have been used for the computation./ The flow diagram of the hill-climbing optimization is shown in Fig. 4. The convergence of the control vector toward the real locus of optimum / 0.5 for all the control variables / in a few steps are shown in Fig. 5. for a 6 dimensional paraboloid. For a model surface of the type:

$$z = a - b/y_2-c/^2-\left[1 - /y_3-d/^2\right]/y_1- e/^2$$

the steps of the optimization are shown in Fig. 6. Some of the results ro far achieved indicate the efficiency of the learning optimization in comparison with other methods reviewed in Fletcher's paper [9] /Table 2./. In this case the model was Rosenbrock's banana-shaped valley and our aim was to investigate the convergence in a given number of steps.

Some other experiments have been carried out with the second type of learning optimization according to the simple flow-diagram of Fig. 7. The results of these have shown the expected convergence.

CONCLUSIONS

The analysis of the living adaptive system reveals some very important principles of the optimal control of high perfection such as the learning through building up associations, the way in which the earlier experiences are used for the

Table 2
The minimization on Rosenbrock's banana-shaped valley.

Methods applied by						Learning optimization /4 experiments starting from different starting points/							
Davies, Swann, Campey		Powell		Smith									
z	f	z	f	z	f	z	f	z	f	z	f	z	f
24	1	24	1	24	1	20	1	100	1	37	1	100	1
3.9	12	4.0	13	3.8	14	3.3	8			27	8	28	8
2.8	21	3.3	25	3.3	28	0.48	12	1.3	8	3	12	0.32	12
2.3	28	2.6	35	2.2	44	0.14	20	0.97	12	0.23	20	0.68	20
1.8	36	1.9	46			0.14	28	0.30	20	0.12	31	0.088	28
1.4	43							0.27	28				

z the value of the function
f the number of function evaluation

evaluation of actions to be done, the hierarchic structure and the use of the same method for the static and dynamic optimisation. Simulating the laws of the living adaptivity, effective procedures can be developed for the optimization of large - scale, time varying processes. The first model - experiments give the evidence of the correctness of these assumptions.

ACKNOWLEDGEMENT

I wish to express my gratitude towards Dr. T. Vámos for his kind help given by reading thoroughly the manuscript of my paper.

REFERENCES

/1/ Prof. D. Gabor, W. P. L. Wilby, R. Woodcock: A Universal Non-Linear Filter, Predictor and Simulator which Optimizes Itself by a Learning Process The Proceedings of the Institution IEE. July 1961.

/2/ G. M.Jenkins: Adaptive Optimization of Reactors Using Sine-Wave Probes. Imperial College of Science and Tech.

/3/ G. E. P. Box, G. M. Jenkins: Further Contributions to Adaptive Quality Control: Simultaneous Estimation of Dynamics; Non-Zero Costs. Statistics in Physical Sciences 1. 1964.

/4/ P. H. Hammond, M. J. Duckenfield: Automatic Optimization by Continuous Perturbation of Parameters. Automatica Pergamon Press 1963.

/5/ G. E. P. Box, G. M. Jenkins: Some Statistical Aspects of Adaptive Optimization and Control. The Journal of the Royal Statistical Society; Series B /Methodological/ 1962.

/6/ H. H. Rosenbrock: An Automatic Method for Finding the Greatest or Least Value of a Function. The Computer Journal 1960.

/7/ A. A. Feldbaum: Rechengeräte in automatischen Systemen. R. Oldenbourg.

/8/ J. M. Nightingale: Parameter-perturbation adaptive control systems with imposed constraints. The Institution of Electrical Engineers Monographs. No. 158.

/9/ R. Fletcher: Function minimization without evaluating derivatives - a review. The Computer Journal April 1965.

/10/ M. Orbán: Obutchaiushtshaiasja optimizatsia mnogomernih processov; Trudi Mezsdunarodnoj Conferentsii po Mnogomernim Systemam Avtomaticheskovo Upravlenia. Prague 1965.

Discussion

Prof. K.S. Fu (USA) agreed with Dr. Orban's idea of using learning to modify a search procedure and suggested that storage of best parameter settings or control actions to date was the best way of achieving this. He mentioned work at Purdue University where this technique had been tried with the addition of quantisation noise and measurement error. Learning time has been estimated and multimodel hills considered.

Dr. Orban thanked Prof. Fu for his information but indicated that the method described in his paper is different from the techniques mentioned by Prof. Fu.

HILLCLIMBING ON HILLS WITH MANY MINIMA

by J. C. Hill   and J. E. Gibson
    Associate Professor of Engineering   Dean of Engineering
    Oakland University   Oakland University
    Rochester, Michigan   Rochester, Michigan

ABSTRACT

Three distinct types of hillclimbing
methods are studied in this paper; variations
on the conventional discrete search as proposed by Bocharov and Fel'dbaum [2], Kushner's
method [3], and the piecewise cubic method [8],
presented herein. The concept of testing
various methods on a large number of randomly
selected hills is proposed, and a class of
such test hills is presented. This approach
is then used to evaluate the performance of the
three types of hillclimber with respect to
five criteria.

It is found that the piecewise cubic
method and Kushner's method, which use a
global model-building approach, offer significant advantages over the gradient methods.

INTRODUCTION

When faced with the problem of designing
a control system or any other assembly of
various components, the engineer is strongly
motivated to build the best system possible
within the given constraints.

When "best" is taken in a well-defined
mathematical sense, the problem becomes one
of the optimal control problems that are
currently of such great interest. These problems are usually attacked from the point of
view of the classical calculus of variations
or related techniques such as the maximum
principle or dynamic programming. A typical
problem of this type would be the following:
Given the differential equation of the plant
to be controlled, find the forcing function
u(t) which will transfer the plant from any
initial point in the state space to the origin
in a certain fixed time T with minimum expenditure of energy. If such an optimum u(t) can
be found, one normally would like to express
u(t) as a function of the state variables
$(y_1,\ldots,y_n)$ of the system

$$u(t) = g(y_1,\ldots,y_n) \qquad (1)$$

so as to have a feedback type of control. The
problem of implementing exactly the control law
g can lead to hardware whose complexity is
clearly out of keeping with the engineering
nature of the problem.

If the attempt to build a practical optimum
system fails, the engineer is likely to pose
the following problem: Given a certain fixed
system configuration made up of components known
to be practical, what are the best possible
choices for the free parameters of the system?
Stated in this way, the problem of fixed configuration optimization is seen to be fundamental to the design of any system to perform a
specified task, whether it be a business organization, a control system, a chemical process, or
an electronic circuit. The majority of the engineering design problems come under this
category.

A related problem has been caled the
"specific optimal control problem" by Bellman,
Kalaba, and Sridhar [6], wherein it is assumed
that the plant differential equation is known.

The fixed configuration optimization problem
is defined as follows: Given a fixed configuration system with n variable parameters
$x_1, x_2,\ldots,x_n$, arbitrarily chosen bounded by
[0, 100], and an <u>index of performance</u> which
is some unknown continuous function $f(x_1,\ldots,x_n)$
of the variable parameters,

$$y = f(x_1,\ldots,x_n) \qquad 0 \le x_j \le 100 \qquad (2)$$

specify an algorithm for determining the arguments $x_1,\ldots,x_n$ which will minimize (or maximize)
the function f by observing the functional
values $y_i = f(x_{1i}, x_{2i},\ldots,x_{ni})$ for a sequence
{i} of parameter settings. The surface defined
by Eqn. 2 is called the IP surface.

In other words, it is desired to <u>extremize</u>
the function f with respect to the n <u>variables</u>
$x_1,\ldots x_n$.

Systems mechanizing such algorithms are
called optimizers, automatic optimalizers,
extremal control systems, or hill-climbers, and
the majority of them have one major drawback:
they will work satisfactorily only if the function
f has but a single minimum (the relative minimum
problem). A problem of slightly less practical
importance is the tendency of most methods to
stagnate in regions where f is relatively insensitive to the parameter variations (the plateau
problem). A third problem with the use of
current algorithms is that most of them are
efficient when there is one variable parameter
(n = 1) but their extension to higher dimensional problems is not satisfactory. A fourth

"Superior numbers refer to similarly-numbered references at the end of this paper."

322

problem is that of extremizing the function f when observations of the functional values are corrupted by noise. In this paper the major emphasis is on the first two problems.

GLOBAL OPTIMIZATION

The first paper to concern itself with practical methods for optimization of functions with more than one minima was that of Bocharov and Fel'dbaum[2]. Three algorithms are proposed that are basically variations on the conventional discrete search techniques as presented in A. A. Fel'dbaum[1].

An interesting technique for the experimental optimization of the performance of a system with a very general multipeak performance function when the only available information is noise-disturbed samples of the function was presented by H. J. Kushner[3] at the 1963 Joint Automatic Control Conference. Kushner's method is one of the first to use global properties of the hill to determine the location of the minimum, a concept that is reinforced in this paper.

Kushner assumes that the unknown hill is a Gaussian random function, a sample of the classical Brownian motion process. Given j observations $f(x_1)$, $f(x_2)$,..., $f(x_j)$ (dispensing with the double subscript notation, since we now have a one-dimensional hill) on a one-dimensional hill $y = f(x)$, the conditional expectation

$$\overline{f(x)} = E[f(x)|f(x_1), f(x_2), ..., f(x_j)] \quad (3)$$

and conditional variance Var f(x) are computed with reference to the assumed hill model. A particularly interesting property of the model is that $\overline{f(x)}$ is piecewise linear, and the breakpoints are the locations of the observations, i.e., $\overline{f(x)}$ is a piecewise linear approximation to f(x). The variance Var f(x) is quadratic between observations. The piecewise linearity of $\overline{f(x)}$ and the quadratic nature of Var f(x) facilitate the visual interpretation of the data and the intuitive relation between the data and the true curve.

GENERAL APPROACH TO A MODEL BUILDING HILLCLIMBING TECHNIQUE

As was stated in the Introduction, the two major problems of hillclimbing techniques based on the gradient method are the unsatisfactory performance on hills with more than one minimum (the relative minimum problem) and the tendency to stop entirely or to move very slowly on portions of the hill that have very small slopes (the plateau problem).

These difficulties can be seen to be a consequence of the fact that only a local property of the hill is measured at each stage, the gradient at the operating point.

The work of Bocharov and Fel'dbaum[2] attempts to provide satisfactory performance by performing a sequence of hillclimbing maneuvers so organized as to cover the entire parameter space, at least on a statistical basis.

It seems that a more radical departure from the gradient techniques might be fruitful, In particular, we assert that a method that makes global measurements on the hill can form a general macroscopic picture (model) of the hill, refine it as necessary by the use of additional measurements (either in the vicinity of the anticipated minimum to improve immediate performance or on other parts of the hill to improve confidence in the model), and provide adequate performance with fewer measurements on the hill.

This is also the viewpoint adopted by Kushner[3].

The intent of this paper will be to provide experimental evidence supporting the above assertion by comparison of the performance characteristics of the Bocharov and Fel'dbaum algorithms, Kushner's method, and a method developed in this paper, the piecewise cubic hillclimber.

The general strategy is the following:

1) make global measurements on the hill.

2) curve-fit these measurements to provide an internal digital model of the IP surface.

3) evaluate the accuracy of the digital model, take more measurements if the accuracy is not satisfactory.

4) minimize (or maximize) the resulting digital model.

The major problem in the implementation of this general strategy is the selection of a suitable curve-fitting method from the plethora of available techniques. Interpolation and approximation problems are the core of classical numerical analysis. In Gibson, Fu, and Hill[8], four methods for modeling were studied with regard to the accuracy obtained and their interaction with the other problem areas; in particular, the computational problem of minimizing the digital model once it has been obtained is influenced strongly by the type of model chosen.

A fundamental point is involved in checking the model accuracy. To check the accuracy of the model it is necessary to make new observations and compare the results with predicted values from the model. Thus the penalty assigned to the cost of measurement is involved in checking the model.

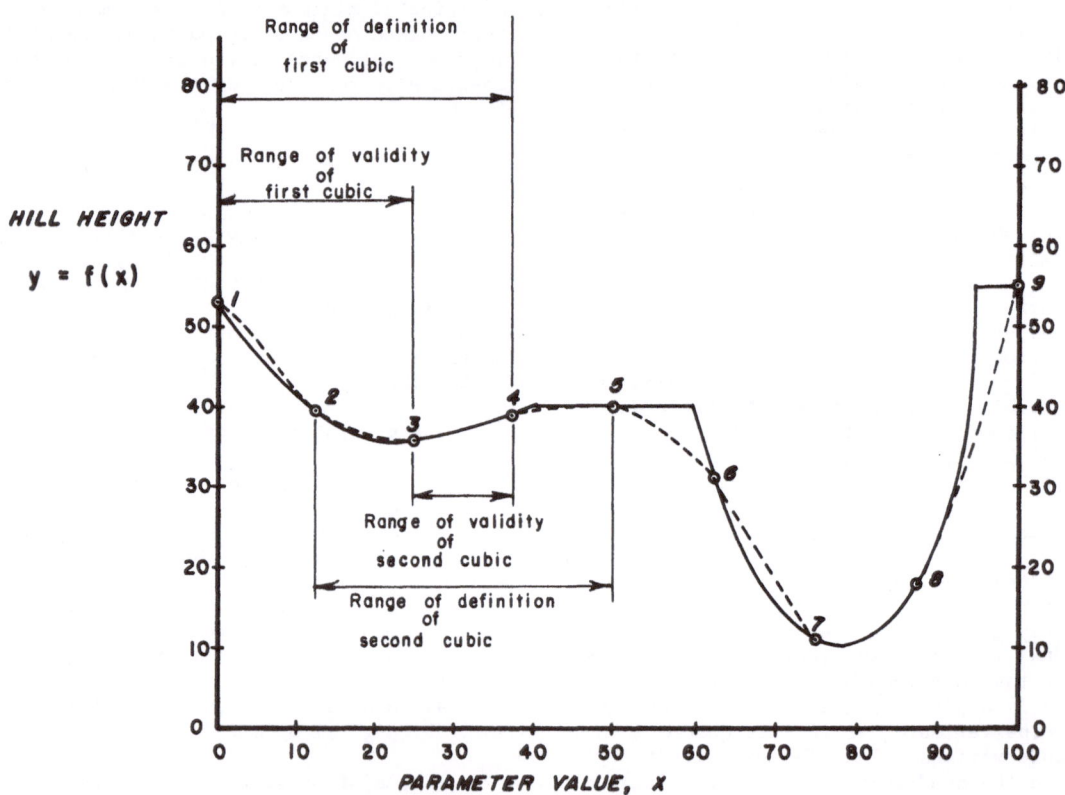

FIGURE 1. PIECEWISE CUBIC FIT

Once a sufficiently accurate model is obtained, the minimum value must be found. The computational problem here is strongly dependent on the type of model used. If the digital model is a 15th degree polynomial approximation to the hill, the computer must minimize digitally a 15th degree polynomial, and little advantage would be gained. Similar objections apply to Fourier approximations.

PIECEWISE CUBIC FIT

The following algorithm (discussed for the case of 9 equally spaced data points) is proposed:

1) In the interval $0 \leq x \leq 25$, use a cubic approximation obtained from $y(0)$, $y(12.5)$, $y(25)$, $y(37.5)$.
2) In the interval $25 \leq x \leq 37.5$, use a cubic approximation obtained from $y(12.5)$, $y(25)$, $y(37.5)$, $y(50)$.
3) In the interval $37.5 \leq x \leq 50.0$, use a cubic approximation based on $y(25)$, $y(37.5)$, $y(50)$, $y(62.5)$.
4) Etc.
5) In the interval $75 \leq x \leq 100$, use a cubic approximation based on $y(62.5)$, $y(75)$, $y(87.5)$, $y(100)$.

Thus the approximating polynomial in each interval will be based on the values obtained at the ends of the two bracketing intervals, except for the special cases at the ends. This is equivalent to assuming that portions of the curve $y = f(x)$ may be adequately fit by a 3rd degree polynomial, although no single polynomial of reasonable degree would be adequate over the entire parameter space[4].

An example of a piecewise cubic fit to a hill is shown in Figure 1.

CONVERGENCE

A means of stopping the successive subdivisions once satisfactory agreement between the model and the hill is obtained must be incorporated. This is done by comparing the new data points obtained at each subdivision with the same data points evaluated from the old model. The maximum absolute error (EMAX) is compared with an allowable error (EALL); if EMAX $\leq$ EALL, the subdivision is stopped.

The piecewise cubic fit is relatively simple computationally but requires a large number of parameters to describe the hill. The method is readily adaptable to describing the hill with more and more detail. In short, it is a balanced technique for one-dimensional optimization.

Notice that this method is "nonintelligent."

That is, it takes no notice of the specific hill being examined, since the points at which data are taken are fixed in advance. It is surprising that it performs as well as it does. Notice that if an attempt is made to alter the sampling sequence based on the observed data, such an attempt must be made in terms of more quantitative (hence more restrictive) assumptions as to the type of hill to be encountered. In other words, more a priori knowledge would be required.

MINIMIZING ROUTINE

The model must now be minimized. Here the advantage of having a model that is piecewise cubic becomes apparent. The procedure for finding the minimum of the model is:

1) Differentiate each polynomial, obtaining a set of quadratics, $df_j(x)/dx=0$, $j=1, \ldots, k-3$.
2) Factor jth polynomial by the quadratic formula obtaining the roots R_{1j} and R_{2j}.
3) Check to see if R_{1j} and/or R_{2j} lie in the range of validity of the jth polynomial.
4) If so, check for the minimum of $f_j(R_1)$, $f_j(R_2)$, f_j (left boundary of jth segment), f_j (right boundary of jth segment). If not, check only right and left boundary points.

Thus in general four candidates for the minimum are obtained from each polynomial.

A CLASS OF TEST HILLS

The intent of this paper is to provide a comparison of the relative performance of several hillclimbing methods, using experimental data from a large number of test hills.

There is no general agreement on what constitutes a model of a typical hill. In particular, there are no satisfactory hills of a random nature proposed in the literature (perhaps with the exception of Kushner's model[3]). The main features required of a class of test hills are the following:

1) The hills must be capable of exhibiting many relative minima.
2) The relative smoothness of the hills must be controllable.
3) The hills must be bounded.
4) The hills must be continuous.
5) The hills must provide extended regions of very small or zero slope.
6) The hills must not be of the exact type assumed by a hillclimbing algorithm.
7) The type of hill should readily apply to more than one dimension.
The test hills used were obtained from a finite Fourier series:

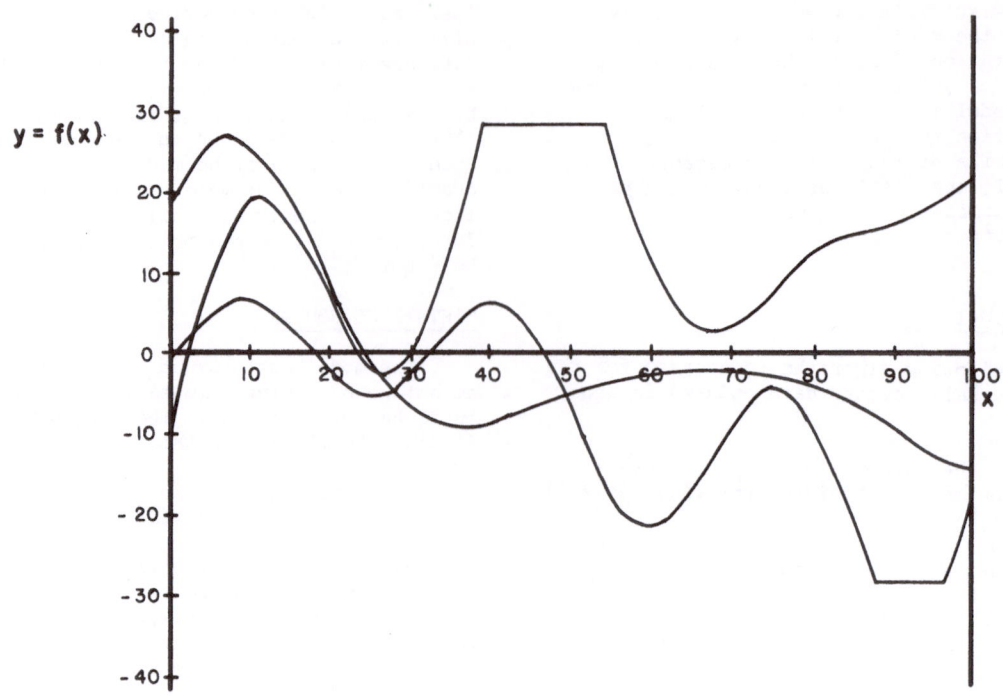

FIGURE 2.   SOME REPRESENTATIVE 6-HARMONIC HILLS

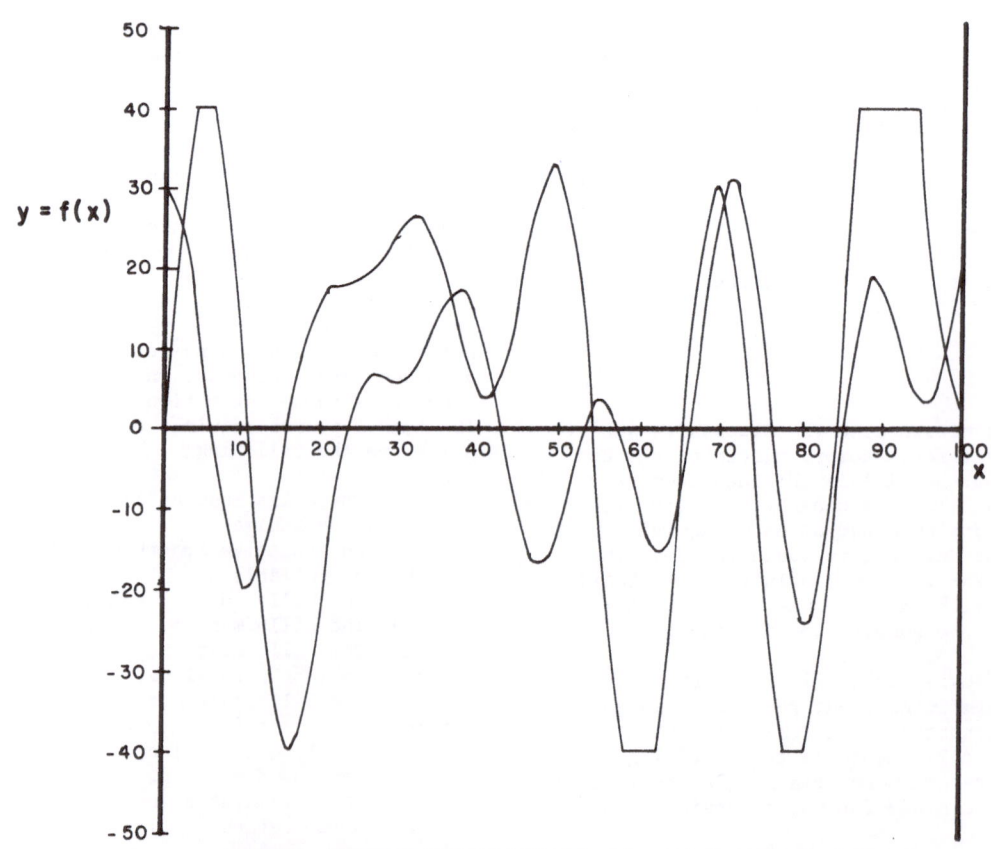

FIGURE 3.   SOME REPRESENTATIVE 12-HARMONIC HILLS

$$y=f(x)=a_1+\sum_{k=2}^{n}\{a_k \cos \frac{(k-1)\pi x}{100} + b_k \sin \frac{(k-1)\pi x}{100}\} \tag{4}$$

where the a_k and b_k are uniformly distributed random variables. As given here, the hills are periodic with period 200, so that a different hill height may be obtained at each endpoint. Control over the number of minima expected and the relative smoothness is exerted through the parameter n. All requirements are then met except 5), the requirement on extended regions of very small or zero slope. This is met by saturating the hills at the ±2σ level to provide plateaus. Some typical 6 and 12 harmonic hills are shown in Fig. 2 and Fig. 3.

For some purposes, it is desirable that the hill always be positive. This is difficult if the hill amplitude (variance) changes as a function of n. Accordingly the hills are modified by:

1) Dividing $f(x)$ by σ, the hill standard deviation to get a unit sigma hill
2) Multiplying by 20; now σ = 20.
3) Clipping at the ±2σ = ± 40 volt levels
4) Adding 50.

The result is a class of test hills with σ = 20 clipped at +10 and +90 levels, independent of the number of harmonics used. Thus n now controls only the relative smoothness of the hills and has no effect on the peak-to-peak amplitude.

For test hills in two dimensions, the preceding approach carries over into a two dimensional finite Fourier series [5]:

$$z=f(x,y)=\sum_{j=1}^{n}\sum_{k=1}^{n}\{a_{jk} \cos [\frac{(j-1)\pi x+(k-1)\pi y}{100}]$$

$$+ b_{jk} \sin [\frac{(j-1)\pi x+(k-1)\pi y}{100}]\} \tag{5}$$

where, as before,

$$a_{jk}, \quad j, k=1, 2, \ldots, n$$

$$b_{jk}, \quad j, k=1, 2, \ldots, n$$

are statistically independent random variables, identically uniform over [-10, 10], and the hills are again normalized and dipped at the ±2σ levels.

Very little data is available on the two dimensional hills because of the expense in generating them. For example, a 100x100 grid of a 12 harmonic two dimensional hills takes approximately three hundred hours to compute on the IBM 1620 Model I.

SELECTION OF A VALUE FOR THE ALLOWABLE ERROR (EALL)

For the piecewise cubic method the only free parameter to be chosen is EALL (cf. CONVERGENCE, above). In order to select a good value for EALL, a computer run was made with a range of values of n and EALL. The number of harmonics of the test hills varied from 6 to 12 in increments of 2, and EALL varied from 10.00 to 30.00 in increments of 5.00. (Note that n=1 + number of harmonics desired). Ten hills were minimized at each parameter value. The "actual" minimum was determined by direct scan to 1 per cent resolution to serve as a check on the accuracy of the piecewise cubic minimization technique. The results of this run are presented in Table 1.

For the range of hill complexities considered, EALL does not seem to be a particularly sensitive parameter. EALL = 15.0 seems a good choice, independent of the number of harmonics, with EALL = 10.0 leading to a significantly greater average number of measurements.

CRITERIA FOR EVALUATION OF HILLCLIMBING METHODS:

Five criteria will be used for evaluation of a hillclimbing method:

1. Average Minimal Hill Height. On the average, the best method should find lower minima than its competitors.
2. Average Number of Measurements. If the problem is such that each measurement has essentially a fixed cost (for example, if the problem is a digital solution of a two-point boundary value problem) then the quantity of interest is the number of measurements needed to optimize the solution.
3. Average Integral of the Index of Performance. If the problem is such that each measurement has a cost proportional to the hill height measured (for example, in process optimization it costs more to make measurements in regions of unsatisfactory performance) then the natural quantity of interest is the sum of the hill heights at all the measurements taken (integral of the IP with respect to time in the continuous case). This function will be called CUMF (cumulative function), and it can be reduced by one method taking measurements lower on the hill than another method, or by the first method taking fewer measurements, or a combination of these factors.
4. Reliability. Although one is interested in such measures of hillclimber performance as the average number of measurements and average CUMF on hills of a given complexity, there is a consideration of more interest to the normal user: What assurance does he have that a true minimum has been found when the minimization is completed? A method with high reliability would be preferable to one of low reliability, even if the low reliability method were to cost less to

Table 1. Performance of Piecewise Cubic Hillclimber as a
Function of the Number of Harmonics

EALL	No. of har- monics	Number of Hills on Which the Indicated Number of Measurements were Made				Average No. of Measurements	No. of Missed Minima
		9	17	33	65		
10.0	6	0	10	0	0	17.0	0
10.0	8	0	0	10	0	33.0	0
10.0	10	0	0	10	0	33.0	0
10.0	12	0	0	7	3	42.6	0
15.0	6	4	6	0	0	13.8	0
15.0	8	3	7	0	0	14.6	0
15.0	10	0	1	9	0	31.4	0
15.0	12	0	0	10	0	33.0	0
20.0	6	4	6	0	0	13.8	0 Satisfactory
20.0	8	4	6	0	0	13.8	1 Unsatisfactory
20.0	10	1	3	3	3	35.4	1
20.0	12	1	1	7	1	32.2	1
25.0	6	8	2	0	0	10.6	0
25.0	8	4	6	0	0	13.8	0
25.0	10	1	7	2	0	19.4	1
25.0	12	2	0	8	0	27.2	1
30.0	6	9	1	0	0	9.8	0
30.0	8	4	6	0	0	13.8	1
30.0	10	2	6	2	0	18.6	1
30.0	12	1	1	8	0	24.0	1

use on the average. One would prefer not to be wildly misled by the results of the minimization.

5. Noise Sensitivity. The Bocharov and Fel'dbaum methods and the piecewise cubic method have not been designed for use when the observation of the hill height is corrupted by noise. Kushner has two versions: one for noise-free situations and the other for noisy situations. The emphasis of this paper has been stated to be on the relative minimum problem and the plateau problem, thus only the methods designed for noise-free operation have been studied experimentally. However, practical systems will never be entirely noise-free, so that a question as to the sensitivity of the noise-free algorithms to small amounts of observation noise is raised. In other words, how small must the noise level be for satisfactory operation of the noise-free algorithms?

EXPERIMENTAL RESULTS

Preliminary studies of the three algorithms of Bocharov and Fel'dbaum indicated that Algorithm No. 2 was superior to the other two algorithms; further study then focused on this method, which will be denoted by B/F No. 2. A major determining factor in the performance of the B/F No. 2 algorithm is the means by which the hillclimbing phases are carried out. Two such methods are studied. The first, B/F No. 2 DISCRETE takes steps of fixed length and compares the resulting hill heights. The second method, B/F No. 2 GRADIENT, determines the gradient at the operating point and then takes a step proportional to the magnitude of the gradient.

For this data, Kushner's method was arbitrarily set to use 35 measurements; for measurements 1-10 ϵ_n = 40.0; for measurements 1-24 ϵ_n = 20.0; and for measurements 25-35 ϵ_n = 10.0. For the piecewise cubic method, EALL = 15.0.

The test hills used were 12 harmonic one-dimensional hills. The σ_n=0.0 averages are over 100 hills; σ_n=1.25%, 2.5%, 5%, and 10% averages are over 25 hills (cf. APPENDIX).

1. Average Minimal Hill Height vs. σ_n

In Fig. 4 the average minimal hill height as found by the four methods is plotted vs. σ_n.

The B/F discrete method has the worst performance according to this criteria, with the average minimal hill height being about 18 for σ_n=0 and varying between 18 and 19 for $0 \leq \sigma_n \leq 5\%$.

The B/F gradient routine finds intermediate minima on the average, and with regard to this criterion this method appears to be least sensitive to noise.

2. Average Number of Measurements vs. σ_n

In Fig. 5 the average number of measurements used is plotted vs. σ_n.

The piecewise cubic hillclimber takes approximately 33 measurements until $\sigma_n \approx 5\%$, where the number of measurements necessary increases drastically. This is because the noise has a level comparable to EALL = 15.0, thus interfering with the error evaluation of the model. The number of measurements necessary is more sensitive to noise than is the average minimal hill height.

For the B/F discrete method, the average number of measurements is independent of the noise level for the range of σ_n studied. The average number of measurements necessary is considerably less than that used by the piecewise cubic hillclimber and Kushner's method. Apparently there is a simple trade-off between, say, the piecewise cubic hillclimber and the B/F discrete method: the piecewise cubic hillclimber has a lower average minimal hill height and the B/F discrete method takes fewer measurements.

To test this conclusion, μ was increased from 10 to 20 in an attempt to force the B/F discrete and gradient methods to find an average minimal hill height competitive with the piecewise cubic method. The result of the change was to increase the average number of measurements taken by the B/F discrete method to 41.20, and by the B/F gradient method to 64.20, while of course the piecewise cubic method remains unchanged at 31 - 32. It is interesting to notice that the large increase in the number of measurements does not produce a sufficient change in the average minimal hill height found by these two methods to make them better than the piecewise cubic method. Thus the curves presented in Fig. 4 for average minimal hill height are not strongly dependent on the value of μ used, while the curves of the two B/F methods in Fig. 5 for average number of measurements may be moved up and down at will.

3. CUMF vs. σ_n

In Fig. 6 the average CUMF accumulated is plotted vs. σ_n.

For the piecewise cubic method, CUMF increases rapidly as σ_n increases, reflecting the increase in the number of measurements taken (see Fig. 5).

For the B/F discrete method, CUFM is independent of the noise level. The similarity between the curves in Fig. 5 and Fig.6 leads to the conclusion that the CUMF of the B/F discrete method is lower than that of the piecewise cubic method because the latter took more measurements, not because the former took its measurements lower on the hill(which would be a desirable

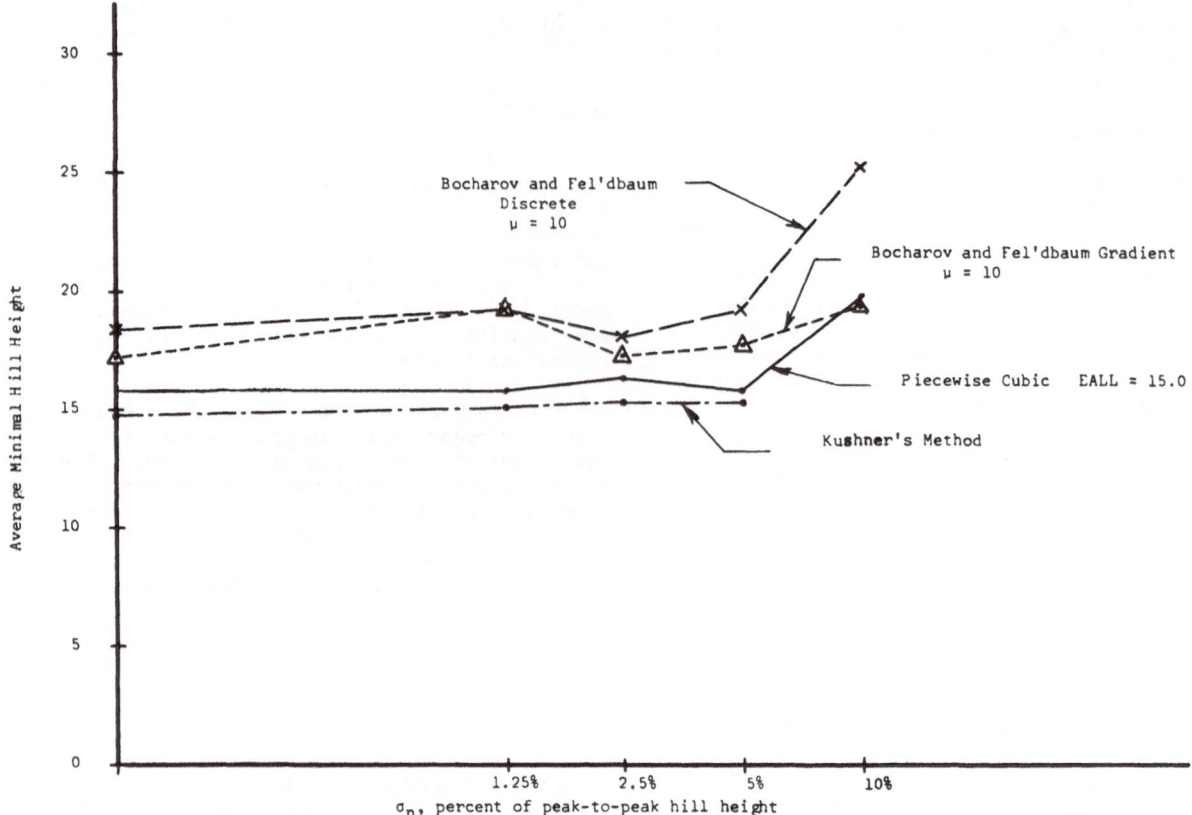

FIGURE 4. Average Minimal Hill Height vs. σ_n for 12 Harmonic One Dimensional Hills

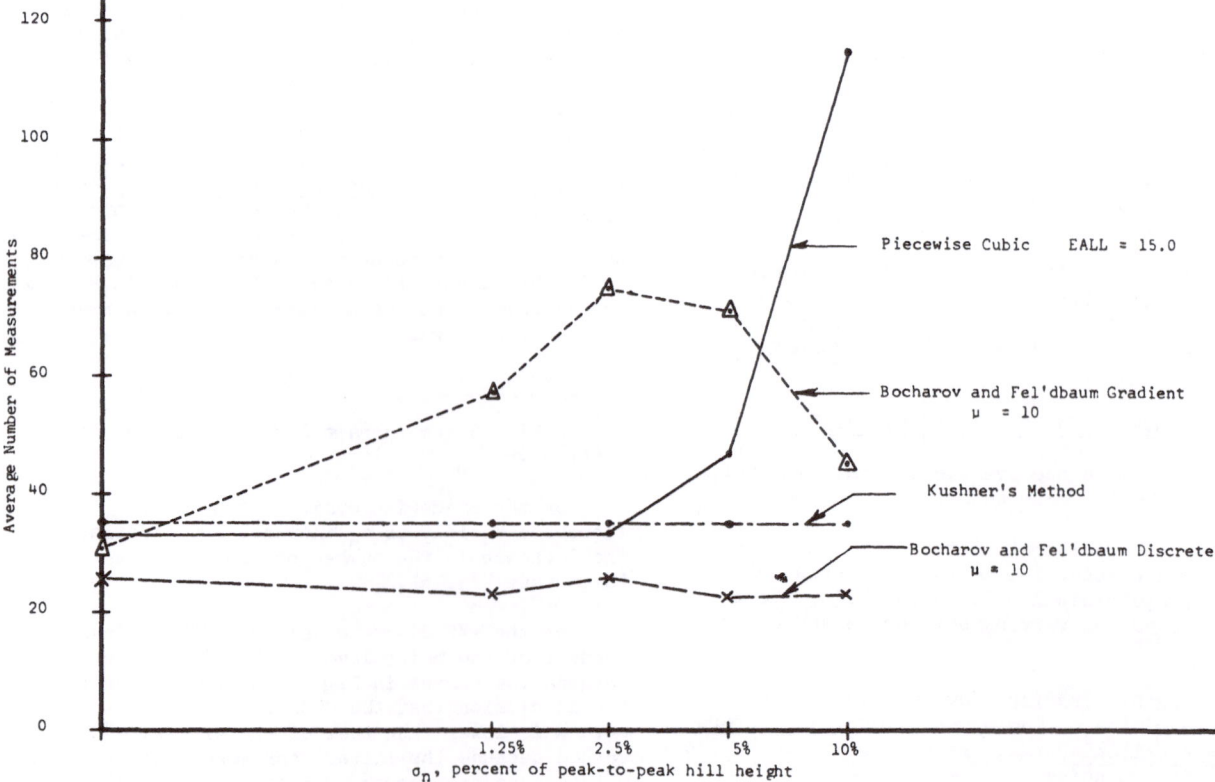

FIGURE 5. Average Number of Measurements vs. σ_n for 12 Harmonic One Dimensional Hills

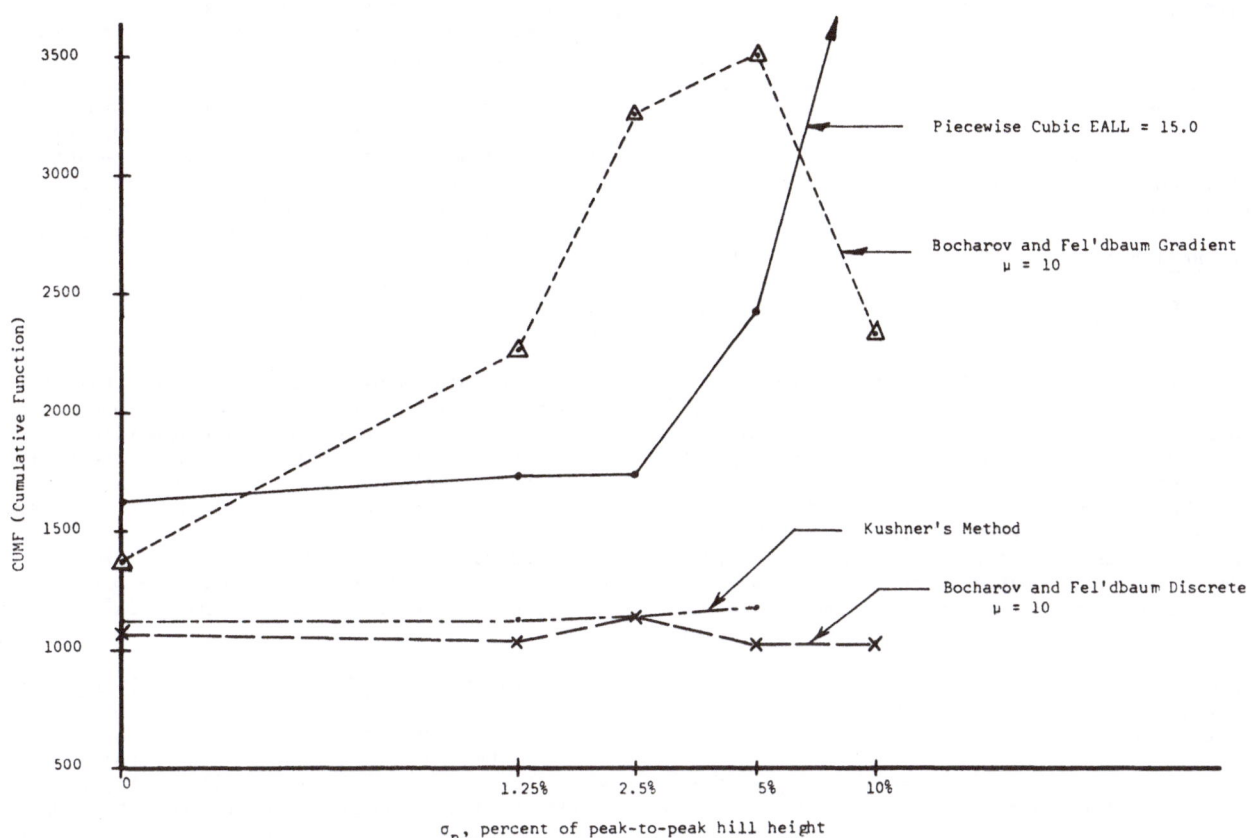

FIGURE 6.   CUMF vs. σ_n for 12 Harmonic One Dimensional Hills

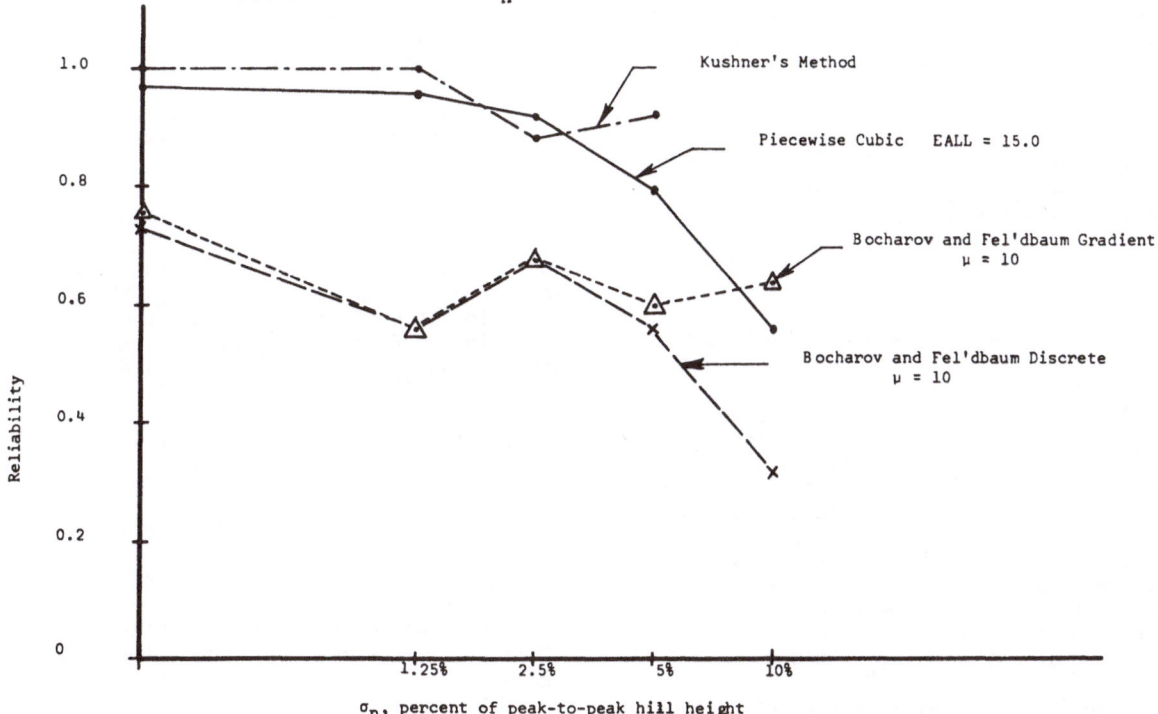

FIGURE 7.   Reliability vs. σ_n for 12 Harmonic One Dimensional Hills

strategy with respect to this criterion).

The B/F gradient method fares in the same way as it did with respect to the average number of measurements; the same discussion applies.

Here Kushner's method possesses a distinct advantage because of its tendency to take measurements low on the hill.

4. Reliability vs. σ_n

In Fig. 7 the reliability of the methods is plotted vs. σ_n.

The reliability figure is obtained by comparing the minimum hill values found by the four methods. For example, suppose the piecewise cubic method found FXMIN 1 = 13.96; the B/F discrete method found FXMIN 2 = 13.98; and the B/F gradient method found FXMIN 3 = 18.41. Thus the piecewise cubic and B/F discrete methods found the same (or equivalent) minima, while the B/F gradient method gave an indication significantly worse. The B/F gradient method is then credited with a failure on this hill (notice that it is not known whether the actual minimum of the hill has been found, all that is known is that two methods beat the third).

The failures are then assessed by noting which method found the lowest minimum on a given hill, and if another method does not find a minimum within one volt (1.25% of peak-to-peak hill amplitude) of this value, it is credited with a failure. The total number of correctly found minima are then expressed as a fraction of the total number of hills to evaluate the reliability.

From Fig. 7 the major advantage of the piecewise cubic method and Kushner's method becomes apparent: they are significantly more reliable than the other methods.

17 MEASUREMENT RESULTS

The performance obtained by Kushner's method using 35 measurements suggested that a much smaller number of measurements might be used to achieve satisfactory results, both from Kushner's method and the piecewise cubic method, if the complexity of the hill is known beforehand. Kushner's method was then allowed 17 measurements; for measurements 1-6 $\varepsilon_n = 40.0$; for measurements 7-12 $\varepsilon_n = 20.0$; and for measurements 13-17 $\varepsilon_p = 10.0$. The piecewise cubic method was also locked out at 17 measurements, and these two were compared with the best Bocharov and Fel'dbaum method, B/F No. 2, on 25 12 harmonic hills with $\sigma_n = 0.0\%$.

With 17 measurements both model-building methods miss 3 out of the 25 hills for a reliability of 0.88. The average minimal hill

heights are equivalent, and Kushner's method achieves a lower CUMF. The B/F No. 2 discrete is bested on all counts.

From this data it is concluded that the piecewise cubic method takes more measurements than it actually needs to do a reasonable job of finding minima; the fault is seen to lie in the error checking procedure. It is overly pessimistic since it tries to fit the sharp corners at the saturation levels.

CONCLUSION

In the preceding, we have proposed that the performance of various hillclimbing methods be compared, and have suggested five practical criteria for making the comparison. A necessary element for making such a comparison is a class of test hills, and a suitable class was presented.

This technique was then applied to provide experimental evidence for the use of a global viewpoint when hillclimbing on multimodal hills.

APPENDIX: ERROR ANALYSIS OF PERFORMANCE AVERAGES

In the preceding, the basic experimental data was obtained in tabular form as the results of minimizing a set of 25 random hills. For purposes of discussion and presentation, these data were reduced to averages across the 25 hills and then plotted vs. the observation noise level. Since most of the comparisons between methods are made on the basis of these averages, the question arises: Is 25 hills a sufficient number for these averages to be meaningful?

Following Cramer[7] on the p% level of significance we have the confidence limits

$$\bar{x} \pm \frac{\lambda_p}{\sqrt{n}} \sigma = \frac{\Sigma x_i}{n} \pm \frac{\lambda_p}{\sqrt{n}} \sigma \qquad (6.1)$$

for the mean m of a random variable x, of which we have n samples $x_1, x_2, \ldots, x_n$. The standard deviation of σ of the x_i is not known and must be estimated from the experimental data, and λ denotes the p% value of the normal distribution (cf. Cramer[7], Table II, p.274.)

Quantitatively, on the 10% level of significance, $\lambda_p = 1.6449$, so that for n = 25 we have

$$\bar{x} \pm \frac{1.6449}{\sqrt{25}} \sigma = \bar{x} \pm 0.329 \sigma , \qquad (6.2)$$

so that any such average quoted has been determined to within one-third of the standard deviation of the particular quantity averaged. This is sufficient to indicate trends in the data.

As an example, the standard deviation for the data presented in Fig. 4 is found to be typically s.d. ≃ 7, so that the confidence limits for each point are approximately ±7/3 = ±2.3. Thus one cannot with certainty separate the individual curves shown; however, the upward break beyond σ = 4 is significant. This inability to make definite statements about the relative performance of the three methods with respect to average minimal hill height may be interpreted as indicating that average minimal hill height is not a good criterion for selection of one method over another.

A different situation exists with regard to Fig. 5 where the average number of measurements is plotted vs. observation noise level for the three methods. For example, for σ_n = 1.25%, the data for the piecewise cubic hillclimber has an s.d. ≃ 7.99, hence a confidence limit of ± 7.99/3 ≃ 2.3. The data for the Bocharov and Fel'dbaum Discrete method has an s.d. ≃ 5.72, hence a confidence limit of ± 5.72/3 ≃ ± 1.9. The data for the Bocharov and Fel'dbaum Gradient method has an s.d. ≃ 37.11, hence a confidence limit of ±37.11/3 ≃ ±12.4. Thus the averages of Fig. 5 are sufficiently separated to distinguish between the various methods at the 10% level of significance. The similarity of Fig. 6 to Fig. 5 allows the same conclusion to be applied to Fig. 6.

The above analysis is expected to be pessimistic with respect to drawing relative conclusions between the three methods since all three methods performed on the same sequence of 25 hills.

ACKNOWLEDGMENT

The research reported in this paper was sponsored by the United States Air Force Office of Scientific Research, Contract AF AFOSR 62-351, at the School of Electrical Engineering, Purdue University, Lafayette, Indiana.

REFERENCES

(1) A. A. Fel'dbaum, "Automatic Optimalizer," A.i.T. 19, 8(1958).

(2) I. N. Bocharov and A. A. Fel'dbaum, "An Automatic Optimizer for the Search of the Smallest of Several Minima (A Global Optimizer)," A. i. T. 23, 3(1962).

(3) H. J. Kushner, "A New Method of Locating the Maximum Point of an Arbitrary Multipeak Curve in the Presence of Noise," 1963 JACC Preprints, pp. 69-79, (June 1963).

(4) F. B. Hildebrand, Introduction to Numerical Analysis, pp. 295 ff., McGraw-Hill, 1956.

(5) R. Courant and D. Hilbert, Methods of Mathematical Physics, Vol. I, pp. 73-74, Interscience, 1953.

(6) R. Bellman, R. Kalaba, and R. Sridhar, "Adaptive Control via Quasilinearization and Differential Approximation," the RAND Corporation, RM-3928, November 1963.

(7) H. Cramer, The Elements of Probability Theory and Some of Its Applications, pp. 207-208, Wiley, 1955.

(8) J. E. Gibson, K. S. Fu, and J. C. Hill, "A Hillclimbing Technique Using Piecewise Cubic Approximation," TR EE-64-7, Purdue University School of Electrical Engineering, April 1964.

Discussion

Dr. D. Graupe (UK) discussed the use of the term "parameter adjustment" in hill climbing, pointing out that in real plants some parameters are not available for direct adjustment. He considered the example of a heat exchanger where a major parameter may be a heat transfer coefficient which can only be adjusted by modifying flow rates or temperatures. He thought that the term "input variable adjustment" was preferable to "parameter adjustment" in the context of the paper.

Dr. Hill replied that the difference between the two terms was concerned more with the properties of their variations than in their physical significance. A system with slowly varying parameters has a different level of compexity to one in which all variables are treated equally.

Mr. R.J. High (UK) asked the authors how they proposed to select information for tracking a hill, once having reached the summit. He also asked how noise on the observations affected the accuracy of the model.

Dr. Hill noted that the paper dealt strictly with the acquisition of a peak and he did not suggest the same method for peak tracking. As regards noisy measurements the author had data showing the method effective with noise of up to 10% peak to peak hill height. With curve fitting methods there was the difficulty of getting continuity between the piecewise cubics. For one dimensional cases Kushner's method had many advantages in the presence of large amounts of noise.

Dr. M.H. Hamza (Switzerland) recalled that his method can be used to find the absolute minima of a hill with several minima. Scanning can be done very quickly in one direction. Jumps in the curvature function give a mapping of the sign of the slope over the hill and hence the position of the minima. As time progresses these minima may change their values and this can be readily observed.

Dr. Hill was doubtful about the ability of
Dr. Hamza's method to speed up the hill climbing
process.

Mr. I. H. Rowe (U.K.) recalled that both
Kushner's method and that due to Bucharov and
Feldbaum can be extended to multivariable situations. He asked the authors whether their test
functions could be extended to make the method
applicable to more than one dimension.

Dr. Hill said that the piecewise cubic
method had been extended to two dimensions and had
been tested on one specially generated test hill,
together with the extension of Bucharov and
Feldbaum's method. Results from one hill did not
give a clear indication. To generate a large
ensemble of test hills, by two dimensional Fourier
series, was expensive in computer time, but this
could be done if someone was willing to pay for it.

Dr. O. L. R. Jacobs (U.K.) asked what was
the relation between these three methods and the
primitive one of sweeping through all possible
values of the argument.

Dr. Hill emphasised that the modelling techniques needed far fewer measurements than scanning methods. In checking the piece-wise cubic
method, 100 scanning measurements were needed,
against 17 for modelling, a pay off of 5:1.

Adaptive Optimisation of a Water-Gas Shift Reactor

by A.J. Kisiel* and D.W.T. Rippin
Department of Chemical Engineering &
Chemical Technology, Imperial College,
London, S.W.7

*Present address: International Systems
 Control Ltd., East Lane, Wembley, Middx.

ABSTRACT

The use of a perturbation method to optimise the performance of a water gas shift reactor with respect to steam flow rate is described.

The effect of the parameters of the optimiser on the overall system performance is studied both experimentally and by a digital computer simulation. The results of these two studies are compared and recommendations of optimiser settings for best performance are made.

1. INTRODUCTION

Maximum seeking control systems have been widely discussed since the early studies of Draper and Li[1] but much of the work reported has been either on the theoretical analysis of idealised systems or on the performance of optimisers working on computer simulations. Relatively few experimental studies of optimiser performance on actual plant in the chemical and process industries have been reported. The present study investigates the performance of an on-line optimiser which seeks to reach and maintain by a perturbation method the best steady operating conditions of a small experimental water gas shift reactor.

A number of parameters of the optimiser are available for adjustment to maximise the optimiser peformance. The best settings of these parameters will depend on factors such as the shape of the response surface relating reactor performance to process variables, the dynamic behaviour of the process and the nature and magnitude of the inherent noise. However, if disturbances affecting the process cause changes in the position and height of the maximum in the response surface without substantially affecting the general shape of the surface or the dynamics of the process then the best settings of the optimiser parameters are not expected to change significantly. It is believed that this situation may occur in many chemical processes so that values of the optimiser parameters once established can be used over a range of values of the input variables with up-dating at only relatively infequent intervals.

The method advocated is to develop a model for the process together with the optimising system and to use this off line to establish best values for the optimiser parameters. These values are confirmed by experimentation on the plant. The procedure has the advantage that only a small amount of on-line computing equipment is required.

Superior numbers refer to similarly-numbered references at the end of this paper.

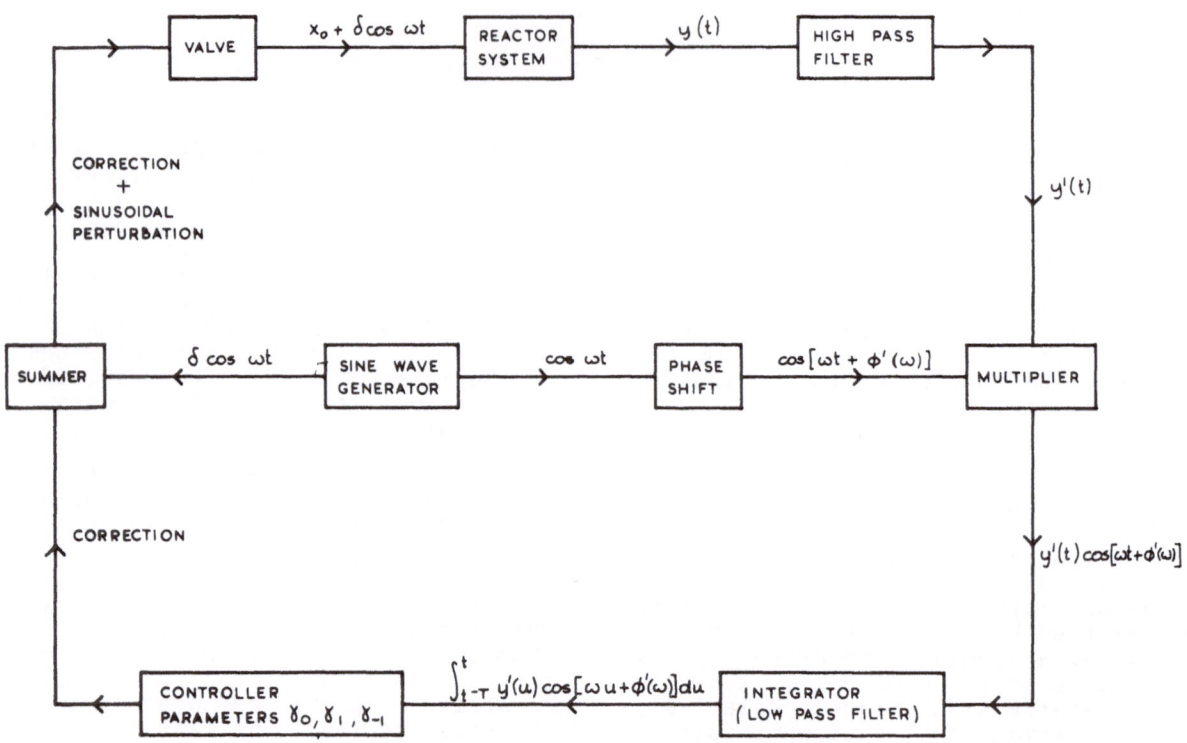

FIG.1. CONTROL LOOP FOR ONE VARIABLE

2. THE OPTIMISER

The perturbation type optimising procedure was described first by Draper and Li[1] and subsequently by, amongst others, Box and Chanmugan[2]. The procedure can be applied to a system with several variables but optimisation was carried out with respect to only one variable in the present study.

The optimising procedure is illustrated in Fig.1 A sinusoidal perturbation in the process variable at the input to the reactor produces a corresponding variation in the objective function computed from reactor outputs which measures the reactor performance. The component of the objective function at the frequency of the perturbation signal contains information about the slope of the objective function with respect to the perturbed process variable at the current operating point. This information is isolated by multiplying the objective function signal by a signal at the perturbation frequency, after phase adjustment to compensate for the dynamics of the process, and subsequently integrating the resulting signal over one or more complete periods of the perturbation. The output of the integrator is proportional to the slope of the objective function and is used to adjust the set point of the process variable about which the perturbations are made.

The best settings of the optimiser parameters are influenced by the criterion used to assess optimiser performance. Two conflicting factors have to be considered - the time taken or the loss incurred in reaching the current maximum, and the loss incurred at the maximum by the continuing perturbations.

Primary adjustable parameters of the optimiser are the amplitude δ and frequency ω of the perturbation signal, and the phase shift adjustment ϕ made to this signal before the multiplier and the gain γ by which the integrator output is multiplied before being used to adjust the set point of the process variable.

The choice of the amplitude δ of the perturbation signal and the gain γ of the slope signal is closely related to the criterion of optimiser performance. A high value of the product $\delta\gamma$ will give a rapid approach to the optimum operating point but a high value of δ will give a large loss at the optimum; further a high value of γ/δ will result in a large amplification of noise in the process[3].

The phase shift should be chosen to maximise the magnitude of the slope signal produced by the integrator. If the transfer function of the process linearised about the current operating point is of the form $\dfrac{\text{constant}}{\sum_{i=0}^{n} a_i p^i}$ the amplitude of the fundamental component of the objective function signal is proportional to the steady state slope. The maximum value of the slope signal from the integrator is obtained when the perturbation signal to the multiplier is in phase with the fundamental component of the objective function signal. If the transfer function is of the form

$$\frac{\sum_{j=0}^{m} b_j p^j}{\sum_{i=0}^{n} a_i p^i}$$

the amplitude of the fundamental component of the objective function is proportional to the steady state slope plus a function of the frequency. A phase shift can in general be found which will produce at the output of the integrator a signal proportional to the steady state slope but careful matching of the two phase shifts becomes more important than in the previous case. In the present study the dynamics of the process could be represented by transfer functions of the first type.

The higher the frequency of the perturbation signal the more rapidly can information about the slope of the response surface be generated. However this frequency cannot be increased indefinitely owing to the increasing attenuation of the output signal from the process - the signal to noise ratio must be maintained at an acceptable level. Further any frequencies must be avoided at which the noise in the process has substantial magnitude since noise components at the perturbation frequency will, of course, not be filtered out by the multiplication and integration procedure.

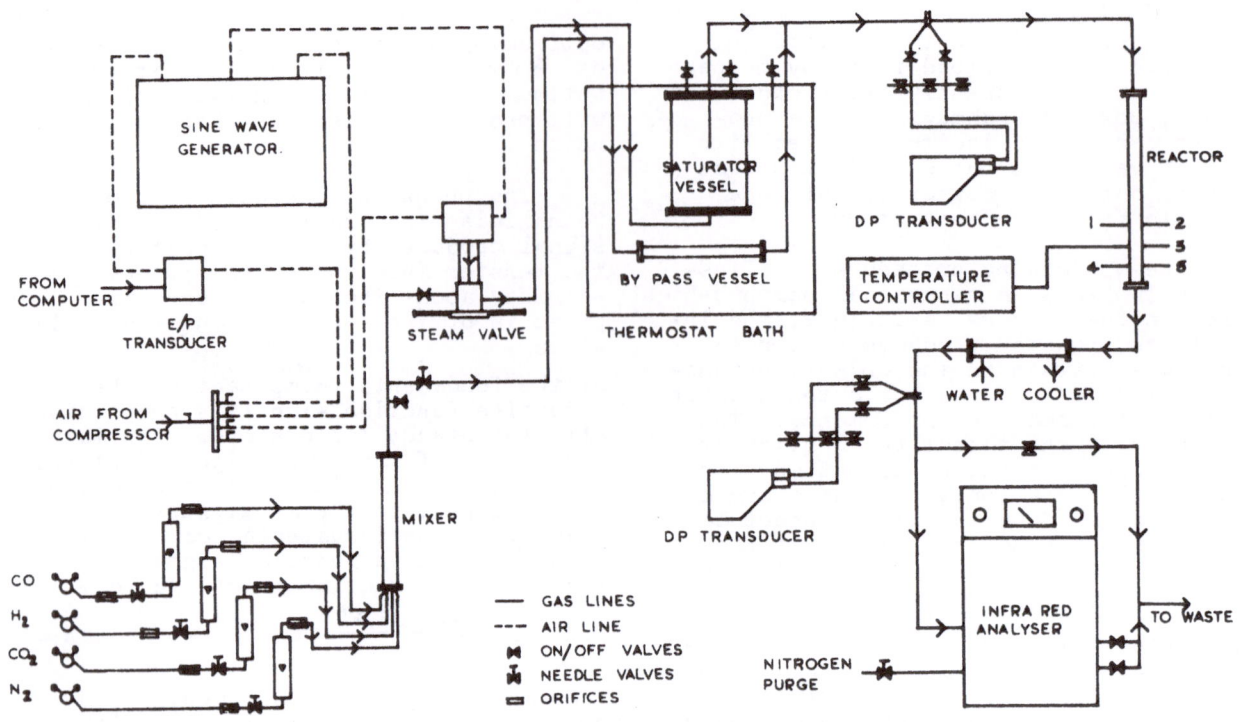

FIG.2. BLOCK LAYOUT OF APPARATUS

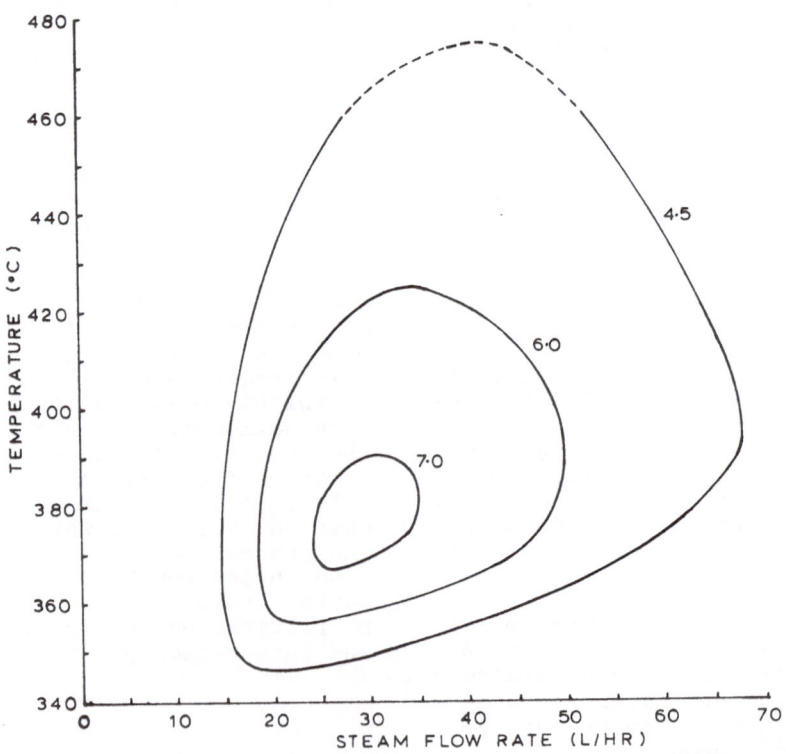

FIG.3. OBJECTIVE FUNCTION RESPONSE SURFACE

3. THE PROCESS

The process studied was the water gas shift reaction between carbon monoxide and steam producing carbon dioxide and hydrogen. The reaction is of considerable commercial interest and is relatively easily carried out on a small scale. The performance of the reactor expressed as the amount of carbon monoxide converted to carbon dioxide exhibits a maximum with respect to variations in steam flow at constant temperature. There is also a maximum with respect to variations in temperature which will be used at a later stage of the investigation when two variable optimisation is studied.

The experimental equipment is shown in Fig 2. The gases other than steam, after measurement of their flow rates by rotameters are mixed. Steam is introduced by passing part of this mixture of gases through a saturator vessel containing water maintained at a constant temperature close to $98°C$. The gases passing through the saturator pick up an amount of of steam proportional to their volume and are then mixed with the remaining gases before entering the reactor which contains 40 cc of ICI high temperature shift catalyst and is operated isothermally, the temperature being maintained by electrical heating. After the reaction the gases are dried and the residual carbon monoxide measured by infrared analysis. Changes in the steam flow are made by means of a pneumatically operated valve which controls the flow through the saturator. Flow rates required in the optimiser for the evaluation of the objective function are measured by differential pressure transducers.

The optimisation experiments were carried out at a reactor temperature of $405°C$ and atmospheric pressure. The flow rates of dry gases measured at room temperature were: carbon monoxide 9.0 l/hr, hydrogen 23.8 l/hr, nitrogen (an inert carrier gas) 21.2 l/hr. Steam flow varied from 10 to 100 l/hr and the mean residence time of gases in the catalyst bed from 0.4 to 0.8 secs.

It was found that under these conditions the model of the chemical kinetics of the water gas shift reaction due to Mars[4] gave a satisfactory account of the steady state performance of the

reactor. Using this model the fraction s of the entering carbon monoxide converted to carbon dioxide is

$$s = \Delta y \left(1 - \exp\left(- \frac{kx_o}{u} \right) \right)$$

where k is the velocity constant of the reaction calculated at the reaction temperature and which is constant at constant temperature

x_o is the length of the reactor

u is the linear velocity of gases in the reactor

Δy is given by the solution of the equation

$$K_T = \frac{(y_{CO_2} + \Delta y)(y_{H_2} + \Delta y)}{(y_{CO} - \Delta y)(y_{H_2O} - \Delta y)}$$

K_T, the equilibrium constant of the reversible reaction is a function only of temperature

y_i is the mole ratio of component i to CO at entry to the reactor (Thus $y_{CO} = 1$)

It was found that the relationship between fractional conversion s and steam flow exhibited a maximum as expected but the slope of the surface was steep for values of steam flow before the maximum and very shallow beyond the maximum. A more suitable surface for testing the initial performance of the optimiser was obtained by subtracting from the conversion a quantity proportional to the current value of the steam flow. The resulting response surface is illustrated in Fig 3. The inclusion of steam flow can be justified on economic grounds as the cost of extra steam required to increase the conversion of carbon monoxide. The form of the objective function used in the experiment was

$$c = 0.3 \text{ (conversion (\%) - 0.4 steam flow (l/hr) - 53.3)}$$

This value is scaled for use on a ± 10 v analogue computer.

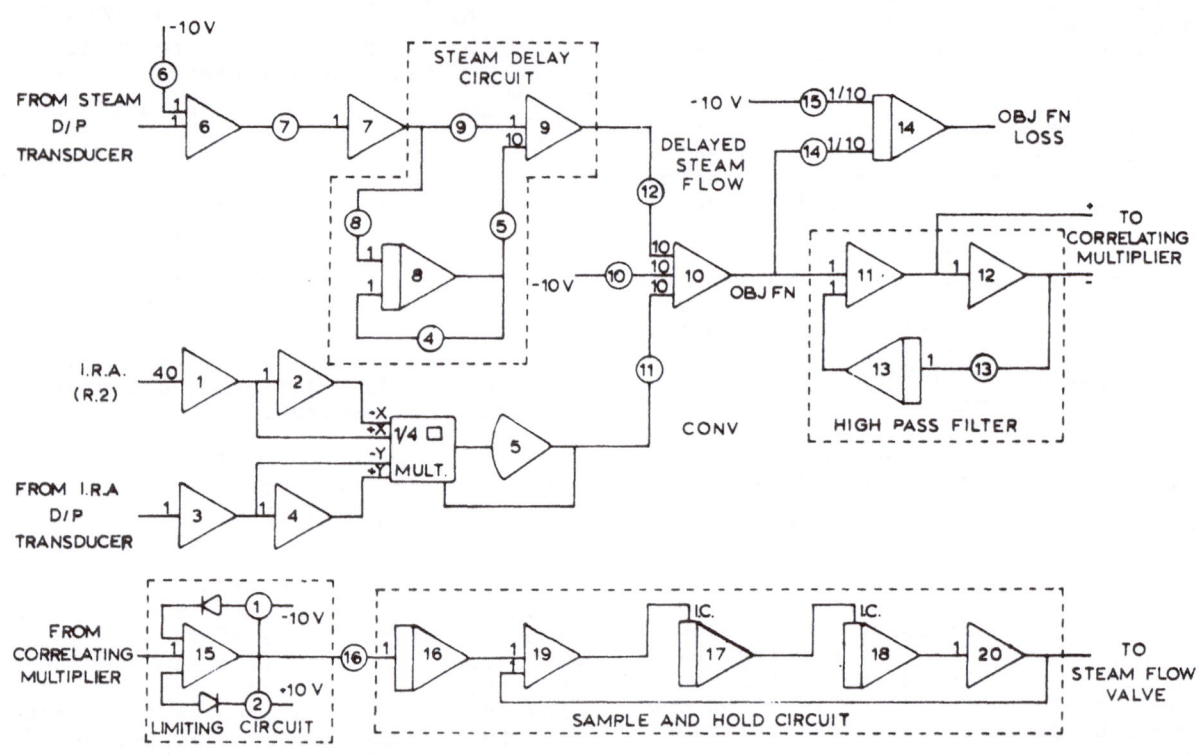

FIG.4. PATCHING DIAGRAM FOR THE ANALOGUE COMPUTER

Due to the small residence time of gases in the reactor (0.4 to 0.8 secs) any dynamics associated with the reactor were negligible compared with the combined dynamics of the saturator, the infra red analyser and other equipment in the optimising loop. It was thus legitimate to construct from experimental measurements a linearised model of the process dynamics and to use this in series with the steady state model of reactor performance already discussed. Frequency response measurements were made of the changes in conversion and steam flow resulting from adjustments of the air pressure signal supplying the steam valve. It was found that the transfer function of each response could be represented by

$$\frac{\text{constant } e^{-pT_2}}{(1 + pT_1)^2} .$$

The values of the time constants were:

steam flow $T_1 = 6.1$ secs $\quad T_2 = 2$ secs

conversion $T_1 = 8.0$ secs $\quad T_2 = 30.5$ secs

In order to obtain a true representation of the effect of the adjustments in steam valve position and the objective function, c, it was essential that the varying components comprising c should be in phase with each other. Therefore the measured value of steam flow was delayed by an amount determined from the transfer functions of steam flow and conversior to bring it into phase with the conversion signal.

The analogue computing equipment required for this and other manipulations is indicated in Fig 4. The steady state and other low frequency components of the objective function were not required for evaluation of the slope of the response surface and were removed by a high pass filter. The correlating multiplier was a sin - cos potentiometer mounted on a rotating shaft of the pneumatic sine wave generator. The phase relationship between the pneumatic signal applied to the steam valve and the signal produced from the sin - cos potentiometer could be varied by adjusting relative position of this shaft and that carrying the swash plate which varied the air pressure signal. The diode limiting circuit was found to be needed after the correlating multiplier to maintain stable operation of the optimiser in face of large changes in operating conditions. The sample and hold circuit which was tri-ggered by a micro switch on the shaft of the sine wave generator was used to integrate the correlated signal over a complete cycle of the perturbation. The resulting measure of slope was used to adjust the setting of the steam flow value at the end of each cycle of the perturbation.

4. EXPERIMENTAL INVESTIGATIONS

The effect of optimiser parameters on the performance of the system was studied with a stationary response sur-face and all runs starting from a fixed value of steam flow.

Experimental results are presented in terms of the following criteria of optimiser performance:

A. Time required to first reach the point at which the objective function has its maximum value.

B. Average loss incurred by perturbation during steady operation about the maximum expressed as a percentage of the maximum value of the objective function.

C. Loss in twenty minutes

D. Loss in one hour

The loss in criteria C and D during time T is normalised by evaluating

$$\frac{\int_o^T (P_{max} - P)dt}{[P_{max} - P_{in}]T} \quad \text{and}$$

presented as a percentage. P is the value of the objective function at time t and P_{max} and P_{in} are its maximum and initial values respectively.

The period of the perturbation sig-nal in steam flow was fixed at about 95 secs, when the gains of steam flow and conversion used in the objective function were both approximately 0.8 times their steady state value. The dry gas flow rates were set at the values already given which produced a maximum in the objective function at

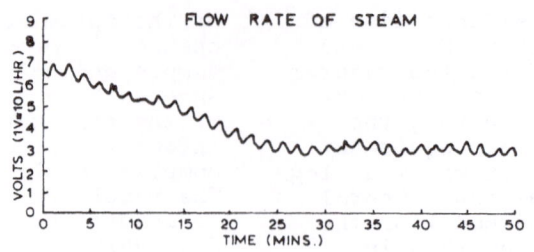

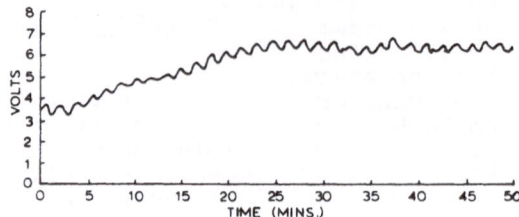

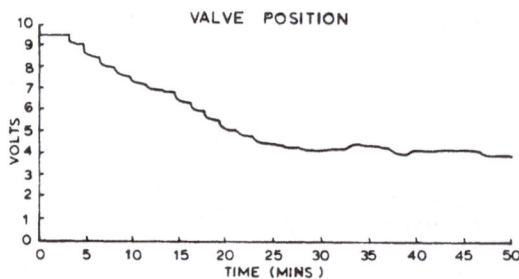

FIG. 5. DIAGRAM OF A TYPICAL OPTIMISATION RUN

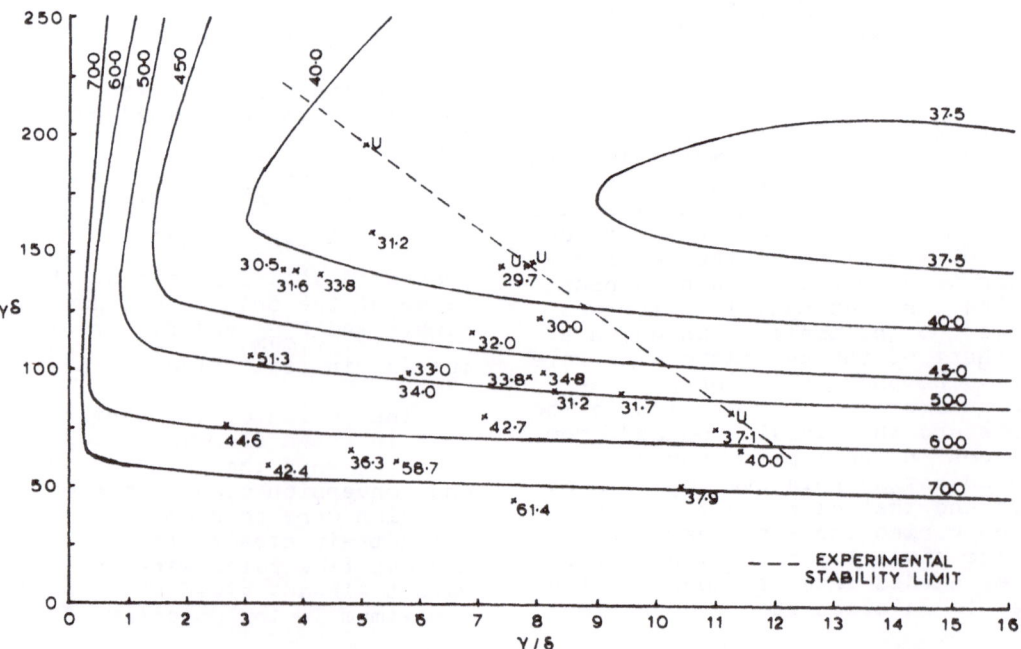

FIG. 6. VARIATION OF THE %AGE LOSS IN 20 MINUTES WITH γδ AND γ/δ

HIGH PASS FILTER CONST = 10 SECS.
PHASE SHIFT ANGLE = 94 DEGS.

a steam flow rate of approximately 30 l/hr. In each run the initial value of steam flow was set at 60 l/hr. After an initial period of operation on open loop the optimising loop was closed and the system run until some time after the maximum value was first attained. The valve position about which perturbations were made, steam flow and objective function were recorded during each run as was the cumulative value of the objective function. Records from a typical run are shown in Fig 5. An initial overshoot can be seen as well as a certain degree of oscillation about the optimum point with a frequency much lower than the perturbation frequency.

Four optimiser parameters were investigated in a 2^{4-1} fractional factorial experiment run in random order with three centre points for replication. The variables were:

1 gain γ of slopesignal
2 amplitude δ of perturbation sig-nal
3 phase shift ϕ
4 high pass filter time constant (H)

The results of these experiments in terms of the several objective functions are shown in Table 1 and a statistical analysis of these results is given in Table 2.

The analysis shows that for all criteria except B the dominating effects were the gain γ and amplitude δ and these should both be increased to improve performance. However, as anticipated, for criterion B the gain has no significant effect but performance can be improved by decreasing the perturbation amplitude.

Further investigations were made of the effect of changes in gain and amplitude on criterion C for fixed values of phase shift and high pass filter constant. In order to locate the parameter settings of γ and δ giving best performance a factorial strategy was used supplemented by a grid around the optimum. The results of these experiments are plotted as points in Fig 6. The coordinates $\gamma\delta$ and γ/δ are chosen since the first controls the time to the optimum and the second the amplification of noise in the system[3]. The system proved unstable at high values of γ

5. DIGITAL SIMULATION

The model of the process described in Section 2 and the components of the optimiser were simulated on a digital computer. The results obtained are plotted as contour lines of criterion C in Fig 6. The simulation gives rather higher losses than the experimental results but the effects of variations in parameter settings are similar[5]. Some nonlinear process components were not simulated accurately and this may account for the discrepancies. The stability limit observed experimentally is not reproduced in the simulation probably because the latter was noise-free.

The minimum loss observed in the simulation at the specified values of $\phi=94^{\circ}$ and H=10 secs was 36.9%. This value could be reduced by further adjustment of the parameters to 32.9% at $\phi=115^{\circ}$, H=11.6 secs $\gamma\delta$=175 γ/δ = 16.8.

6. CONCLUSIONS

The experimental study demonstrated the satisfactory performance of an on-line perturbation optimiser. Points requiring particular attention in the design of such a system are the phase matching of the components of the objective function and the incorporation of a limiting circuit to extend the value of gain which can be used without producing instability.

The digital simulation correctly predicted trends in optimiser performance with respect to the optimiser parameters but assessment and simulation of the noise in the process must be made before a realistic representation of the stability bounds of the system can be produced.

7. ACKNOWLEDGEMENTS

Equipment for this investigation was provided by a research grant from the Department of Scientific and Industrial Research.

Water gas shift catalyst was provided by Imperial Chemical Industries Limited, who also loaned the pneumatic sine wave generator.

343

TABLE 1

2^{4-1} factorial experiment for optimisation of optimiser parameters

Run No.	Gain γ	Ampli- tude 1/hr δ	Phase Shift degrees ϕ	High Pass Filter Content secs H	Perturb -ation Period secs	Time to Maximum mins A	Loss at Maximum % B	Loss in 20 mins % C	Loss in 1 hr. % D
10	29.7	4.0	114	10	99	6.9	7.7	32.9	20.2
6	9.7	5.4	114	6	97	62.9	8.1	56.2	52.0
7	30.0	1.4	114	6	100	58.8	3.9	73.2	41.6
11	9.2	1.4	114	10	92	70.7	3.3	78.2	48.2
8	29.7	4.6	94	6	99	13.4	5.1	34.5	20.1
2	9.7	4.9	94	10	97	26.6	7.8	50.0	29.4
4	28.5	1.1	94	10	95	11.3	6.4	28.0	18.3
5	9.5	1.4	94	6	95	101.8	3.7	82.6	63.8
1	19.6	2.9	104	7.5	98	23.8	5.4	55.1	28.4
3	19.4	2.6	104	7.5	97	20.0	4.9	49.4	24.0
9	19.6	3.2	104	7.5	98	20.0	4.8	46.2	22.2

TABLE 2

ANALYSIS OF THE RESULTS IN TABLE 1

SOURCE OF VARIATION	OBJECTIVE FUNCTION									
	Time to Reach Maximum		Loss in Reaching Max		Loss at Maximum		Loss in 20 Minutes		Loss in 1 hour	
	EFFECT	MEAN SQUARE	EFFECT	MEAN SQUARE	EFFECT	MEAN SQUARE	EFFECT	MEAN SQUARE	EFFECT	MEAN SQUARE
GAIN	-42.9	1840	-426	181500	0.05	0.00	-24.6	605	-23.3	543
AMPL	-34.2	1170	-348	121100	2.85	8.12	-22.1	488	-12.5	156
PHASE SHIFT	8.3	69	120	14400	1.10	1.21	-14.3	205	-15.3	234
H.P.F. CONST	11.5	132	80	6400	0.00	0.00	11.3	128	7.6	58
	-29.4	864	-297	88200	-1.60	2.56	5.4	29	-2.7	7
	9.0	81	129	16600	0.05	0.00	10.4	108	4.1	17
	3.1	9	68	4600	1.40	1.96	- 9.0	81	3.7	14
AVERAGE	44.0	1940	440	193600	5.75	33.06	108.9	11860	73.4	5388
ESTIMATED VARIANCE	4.81		1567		0.10		20.3		10.2	
S.S. AT 95% SIGNIFICANCE LEVEL	89		29000		1.85		376		189	
S.S. AT 99% SIGNIFICANCE LEVEL	474		154200		9.85		2000		1005	

One of the authors (A.J.K.) acknowledges the support provided by an Esso Studentship.

8. REFERENCES

1. Draper C.S. and Li Y.T. Principles of optimalising control systems. A.S.M.E. special publication September 1951

2. Box G.E.P. and Chanmugan J., I.E.C. Fundamentals $1,2$, 1962

3. Hammond P.H. and Duckenfield M.J. Automatica $\underline{1}$, 147, 1963.

4. Mars P. Chem.Eng.Sci. $\underline{14}$,375, 1961

5. Kisiel A.J., Ph.D. Thesis, Imperial College, London University, 1965.

DISCUSSION OF PAPER BY A. J. KISIEL AND D. W. T. RIPPIN

Mr. R. J. High (U.K.) compared the experimental water gas shift reactor at the Warren Spring Laboratory, Stevenage with that of the authors. In particular he asked whether the authors perturbation technique would be handicapped by adiabatic operation of the reactor and the high thermal capacity of the reaction bed. These factors led to a time constant of about one hour in the Warren Spring installation.

Dr. Rippin replied that clearly perturbation frequencies used and phase shifts encountered are strongly dependent on system lags. However, apart from a resulting decrease in speed of optimisation he saw no fundamental difficulty in applying his technique.

Mr. I. H. Rowe (U.K.) described a quadrature correlation method which he claimed would eliminate the box labelled "phase shift" in the authors figure 1. He considered sinusoidal perturbation of a quadratic hill followed by a simple low order dynamic system, possibly with transport delay. Input x is perturbed about a mean value x_i at a frequency w. The output (inaccessible) of the static part of the system is $J(t)$ and of the dynamic part $y(t,\tau)$ where τ is the time delay (ref. 1).

It is required to determine $\frac{\partial J}{\partial x}$ and the usual method is to take the mean value of the product of the fundamental in the output with the perturbing signal over as many cycles in as short a time as possible.

High perturbation frequencies normally lead to both phase shift and attenuation difficulties.

If the reference channel is phase shifted, as suggested by the authors, only the attenuation problem remains, provided that the system phase shift at frequency w is known and constant. A secondary adaptive loop has been suggested to take care of non constant phase shift but this is unnecessary using quadrature detection as shown in the figure. This gives the sign and magnitude of the gradient as outputs independent of τ

REFERENCES

1. FREY, DEEM and ATTPETER, JACC 1965.

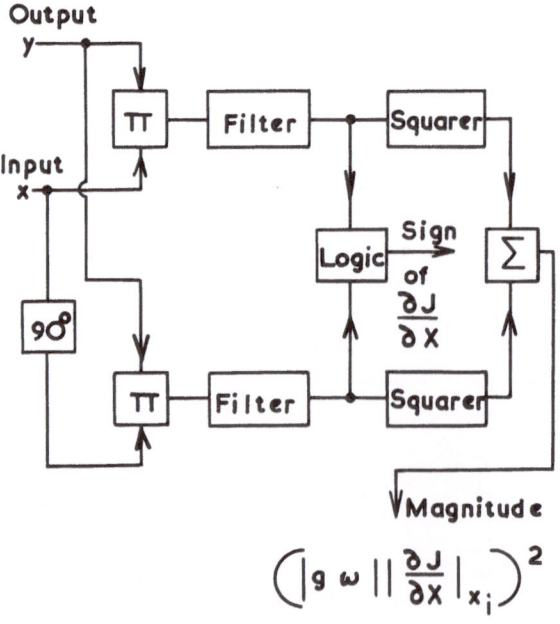

FOR DISCUSSION BY I.H. ROWE

DECOUPLING IN MULTIDIMENSIONAL EXTREMA-SEARCHING ADAPTIVE CONTROL SYSTEMS.

by Madhav Dinkar Khadkikar
Alexander von Humboldt-Research Fellow
Institut für Elektrische Anlagen an der
Technischen Hochschule Stuttgart
STUTTGART, West Germany.

ABSTRACT

The problem of Multidimensional Extrema-Searching Adaptive Control Systems
is discussed and it is pointed out that
for rapid and successful adaptation it
is necessary to achieve 'decoupling'
between various adaptive loops. The extent to which decoupling is possible
depends on our ability to select out
information from an amplitude modulated
sinusoidal oscillation in presence of
other disturbing signals lying in the
same frequency range. A new method of
signal selection to achieve this goal is
presented. The method is based on obtaining orthogonal correlation products,
phase adjustments and subtraction, leading to a minimum of delay and deformation of information signal.

INTRODUCTION

In recent years much attention has
been paid to a class of Self-Optimising
Control Systems known as Extrema-Searching Systems. The study of such systems
is of great practical importance, for it
leads to economic utilisation of fuel in
space vehicles and also leads to cutting
down the production costs in various
industries such as petroleum, chemical
etc.

For searching the extremum of a certain function which defines the quality
of performance, some type of perturbation test signal is required. Draper
and Li[1] in their pioneering study on
optimalisation in connection with an
internal combustion engine have used
sinusoidal oscillations as test signal.
The sinusoidal oscillations as test signal appear to be a very promising solution to this problem, because of the

----
Superior numbers refer to similarly-
numbered references at the end of
this paper.

simplicity in generation, selection,
delay and detection of such signals.
Various authors have further developed
different aspects of this problem. (For
example see Feldbaum[2,4]). In a recent
analysis Eveleigh[3] has developed an
equivalent linearised model of such an
Extrema Searching System and has given
reasonable estimates of the stability
and limit cycles arising. This and other
recent research papers point out the
limitations of the present state of the
theory. A severe limitation comes up in
generalisation of the problem to
n-dimensional case, where the intercoupling between various adjustments demands
special attention. Decoupling of various
adaptive loops is possible to a certain
extent, by introducing sharply tuned
bandpass filters, which all the same
affect the rate of adaptation adversely.

A promising solution of the decoupling
of various adaptive loops depends on our
ability to separate different signals,
which lie essentially in the same frequency range. The author has analysed a
similar situation in case of identification problem in plant adaptive systems
and has developed a method of signal
separation in case of amplitude modulated
sinusoidal oscillations.[5] This method
lends itself to the present problem of
decoupling in multidimensional extrema-
searching systems.

ANALYSIS OF A SINUSOIDAL PERTURBATION EXTREMA-SEARCHING ADAPTIVE SYSTEM

Let the function whose extremum we are
searching be the output Y of an object O
subjected to n-input variables $x_1 \cdots x_n$.
It is assumed that the manipulated input
variables $x_1 \cdots x_n$ are independent
of one another, that is they form a set
of orthogonal co-ordinates. The relationship between Y and any one of the input
variables x_i be a parabola expressed by
the equation :

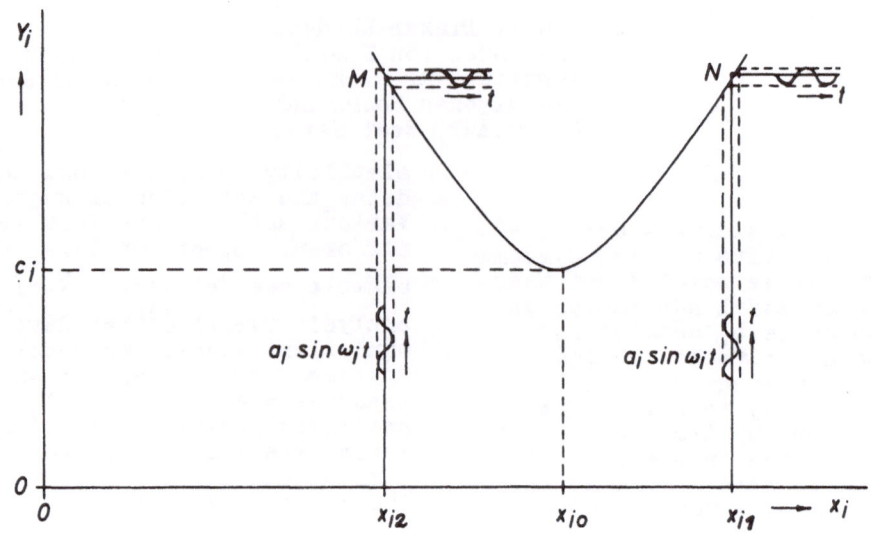

Fig.1 Extremum (Minimum) Characteristic

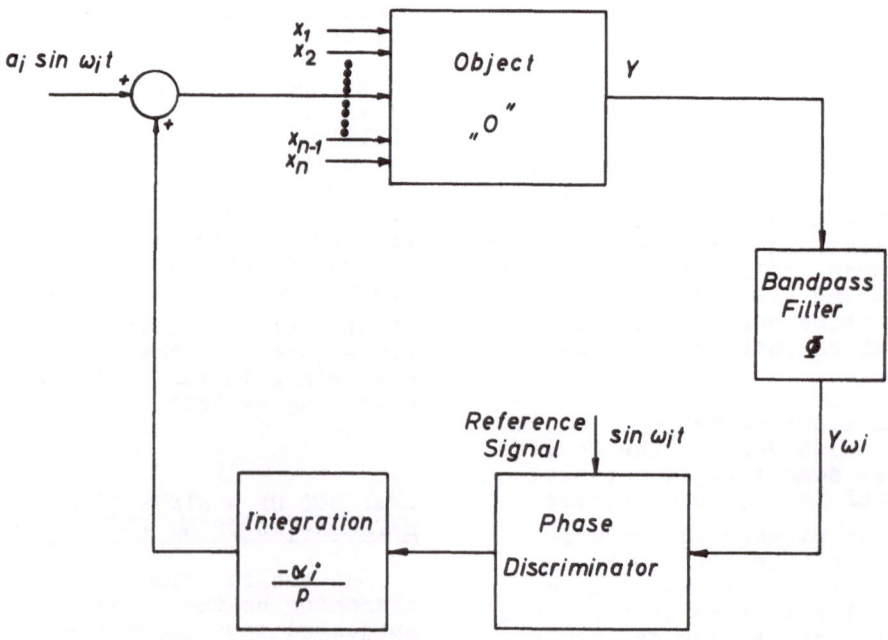

Fig.2 i-th Adaptive loop

$$y = b_i(x_i - x_{io})^2 + c_i \; ; \; i = 1 \cdots n \tag{1}$$

This equation is graphically represented in fig. 1. c_i denotes the minimum value of y (in x_i-y plane) and occurs at $x_i = x_{io}$. In this case we have assumed the extremum to be a minimum but the analysis is valid for both minima and maxima.

Due to random external influences on the object, the location of the minimum (x_{io}, c_i) is subject to random variations. It is our aim to adjust the manipulated inputs x_i such that y attains its minimum value in face of those variations. Suppose at a particular instant $x_i = x_{i_1}$. Now we must know the direction in which x_i should be changed to achieve the minimum. To obtain the direction of the required change, a sinusoidal test signal $a_i \sin \omega_i t$ (a_i-small) is superposed on the manipulated variable x_i. The fact that such a signal gives us the direction of change can be readly verified from fig. 1. The point x_{i_1} is on the right of the minimum and corresponding to this, the output of is 180° out of phase with the input. When we are on the left of the minimum at a point x_{i_2}, the test signal and the corresponding output are in phase. As such the phase of the test signal response gives us sufficient information required for adjusting x_i.

The strategy of searching the minimum can be mathematically expressed as follows. Let the ith input consisting of the manipulated variable and the superposed test signal be expressed as : ($x_i + a_i \sin \omega_i t$). The corresponding component in the output will then be :

$$y_i = b_i(x_i + a_i \sin \omega_i t - x_{io})^2 + c_i$$

$$= b_i[(x_i - x_{io})^2 + a_i^2 \cdot \sin^2 \omega_i t$$

$$+ 2(x_i - x_{io}) \cdot a_i \cdot \sin \omega_i t] + c_i$$

$$= b_i[(x_i - x_{io})^2 + \frac{a_i^2}{2} - \frac{a_i^2}{2} \cos 2\omega_i t$$

$$+ 2(x_i - x_{io}) \cdot a_i \sin \omega_i t] + c_i \tag{2}$$

From this we select out the component of frequency ω_i for our further analysis.

$$y_{\omega_i} = b_i \cdot 2 \cdot (x_i - x_{io}) \cdot a_i \sin \omega_i t \tag{3}$$

To determine the phase of this component, some type of phase-discriminator is used. One way of obtaining this is to multiply y_{ω_i} with a reference or control signal of the same frequency : $\sin \omega_i t$.

The output of the phase-discriminator will then be:

$$= b_i \cdot 2 \cdot (x_i - x_{io}) \cdot a_i \cdot \sin^2 \omega_i t$$

$$= a_i b_i (x_i - x_{io}) - a_i b_i (x_i - x_{io}) \cos \omega_i t$$

From this expression we make use of the component $a_i b_i (x_i - x_{io})$ for adjustment of x_i, which is carried out by means of an integrator.

i.e. $$x_i = -d_i \int a_i b_i (x_i - x_{io}) \cdot dt \tag{5}$$

$$\text{where } d_i = \text{Const. of integrator.}$$

where the -ve sign before the integrator takes into account the proper sense of adjustment.

$$\frac{d(x_i - x_{io})}{dt} = -d_i a_i b_i (x_i - x_{io}) \cdot$$

$$= -T(x_i - x_{io}) \tag{6}$$

$$\text{where } T = d_i a_i b_i$$

Or the solution would be :

$$(x_i - x_{io}) = (x_i - x_{io})_{t=0} \cdot e^{-t/T} \tag{7}$$

That is x_i approaches x_{io} exponentially with the time constant T.

The mechanisation of the above mathematical operations required for searching the minimum is shown in the block-diagram of fig. 2.

The output y of the object O consists of the system signals, superposed with response to n-test signals and system noise. The small-band filter ϕ selects out the component y_{ω_i} from the output y. The phase discriminator used is a multiplier. The frequency component $2\omega_i$ in the output of the discriminator is sufficiently damped by the following integrator, therefore no additional low-pass filter is required to remove this component.

The adaptive loops for other manipulated variables have similar structure.

PRACTICAL LIMITATIONS AND DIFFICULTIES IN REALISING THE ADAPTIVE LOOPS

An assumption which is implicitly made in the above analysis is that the corrective-action in x_i takes place slowly so that $(x_i - x_{io})$ can be considered as constant within a period $T_i (= 2\pi/\omega_i)$ of the test signal. Therefore for rapid adaptation the choice of test frequencies $\omega_i (i = 1 \cdots n)$ should be as high as is consistent with the object dynamics. That is the test frequencies $\omega_i (i = 1 \cdots n)$ should necessarily lie very near to one another, that is, essentially in the same frequency range. This makes the selection of the component $y_{\omega i}$ from the output y (see eqns (2) and (3)) extremely difficult. The presence of other frequency components $\omega_j (j \neq i)$ in the i-th adaptive loop is called as 'intercoupling' and it can under certain circumstances, have a very harmful effect on the adaptive action, in that x_i may be driven away from the minimum.

The use of sharply tuned band-pass filters means a large time-delay[3]. Moreover the term $(x_i - x_{io})$ is a function of time, therefore the signal $y_{\omega i}$ is an amplitude modulated signal having a certain bandwidth. This puts up a limit below which we can not reduce the bandwidth.

The factors which speak for reducing the bandwidth of the band-pass filter are the desired 'decoupling' between various loops and the presence of the system noise. On the contrary the requirements on rapid adaptation necessarily lead to the choice of sufficiently large band width. A compromise between those contradictory conditions on the choice of band-width is very difficult to achieve, particularly in case of multidimensional systems for $n = 2$ or more. This problem in its ultimate analysis crystalises in the problem of detecting the information in amplitude modulated sinusoidal oscillations when the superposed noise lies in the same frequency range. The method of detection should be such that the delay and deformation of information should be minimum. How this can be achieved is demonstrated in the next section.

THE METHOD OF DETECTION

For this section we rewrite the eqn. (3) as follows :

$$y_{\omega i} = A_i \cdot \sin \omega_i t + A_j \cdot \sin \omega_j t \quad —(8)$$

$$\text{where} \quad i \neq j$$

$$\text{and} \quad A_i = A_i(t),$$

$$A_j = A_j(t).$$

A_i is the desired information, which comes up as modulation of the sinusoidal signal of frequency ω_i. The term $A_j \cdot \sin \omega_j t$ is the effect of j-th test signal entering the i-th loop. As the effect of the system noise entering the adaptive loop is similar to the effect of this (j-th) component, no additional term is considered in this equation.

To achieve 'decoupling' between adaptive loops, it is aimed that the detector output should be proportional to A_i and should be independent of A_j.

We build up two correlation products P_1 and P_2 multiplying $y_{\omega i}$ with two reference signals, which are orthogonal to one another.

$$P_1 = [A_i \cdot \sin \omega_i t + A_j \cdot \sin \omega_j t] \cdot \sin \omega_i t$$

$$= \underbrace{\frac{A_i}{2}}_{\text{Information}} - \underbrace{\frac{A_i}{2} \cos 2\omega_i t + \frac{A_j}{2} \cos(\omega_i - \omega_j)t}_{\text{Undesired components}}$$

$$\underbrace{- \frac{A_j}{2} \cos(\omega_i + \omega_j)t}_{\text{"Noise"}} \quad ——(9)$$

$$P_2 = [A_i \cdot \sin \omega_i t + A_j \cdot \sin \omega_j t] \cdot \cos \omega_i t$$

$$= \underbrace{\frac{A_i}{2} \cdot \sin 2\omega_i t + \frac{A_j}{2} \cdot \sin(\omega_i + \omega_j)t}_{\text{Components corresponding to}}$$

$$\underbrace{- \frac{A_j}{2} \cdot \sin(\omega_i - \omega_j)t}_{\text{noise in } P_1} \quad ——(10).$$

A close observation of the products P_1 and P_2 reveals the following facts :

350

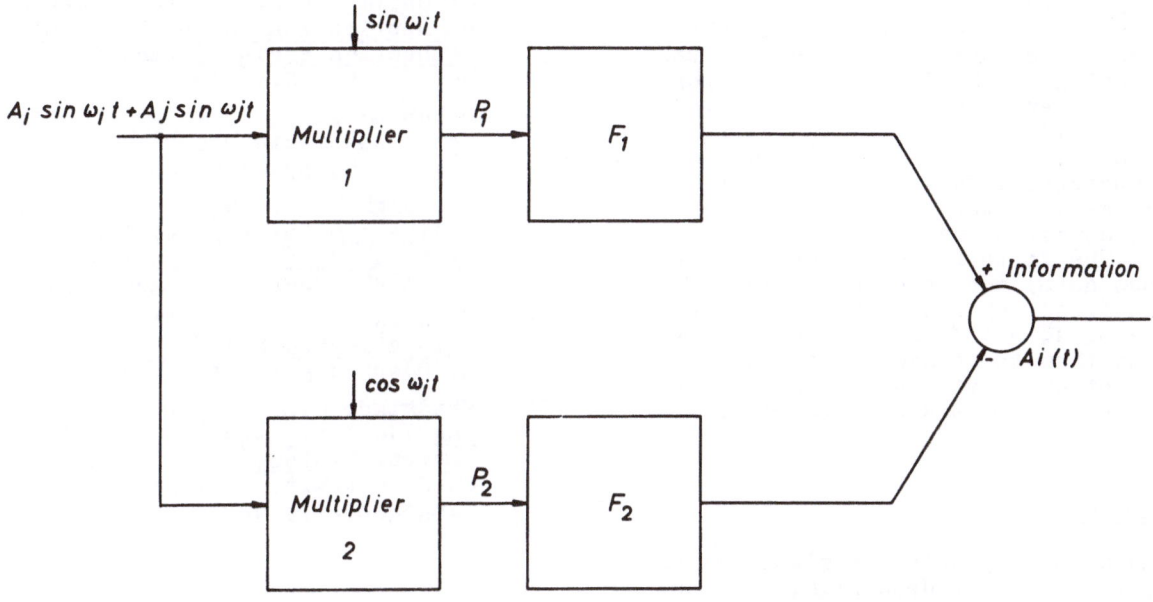

Fig. 3   Improved Phase Discriminator

(<u>1</u>) The product P_1 contains the desired information signal superposed with 'noise components'.

(<u>2</u>) The information component in P_1 is absent in the product P_2. The components of P_2 correspond to the noise components of P_1 having same frequencies and amplitudes but differing in phase.

As such the noise components in P_1 can be compensated by subtraction. For such compensation, it is however necessary that the noise components in products P_1 and P_2 are adjusted to have the same phase. The block-diagram of the compensation arrangement is shown in fig. 3.

To achieve the required phasing in the components to be compensated, two filters F_1 and F_2 are used. The reason for introducing the filter F_1 in the information channel is to simplify the realisation of the filter F_2. Suppose F_1 was absent. Then F_2 should introduce a phase shift of $90°$ over the entire range of noise frequencies which are to be compensated. This is very difficult to achieve. Filter F_1 relaxes this severe condition on the phase shift of F_2. The filter F_1 is such that it has a minimum delay and deformation in information frequency range. The characteristics of F_1 and F_2 are so chosen that both of them to-gether satisfy the compensation condition for phasing, and at the same time are easy to realise.

CONCLUSION

The complex problem of signal selection, required for decoupling, can not be solved by using a band-pass filter, because the signals lie essentially in the same frequency range. Under such difficult conditions, the suggested method of compensation appears to be a promising solution, in which the property of orthogonal correlation products is made use of in generating the compensation signals in a parallel channel.

ACKNOWLEDGEMENT

The author takes this opportunity to express his sincere thanks to Professor Dr. Ing., Dr. tech.E.h. A. Leonhard for his encouragement and helpful guidance. His thanks are also due to Alexander von Humboldt-Foundation (West-Germany) for the grant of a research fellowship.

REFERENCES

(1) Draper C.S. and Li Y.T., '<u>Principles of Optimalising Control Systems and an Application to the Internal Combustion Engine</u>', Ame. Soc. of Mech. Engs. N.Y. 1951

(2) Feldbaum A.A., '<u>Automatic Optimaliser</u>', Automatic and Remote Control, Aug. 1958, pp. 718-728.

(3) Eveleigh V.W., '<u>General Stability Analysis of sinusoidal Perturbation Extrema-searching Adaptive Systems</u>' 2nd IFAC Congress, August 1963,Basel.

(4) Feldbaum A.A., '<u>Rechengeräte in Automatischen Systemen</u>' (Book), R. Odenbourg Verlag, München, 1962.

(5) Khadkikar M.D., '<u>A Method of Reducing Identification Time in Adaptive Control Systems</u>', IFAC Symposium on Sensitivity Analysis, Dubrovnik, (Yugoslavia), September 1964.

STEADY-STATE OSCILLATIONS DUE TO FALSE SWITCHING IN EXTREMUM CONTROL SYSTEMS

by H. L. Burmeister - Institut für Regelungs- und Steuerungstechnik
Dresden der Deutschen Akademie der Wissenschaften
zu Berlin

1. INTRODUCTION

Step-type extremum controllers frequently use the difference of consecutive sampled values of the process output to determine the sign of the following step of the process input. It is well known that this control law may cause false sign reversals due to dynamic lag of the process. These may lead the system away from the extremum [6, 7]. However, quantitative investigations of this phenomenon are rare [3].

In this paper a simple model of an extremum control system is analysed. Oscillations due to false reversals in the open-loop system are calculated exactly and conditions for their existence and stability in the closed-loop system are given. The results make it possible to avoid oscillations due to false switching and thus to reduce the hunting loss. Main features of the analysis are the transformation to normal coordinates [5] and the description of the system dynamics by mapping the phase space into itself.

2. THE EXTREMUM CONTROL SYSTEM

The system considered is shown in Figure 1. The steady-state characteristic of the process is assumed to be symmetric and to have a single extremum (maximum) at $x = 0$. The dynamic lag of the process is taken into account by a stable linear n-th order element proceding the non-linearity, the measurement lag of the process output being neglected. The equations of the system are:

$$x(t) = \frac{c}{p} G(p) u(t) = F(p) u(t) \; ; \quad p = \frac{d}{dt} \quad \cdots (1)$$

$$y(t) = f[x(t)] \qquad \cdots (2)$$

$$u(t) = \sum_{m=0}^{\infty} u[m] \, \delta(t-mT) \qquad \cdots (3)$$

$$u[m] = u[m-1] \, \text{sgn}\{y[m] - y[m-1]\} \; ; \quad y[m] = y(mT) \qquad \cdots (4)$$

$G(p)$ is a proper rational algebraic fraction with first-order poles $a_k < 0$ $(k = 1,2,...,n)$ and arbitrary zeros. Obviously the dynamic

behaviour of the closed loop system can be completely described by the sampled values $x(mT) = x[m]$ $(m = 0,1,2,...)$ of $x(t)$. They satisfy the $(n+1)$-th order difference equation[*]

$$A^*(z) \, x[m] = B^*(z) \, u[m] \qquad \cdots (5)$$

where

$$\frac{B^*(z)}{A^*(z)} = F^*(z) = \sum_{k=0}^{n} \frac{\beta_k}{z-\alpha_k} = \sum_{\nu=1}^{\infty} \gamma_\nu z^{-\nu} \qquad \cdots (6)$$

with

$$A^*(z) = \prod_{k=0}^{n} (z-\alpha_k) , \qquad \alpha_k = e^{a_k T} \qquad \cdots (7)$$

$$(\alpha_0 = 1, \; 0 < \alpha_k < 1 \; \text{for} \; k \geqq 1)$$

is the z transfer function [4] of the linear part of the system. Introducing normal coordinates as phase (state) variables by

$$x_k[m] = \frac{1}{A^{*\prime}(\alpha_k)} \left[\frac{A^*(z)}{z-\alpha_k} x[m] - \frac{B^*(z)-B^*(\alpha_k)}{z-\alpha_k} u[m] \right] \qquad \cdots (8)$$

$$(k = 0,1,2,...,n)$$

leads to a system of uncoupled first order difference equations

$$x_k[m+1] = \alpha_k x_k[m] + \beta_k u[m] . \qquad \cdots (9)$$

The inverse of transformation (8) is [1]

$$z^k x[m] = \sum_{j=0}^{n} \alpha_j^k x_j[m] + \sum_{j=0}^{k-1} \gamma_{k-j} z^j u[m] \qquad \cdots(10)$$

and for $k = 0$:

$$x[m] = \sum_{j=0}^{n} x_j[m] . \qquad \cdots(11)$$

[*]In equations (5), (8) and (10) the variable z is to be considered as a shift operator with the property $zx[m] = x[m+1]$.

Bracketted numbers refer to similarly-numbered references at the end of this paper

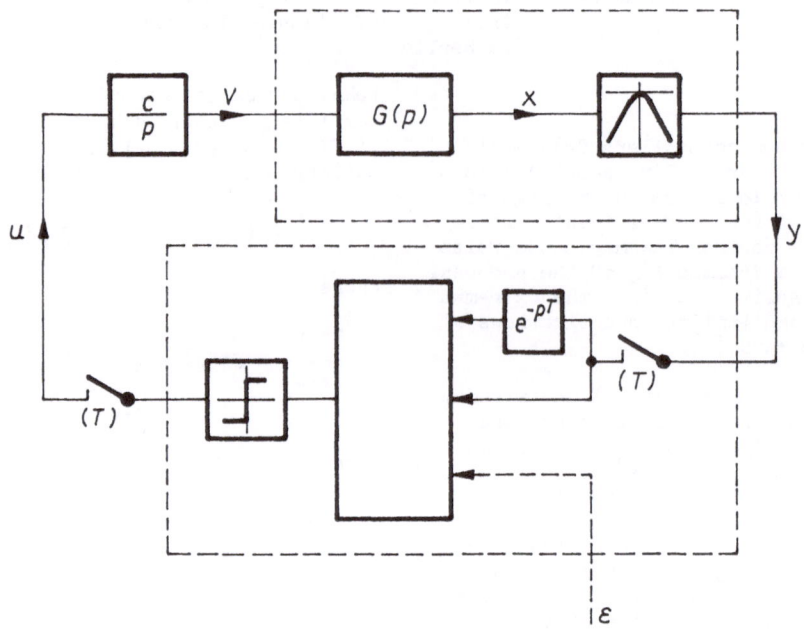

FIGURE 1

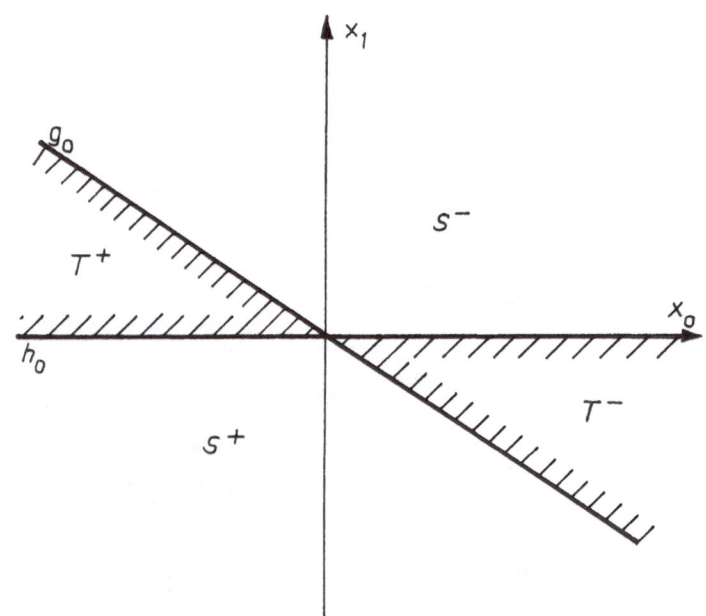

FIGURE 2

354

5. SECOND-ORDER SYSTEMS

Equations (9) and (15) describe the transition between the states $\underline{x}[m] = \{x_0[m], x_1[m]\}$ and $\underline{x}[m+1]$ in consecutive sampling instants as a piecewise linear mapping of the phase plane into itself, coinciding with

$$\Phi^+ : x_k[m+1] = \alpha_k x_k[m] + \beta_k, \quad \text{if} \quad u[m] = 1,$$

and with

$$\Phi^- : x_k[m+1] = \alpha_k x_k[m] - \beta_k, \quad \text{if} \quad u[m] = -1.$$

In order to visualize this mapping, the following decomposition of the phase plane is carried out.

(a) The straight lines

$$g_0 : 2x_0 + \frac{1 + \alpha}{\alpha} x_1 = 0$$

and

$$h_0 : \qquad\qquad x_1 = 0$$

divide the phase plane into four sectors S^+, S^-, T^+, T^- (Figure 2), to which is assigned a rule for the sign change of the control impulse:

$$S^+ \text{ and } S^- : u[m] = u[m-1]$$

$$T^+ \text{ and } T^- : u[m] = -u[m-1] ;$$

(b) the sectors S^+ and S^- are divided by the straight lines (Figure 3)

$$g_k^+ = (\Phi^+)^{-k} g_0 \quad \text{and} \quad g_k^- = (\Phi^-)^{-k} g_0$$

into the regions S_k^+ and S_k^- ($k = 1,2,\ldots$);

(c) the sectors T^+ and T^- are divided by the straight lines - parallel to the x_0-axis -

$$h_0^+ = (\Phi^+)^{-1} h_0 , \quad h_k^+ = (\Phi^- \Phi^+)^{-k} h_0^+$$

and

$$h_0^- = (\Phi^-)^{-1} h_0 , \quad h_k^- = (\Phi^+ \Phi^-)^{-k} h_0^-$$

into the parallel strips T_k^+ and T_k^- ($k = 0,1,2,\ldots$);

(d) the strips T_0^+ and T_0^- are divided by the straight lines g_k^+ and g_k^- into the regions T_{ok}^+ and T_{ok}^- ($k = 1,2,\ldots$).

It is easily proved that, starting from an arbitrary initial state $\underline{x}[0]$ with an arbitrary $u[0]$, a state $\underline{x}[m_0] \in T^+$ with $u[m_0] = 1$ is

reached after a finite number m_0 of steps. Hence it may be assumed, without loss in generality, that

$$\underline{x}[0] \in T^+ , \quad u[0] = 1$$

and

$$\underline{x}[-1] \in S^- , \quad u[-1] = -1$$

for each periodic motion. Then the following assertions hold, which can be proved by elementary calculation:

(a) Each phase point $\underline{x}[0] \in T_{ok}^+$ is mapped by $(\Phi^+)^l$ ($1 \le l < k$) into S_{k-l}^+ and by $(\Phi^+)^k$ into T^- (no false switching).

(b) Each phase point $\underline{x}[0] \in T_k^+$ ($k > 0$) is mapped by Φ^+ into T^+, by $\Phi^- \Phi^+$ into T_{k-1}^+ and hence by $(\Phi^- \Phi^+)^k$ into T_0^+ (k false switchings).

(c) The trajectory of an oscillation passes

from $\underline{x}[0] \in T_k^+$ to $\underline{x}[2k] \in T_{0M_1}^+$

by means of $(\Phi^- \Phi^+)^k$,

from $\underline{x}[2k]$ to $\underline{x}[2k+M_1] \in T_1^-$

by means of $(\Phi^+)^{M_1}$,

from $\underline{x}[2k+M_1]$ to $\underline{x}[2k+M_1+2l] \in T_{0M_2}^-$

by means of $(\Phi^+ \Phi^-)^l$ and

from $\underline{x}(2k+M_1+2l)$ to $\underline{x}[2k+M_1+2l+M_2] \in T^+$

by means of $(\Phi^-)^{M_2}$.

Now we have either $\underline{x}[2k+M_1+2l+M_2] = \underline{x}[0]$, or one or more cycles of this kind follow. The further treatment is confined to the first case. Obviously the numbers M_1 and M_2 must be equal on account of $\displaystyle\sum_{m=0}^{N-1} u[m] = 0$. A further necessary condition is found by adding the inequalities

$$x_1[2k-1] > 0, \quad x_1(2k+1) < 0, \quad x_1[2k+M+2l-1] < 0,$$

$$x_1[2k+M+2l+1] > 0 :$$

$$x_1[2k-1] + x_1[2k+M+2l+1] > 0$$

$$x_1[2k-1] + x_1[2k+M+2l-1] < 0$$

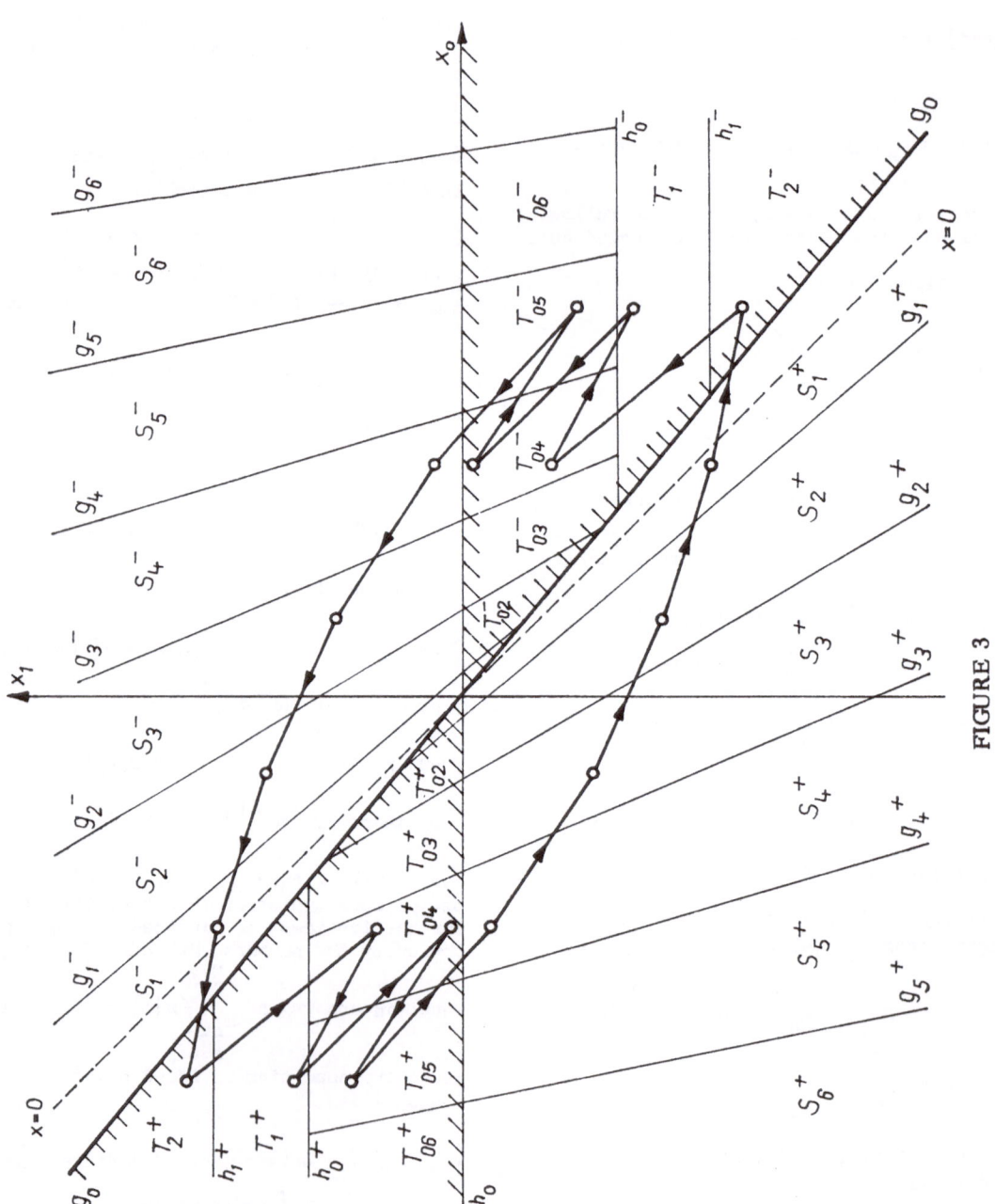

FIGURE 3

Equations (9), (4), (2) and (11) give a complete mathematical description of the sampled-data extremum control system.

3. OSCILLATIONS IN THE OPEN-LOOP LINEAR SYSTEM

Since a steady-state oscillation in the closed-loop system is sustained only by a periodic series of control impulses at the input of the linear system, its period is a multiple of the sampling period T.

Hence to each oscillation corresponds a periodic solution of equations (9), (4), (2) and (11), which is found in two steps:

(a) Calculation of the steady-state oscillations arising in the open-loop linear system when excited by a periodic series of control impulses,

(b) derivation of the conditions that these oscillations should continue to exist in the closed-loop system when the equations of the non-linearity and the extremum controller are taken into account.

The first step is carried out immediately. To each series of impulses

$$u[0], u[1], \ldots, u[N-1]; \quad u[m+N] = u[m] = \pm 1 \quad \ldots(12)$$

$$(m = 0, 1, 2, \ldots)$$

corresponds a family of periodic solutions of equations (9)

$$x_0[m] = x_0[0] + \beta_0 \sum_{\ell=0}^{m-1} u[\ell]; \quad x_0[0] \text{ arbitrary}$$

$$\ldots(13)$$

$$x_k[m] = \frac{\beta_k}{1-\alpha_k^N} \sum_{l=0}^{N-1} u[l+m] \, \alpha_k^{N-1-l} \quad (k = 1, 2, \ldots, n)$$

only when $\sum_{m=0}^{N-1} u[m] = 0$. Hence $N = 2M$ is an

even integer. From (11) and (12) follows by contour integration [1]

$$x[m] = C + \sum_{r=1}^{N-1} \vartheta_r z_r^m F^*(z_r) \quad (m = 0, 1, 2, \ldots),$$

$$\ldots(14)$$

where z_r are the N-th roots of unity different from 1, and

$$\vartheta_r = \frac{1}{N} \sum_{l=0}^{N-1} u[l] z_r^{-l} \quad (r = 1, 2, \ldots, N-1)$$

are constants depending only on the series of impulses $u[m]$. For solution (13) and the frequently assumed characteristic $y = -ax^2$ $(a > 0)$ the hunting loss becomes

$$\delta = a \left[C^2 + \sum_{r=1}^{N-1} |\vartheta_r F^*(z_r)|^2 \right].$$

The solution with least hunting loss, $C = 0$, will be called a symmetric oscillation. To any periodic series of impulses with the property $u[m+M] = -u[m]$ $(2M = N)$ corresponds a symmetric oscillation with $x_k[m+M] = -x_k[m]$.

An oscillation due to false switching is characterized by the property that two consecutive sign reversals of $u[m]$ occur at the same side of the extremum, i.e. that $x[N_r] = x[N_{r+1}]$ for at least one r, if $0 = N_0 < N_1 < \ldots < N_{s-1} < N_s = N$ are the sampling instants where $u[m]$ changes sign. A series (12) being given, this condition can be checked by making use of (11) and (13) or (14).

4. EXISTENCE CONDITIONS FOR OSCILLATIONS IN THE CLOSED-LOOP SYSTEM

An oscillation (13) is sustained in the closed-loop system without external excitation, if the impulses (12) coincide with the impulses (4) produced by the extremum controller. Since

$$\text{sgn}\{y[m] - y[m-1]\} = -\text{sgn}\{x^2[m] - x^2[m-1]\}$$

for each characteristic satisfying the assumptions of section 2, the shape of the characteristic in detail is irrelevant, and the existence conditions take the form

$$u[m] = u[m-1] \, \text{sgn} \sum_{k=0}^{n} \frac{1+\alpha_k}{\alpha_k} x_k[m] \, .$$

$$\text{sgn} \sum_{k=1}^{n} \frac{1-\alpha_k}{\alpha_k} x_k[m] \qquad \ldots(15)$$

$$(m = 0, 1, 2, \ldots, N-1) \, .$$

After insertion of (12) and (13), these equations constitute a system of inequalities defining the existence region of the oscillation under investigation in the space of parameters α_k, β_k and $x_0[0]$. The numerical or graphical evaluation of these inequalities may be laborious if the order of the linear system and the number of free parameters are large. For second-order systems the complete solution is obtained by means of simple phase plane considerations.

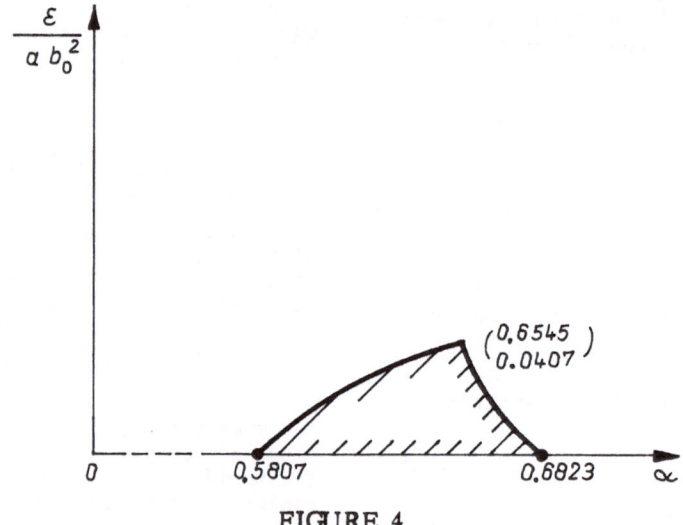

FIGURE 4

and inserting the values (13):

$$\alpha^{2(k-1)} - \alpha^2 > \alpha^{2k+M}(1 - \alpha^2) > 0$$

$$\alpha^{2(1-k)} - \alpha^2 > \alpha^{21+M}(1 - \alpha^2) > 0 .$$

The first inequality can be satisfied only for $k \leq 1$, the second one only for $1 \leq k$, hence $k = 1$. Thus it has been proved, that false sign reversals of $u[m]$ occur in equal numbers at the beginning of both half periods.

Now the necessary and sufficient conditions for the existence of a symmetric oscillation with M regular steps and k false switchings per half period can be formulated:

$$\underline{x}[0] \in T_k^+ ; \quad \underline{x}[2k] \in T_{OM}^+ . \qquad \ldots(16)$$

These conditions can be summarised by the chain of inequalities

$$\frac{1}{\alpha^{2k-1}} < \frac{2\alpha(1-\alpha^{M-1})}{(1-\alpha)(1+\alpha^{2k+M})} < M - 1 < \frac{2(1-\alpha^{M-1})}{(1-\alpha)(1+\alpha^{2k+M})} < \frac{1}{\alpha^{2k+2}} \qquad \ldots(17)$$

which are valid in the following intervals (τ = time constant of the linear term):

M	k	α	$\frac{\tau}{T}$
1	$\geqq 0$	no α	no $\frac{\tau}{T}$
2	0	$0 < \alpha < 1$	$0 < \frac{\tau}{T} < \infty$
2	$\geqq 1$	no α	no $\frac{\tau}{T}$
3	0	$0 < \alpha < 0.618$	$0 < \frac{\tau}{T} < 2.08$
3	1	$0.580 < \alpha < 0.682$	$1.84 < \frac{\tau}{T} < 2.61$
3	$\geqq 2$	no α	no $\frac{\tau}{T}$
4	0	$0.385 < \alpha < 0.543$	$1.05 < \frac{\tau}{T} < 1.64$
4	1	$0.529 < \alpha < 0.710$	$1.57 < \frac{\tau}{T} < 2.92$
4	2	$0.701 < \alpha < 0.719$	$2.81 < \frac{\tau}{T} < 3.04$
4	$\geqq 3$	no α	no $\frac{\tau}{T}$
5	0	no α	no $\frac{\tau}{T}$
5	1	$0.554 < \alpha < 0.682$	$1.69 < \frac{\tau}{T} < 2.61$
5	2	$0.687 < \alpha < 0.762$	$2.67 < \frac{\tau}{T} < 3.68$
5	$\geqq 3$	no α	no $\frac{\tau}{T}$

The table contains, in addition to the oscillations due to false switchings, the simple oscillations without false switching (k = 0). Comparing the results leads to the conclusion that oscillations of both types may coexist in the same system. Which of these oscillations arises as the steady-state operation, depends on the initial conditions. Hence it is incorrect to identify the boundary for the occurrence of false reversals with the boundary for the existence of simple oscillations [3,6].

Figure 3 shows as an example the trajectory of a symmetric oscillation with 5 regular steps and 2 false reversals per half period for $\alpha = 0.7$.

From such a trajectory result, by a translation parallel to the x_0-axis without any phase point crossing the straight line g_0, the trajectories of all the unsymmetric oscillations corresponding to the same series of control impulses.

6. A MODIFIED EXTREMUM CONTROLLER

In order to make the extremum controller less sensitive to random disturbances and to facilitate the implementation of the control law in practice, a threshold is introduced into the control law:

$$u[m] = u[m-1] \, \text{sgn} \, \{y[m] - y[m-1] + \epsilon\} \quad (\epsilon > 0) . \qquad \ldots(18)$$

The method of the preceding sections can be taken over with slight modifications. The existence conditions, however, now depend essentially on the shape of the nonlinear characteristic. For the parabolic characteristic $y = -ax^2$ ($a > 0$) Figure 4 shows the existence region of an oscillation with three regular steps and one false switching per half period in the $\left(\alpha, \frac{\epsilon}{ab_0^2}\right)$-plane, where $b_0 = cG(0)$ is the gain of the linear chain. Obviously false switching can be excluded merely by introducing a relatively small threshold ϵ.

REFERENCES

[1] Burmeister, H. L.; Zur exakten Bestimmung von einfachen Schwingungen in Schritt-Extremalsystemen. III. Konferenz über nichtlineare Schwingungen, Berlin 1964, Teil II.

[2] Burmeister, H. L.; Einfache und komplizierte stationäre Schwingungen in Extremalsystemen. Zeitschr. f. Messen, Steuern, Regeln (to appear).

[3] Fitzner, L. N.; O principach postroenija i metodach analisa nekotorych tipov ekstremal'nych sistem. (On the principles of synthesis and analysis of some types of extremum systems. Theory and applications of discrete automatic systems.) Proceedings of a Conference in Moscow, 1960, pp.486-504.

[4] Jury, E. I.; Sampled-data control systems. New York 1958.

[5] Kalman, R. E.; Nonlinear aspects of sampled-data control systems. Proc. Symp. Nonl. Circ. Anal. VI (1956), 273-313. New York.

[6] Manczak, K.; O warunkach istnienia drgan prostych i zlozonych w przekaznikowo - impulsowych ukladach regulacji ekstremalnej. (On the conditions for the existence of simple and complex oscillations in relay pulse schemes of extremum control.) Arch. Autom. i. Telem. 7 (1962), 71-88.

[7] Popkov, Ju.S.; Perechodnye i ustanovivsiesja rezimy v impul'snych ekstremal'nych sistemach s nezavisimym poiskom. (Transient and steady state regimes in pulse extremum systems with independent search.) Aut. i. Telemech. 24 (1963), 1437-1500.

GENERAL CONTRIBUTION ON THE PAPERS ON "CONTROL

USING PERTURBATION TECHNIQUES"

Dr. P. T. Priestly (U.K.)

It was of some surprise, after listening to all the contributions, to conclude that nothing had been said about the theory and application of extremum-seeking devices based on Evolutionary Operation EVOP and its associated modifications such as SIMPLEX and COMPLEX. In many real control systems the values of the measured variables are obtained at discrete intervals of time, which may be of the order of minutes, and consequently there may be considerable delay in the evaluation of the Performance Index. Many of the continuous-perturbation methods proposed at the Symposium do not lend themselves to systems of the former type, and it is thus desirable to be aware of alternative methods of control.

We propose to use in a pilot-plant chemical system a SIMPLEX method of control involving two control variables, with an associated Cost Function as the Performance Index to be minimised. In this case the simplex will be a triangle, in the two dimensions of control with the cost function evaluated at three different combinations of the control variables, corresponding to the corners of the triangle. Evaluation of the cost function will be carried out using analogue elements while a digital system will be used to decide which points are to be discarded or used. Obviously, this system could be controlled by a digital computer alone, if one were available. The latter would be essential if control were to be attempted in three or more control variables.

It may be argued that such a system is never stationary and is therefore not an efficient or optimal extremum controller, but in the presence of considerable noise, this is equivalent to say the hunting loss in a continuous-perturbation system when the amplitude of perturbation is made high in order to reach the extremum in a short time. The writer tentatively suggests that a two-level hierarchical system can be more efficient in respect of these losses. Thus, for example the amplitude of perturbation could be made fairly large until the system reached the extremum, followed by a phase in which the amplitude were kept small in order to reduce the hunting loss, but still sufficiently large to maintain the system at the optimum.

As a newcomer to the field of process control, the writer feels unqualified to comment on the various papers, but it seems quite essential to standardise on certain types of "hill" so that the progress of hill-climbers set loose on them may be compared. Such standardisation was mentioned by Dr. J. C. Hill and research of this nature is also being carried out by Dr. M. Box in the United Kingdom.

AUTHOR INDEX

363